APPLIED OPTIMIZATION

For further volumes:
http://www.springer.com/series/5634

Applied Optimization

HANDBOOK OF MULTICRITERIA ANALYSIS

Edited by

CONSTANTIN ZOPOUNIDIS
Technical University of Crete, Chania, Greece

PANOS M. PARDALOS
University of Florida, Gainesville, FL, USA

Springer

Prof. Constantin Zopounidis
Technical University of Crete
Department of Production Engineering
and Management
Financial Engineering Laboratory
University Campus
731 00 Chania
Greece
kostas@dpem.tuc.gr

Prof. Panos M. Pardalos
Department of Industrial and
Systems Engineering
University of Florida
303 Weil Hall
P.O. Box 116595
Gainesville, FL, 32611-6595
USA
pardalos@ufl.edu

ISSN 1384-6485
ISBN 978-3-540-92827-0 e-ISBN 978-3-540-92828-7
DOI 10.1007/978-3-540-92828-7
Springer Heidelberg Dordrecht London New York

Library of Congress Control Number: 2010927686

Cover design: WMXDesign GmbH, Heidelberg, Germany

Printed on acid-free paper

Springer is part of Springer Science+Business Media (www.springer.com)

To our families

Preface

The Multicriteria Analysis Paradigm

During the past decades, operations research (OR) has come a long way as a field that supports scientific management. Within the OR field, various interconnected areas have been developed on the basis of different decision-making paradigms and problem contexts. OR is mainly involved with model building and algorithmic optimization procedures that facilitate the analysis of complex real-world problems. This complexity can be due to the dimensionality of a given problem (e.g., the number of available options and actions), the uncertainty that prevails in most real-world situations, the nature of the available data which are often imprecise, as well as the multiple stakeholders that are often involved.

An important implication of the above issues involves the multidimensional character of real-world decision-making problems, which requires the consideration of multiple conflicting points of view, even in situations where a single decision maker is involved. Nowadays, economic, social, and environmental criteria are nowadays involved in practically all decision situations, in order to describe the diverse outcomes of the existing options. Within this context, the decision process should naturally explore the conflicting nature of the criteria, the corresponding tradeoffs, the goals set by the decision makers, and of course the way that these can be introduced in an appropriate decision model that takes into account the subjectivity of the decision process and the preferences of the decision makers.

Nevertheless, with the introduction of multiple points of view, criteria, and factors, universally acceptable (objective) solutions are no longer feasible. While this may be cumbersome, it highlights the difficulty of decision-making in a realistic context. The well-known theorem of Arrow [1] is indicative of these difficulties. In a social choice setting, Arror's axiomatic system defines the necessary conditions for democracy, and the paradox is that it leads to dictatorship. Arrow and Raynaud put this argument in a decision-making context (see [2], p. 21):

> You want to make a real, wise, *multicriterion* decision, and the simplest and most natural axioms drive you toward a *monocriterion* one!

In explaining this paradox Vincke [9] notes that Arrow's axioms are incompatible when one tries to select a single (right) solution (preorder of some alternatives) from a set of plausible ones defined by the information that multiple criteria provide. Vincke emphasizes that the final choice requires more information or implicit assumptions about the preferences of the individuals involved in the decision process (i.e., the decision-makers).

With these issues in mind, multicriteria analysis has become an important and active discipline in OR, focusing on providing the theory and methodologies needed for supporting the decision-making process in complex and ill-structured problems, within a realistic context taking into account all the multiple points of view, criteria, and stakeholders involved. Among others multicriteria analysis focuses on issues such as: (1) the resolution of the conflicting nature of the criteria, (2) the modeling of the decision-makers' preferences, (3) the identification of compromise solutions and the analysis of the consequences of multicriteria solutions, and (4) the development of decision-making models.

The Evolution of Multicriteria Analysis and Current Status

The increasing complexity the economic, technological, and business environment, have contributed to the establishment of multicriteria analysis as an important field of OR and management science. Actually, however, the field has a long history, which can be traced back to the works of Jean-Charles de Borda and Marquis de Condorcet on voting systems in the late 18th century. About a century after these works, Vilfredo Pareto introduced the concept of dominance, which is fundamental in modern theory of multicriteria analysis, which was later on extended by Koopmans [4]. During the 1940s and the 1950s, von Neumann and Morgenstern [10] as well as Savage [8] introduced utility theory for normative decision-making, which set the grounds for multiattribute utility/value theory, one of the major methodological streams of multicriteria analysis. These pioneering works inspired several researchers during the 1960s. Charnes and Cooper [3] extended the traditional mathematical programming theory through the introduction of goal programming. By the end of the 1960s, multicriteria analysis attracted the interest of European OR too. Roy [7], one of the pioneers in this field, introduced the outranking relation approach; he is considered as the founder of the "European" school of multicriteria analysis.

During the recent years, multicriteria analysis has continued its growth through:

- New theoretical developments on new techniques and the characterization of existing decision models.
- The implementation of multicriteria methodologies into integrated decision support systems.
- Innovative applications into new areas, including among other management, economics and finance, environment and energy planning, telecommunications, transportation, etc.

- The exploration of the interactions with other disciplines such as artificial intelligence, evolutionary computation, fuzzy sets theory, and soft computing.

Outline of the Book

Aims and Scope

Research works on multicriteria analysis is often published in premier OR journals, special issues, conference proceedings, and textbooks. Nevertheless, a publication such as this edited volume provides the unique opportunity to present in a unified and comprehensive way the foundations of multicriteria analysis, its core concepts, and the recent advances in the field. To this end, the book covers the theory of multicriteria analysis on discrete problems and multiobjective optimization, the connections of multicriteria analysis with other disciplines, and applications. All chapters are written by leading experts in the field, in an expository yet scientifically vigorous manner. In this way, we think that a broad readership including among others researchers, graduate students, and practitioners who are interested in management science, operations research, and decision analysis, will find in this book a complete coverage of the recent advances in the different aspects of MCDA and the state-of-the-art research in this field.

The book is organized into four main parts, covering all aspects of multicriteria analysis, including issues in decision aiding and support, discrete problems, multiobjective optimization, and applications. Below we provide an outline to the contents of the book.

Organization

Issues in Decision Aiding

The first part of the book is devoted to some important issues that analysts and decision-makers should bear in mind during the decision process. Irrespective of the methods used, it is always important to have in mind issues related to problem modeling and structuring, as well as the robustness and sensitivity of the obtained solutions and recommendations. The careful consideration of such points ensures that the decision aiding process is well designed and implemented, which is fundamental for the quality of the results, their acceptability by the decision-maker, and their actual applicability.

In the first chapter of this introductory part, Roy discusses the important concept of robustness. All decision aiding models are naturally based on a set of assumptions of the real world, which by itself implies that such models provide (by definition) an approximation of reality. This raises the question on whether the results and recommendations obtained by decision models are actually as good as analysts and decision-makers assumed during the analysis process. Robustness analysis enables the consideration of this issue which is crucial for decision aiding. Roy analyzes the concept of robustness in decision aiding and discusses four proposals that clarify the broad and subjective nature of robustness, its importance for decision aiding, and its implementation in multicriteria methodologies.

The second chapter by Montibeller and Franco, involves the use of multicriteria analysis for strategic decision-making. Strategic decisions are taken at the top level of an organization, require a vast amount of resources, and the magnitude of their results is decisive for the future of the organization. Multicriteria methods are particularly well-suited to this type of problems, but there is a number of issues involved for their successful application. Montibeller and Franco discuss the complexity of the strategic decision-making process as far as its technical and social aspects are concerned and outline the multicriteria modeling framework for strategic decisions including the role of multiple objectives, the uncertainties involved, the identification of robust options and their evaluation, as well as the analysis of their long term consequences. The authors provide suggestions on how to implement a multicriteria modeling approach, focusing on the design of decision support processes to tackle the complexity of strategic decisions.

Multicriteria Decision Aid Methodologies

The second part of the book covers issues related to discrete multicriteria problems. By "discrete problems", we refer to decision situations involving the evaluation of a finite set of alternatives and actions over a predefined set of evaluation criteria. Such problems are encountered in numerous cases, including among others finance and economics, strategic planning, human resources management, marketing, engineering, etc. The chapters in this part cover the main multicriteria paradigms and methods to address such problems.

The first chapter in this part involves the family of ELECTRE methods. Founded on the theory of outranking relations, the ELECTRE methods have played a prominent role in the development of multicriteria analysis during the past 40 years. During this period several extensions and variants of the original ELECTRE method have been developed for decision aiding in different kinds of problem contexts. In this chapter Figueira, Greco, Roy, and Słowiński provide a comprehensive discussion of the main features of the ELECTRE methods, their weak and strong points, and present an overview of all the recent advances of this modeling approach.

In the next chapter, Saaty provides a comprehensive discussion of the analytic hierarchy process (AHP) and its generalization to dependence and feedback, the analytic network process (ANP). AHP/ANP implement a completely different approach to multicriteria analysis compared to the outranking relations approach of the ELECTRE methods. Both methods have become very popular techniques with numerous applications in various fields, since the introduction of AHP by Saaty in

1980. In this chapter Saaty describes the foundations of AHP/ANP and their assumptions, using several detailed examples and real-world applications. Extensions to group decision-making are also discussed together with the well-known issue of rank preservation and reversal.

The third chapter, by Salo and Hämäläinen, introduces preference programming methods, which include techniques that model incomplete preference information through set inclusion, provide decision recommendations based on dominance concepts and decision rules, and support the iterative exploration of the decision-maker's preferences. The chapter presents the key concepts of this approach, the preference elicitation techniques employed, the existing methods, as well as their implementation into decision support systems, and their applications.

The next chapter is devoted to the aggregation-disaggregation approach. Similarly to preference programming, disaggregation techniques facilitate the preference elicitation process during model building. This approach is particularly helpful when the decision-maker has difficulties in specifying the detailed preferential information that is needed for the decision aiding process. Disaggregation analysis uses regression-like techniques to infer this information from a set of decision examples. In this chapter Siskos and Grigoroudis present the new research developments on aggregation-disaggregation models, including among others issues such as post-optimality analysis, robustness analysis, group and collective decision-making.

In last chapter in the second part of the book, Doumpos and Zopounidis explore the connections between the disaggregation paradigm of multicriteria analysis with statistical learning and data mining. Similarly to disaggregation analysis, statistical learning/data mining is also involved with the problem of inference from data. However, the scope and context of the two fields is substantially different. This chapter discusses the similarities and differences between the two fields, and explores the ways that they can be combined to provided integrated decision support. To this end, a comprehensive literature review is presented.

Multiobjective Optimization

The third part of the book is devoted to multiobjective optimization, which extends the traditional single objective toward the consideration of multiple (conflicting) objectives and goals. Multiobjective optimization has been a very active research field in multicriteria analysis. Its development has been motivated by the diversity and complexity of business and engineering problems, which led to important developments such as interactive and iterative algorithms, goal programming formulations, and (more recently) multiobjective evolutionary optimization techniques.

This part of the book begins with the chapter of Zeleny, who explains in the simplest possible terms what multiobjective optimization is and discusses the role of tradeoffs, distinguishing between a tradeoffs-based approach and tradeoffs-free thinking. This distinction leads to two different optimization paradigms: the optimization of a given system versus the design of the optimal system. On the basis of the latter approach, Zeleny discusses the foundations of De novo programming and provides examples optimal design and multiobjective optimization can be used in areas such as risk management, conflict dissolution, and product pricing.

In the next chapter, Korhonen and Wallenius focus on interactive multiobjective optimization methods. Interactive methods are widely used in multiobjective optimization to facilitate the search for an appropriate solution in accordance with the decision-maker's preferences. Korhonen and Wallenius present the main ideas and principles of interactive methods, their implementation in different methodological approaches, together with a discussion of the context within which such methods can be used efficiently for decision-making and support. Examples of two software implementations are also given.

The third chapter of this part is devoted to evolutionary multiobjective optimization, which has a emerged during the past decade as an important development in the theory and practice of multicriteria analysis in computationally complex problems that are hard to solve to optimality with traditional optimization techniques. In this chapter Fontes and Gaspar-Cunhathe introduce the concepts on which evolutionary multiobjective optimization algorithms are based, together with a summary of the main algorithmic implementations and their applications in engineering, industry, economics, and management, among others.

Goal programming is the subject of the next chapter. Since the 1950s, goal programming formulations have constituted an important modeling approach in multicriteria analysis and OR. Instead of considering multiple objectives as in a multiobjective optimization context, goal programming is based on the definition of goals, which represent targets that the decision-maker's would like the obtained solution to reach on a number of measures (criteria). González-Pachón and Romero, begin with the discussion of the connections between goal programming and Simon's bounded rationality theories, and present several types of goal programming formulations. Important issues for the successful use of this approach are also discussed along with several extensions.

In the last chapter of this part, Engau discusses the use of decomposition-coordination methods in multiobjective problems. Such methods are very useful in complex decision problems involving a large number of decision variables and/or objectives. Decomposing such problems into smaller ones not only improves the computational aspects of the solution process, but it may also facilitate the active participation of the decision-maker. The overview in this chapter collects and reports on some of the most important methods and results, including but not limited to multidisciplinary design optimization, multiobjective hierarchical analysis, tradeoff-based decision-making, and multiscenario multiobjective optimization.

Applications

The fourth part of the book is dedicated to applications of multicriteria analysis. We consider applications as being of equal importance to the advances in theory. Applications enable the testing of new multicriteria methodologies in real-world situations, thus highlighting the practical contributions of this field, the existing limitations of the methods, and pinpointing new issues that require new methodologies or further validation [5, 6, 11].

In the first chapter of this last part, André and Roy present a multicriteria approach, based on the ELECTRE TRI method, designed to evaluate non-financial performance in companies. The methodology implements a hierarchical evaluation scheme, in accordance with a hierarchy of responsibilities in an organization. The chapter discusses in detail the modeling framework and discusses the application of the proposed methodology in several companies for the evaluation of sponsorship projects.

The next chapter, by Schauten and Spronk, deals with the optimal capital structure problem, which is of major importance in corporate finance, which emphasizes the objective of shareholder wealth maximization. However, as Schauten and Spronk show the management of the firm's capital structure is much more involved, as the capital structure decision process involves multiple considerations of diverse nature. Within this context, the authors propose to translate some of these considerations as separate criteria, which can be traded off against the hard and quantifiable criterion of market value, and present an overview of the different objectives and considerations that have been proposed in the literature.

The book ends with the chapter of Tsafarakis, Lakiotaki, and Matsatsinis on the applications of multicriteria analysis in marketing and e-commerce. In the marketing domain, the chapter illustrates the uses of multicriteria analysis in issues such as consumer preference measurement, market segmentation, targeting, and simulation. In the e-commerce domain business-to-consumer tools and recommender systems are discussed among others. Examples of multicriteria applications and decision support methodologies are also discussed.

Acknowledgements

Sincere thanks must be expressed to all the authors, who have shown increased interest in contributing to this edited volume. They have devoted considerable time and effort to prepare excellent comprehensive works of high scientific quality and value. Without their help it would be impossible to prepare this book in line with the high standards that we have set from the very beginning of this project. We are also grateful to Michael Doumpos, Aggeliki Liadaki, and Dimitrios Niklis for their valuable help with the editorial process and the preparation of the final manuscript.

Constantin Zopounidis
Panos M. Pardalos
February 2010

References

1. K.J. Arrow. *Social Choice and Individual Values*. John Wiley & Sons, New York, 1951.
2. K.J. Arrow and H. Reynaud. *Social Choice and Multicriterion Decision-making*. MIT Press, Cambridge, MA, 1986.
3. A. Charnes and W.W. Cooper. *Management Models and Industrial Applications of Linear Programming*. John Wiley & Sons, New York, 1961.
4. T.C. Koopmans. *Activity Analysis of Production and Allocation*. John Wiley & Sons, New York, 1951.
5. P.M. Pardalos and M.G.C. Resende. *Handbook of Applied Optimization*. Oxford University Press, Oxford, 2002.
6. P.M. Pardalos, Y. Siskos, and C. Zopounidis. *Advances in Multicriteria Analysis*. Springer, Dordrecht, 1995.
7. B. Roy. Classement et choix en présence de points de vue multiples: La méthode ELECTRE. *R.I.R.O*, 8:57–75, 1968.
8. L.J. Savage. *The Foundations of Statistics*. John Wiley & Sons, New York, 1954.
9. Ph. Vincke. Arrow's theorem is not a surprising result. *European Journal of Operational Research*, 10:22–25, 1982.
10. J. von Neumann and O. Morgenstern. *Theory of Games and Economic Behavior*. Princeton University Press, Princeton, NJ, 1944.
11. C. Zopounidis, P.M. Pardalos, and G. Baourakis. *Fuzzy Sets in Management, Economics and Marketing*. World Scientific, Singapore, 2001.

Contents

Part III Multiobjective Optimization

List of Contributors

Stéphane André
Université Paris-Dauphine, LAMSADE, 120/122 rue Réaumur, 75002 Paris, France, e-mail: stephane.andre@dauphine.fr

Michael Doumpos
Technical University of Crete, Dept. of Production Engineering and Management, Financial Engineering Laboratory, University Campus, 73100 Chania, Greece, e-mail: mdoumpos@dpem.tuc.gr

Alexander Engau
Department of Mathematical and Statistical Sciences, University of Colorado Denver, USA
Department of Management Sciences, Faculty of Engineering, University of Waterloo, Canada, e-mail: aengau@alumni.clemson.edu

Jose Rui Figueira
CEG-IST, Center for Management Studies, Instituto Superior Técnico, Technical University of Lisbon, Tagus Park, Av. Cavaco Silva, 2744-016, Porto Salvo, Portugal, e-mail: figueira@ist.utl.pt

Dalila B.M.M. Fontes
LIAAD - INESC Porto L.A. and Faculdade de Economia, Universidade do Porto, Rua Dr. Roberto Frias, 4200-464 Porto, Portugal, e-mail: fontes@fep.up.pt

Alberto Franco
Warwick Business School, University of Warwick, Coventry CV4 7AL, UK, e-mail: alberto.franco@wbs.ac.uk

António Gaspar-Cunha
IPC/I3N - Institute of Polymers and Composites, Department of Polymer Engineering, University of Minho, Campus de Azurém, 4800-058 Guimarães, Portugal e-mail: agc@dep.uminho.pt

Salvatore Greco
Faculty of Economics, The University of Catania, Corso Italia, 55, 95 129 Catania, Italy, e-mail: salgreco@unict.it

Evangelos Grigoroudis
Technical University of Crete, Dept. of Production Engineering and Management, Decision Support Systems Laboratory, University Campus, 73100 Chania, Greece, e-mail: vangelis@ergasya.tuc.gr

Raimo P. Hämäläinen
Helsinki University of Technology, Systems Analysis Laboratory, P.O. Box 1100, FIN-02015 TKK, Finland, e-mail: raimo@tkk.fi

Pekka Korhonen
Helsinki School of Economics, Department of Business Technology, Runeberginkatu 22-24, 00100 Helsinki, Finland, e-mail: Pekka.Korhonen@hse.fi

Kleanthi Lakiotaki
Technical University of Crete, Dept. of Production Engineering and Management, Decision Support Systems Laboratory, University Campus, 73100 Chania, Greece, e-mail: klio@ergasya.tuc.gr

Nikolaos Matsatsinis
Technical University of Crete, Dept. of Production Engineering and Management, Decision Support Systems Laboratory, University Campus, 73100 Chania, Greece, e-mail: nikos@ergasya.tuc.gr

Gilberto Montibeller
Department of Management, London School of Economics, Houghton Street, London, WC2A 2AE, UK, e-mail: g.montibeller@lse.ac.uk

Jacinto González-Pachón
Universidad Politécnica de Madrid, Departamento de Inteligencia Artificial, Facultad de Informática, Campus de Montegancedo s/n 28660-Boadilla del Monte, Madrid, Spain, e-mail: jgpachon@fi.upm.es

Carlos Romero
Universidad Politécnica de Madrid, ETS Ingenieros de Montes, Ciudad Universitaria s/n, 28040 Madrid, Spain, e-mail: carlos.romero@upm.es

Bernard Roy
Université Paris-Dauphine, LAMSADE, Place du Maréchal de Lattre de Tassigny, 75775 Paris Cedex 16, France, e-mail: roy@lamsade.dauphine.fr

Thomas L. Saaty
University of Pittsburgh, The Katz Graduate School of Business, Pittsburgh, PA 15260, USA, e-mail: saaty@katz.pitt.edu

Ahti Salo
Helsinki University of Technology, Systems Analysis Laboratory, P.O. Box 1100, FIN-02015 TKK, Finland, e-mail: ahti.salo@tkk.fi

Marc B.J. Schauten
Erasmus University, P.O. Box 1738, 3000DR Rotterdam, The Netherlands, e-mail: schauten@ese.eur.nl

Yannis Siskos
University of Pireaus, Department of Informatics, 80 Karaoli & Dimitriou str., 18534 Pireaus, Greece, e-mail: ysiskos@unipi.gr

Roman Słowiński
Institute of Computing Science, Poznań University of Technology, Street Piotrowo 2, 60-965 Poznań, Poland, e-mail: roman.slowinski@cs.put.poznan.pl

Mariya Sodenkamp
University of Paderborn, Germany, e-mail: msodenk@mail.uni-paderborn.de

Jaap Spronk
Erasmus University, P.O. Box 1738, 3000DR Rotterdam, The Netherlands, e-mail: jspronk@rsm.nl

Stelios Tsafarakis
Technical University of Crete, Dept. of Production Engineering and Management, Decision Support Systems Laboratory, University Campus, 73100 Chania, Greece, e-mail: tsafarakis@isc.tuc.gr

Jyrki Wallenius
Helsinki School of Economics, Department of Business Technology, Runeberginkatu 22-24, 00100 Helsinki, Finland, e-mail: Jyrki.Wallenius@hse.fi

Milan Zeleny
Graduate School of Business, Fordham University, New York, USA, e-mail: mzeleny@fordham.edu

Constantin Zopounidis
Technical University of Crete, Dept. of Production Engineering and Management, Financial Engineering Laboratory, University Campus, 73100 Chania, Greece, e-mail: kostas@dpem.tuc.gr

Part I
Issues in Decision Aiding

Chapter 1
To Better Respond to the Robustness Concern in Decision Aiding: Four Proposals Based on a Twofold Observation

Bernard Roy

Abstract After reviewing what the adjective "robust" means in decision aiding [1] (DA) and explaining why it is important to be concerned about robustness in DA, I present a twofold observation (section 1.2), which leads me to make four proposals in order to better respond to robustness concern in decision aiding. With the first two proposals (sections 1.3 and 1.4), I show that, in many cases, the vague approximations and the zones of ignorance against which robustness helps to prevent, must be considered in terms of substituting the concept of version for the usual concept of scenario and focusing on the diverse processing procedures that must be used to study the decision aiding problem as it was formulated. Next, I show (section 1.5) that the traditional responses formulated in terms of "robust solutions" limit the meaning of this concept. I briefly describe a certain number of avenues for research that could be explored further, not only in order to otherwise conceive the solutions that could be qualified as robust in another way, but also to better interact with decision-makers to make them aware that the adjective "robust" can be subjective. Finally, the fourth proposal is related to forms of responses that lead to stating "robust conclusions", which do not necessarily refer to solutions characterized as robust. After defining what I mean by robust conclusions and giving some examples, I mention the rare approaches that have been proposed for obtaining such conclusions.

1.1 The Robustness Concern in Decision Aiding

Robustness in decision aiding (DA) is a subject that has given rise to many publications over the last several years (see, [3, 4, 11, 12]). After reviewing what the adjective "robust" means in terms of decision aiding and explaining why it is im-

Université Paris-Dauphine, Paris, France e-mail: roy@lamsade.dauphine.fr

[1] I use "decision aiding" rather than "decision support", "decision making" or "decision analysis" to avoid simplistic assimilations

C. Zopounidis and P.M. Pardalos (eds.), *Handbook of Multicriteria Analysis*, Applied Optimization 103, DOI 10.1007/978-3-540-92828-7_1, © Springer-Verlag Berlin Heidelberg 2010

portant to be concerned about robustness in DA, I present a twofold observation, which leads me to make four proposals designed to facilitate a better response to the robustness concern in decision aiding.

When proposing solutions and making recommendations to a decision-maker, the analyst must base these solutions and recommendations on models. However, the inevitable presence of vague approximations and zones of ignorance (explained in more detail later on) affect the modeling. The models must try to take into account these vague approximations and zones of ignorance so that the solutions and the recommendations based on the modeling do not risk ending up in results that are far worse than was anticipated when they will be implemented. Paying attention to robustness means taking robustness concern into account throughout the decision aiding activity (see [15]).

I have proposed (see [38, 39]) that, by definition, paying attention to robustness in decision aiding means seeking to be able to **withstand "vague approximations" and "zones of ignorance", in order to prevent undesirable impacts**, such as goals attained that are much worse than those anticipated or degradation of some of the properties to be maintained.

With respect to the decisions in question, the presence of vague approximations and zones of ignorance intervene at two levels of the decision aiding process:

1. the model on which the decision aiding is based may not conform exactly to the context in which the decisions will be implemented; and
2. the value system used to conceive and process the model may not conform exactly with the value system that will be used to judge these decisions.

These are two clear-cut facts which are the source of the inevitable gaps that exist between what I called the formal representation and the real-life context (see [40]):

- *formal representation (RF)*: the model and the processing procedures that are applied to highlight the results on which decision aiding is based);
- *real-life context (RLC)*: the context in which decisions will be made, executed and judged. [2]

Assuming that FR conforms exactly to RLC may lead the analyst to make recommendations that are likely to have extremely undesirable impacts. I will refer here only to one example in the field of linear programming. Believing that this conformity, without being perfect, could be judged excellent would justify the analyst recommending that the decision-maker adopt the optimal solution. In this type of model, it is the extreme point of a polyhedron that characterizes the optimal solution. An extreme point corresponds to a point on the boundary that separates that which has been judged acceptable from that which has been judged unacceptable for each of the constraints that define the extreme point. If only one of these boundary constraints has been poorly apprehended, the recommended solution may be quite

[2] *State of nature* could be used instead of *real-life context*, but because the latter expression refers to real life, it is, in my opinion, more appropriate to decision aiding than the former, which refers to nature.

different from the solution that would really be optimal, and may even be unfeasible. The optimization criterion is often defined as a synthesis of the elementary partial criteria. If these criteria are taken into account in a way that differs from the way that they are taken into account in the RLC, the recommended solution could be much worse than was expected.

In a given decision-making context, the analyst must thus seek to identify the vague approximations and zones of ignorance because they can engender a lack of conformity between FR and RLC. If the analyst ignores them, it could lead to proposing solutions and/or to making recommendations that would not protect the decision-maker from the impacts judged undesirable.

These brief considerations[3] are sufficient to make understand that the analyst who is concerned about robustness must formulate and process the problem in a way that responds appropriately to the following two requirements:

- *Requirement 1*: Inventory carefully what I have proposed calling **frailty points** and appropriately take these points into account in FR (see [39]).
  By definition, these frailty points are the place in the FR where the vague approximations and zones of ignorance are situated. I will give further examples later in the text. For the moment, I just want to draw attention on the fact that these frailty points are not only located in the model; they can also affect the processing procedures.
- *Requirement 2*: Develop forms of responses that are capable of helping decision-makers protect him/herself from the undesirable impacts that can result from the presence of vague approximations and zones of ignorance.
  These forms of responses must take into account what the decision-maker expects from this protection: are there levels of impact that he/she is willing to tolerate in certain circumstances and others that he/she judges unacceptable with respect to what he/she wants to be protected from, regardless of what happens and regardless of the cost? Depending on the RLC, the solution adopted by the decision-maker could, due to the presence of vague approximations and zones of ignorance in FR, reconcile relatively well the desire for protection and the hope to optimize the performance criterion/criteria that serve to evaluate this solution. The forms of responses elaborated will be really useful to the decision-maker only if these forms give him/her enough information to be able, given his/her own subjectivity, to decide between two conflicting risks: being poorly protected with regard to very poor performances that could result from undesirable impacts and/or abandoning all hope of good, even very good, performances.

The two above requirements above are the starting point for the twofold observation presented below.

[3] For more details, see the references cited above, as well as Roy [35] and Roy and Bouyssou [41].

1.2 A Twofold Observation

In the literature (see the enclosed references, particularly the bibliographies in these references), I have observed that many researchers who study the robustness concern perceive the two requirements stated above in a way that leads them to concentrate their research in two "special channels". By referring to these channels, I do not mean to minimize the importance of this research. I simply want to emphasize the following point: if, at this point in time, most researchers conduct their research exclusively on these two favored channels, other paths risk being ignored, despite the fact that they are equally fruitful on both the theoretical and practical levels.

- *First observation (first channel)*: The frailty points generally taken into account concern solely the **data** whose purpose is to bring the RLC aspect qualified as **uncertainty** into play in the FR. This conception leads to defining a set of **scenarios** that are supposed to describe all the realities likely to occur, which the robustness concern requires be taken into account.
  Although appropriate for responding to the first of the two requirements presented above, this conception may, in some cases, prove insufficient for identifying all the frailty points. I will show in the following two sections that this conception may lead to ignoring some vague approximations and zones of ignorance that should be taken into consideration.
- *Second observation (second channel)* : Most researchers focus on only one particular form of response, which makes the concept of **robust solution** play a central role. To qualify a solution as robust, these researchers assign a privileged role to the solution's performance in the **worst scenario**(s).In addition, this performance is modeled in the FR with a single criterion.
  This form of response leads to proposing to the decision-maker one or more solutions, which, depending on the vague approximations and zones of ignorance taken into account, optimize the most undesirable impact that the decision-maker faces. This impact can be understood either in terms of absolute performance or in terms of regret with respect to the optimal solution in the scenario that will be implemented. Such a solution can very well not correspond to what the decision-maker means by robust solution. In any case, in order to adopt such a solution, the decision-maker may need more information, particularly about what could happen if the scenario that occurs is not the worst case. These brief considerations show that the second of the two requirements presented above may not be taken into account in a satisfactory manner by this form of response. In sections 1.5 and 1.6, I will show that the robustness of a solution can be defined otherwise, not only by referring to a single performance criterion, and that, more generally, other forms of responses deserve to be studied.

1.3 First Proposal: Move Beyond the Scenario Concept

In some cases, the concept of scenario can be too reductive to allow an appropriate response to the first requirement (see section 1.1). This can lead to the desire to move beyond this concept. This reductive nature is due to two reasons:

The first reason is semantic. For a decision-maker, "scenario" stands for a description of a potential future (sometimes present) situation. It is thus this meaning of the term that the decision-maker will understand when the analyst employs "scenario" when speaking of the robustness concern. Therefore, the analyst must be aware of the risk of misunderstanding.

The second reason is related to the fact that "scenario" is generally associated with the notion of **uncertainty**. This uncertainty can lead to a belief that, somewhere, certainty exists. This implicit belief can lead to thinking that there is an objective state of reality that needs to be reproduced as closely as possible, either in qualitative or quantitative terms, in order to make the FR conform as closely as possible to the RLC.

It is thus with the aspects of reality said to be "**first order**"[4],which by definition are linked to objectively significant physical properties of things. By nature, such aspects give rise to repeated experimental verifications. Thus, speaking of approximation, either in qualitative or quantitative terms, in relation to such aspects of reality makes sense. For example, this is the case with the length of a path in kilometers, the quantity of CO2 emitted by a motor, or the price of goods available today on the market.

However, in decision aiding, aspects of reality that are not of this nature are also present. These aspects of reality are said to be "**second order**". By definition, these aspects of reality involve a connotation, the essentially subjective value that is assigned to the content of this reality. They can thus not give rise to repeated experimental verifications to reach an agreement on the way that they should be taken into account. It does not seem to me very appropriate to speak of uncertainty, and even less of approximation, when describing them. This is the case when speaking of, for example, objectives, standards, attitudes towards risk, preferences and understanding of a future situation.

It is thus not only uncertainty that leads to being concerned about robustness. More generally, it is the presence of vague approximations and zones of ignorance. Some of them cannot stem from the presence of uncertainties, as is shown in the following examples.

Vague approximations stemming from a part of arbitrariness in the modeling: It can come from the presence of a form of simplification that can be envisioned in two or three different ways, or from the introduction of two or three probability distributions representing a random phenomenon with the choice of these distributions being highly dependent on the ease of calculation.

Vague approximations stemming from the way that qualitative information is coded: : Consideration of such information (notably when modeling preferences)

[4] I borrow this terminology from Watzlawick [49].

can lead to several possible interpretations: several techniques, particularly normalization, can be envisioned to quantitatively process data that, in the form in which they are gathered, are not really quantitative.

Zones of ignorance stemming from imperfect knowledge about a complex phenomenon: This imperfect knowledge can lead to introducing different variants to describe the phenomenon.

Zones of ignorance stemming from the subjective nature of certain constraints: The boundary between possible and impossible or acceptable and unacceptable can remain very poorly defined and thus can lead to a set of options, which can be defined either as the set of all the real values that a parameter in a given interval can have or as a list of all the possible values in such an interval.

Zones of ignorance stemming from the need to differentiate the role that the criteria must play in a multicriteria aggregation model: This differentiation is most often made by means of parameters that are called, depending on the case, importance coefficients, scaling constants, substitution rates or weights, for example. Frequently, these parameters have no real existence in the mind of the decision-maker, who may even have a very superficial comprehension of the role that these parameters play in the model's functioning. Depending on the available information, the model's aggregation logic and the desired objectives, several not necessarily discrete set of values may be appropriate for assigning a value to these parameters.

Chosen among many others, these examples show that the notion of uncertainty covers rather imperfectly all the vague approximations and zones of ignorance that can cause frailty points. Consequently, using uncertainty to inventory these frailty points can lead to ignoring some that should be taken into account by the robustness concern.

It is important not to underestimate the impact of vocabulary. The words used often play a decisive role in the direction that research takes, as well as in the way that applications are conducted. In order to reduce the risk of decision-maker misunderstandings, and especially to avoid the often unconscious restrictive influence that vocabulary choices have on thought processes, at the 56th meeting of the European work group "Multiple Criteria Decision Aiding" in 2002 (see [36]), I proposed to replace the term "scenario", with the more general term **version**, The term "version" does not emphasize the notion of uncertainty. In addition, this term is more clearly associated than "scenario" with a decision aiding problem formulation (DAPF). A DAPF version is characterized by the selection of a specific option for each identified frailty point. This option is one of the items in the (finite or infinite) set of possible options associated with the frailty points under consideration. As the examples given above show, the objective of this option is different according to the nature of the vague approximations and zones of ignorance that have caused a given frailty point. This option can consist of, for example, setting the numerical value assigned to a parameter, determining the analytic form retained for a probability distribution or a utility function, choosing the coding technique, selecting the model variant adopted to describe a complex phenomenon or defining the way a trend is described.

In a given decisional context, the analyst must define a DAPF version set $\widehat{V}$ that is likely to be appropriate for responding to the robustness concern. To do so, this analyst must carefully inventory the frailty points connected to both the way the decision aiding problem was formulated and the way the problem was modeled. Questions about how this inventory can be completed and how $\widehat{V}$ can be defined so that it can be taken into account effectively without being so reduced that it no longer responds to robustness concern are not within the scope of this article. The interested reader can find some paths that can be considered to answer these questions in Roy [37, 38, 39]. Their level of difficulty is fairly similar regardless of whether version or scenario is used.

1.4 Second Proposal: Take the Way That the Different Versions are Processed Explicitly into Account

Whether the goal is finding robust solutions or developing other forms of responses related to the robustness concern, it is necessary to apply at least one processing procedure to each of the versions in $\widehat{V}$. It is also possible to apply several procedures that may, or may not, be part of the same method. Before explaining my second proposal, I need to explain briefly my meaning for the terms "procedures" and "methods". (For more details, please consult Roy [37, 39].)

A **procedure** P designates a set of execution instructions for handling a problem that will produce a $R(P,V)$ result when applied to a version V of a DAPF. Depending on the case, this result takes diverse forms. It can be a solution (e.g., an optimal solution) or a bundle of solutions (e.g., all non-dominated solutions). It can also be a statement: "there is no feasible solution", "according to the criterion g, the deviation from the optimal solution $\mathbf{x}$ is bounded by such-and-such value", or "action a is outranked by all the other actions, with the exception of the following, ...".

A **method** M designates a family $\widehat{P}$ of similar procedures that satisfy the following two requirements:

- the procedures in the family have enough similar features (e.g., structures, group of concepts, axiomatic corpus, types of approach) that considering them as belonging to the same class is justified; and
- the procedures in the family can only be differentiated by the options chosen with regard to certain frailty points in the method (e.g., concordance levels or cut thresholds in the ELECTRE methods, thresholds that make some inequalities strict in mathematical programming, the various parameters involved in a metaheuristic method, individual subjectivity in a process that calls for expert intervention at certain steps).

Wanting to apply one single processing procedure P to each of the versions in $\widehat{V}$ can, in many cases, prove to be insufficient for correctly understanding the links between FR and RLC. This is the case when distinct procedures warrant being taken into account. Even if a single procedure P may appear sufficiently appropriate, it is actually quite rare that this procedure will not have any frailty points at all. It can be the value assigned to a purely technical parameter or related to a second-order reality, the arbitrariness involved in the way certain arbitrages are conducted, especially between really similar solutions, the manner that some syntheses are conducted, particularly for taking into account the solutions that are close to the optimum in the case of a "flat" optimum.

Whatever the procedure P considered, it is necessary to inventory its frailty points and to identify the one that could significantly affect the result $R(P,V)$, For such frailty points, several options must be introduced. This leads to a multiplication of the procedures. Here again, it is important to choose wisely in order to not retain too many procedures, which would lead to processing times that are too long to adequately respond to the robustness concern.

In the following, $\widehat{P}$ designates the set of procedures retained. These procedures may, or may not, be part of a same method. Each one can be applied to each of the versions in $\widehat{V}$. It may also be that certain procedures will appear inappropriate for processing a specific version in $\widehat{V}$. In general, this leads to defining the set S of pairs (procedure P, version V) judged compatible: $S \subseteq \widehat{P} \times \widehat{V}$.

As the above observations show, basing the response on a single version set alone, or even more, on a scenario set, can in many cases not be enough to adequately respond to the robustness concern. I propose a larger conception that begins by defining the three sets - $\widehat{V}$, $\widehat{P}$ and S - which leads to basing the responses to robustness concern on the set of results: $R(P,V)$, $\forall(P,V) \in S$. Nevertheless, basing the responses on a subset $\widehat{S} \subset S$ is sometimes enough.

Before examining the different possible forms of response to the robustness concern (see sections 1.5 and 1.6), I need to make one thing clear. There are frailty points for which it is possible to wonder if the points are due to the way the problem was modeled or to certain characteristics of the processing procedures considered. This is the case, for example, with the parameters that define the role the different criteria must play in preference modeling. The way that this question is treated modifies the definition of the sets $\widehat{P}$ and $\widehat{V}$. The set of results $R(P,V)$ remains invariant facing such modifications. This comes from that the answer brought to the question does not modify the definition of S. Every element $s \in S$ is defined by a particular option retained for each of the frailty points, regardless of whether a given frailty point is assigned to procedures or to versions. To refer to this sequence of options that defines an element $s \in S$, I will use the term **variable setting**.

1.5 Third Proposal: Look for "Other" Definitions for *Robust Solutions*

In this section, I focus exclusively on the forms of response used to propose to the decision-maker one or more solutions qualified as robust. In these conditions, abandoning the second channel (see section 1.2) consists of proposing definitions for robust solutions that, one way or another, go beyond the usual schema to which most researchers currently conform. This schema seems to me to be characterized by the following three features:

Characteristic feature #1: In the FR model, preference is modeled using a single performance criterion (e.g., gain, cost); the robustness concern is not taken into account in any way in this single criterion.

Characteristic feature #2: A single robustness measure is defined to give meaning to the statements "solution **x** is at least as robust as solution **y**". This measure is the only criterion used to define robust solutions.

Characteristic feature #3: The single robustness measure assigns a preponderant role to the performance of the solution in the worst case(s), for example, the worst scenarios or the worst pairs (P,V) in S.

The basic and innovative work of Kouvelis and Yu [27] has probably contributed the most to the fact that research is mainly conducted within the framework of the above schema. A robust solution is a solution **x** that, by definition, optimizes a single robustness measure $r(\mathbf{x})$, which is one of the three measures on which Kouvelis and Yu primarily focused. Below, I remind the reader of their definitions.

These three measures insure that the optimization criterion v comes into play in a given FR model (Characteristic feature #1). This criterion assigns a value $v_s(\mathbf{x})$ to **x** in scenario s, assuming that "optimum" means "maximum".

Absolute robustness. The robustness measure that must be maximized is defined by the value of the solution in the worst scenario:$r(\mathbf{x}) = \min_s v_s(\mathbf{x})$.

Absolute deviation. The robustness measure that must be minimized is defined by the value of the absolute regret in the worst scenario, due to the fact that the solution differs from that which would be optimal in this scenario:$r(\mathbf{x}) = \max_s[v_s^* - v_s(\mathbf{x})]$, where v_s^* is the value taken by the optimal solution in scenario s.

Relative deviation. The robustness measure that must be minimized is defined by the value of the relative regret in the worst scenario, due to the fact that the solution is not optimal in this scenario:

$$r(\mathbf{x}) = \max_s \frac{v_\mathbf{x}^* - v_s(\mathbf{x})}{v_\mathbf{x}^*}$$

When, further on, I need to refer to these measures, I will speak of the "three measures of Kouvelis and Yu". I remind the reader that these measures correspond to the classic Wald and Savage criteria in decision under uncertainty.

In the sub-sections below, without any pretentions of exhaustivity, I evoke the directions that have begun to be explored but that I think deserve to be more explored. All aim to more or less free themselves from the schema that I have described above with its three characteristic features. This exploration should lead to new definitions of "robust solutions", which because they abandon the second channel (see section 1.2) will permit, at least in certain cases, to better respond to the robustness concern. To accomplish this aim, these new definitions must consider a double goal:

- take into account other conceptions of robustness than those taken into account in the schema described above;

- be appropriate for organizing a discussion with the decision-maker to help him/her understand the subjective aspect of robustness.

1.5.1 Abandoning the Third Characteristic Feature Without Necessarily Abandoning the First Two

Three directions at least seem to deserve more exploration:

a) Involve the neighborhood of the worst case

Once one of the three robustness measures of Kouvelis and Yu is chosen, it is possible to consider as robust not only the solutions that optimize this measure, but also all those that are in the neighborhood of the optimum. This neighborhood can be defined in several ways.

For example, let us assume that the measure $r(\mathbf{x})$ chosen is the absolute robustness of Kouvelis and Yu. It is then possible to qualify as robust every solution $\mathbf{x}$ that verifies $r(\mathbf{x}) \geq w$, where w is a value defined *a priori* by the decision-maker. When the criterion v represents a gain, the value w is interpreted as the minimum gain that the decision-maker wants to guarantee. The neighborhood of the optimum thus taken into account does not allow the robust solutions to be highlighted unless $w \leq w^* = \max_{\mathbf{x}\in X} r(\mathbf{x})$.

Among others, Aubry [6] and Rossi [33] have focused on this way of qualifying a robust solution. Clearly, the greater the difference $w^* - w$ the greater the number of solutions qualified as robust. Limiting the definition of the neighborhood is possible and allows a solution to be qualified as robust by retaining only the solutions that maximize the number or the proportion of scenarios that permit an objective b to be reached. This objective can be defined either in terms of absolute gain, or in terms of absolute regret, or even in terms of relative regret (again, with the meaning of the three criteria of Kouvelis and Yu). It was precisely this way of defining that led me to define three new measures: (b,w)-absolute robustness, (b,w)-absolute deviation, (b,w)-relative deviation (see [38, 40]). In these articles, I particularly emphasized the role that the analyst could assign to parameters b and w to help the decision maker determine the precise meaning that he/she wants to give to *robust solutions*. The choice of the values that he/she assigns to these parameters determines the compromise between guaranteed gain and the hope to reach a goal.

It is also possible to propose to qualify as robust the solutions that verify:

$$r(\mathbf{x}) \geq \left(1 - \frac{p}{100}\right) \max_{\mathbf{x}\in X} r(\mathbf{x}),$$

where $r(\mathbf{x})$ denotes absolute robustness and p, a percentage of deviation to the worst case that the decision-maker gives *a priori*. This is the p-robustness (see [1]; [45]). By proceeding in this manner, regardless of the value of the percentage p, the set of solutions qualified as robust is never empty. Diverse variants have been proposed to limit the set of solutions thus qualified as robust.

It is possible to propose similar definitions for robust solutions by adopting, for $r(\mathbf{x})$, the absolute regret or the relative regret of Kouvelis and Yu. However, it is not certain that such definitions would be so easily interpreted by the decision-maker.

b) Design a robustness measure that, while assigning a role to the worst case, weakens its influence

With the exception of research limited to very specific concrete cases, only one proposal seems to have been made in this direction: "lexicographic α-robustness" (see [5, 25]). This proposal is based on any of the three robustness measures of Kouvelis and Yu, and it assumes that a finite set S of scenarios has been defined. It takes into account all the gains or regrets lexicographically, from the worst case to the best case. This lexicographic ranking brings into play a reference point and an indifference threshold α, which according to the authors is supposed to reflect "the subjective dimension of robustness". The set $A(\alpha)$ of the solutions thus qualified as robust is such that $\alpha' > \alpha \Rightarrow A(\alpha) \subseteq A(\alpha')$. With the values of α, $A(\alpha) = \varnothing$ can be obtained. The conditions for solution $\mathbf{x}$ belonging to, or not belonging to, $A(\alpha)$ are highly complex. Thus, it seems to me that it would be difficult to make a decision-maker understand the general meaning of "lexicographic α-robustness". Still, as the authors have underlined, this approach can also be seen as a source for developing robust conclusions (see section 1.6).

c) Limit the option combinations appropriate for defining a worst case

Exploring this direction consists of seeking to benefit from the following empirical judgments. Let us assume that, for certain "critical" frailty points, the identity of the option that would most negatively affect the performance $v(\mathbf{x})$ of a solution $\mathbf{x}$ can be identified. In general, these options are "extreme". A scenario, or more generally a variable setting (in the sense given at the end of section 1.4), which combines a very large number or a very large proportion of such options, reflects a real-life context that could be judged highly unrealistic. Under these conditions, it can seem justifiable either to totally eliminate from S the scenarios or variable settings that are not very meaningful for defining the robustness $r(\mathbf{x})$ of a solution or to assign them a separate role that will lessen their impact.

To characterize the elements $s(S)$ judged not very meaningful, many different ways are possible. For example, each of the critical frailty points can be associated to a particular option, called "reference option", that must have a central position or must appear to be the most realistic. Then, it is possible to consider as not very meaningful the element $s(S)$ in which the number or proportion of options that differ from the reference option for which the critical frailty points, exceed a certain threshold. If, for each critical frailty point, a measure of the gap between a given option and the reference option can be defined, and if the aggregation of such a measure has meaning, the result of this aggregation can be used to define the relative meaningfulness of an element s not very meaningful.

The work of Bertsimas and Sim [8, 9] related to linear programming FR models goes in this direction, as does the work of Ben-Tal and Nemirovski [7], Chinneck and Ramadan [14], Gabrel and Murat [20], and Minoux [30].

These three directions are certainly not the only ones. Other directions have already begun to be explored. A non-exhaustive list would include the research of Lamboray [28], Liesiö et al. [29], Perny et al. [31], Salazar and Rocco [44], and Spanjaard [47]. The proposal of Aissi and Roy [3, 4] (sub-section 3.4) also goes in this direction. This proposal leads to (almost) ignoring the role played by the worst case. It takes into account only the position of solution $\mathbf{x}$ in each of the rankings defined by the decreasing performance of $v_s(\mathbf{x}), \forall s \in S$.The values assigned to the two parameters that fix two types of thresholds should help the decision-maker to find a good compromise between the performance desired and the risk tolerated.

1.5.2 Abandoning the Last Two Characteristic Features Without Necessarily Abandoning the Third

Here again, the case in which the preference has been modeled (in the FR model) by a single performance criterion that does not take robustness concern into account is considered. In such conditions, several criteria can be brought into play in order to qualify a solution as robust. Two possibilities can be considered.

a) Take only a single robustness measure into account

Along with the criterion defined by a single robustness measure, one or more other criteria can be used to define the precise conditions in which a solution warrants being qualified as robust. The role of this/these complementary criterion/criteria, which are not based on a robustness measure, is to provide additional information, essentially about performance, in order to help the decision-maker determine what "robust" means to him/her.

For example, a complementary criterion $w(\mathbf{x})$ can be defined in one of the two following ways:

- $w(\mathbf{x})$ is the median performance $v_s(\mathbf{x}), \forall s \in S$ (if S is probalilized, the median can be replaced by the expected value); and
- $w(\mathbf{x}) = v_{s_0}(\mathbf{x})$, where s_0 is a reference scenario or a reference variable setting defined by retaining, for each frailty point, the option that the decision-maker thinks is the most realistic.

Once a robustness measure $r(\mathbf{x})$ has been chosen (not necessarily one of the ones proposed by Kouvelis and Yu), let us denote by Ω the set of solutions that verifies:

$$r(\mathbf{x}) \geq \left(1 - \frac{p}{100}\right) \max_{s \in S} r(\mathbf{x}) \quad \text{if } r(\mathbf{x}) \text{ must be maximized,}$$
$$r(\mathbf{x}) \leq \left(1 + \frac{p}{100}\right) \min_{s \in S} r(\mathbf{x}) \quad \text{if } r(\mathbf{x}) \text{ must be minimized,}$$

where p is a percentage introduced to set the boundaries of the neighborhood Ω of the optimum; outside of this neighborhood, the solutions cannot be considered robust. This leads to associating each solution $\mathbf{x} \in \Omega$ to its robustness measure $r(\mathbf{x})$ and the value corresponding to the performance indicator $w(\mathbf{x})$. Let:

$$w_\rho = \min_{\mathbf{x} \in \Omega . r(\mathbf{x}) = \rho} w(\mathbf{x})$$

Showing the decision-maker the curve (efficient frontier) representing the way that the performance indicator w_ρ varies in relation to the robustness measure ρ can be very useful. This curve can help the decision-maker select the solutions that, in his/her eyes, achieve a suitable compromise between the acceptable risk and the expected performance.

The following are examples of publications that take this approach: Chen et al. [13], Ehrgott and Ryan [16], Kennington et al. [26], and Salazar and Rocco [44]. Interested readers can consult the brief analysis of these publications in Aissi and Roy [3, 4] (section 3.2).They will also find in this article (at the end of section 1.5.2) a proposal dealing with linear programming models in which only the coefficients of the constraint matrix are imperfectly known (the set S that allows the imperfect knowledge to be modeled being assumed finite). In addition, this proposal assumes that the decision-maker can tolerate a robust solution that does not verify all the constraints, if this is true only for a small number of elements $s(S)$ and leads to a low cost solution. It also assumes that a solution that is mathematically infeasible is preferable to a solution that satisfies all the constraints but costs a lot more. This proposal brings into play a robustness measure and a cost criterion. In fact, three possible measures are proposed. Consequently, this approach helps the decision-maker to find a compromise between the solution's performance (here, the cost) and the gaps that reflect the imperfect satisfaction of the constraints.

b) Take several robustness measures into account

It is clear that robustness can be apprehended from various perspectives. The undesirable impacts from which the decision-maker wants to protect him/herself can also vary, including for example objectives that remain unattained or properties that are not maintained. One robustness measure can be associated to each of these objectives and/or to each of these properties.

Let $R = \{r_1(\mathbf{x}), \dots, r_k(\mathbf{x})\}$ be a set of k measures judged appropriate for apprehending robustness. To qualify a solution as robust, it is possible to use a multi-criteria approach based on these k measures. They should, as much as possible, be uncorrelated. It is thus a question of generalizing, for k robustness measures, what can be done with only one measure. Two possibilities can be envisaged. The first consists of bringing into play only the way robustness is measured for qualifying a solution as robust as described in section 1.5.1. The second consists of bringing into play, along with the k criteria of R, one or more other complementary criteria that provide additional information destined to help the decision-maker determine what "robustness" means to him/her, as described in section 1.5.2a).

The interesting attempt of Hites et al. [23] takes the direction of the first possibility. Assuming the set S of scenarios or variable settings to be finite, the authors consider that for each $s \in S$, $v_s(\mathbf{x})$ provides information that will be pertinent for evaluating the relative robustness of solution $\mathbf{x}$. This leads them to state $R = \{v_s(\mathbf{x})/s \in S\}$. They point out the limitations of the traditional multicriteria decision-aiding tools for defining, based on R, the solutions that warrant being qualified as robust. However, they also emphasize that a solution can only be qualified as robust if it is non-dominated by the criteria set R.

I know of only one article that goes in the direction of the second possibility stated above. Jia and Ierapetritou [24] focus on the design phase of a process for synthesizing chemical products in small or intermediate quantities. Interested readers can consult a brief description by Aissi and Roy [3, 4] (section 4.3). They will also find at the end of section 4.3 a general proposal for the case in which S is finite. This proposal is based on using two robustness measures, completed by a performance indicator $w(\mathbf{x})$ of the same type as the one introduced in section 1.5.2a).

1.5.3 Abandoning the First of the Three Characteristic Features

One way to break away from feature #1is not to introduce an initial performance preference in the FR model but rather, from the beginning, to seek to apprehend the preferences in terms of robustness. This assumes that the FR model brings into play a family of several criteria (the case of one single criterion can, it seems to me, be left aside), each of these criteria translating one perspective of robustness. I do not know any work using this approach, other than the one in which I participated (see [32]), which concerned the regulation of dense traffic on a rail network (timetable elaboration and local network improvements). Interested readers can consult section 4.2 of the article by Aissi and Roy [3] for a brief description.

Another probably more natural way to break away from feature #1 leads to focusing on the cases in which the FR preference model is based on a family F of $n(\geq 2)$ criteria. Each criterion v_i models one component of the multidimensional performance of a solution $\mathbf{x}$, evaluated without taking the robustness concern into account. Let $v_{is}(\mathbf{x})$ be this performance when the real-life context is assumed to conform to the scenario or variable setting $s \in S$. Depending on whether one or several robustness measures are used to give meaning to the qualifier "robust", two cases can be distinguished.

a) With a single robustness measure

This single robustness measure must synthesize, for a single dimension characterizing the robustness of a solution **x**, the totality of the performances $v_{is}(\mathbf{x})$, not only $\forall s \in S$ as in section 1.5.1, but also for $i = 1,..,n$. I know of no publication that has proposed such a general measure. It is obviously possible to begin by aggregating the n criteria into a single criterion, which bring us back to the cases mentioned in section 1.5.1. I limit myself below to presenting only a brief description of a different path already presented in sub-section 3.3 of the article by Aissi and Roy [3, 4].

First, a specific robustness measure (see 1.5.1) is associated to each of the n criteria. Second, a way to aggregate the n measures is sought. If these measures are expressed on a common scale, the min or max can be used as a definition of the synthesis measure. If the role played by the different robustness measures needs to be differentiated according to the criterion to which they are associated, such operators as the Ordered Weighed Average (OWA) or Choquet's integral can, with certain precautions, be used.

b) With several robustness measures

The set R of robustness measures that can be brought into play may be conceived in two ways. The first consists of apprehending the robustness of a solution from various points of view, as in sub-section 1.5.2b). The second consists of associating a specific robustness measure to each of the n criteria, as in sub-section 1.5.3a). Like monocriterion preference models, multicriteria preference models do not seem to have been considered in the works that examine the robustness of a solution from a multicriteria perspective. Aissi and Roy [3, 4] (subsection 4.4) have described the only two studies I know of: Bescharati and Azarm [10] and Fernàndez et al. [17].

1.6 Fourth Proposal: Seek to Construct "Robust Conclusions"

Stating a robust conclusion constitutes a form of response to the robustness concern, which generalizes those presented in the previous section. After reviewing several definitions, I present diverse examples of robust conclusions and end this section by evoking some of the approaches that allow this kind of conclusions to be obtained.

1.6.1 Definitions

A ***conclusion*** *is an assertion that take the results $R(P,V)$ into account, either for all pairs $(P,V) = s \in S$, or only for the elements $\widehat{S} \subset S$. These assertions must help the decision-maker to frame, mark out or restrict his/her range of reflection and action.*

A ***robust conclusion*** *is a conclusion that contains in its statement a set of conditions in which its validity has been established.* This means that the robustness of a conclusion is contingent on both the validity range $\widehat{S} \subset S$ to which it belongs and the way that the validity conditions are formulated. Focusing only on the case of $\widehat{S} = S$ would lead to restricting the diversity and the importance of robust conclusion to no useful purpose. It would be the same for requirements that the set $\widehat{S}$ be rigorously defined and that the conditions of validity contain no ambiguities. These considerations led me (see [34, 35]) to distinguish three types of robust conclusions:

A conclusion is qualified as perfectly robust when its validity is rigorously established $\forall s \in \widehat{S}$, $\widehat{S}$ subset of S perfectly identified.

A conclusion is qualified as approximately robust when it has been established without ambiguity for "almost" all $s \in \widehat{S}$, $\widehat{S}$ subset of S perfectly identified. "Almost" means that the exceptions are related to pairs (P, V) that are not necessarily perfectly identified but which can be neglected because they are not very realistic or are almost totally without interest.

A conclusion is qualified as pseudo robust when the conditions that establish its validity contain a degree of ambiguity, and this validity is only established for the elements of a subset of $\widehat{S}$, which is not necessarily perfectly identified but is judged representative of $\widehat{S}$.

Any perfectly robust conclusion is *a fortiori* approximately robust, and any approximately robust conclusion is *a fortiori* pseudo robust.

1.6.2 Examples

The following assertion constitutes the most common example of a robust conclusion: "in S, $\mathbf{x}^*$ is the only solution that optimizes the robustness measure $r(\mathbf{x})$". More generally, since Π defines a set of properties that a solution $\mathbf{x}$ must have to be qualified as robust, the assertion "$\mathbf{x}^*$ is a robust solution because the properties Π are validated without ambiguity for all S" is a robust conclusion.

The concept of robust conclusion would have little interest if it was only applied to assertions that affirm the robustness of a solution on all S in a perfectly well-defined manner. The analyst may be brought to formulate highly interesting robust conclusions for the decision-maker who does not explicitly refer to a definition of a robust solution. In addition, as the following examples show, the analyst may, because he/she lacks time or does not have sufficiently rigorous algorithms available, be led to propose assertions whose validity range is not necessarily all of S and/or which potentially tolerate exceptions.

a) Other examples of perfectly robust conclusions

i) "$\widehat{S}$ being a perfectly identified subset of S, $\forall s \in \widehat{S}$, $\mathbf{x}$ is a solution whose deviation from the optimum according to criterion g never exceeds ε.

ii) "The following solutions... are feasible solutions for all the variable settings of S with the exception of the following..." (the range of validity of this perfectly robust conclusion is S, without the excepted variable settings).
iii) "Solution **x** dominates solution **y**, $\forall s \in S$, except perhaps for the variable settings $s_1,...,s_q$" (the range of validity of this perfectly robust conclusion is S, without the variable settings $s_1,...,s_q$).
iv) "The variable settings of $\widehat{S}$ (perfectly identified subset of S) prove the incompatibility of the following objectives ..." (example of objectives: attain a performance at least equal to b_i according to criterion g_i for $i = 1,...,k$).
v) "$\widehat{S}$ being a perfectly identified subset of S, **x** is a solution that, $\forall s \in \widehat{S}$, has the property P and, $\forall s \notin \widehat{S}$, has the property Q" (for example, with a finite solution set, property P: **x** is always among the α best solutions; property Q: **x** is never among the β worst solutions).

b) Examples of approximately robust conclusions

i) "$\widehat{S}$ being a perfectly identified subset of S, $\forall s \in \widehat{S}$, **x** is a solution whose deviation from the optimum according to criterion g only exceeds ε for certain variable settings that, because they correspond to option combinations that are not very realistic, can be neglected.
ii) "A probability distribution having been defined on S, **x** is a feasible solution with a probability at least equal to $1 - \varepsilon$" (as long as the risk ε of non-feasibility of a solution is judged negligible, this type of conclusion can be considered as approximately robust).
iii) "Solution **x** dominates solution **y**, $\forall s \in \widehat{S}$, $\widehat{S}$ only differing from S by the variable settings that correspond to combinations of procedures and versions that can be judged as not easily compatible."
iv) "$\widehat{S}$ being a perfectly identified subset of S, **x** is a solution that, for *almost* all variable settings $s \in \widehat{S}$, has the property P, where *almost* means that the variable settings for which it could be considered otherwise are judged negligible."
v) "Solution **x** is non dominated, $\forall s \in S$, except for certain variable settings that, because they are *too different* from a reference variable setting, can be neglected."

c) Examples of pseudo robust conclusions

i) "The performance of the solution **x** according to the criterion g can be considered as always being at least equal to W because a simulation using a very large number of random drawings of variable settings of S has shown that $g(\mathbf{x}) \geq W$ was always verified."
ii) "A systematic but non-exhaustive exploration of S showed that the following solutions ... were always feasible solutions."
iii) "Solution **x** dominates (or almost dominates) solution **y**, $\forall s \in \widehat{S}$ (perfectly defined subset of S), where *dominates (or almost dominates)* means that when the

dominance is violated, it is only violated for one or two criteria, with slight deviations."

iv) "$\widehat{S}$ beings a subset of S that was constructed in order to serve as a *representative* sub-set of the variable settings making up S, the assertions validated on $\hat{S}$ can be judged valid on S."

v) "The assignment of $\mathbf{x}_i$ to the category C_j can, $\forall s \in S$, be considered as robust, given the meaning of this term, although it is not completely without ambiguity."

Interested readers will find concrete illustrations of some of the previous examples in works by Pomerol et al. [32], Roy [34, 35], Roy and Bouyssou [41] (chapter 8), Roy and Hugonnard [42, 2], and Roy et al. [43].

1.6.3 How Can Robust Conclusions be Obtained?

The examples in the previous section show that extremely varied forms of robust conclusions warrant consideration. They can encourage thinking that it should be possible to conceive a list of typical statements that are particularly worthy of interest for each of the three types of robust conclusions. However, the statements on such a list would be of interest only for those robust conclusions to which an approach allowing their validation can be associated. This approach would obviously depend on:

- the formal representation (FR): mono or multicriteria models with solutions characterized by a set of variables subjected to constraints or by a set of numbered actions;
- the way that S was conceived: finite or infinite set, nature of the frailty points concerned for $\widehat{V}$ as well as $\widehat{P}$ (see sections 1.3 and 1.4).

Rather than begin by establishing a list of typical conclusions and then wondering about the context (FR, S) in which they could be validated, it is perhaps preferable to proceed in the opposite manner. This is what seems to have been done by the rare authors who sought to respond to the robustness concern by validating more elaborate robust conclusions than those that are limited to affirming than such a solution is robust according to very precise definitions. I will end this section with a brief presentation of the context in which are situated the three approaches of which I am currently aware. All three focus on cases in which:

- the decision-making possibilities are defined by a finite list of what are called *actions* rather than *solutions*;
- the preference model is multicriteria;
- the problematic can be one of choice, of sorting or of ranking.

a) J. Figueira, S. Greco, V. Mousseau and R. Slowinski have focused on the following context (see [18, 19, 21, 22]):

The FR preference model is an additive value function that aggregates several partial monotonic non-decreasing functions. S takes into account a set $\widehat{P}$ of procedures: the procedures of $\widehat{P}$ are all those that allow an additive value function to be built that is compatible with the preference data that the decision-maker provides about a reference set of actions.

There are two types of robust conclusions:

- those that are validated by all the procedures of $\widehat{P}$ that are qualified as **necessary**;
- those that are validated by at least one procedure of $\widehat{P}$ that are qualified as **possible**.

b) T. Tervonen, J. Figueira, R. Lahdelma, P. Salminen and have focused on the following context (see [48]):

- the FR preference model is a relational outranking system of the ELECTRE type, which takes several pseudo criteria into account; and
- the variable setting that define S are related to the possible options in terms of weights, veto thresholds and discrimination thresholds.

The robust conclusions (stated in S) are those that can be validated for a subset $\widehat{S}$ of S built by simulation, which brings into play the probability distributions of all the possible options for each of the frailty points considered. The interested reader can consult Sörensen [46] for another approach to the robustness concern using the Monte Carlo method.

c) H. Aissi and I have focused on the following context (see [3, 4], subsection 5.3):

- the FR preference model is a relational outranking system of the ELECTRE type that uses pseudo criteria; and
- .S, based on two previously defined sets $\widehat{S}$ and $\widehat{V}$, is finite

The robust conclusions are obtained at the end of a three-step procedure. The first step constructs a subset $\widehat{S}$ of S, which must be the result of a compromise between two relatively conflicting requirements, *calculability* and *representativeness*. The second step highlights two categories of frailty points: those that can greatly influence the results, and those whose influence on the results is negligible. On this basis, we proposed replacing $\widehat{S}$ by a set $\widehat{S}'$ that is at least as representative and, as far as possible, much more limited. The third step, based on a careful analysis of the results $R(P,V), \forall (P,V) \in \widehat{S}'$, must allow robust conclusions pertinent to the problem under study to be obtained. This approach generalizes and extends the insufficiently formalized approach that was used to obtain robust conclusions in two concrete cases (see [41], chapter 8; [42, 2].

The types of statements of robust conclusions that each of the three approaches above can validate warrant a more detailed explanation. There are many other contexts (FR, S) than those mentioned above (especially in relation to mathematical programming); it should thus be possible to propose approaches specifically designed to obtain robust conclusions.

1.7 Conclusion

Most publications that propose ways to respond to robustness concern (as defined in section 1.1) remain limited to the two special channels described in section 1.2. This is the two-fold observation that led me to formulate the four proposals presented above. The first two proposals (see sections 1.3 and 1.4) should help the analyst not to overlook frailty points in what I call the formal representation (see section 1.1). If these frailty points are ignored, the analyst could propose solutions or make recommendations that could, when implemented, lead to results much worse than those expected. The last two proposals (see sections 1.5 and 1.6), in conjunction with the first two, present research directions that, even though some have already begun to be explored, should, it seems to me, help to develop new forms of responses to robustness concern.

Using more general definitions of the concept of robust solution or even forms of responses conceived in terms of robust conclusions, it should be possible to take into account a greater variety of decision-makers' expectations, to provide them with a better understanding of the subjectivity of the robustness concern, to help them choose better compromises between the risk of being poorly protected against very bad performances and the abandoning of hope for good, even very good, performances. The research directions that have been outlined above should not only allow a better understanding of the robustness concern in applications, but also stimulate interesting research on more theoretical levels.

References

1. H. Aissi, C. Bazgan, and D. Vanderpooten. Min-max and min-max regret versions of combinatorial optimization problems: A survey. *European Journal of Operational Research*, 197(2):427–438, 2009.
2. H. Aissi and B. Roy. Classement des prolongements des lignes de métro en banlieue parisienne (présentation d'une méthode multicritère originale). *Cahiers du CERO*, 24(2-4):153–171, 1982.
3. H. Aissi and B. Roy. Robustesse en aide multicritère à la décision. 2008.
4. H. Aissi and B. Roy. Robustness in multi-criteria decision aiding. In J. Figueira, S. Greco, and M. Ehrgott, editors, *New Trends in Multicriteria Decision Analysis*. Springer, New York, 2009. In press.
5. M.A. Aloulou and Vanderpooten D. Kalaï, R. Une nouvelle approche de robustesse: α-robustesse lexicographique. *Bulletin du Groupe de Travail Européen Aide Multicritère à la Décision*, 3(12):5–8, 2005.
6. A. Aubry, A. Rossi, and M. Jacomino. A generic off-line approach for dealing with uncertainty in production systems optimisation. In *13th IFAC International Symposium on Information Control problems in Manufacturing (IFAC-INCOM'09)*, pages 1464–1469, 2009.
7. A. Ben-Tal and A. Nemirovski. Robust solutions of uncertain linear programs. *Operations Research Letters*, 25:1–13, 1999.
8. D. Bertsimas and M. Sim. Robust discrete optimization and network flows. *Mathematical Programming B*, 98(1-3):49–71, 2003.
9. D. Bertsimas and M. Sim. The price of robustness. *Operations Research*, 52(1):35–53, 2004.

10. B. Bescharati and S. Azarm. Worst case deterministic feasibility and multiobjective robustnesss measures for engineering design optimization. *International Journal of Reliability and Safety*, 1:40–58, 2006.
11. J.C. Billaut, A. Moukrim, and E. Sanlaville. *Flexibilité et Robustesse en Ordonnancement*. Lavoisier, Paris, 2005.
12. J.C. Billaut, A. Moukrim, and E. Sanlaville. *Flexibility and Robustness in Scheduling*. Wiley-ISTE, New York, 2008.
13. W. Chen, J. Allen, K.L. Tsui, and Mistree F. Procedure for robust design: minimizing variations caused by noise factors and control factors. *Journal of Mechanical Design*, 118(4):478–493, 1996.
14. J.W. Chinneck and K. Ramadan. Linear programming with interval coefficients. *The Journal of the Operational Research Society*, 51(2):209–220, 2000.
15. L.C. Dias. A note on the role of robustness analysis in decision-aiding processes. In B. Roy, M. Aloulou, and R. Kalaï, editors, *Robustness in OR-DA*, pages 53–70. LAMSADE, Paris, 2007. Annales du Lamsade no 7.
16. M. Ehrgott and D.M. Ryan. Constructing robust crews schedules with bicriteria optimization. *Journal of Multi-Criteria Decision Analysis*, 11(3):139–150, 2002.
17. Fernàndez F., S. Nickel, J. Puerto, and A.M. Rodriguez-Chia. Robustness in the Pareto-solutions fort he multicriteria weber location problem. Technical Report no 50/1999, University of Kaiserslautern, 1999.
18. J. Figueira, S. Greco, V. Mousseau, and R. Slowinski. Robust ordinal regression. In J. Figueira, S. Greco, and M. Ehrgott, editors, *New Trends in Multicriteria Decision Analysis*. Springer, New York, 2009. In press.
19. J. Figueira, S. Greco, and R. Slowinski. Building a set of additive value functions representing a reference preorder and intensities of preference: GRIP method. *European Journal of Operational Research*, 195(2):460–486, 2009.
20. V. Gabrel and C. Murat. Robustness and duality in linear programming. *Journal of the Operational Research Society*, 2009. In press.
21. S. Greco, V. Mousseau, and R. Slowinski. Ordinal regression revisited: Multiple criteria ranking with a set of additive value functions. *European Journal of Operational Research*, 191:416–435, 2008.
22. S. Greco, V. Mousseau, and R. Slowinski. Multicriteria sorting with a set of value functions. 2009. Submitted.
23. R. Hites, Y. De Smet, N. Risse, M. Salazar, and P. Vincke. About the applicability of MCDA to some robustness problems. *European Journal of Operational Research*, 191(1):322–332, 2006.
24. Z. Jia and M.G. Ierapetritou. Generate Pareto optimal solutions of scheduling problems using normal boundary intersection technique. *Computers and Chemical Engineering*, 31(4):268–280, 2007.
25. R. Kalaï. *Une nouvelle approche de robustesse: Application à quelques problèmes d'optimisation*. Thèse de Doctorat, Université Paris-Dauphine, 2006.
26. J. Kennington, K. Lewis, E. Olinick, A. Ortynski, and G. Spriride. Robust solutions for the WDM routing and provisioning problem with uncertainty demands and single link failure protection. Technical Report 01-EMIS-07, Southern Methodist University, EMIS Dept., 2001.
27. P. Kouvelis and G. Yu. *Robust Discrete Optimization and its Applications*. Kluwer Academic Publishers, Dordrecht, 1997.
28. C. Lamboray. Supporting the search for a group ranking with robust conclusions on prudent orders. In B. Roy, M. Aloulou, and R. Kalaï, editors, *Robustness in OR-DA*, pages 145–171. LAMSADE, Paris, 2007. Annales du Lamsade no 7.
29. J. Liesiö, P. Mild, and A. Salo. Robust portfolio modelling with incomplete cost information and projects interdependencies. *European Journal of Operational Research*, 190:679–695, 2008.
30. M. Minoux. Duality, robustness, and 2-stage robust LP decision models. In B. Roy, M. Aloulou, and R. Kalaï, editors, *Robustness in OR-DA*, pages 173–190. LAMSADE, Paris, 2007. Annales du Lamsade no 7.

31. P. Perny, O. Spanjaard, and L.X. Storme. Enumeration of robust paths and spanning trees in multivalued graphs. *Annals of Operations Research*, 2009. In press.
32. J.C. Pomerol, B. Roy, C. Rosenthal-Sabroux, and A. Sad. An intelligent DSS for the multicriteria evaluation or railway timetables. *Foundations of Computing and Decision Sciences*, 20(3):219–238, 1995.
33. A. Rossi. *Ordonnancement en milieu incertain - Mise en œuvre d'une démarche robuste.* Thèse de Doctorat, Institut National Polytechnique de Grenoble, 2003.
34. B. Roy. Une lacune en RO-AD - les conclusions robustes. 1997.
35. B. Roy. A missing link in OR-DA - robustness analysis. *Foundations of Computing and Decision Sciences*, 23(3):141–160, 1998.
36. B. Roy. Robustesse de quoi et vis-à-vis de quoi mais aussi robustesse pourquoi en aide à la décision? In C. Hengeller Antunes, J. Figueira, and J. Climaco, editors, *Aide Multicritère à la Décision - Multiple Criteria Decision Aiding*, pages 29–39. INESC, Coimbra, 2004.
37. B. Roy. A propos de robustesse en recherché opérationnelle et aide à la decision. In J.C. Billaut, A. Moukrim, and E. Sanlaville, editors, *Flexibilité et Robustesse en Ordonnancement*, pages 35–50. Lavoisier, Paris, 2005.
38. B. Roy. La robustesse en recherche opérationnelle et aide à la décision: Une préoccupation multi-facettes. In B. Roy, M. Aloulou, and R. Kalaï, editors, *Robustness in OR-DA*, pages 209–231. LAMSADE, Paris, 2007. Annales du Lamsade no 7.
39. B. Roy. Robustness in operations research and decision aiding. In Billaut J.C., Moukrim A., and Sanlaville E., editors, *Flexibility and Robustness in Scheduling*, pages 35–51. Wiley-ISTE, New York, 2008.
40. B. Roy. Robustness in operational research and decision aiding: A multi-faceted issue. *European Journal of Operational Research*, 200(3):629–638, 2010.
41. B. Roy and D. Bouyssou. *Aide Multicritère à la Décision: Méthodes et Cas.* Economica, Paris, 1993.
42. B. Roy and J.C. Hugonnard. Ranking of suburban line extension projects on the Paris metro system by a multicriteria method. *Transportation Research A*, 16(4):301–312, 1982.
43. B. Roy, M. Présent, and D. Silhol. A programming method for determining which Paris metro stations should be renovated. *European Journal of Operational Research*, 24:318–334, 1986.
44. D.E. Salazar and C.M.S. Rocco. Solving advanced multi-objective robust designs by means of multiple objective evolutionary algorithms (MOEA): A reliability application. *Reliability Engineering and System Safety*, 92(6):697–706, 2007.
45. L. Snyder. Facility location under uncertainty: A review. *IIE Transactions*, 38:537–554, 2006.
46. K. Sörensen. Investigation of practical, robust and flexible decisions for facility location problems using tabu search and simulation. *Journal of the Operational Research Society*, 59(5):624–636, 2008.
47. O. Spanjaard. *Exploitation de préférences non-classiques dans les problèmes combinatoires: Modèles et algorithmes pour les graphes.* Thèse de Doctorat, Université Paris-Dauphine, Paris, 2003.
48. T. Tervonen, J. Figueira, R. Lahdelma, and P. Salminen. A stochastic method for robustness analysis in sorting problems. *European Journal of Operational Research*, 192(1):236–242, 2009.
49. P. Watzlawick. *La Réalité de la Réalité (Confusion, Désinformation, Communication).* Editions du Seuil, Paris, 1976.

Chapter 2
Multi-Criteria Decision Analysis for Strategic Decision Making

Gilberto Montibeller and Alberto Franco

Abstract In this chapter we discuss the use of MCDA for supporting strategic decision making, particularly within strategy workshops. The chapter begins by exploring the nature of strategic decisions and the characteristics of the strategic decision making process. Specifically, we examine the technical issues associated with the content of strategic decisions, and the social aspects that characterise the processes within which they are created. These features lead us to propose a number of adaptations to the standard MCDA approach if it were to be used at a more strategic level. We make suggestions on how to implement these proposals, and illustrate them with examples drawn from real-world interventions in which we have participated as strategic decision support analysts.

2.1 Introduction

A strategic decision has been defined as one that is "important, in terms of the actions taken, the resources committed, or the precedents set" [48] (p. 126). Strategic decisions are "infrequent decisions made by the top leaders of an organisation that critically affect organizational health and survival" [18] (p. 17). Furthermore, the process of creating, evaluating and implementing strategic decisions is typically characterised by the consideration of high levels of uncertainty, potential synergies between different options, long term consequences, and the need of key stakeholders to engage in significant psychological and social negotiation about the strategic decision under consideration.

Gilberto Montibeller
Department of Management, London School of Economics, Houghton Street, London, WC2A 2AE, UK, e-mail: g.montibeller@lse.ac.uk

Alberto Franco
Warwick Business School, University of Warwick, Coventry CV4 7AL, UK, e-mail: alberto.franco@wbs.ac.uk

C. Zopounidis and P.M. Pardalos (eds.), *Handbook of Multicriteria Analysis*, Applied Optimization 103, DOI 10.1007/978-3-540-92828-7_2, © Springer-Verlag Berlin Heidelberg 2010

A recent trend within organisations is the employment of strategy workshops as an effective means to engage in the strategic decision making process and ensure the participation of key stakeholders in that process. A recent survey by Hodgkinson et al. [29] suggests, however, that Multi-Criteria Decision Analysis (MCDA) is hardly used for supporting strategy workshops. This is somehow surprising, as facilitated forms of MCDA [21] - where the model is created directly with a group of managers in a decision conference [59] - seem a perfect tool for supporting strategic decisions in a workshop setting.

We believe that discrete-alternative MCDA methods (using the taxonomy suggested by Wallenius et al. [85]) can be useful for supporting a strategy team tasked with designing and selecting high-value strategic options. However, its apparent lack of use in strategy workshops may be due to, we argue, some limitations in the MCDA approach, which may render it unsuitable for supporting strategic decisions and the workshop processes within which they are created, debated and evaluated. We thus propose a number of changes to the standard MCDA approach, so that it can be deployed as an effective strategic decision support tool. Such changes will require the consideration of both *technical* and *social* aspects of strategic decision making. The general purpose of this chapter is, therefore, to suggest these changes and produce a framework for using MCDA to support strategic decision making in strategy workshops. We illustrate these changes with examples drawn from real-world interventions in which we have been involved as strategic decision support analysts.

The remaining of the chapter is organised as follows. In the next section we examine the nature of strategic decisions and the main characteristics of strategic decision making. The discussion will lead to the identification of complex technical and social aspects associated with the provision of decision analytical support at the strategic level. The implications of these characteristics for MCDA interventions are then addressed in the two subsequent sections. The chapter ends with some conclusions and offers directions for further research.

2.1.1 Strategic Decisions and Strategic Decision Making

The popular view of strategic decisions is that they typically involve a high degree of uncertainty, high stakes, major resource implications, and long-term consequences [36]. This view is associated with the traditional conceptualisation of strategic decisions as the product of intentional attempts at rational choice, and context-setters for subsequent strategic action [18, 75].

One of the strengths of this traditional view is that it conceptualises the strategic decision making process in a way that is consistent with the reality faced by practising managers. This conceptualisation, however, has been criticised for assuming a rational and linear relationship between decisions and actions that has not been empirically proven (e.g. [49, 76, 77]). The fundamental point underlying this critique is that organisational decisions are not always decisive, in the sense that they not always imply the presence of "commitment to act". Intentionality and action are difficult to trace and correlate empirically. Sometimes decisions result in action, in the form of a commitment of resources, sometimes they do not [47].

Notwithstanding the above criticisms, we believe like others (e.g. [41]) that the view and quest of intentional decision making is an undeniable aspect of organisational life. In our experience, managers act in accordance to the belief that strategic decisions must be intentional acts and the result of a well-designed rational process. Indeed that is the main reason that they look for our help as decision analysts. There is, therefore, a clear role for Decision Analysis in these contexts, to support strategic decision making.

In the section that follows, we will discuss both the technical and the social complexities associated with strategic decisions and the processes which produce them, respectively. Such articulation will allow us to identify the necessary conditions for the effective deployment of MCDA at the strategic level.

2.1.2 Technical Complexity

From a decision analytical perspective, the two most troublesome challenges in dealing with strategic decisions are the inescapable presence of high levels of *uncertainty* and *decision complexity*. Under these conditions, managers experience great difficulty in choosing how they should act in response to the strategic decision with which they are concerned.

Let us focus on the characteristics of uncertainty first. If an organisation is to decide whether to launch a new innovative product into the market, the choice is difficult because it may be hard to assess whether the new product would be successful or not, given that the product has never been released before. This type of uncertainty is known as "epistemic" [30], and refers to a lack of complete knowledge about an organisation's external environment and its impact on the performances of potential strategies. What is often found in practice is that managers will see a way out in terms of conducting various research activities (e.g. market research, prototype development) to reduce as much as possible the uncertainty.

Another source of uncertainty, which is frequently associated with strategic decisions, is about organisational values. This happens when there is doubt about what strategic objectives, or policy values, should guide the decision or choice of action. Managers need then to undertake activities designed to clarify organisational goals and objectives, or policy guidelines.

A major source of decision complexity is the inter-relationship among choices. Strategic decisions involve different levels of granularity. For example, the strategic decision of entering into a new market will require a major allocation of resources across different parts of the organisation such as marketing, finance, operations, research and development, etc. This in turn will lead to the consideration of other strategic choices associated with the primary strategic decision. Such consideration must include an exploration of the interrelationship among these choices. The challenge for managers is thus to overcome the cognitive burden associated with evaluating a large set of interconnected strategic decisions, and to devote a substantial amount of time working to achieve a holistic and satisfactory strategic focus.

Consequently, if MCDA is to be used for tackling strategic decisions, it must be able to provide support for dealing both with uncertainties about the environment and organisational values and also with decision complexity.

2.1.3 Social Complexity

Having discussed two key technical challenges of strategic decisions, which makes them particularly demanding for decision analysis, we now move on to discuss those aspects associated with the processes that produce them. We will offer below a view of strategic decision making as a socio-discursive process, drawing on the interpretive and strategy-as-practice perspectives of strategic decision making (e.g. [16, 35, 41, 87]).

We posit that strategic decisions are socially produced and reproduced mental frameworks through which managers make sense of their strategic concerns and so are able to act upon them. This conceptualisation is consistent with Laroche's [41] ideas of decisions as "social representations"; Weick's [87] view of decision making as a retrospective sense making process; and Eden's [16] notion of strategy as a social process. The strategic decision making thus provides the cognitive structure within which strategic change takes place in organisations. Under this view, the reality of strategy and change is "socially constructed" [9] in the form of strategic decisions.

Where does the process of producing and reproducing strategic decisions take place? Discursive studies of strategic decision making have strongly (and persuasively) argued that strategic decisions are discussed and debated through different communication channels, e.g. written reports, minutes, speeches, letters to shareholders, or informal conversations (e.g. [27]). Nonetheless, there are other modes of communication within which strategic decision making takes place, such as strategy workshops (e.g. [29]). It is this latter mode of strategic decision making practice which is of interest to the decision analyst.

Strategy workshops typically involve a group of managers representing key organisational stakeholder groups that come together to impose a structure to a decision problem which they perceive as "strategic". Participants bring to the workshop their individual mental frameworks of the issues constituting the problem, how these interrelate, and their perceived implication in relation to the strategic choices open to them. Differences in interpretations of the issues are possible, which in turns creates cognitive conflict. To resolve the conflict, participants engage in negotiation in order to produce a shared mental framework of the strategic decision. However, the negotiation process can be hindered by cognitive limitations. Research on individual and group decision making has vividly shown how cognitive biases and dysfunctional dynamics can produce ineffective and sometimes catastrophic decisions (e.g. [34, 83]).

Furthermore, the internal negotiation process among members of the managerial team does not take place in a political vacuum and political conflict is also possible. Managers will compete to instil their own mental frameworks ([20, 32]). It has been argued that successful agreements regarding strategic decisions may depend on the willingness of managers to engage in open dialogue if the choices they face are not to be dictated by means other than an overt exercise of power (e.g. [33, 62]).

Episodes of strategic decision making such as strategy workshops then involve processes of *psychological* and *political* internal negotiation, where issues of decision structuring, group dynamics and power become critical in building up momentum for strategic action [17]. Traditionally, the focus of decision analysis has been the individual decision maker. However, the now well established management practice of conducting strategy workshops requires a shift in emphasis by decision analysts from the individual to the group, as done in decision conferencing [59].

We contend that the presence or absence of decision analytical assistance might be expected to make a difference to the effectiveness of strategic decision making process, particularly when it takes place within collective fora such as strategy workshops. However, our foregoing discussion suggests that certain adaptations to the methods, tools and processes of decision analysis are required if it is to be effectively applied in such a context . Our suggestions are presented next.

2.2 MCDA for Strategic Decision Making: modelling content

In this section we focus on the technical aspects of modelling strategic decisions using MCDA and suggest ways of tackling these.

2.2.1 Tackling Uncertainty with Future Scenarios

In traditional decision analysis, the standard way of analysing decisions under uncertainty is to represent options and uncertainties as a decision tree and then select the option with the highest expected value (for details, see [11]). For example, if two mutually exclusive options a_1 and a_2 are being considered, their mono-criterion outcomes may vary due to events 1 and 2, respectively (see Figure 2.1). If option a_1 were implemented, event 1 could generate either outcome $o_{1.1}$ (with a probability $p_{1.1}$) or outcome $o_{1.2}$ (with probability $p_{1.2}$). The probabilities of outcomes should sum up to one (e.g.: $p_{1.1} + p_{1.2} = 1$). The option with the highest expected value EV should be selected: $EV(a_i) = \sum p_{i.j} . o_{j.j}$, where j-th is the event index.

If multiple criteria are considered in the evaluation, usually a multi-attribute utility function is employed to aggregate the partial performances; the option selected is the one with the maximum expected utility (for details see Keeney and Raiffa

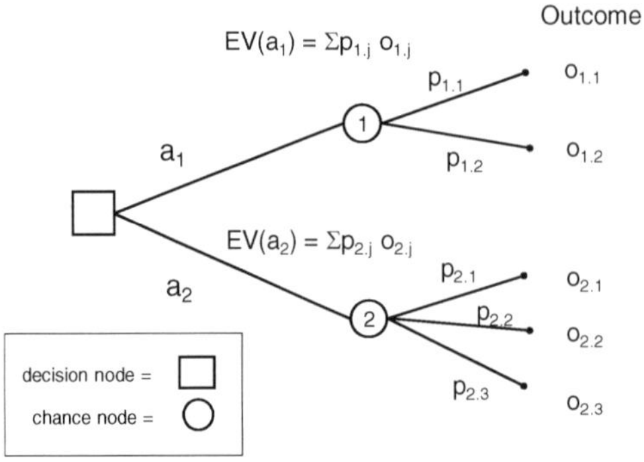

Fig. 2.1 Traditional decision analysis with decision trees

[40]). For example, if there were three criteria (C_1, C_2 and C_3) for assessing the performances of the two options depicted in the previous example, each k-th criterion would have a x_k attribute, measuring the performance of options, an associated u_k partial utility function and a w_k weight, as shown in Figure 2.2. If an a_i-th option were implemented, there would be three outcomes from each branch of the j-th event node: $o_{i.j.k}$. Partial utility functions (u_k) then convert partial performances into partial utility; and an overall utility function can then be calculated for each a_i-th option: $U_{i.j}(a_i) = f[w_k, u_k(o_{i.j.k})]$. The option with the highest expected utility should be selected: $EV(a_i) = \sum p_{i.j} U_{j.j}(a_i)$.

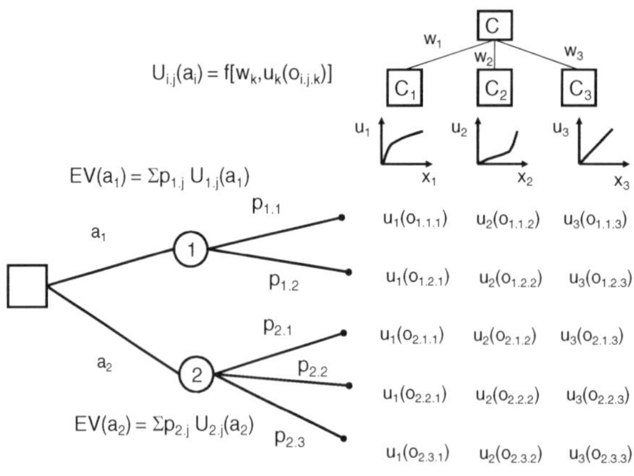

Fig. 2.2 Traditional decision analysis with a multi-attribute utility function

There are three main assumptions in this type of analysis. The first is that the outcomes from a chance node should be mutually exclusive (i.e., only one of them will happen) and collectively exhaustive (i.e., they cover all possible outcomes that may happen in the future). These two conditions make the sum of the probabilities of outcomes equal to one. The second assumption is that it is possible to obtain, in a reliable way, accurate probabilities of outcomes. The third assumption is the use of the expected value rule as a way of selecting the best alternative. We believe that these three assumptions are difficult to hold in strategic decisions.

Let us first analyse the need for a collectively exhaustive set of outcomes. As we discussed earlier, decision makers have to confront epistemic uncertainty, where there is lack of knowledge about the parameters that characterise a phenomenon. Epistemic uncertainty plays a major role in strategic decision making. In particular, most strategic decisions are one-off enterprises with very long term consequences, which makes quite difficult to determine all possible future outcomes. Consequently, it is impossible to assure that the set of outcomes is really exhaustive.

The assessment of reliable probabilities, the second assumption in traditional decision analysis, is also problematic in strategic decisions. Again the inescapable presence of epistemic uncertainty plays an important role here. As historical data is either not available or of little use in forecasting the long-term future, these probabilities are usually provided by experts. Such estimates are difficult to be "accurate", both because of the unavoidable presence of biases during the probability elicitation processes [37] and the impossibility of knowing the likelihood of events in the long term [64, 44]).

With respect to the third assumption, the expected value rule only makes sense in repeated gambles, where the expected value provides a weighted-average outcome. But in one-off gambles, it has shown to be a poor guide for choice (see Benartzi and Thaler [8] and the discussion by Lopes [45] and Luce [46]). Again, strategic decisions are by nature unique and most of them one-off, so it may not be always appropriate to use the expected value rule.

On the other hand, since the 1980s scenario planning has been suggested as an alternative way of considering uncertainty in strategic decisions, instead of traditional forecasting. The idea is to construct a small set of possible future scenarios that describe how the main uncertainties surrounding the problem would behave (e.g., interest rates, prices of commodities, demographic trends). Each scenario presents a coherent story that may happen in the future and is used to explore how different strategies would perform under such circumstances (for details see [71, 72, 82]).

Once scenarios are developed and suitable strategies are devised, a table can be built - which describes qualitatively the outcomes of each strategy under each scenario. For example, Table 2.1 presents the scenarios and strategies for a decision that we supported recently: the strategic direction of an insurance broker in England. The directors of our client company were near retirement and wanted to consider five strategies for the organisation. Three scenarios were developed and the qualitative outcomes of each strategy, under each of these scenarios, were assessed (for details see [52]).

Another advantage of using such scenarios is that scenario planning has been widely employed in practice, and seems to be a tool which managers are comfortable to work with [71, 82]).

Two features distinguish scenarios created with scenario planning from the way that uncertainty is modelled in traditional decision analysis (event nodes). The first

Table 2.1 Strategic options and scenarios for the English insurance broker (from [52], p.10)

		Scenarios		
		Direct future Limited future over next 5 yrs with insurers preferring to deal directly with clients	**Symbiotic future** Relationship-driven, where brokers flourish with no immediate time horizons	**Network-based future** A co-operative set up, (similar to franchise) - all products and contacts are dictated by the network
Strategy	**Maintain existing business** Continue with current business, with marginal increase in profit, sell in 5–7 yrs	• Remain in control • Competition more difficult • Reduced profits • No real future unless environment changes • May be difficult to realise full worth if sold in 5 yrs	• With compliance, broker is viewed as a strong partner by insurers • May obtain higher sales price in 5 yrs • Better working relationships and easier business	• Joining network would relinquish control of business • Increased power for price & products • Can't deal directly with insurers - only through network • % of turnover given to network
	Grow existing business Increase current business and find areas for diversification	• Find other products to sell • Gain a few key clients • May be difficult to sustain growth	• Renewed enthusiasm • Less competition with insurers • Competing mainly with other brokers • Push new product areas to larger client base	• As above • Purchase power of network may allow more growth & diversification
	Buy another business As above by control retained by SIB. Gain larger market share	• Remain in control • Should be easy to find a company but may not be able to compete on purchase price	• Remain in control • Could increase strength & market share • More negotiation power with insurers for better products	• As above with a larger company • In control of new company but still constrained by network • More profits due to increase volume of trade
	Merge with another business Find broker with complimentary competences. Control of business to be negotiated	• Loss of control - culture clashes • With brokers selling up, may be difficult to find appropriate company	• May be difficult to find company to purchase • Loss of control - culture clashes • Could make broker stronger	• As above but with a larger company • More loss of control
	Sell immediately Gain compliance & sell entire business before enforced regulations	• May be difficult to negotiate best price or • Could negotiate better price as more companies compete for purchase • No future job unless secured with purchaser	• Insurers may not be so keen to purchase if they need brokers as a sales force • Could sell to another broker for increased market share • No jobs for Directors	• Uncertain about sales price • No future job unless secured with purchaser

one is that some scenarios may be quite extreme, with a very small possibility of occurrence. While this can be accommodated in a decision tree, with low-probability events, it is our opinion that its structure may inhibit the consideration of extreme cases (not to mention that it tends to produce a low overall utility, given the typically low probability attached to the outcome). Considering an extreme scenario also lead managers to think about how robust the strategy they are assessing is. This may support the creation of new, more robust, strategies and make managers think about the future without merely extrapolating present trends.

The second feature that distinguishes scenarios from outcomes of events in a decision tree, is that the scenarios are not, necessarily, mutually exclusive and exhaustive. This means that one should not attach a probability of occurrence to each scenario. Indeed, most scenario planning proponents are strongly against the use of probabilities in appraising scenarios, as they should not be considered as either states of nature or predictions [70] but, instead, as learning tools [13] to explore the future.

In our own experience of providing strategic decision support to organisations, scenario planning has proved to be a powerful tool to increase awareness about future uncertainties and enhance creativity in thinking about possible strategies. However, the literature on scenario planning is limited in discussing how to identify/design high-quality strategies from a scenarios analysis. It also does not address the need of appraising these strategies taking into account multiple organisational objectives, which we will discuss next.

2.2.2 Considering Multiple Objectives

There is ample evidence in the management literature of the pervasiveness of multiple and conflicting objectives in strategic decision making (e.g. [18]). The fact that strategic decisions typically involve the consideration of multiple strategic objectives suggests the adoption of MCDA as the evaluation tool for strategic choices.

The popularity and advantages of scenario planning, combined with the power of evaluation of MCDA, provides a potent set of decision-support tools for strategic decisions [52]. Indeed, since the 1980s there are suggestions of considering the use of MCDA with scenario planning. Most papers use multi-attribute value analysis, such as Phillips [58] and Goodwin and Wright [25]. However other MCDA methods can also be employed, such as Durbach and Stewart's [14] suggestion of using goal programming with scenario planning. It is also possible to imagine the use of outranking methods such as those advocated by Roy [66], but we are not aware of any published paper adopting this approach.

On a more theoretical level, Belton and Stewart [81] discussed the potential use of MCDA and scenario planning. Stewart [78, 79] presented several technical issues about this integration and provided a thoughtful discussion on how it could be made. Montibeller et al. [52] suggested a framework for conducting a multi-attribute value analysis under multiple scenarios, in the same way as Belton and Stewart [81], but with an emphasis on robustness of strategies. It is this latter approach that we adopt here.

Let a set of n strategic choices be: $A = a_1, a_2, ..., a_n$. There are m criteria: C_1, C_2, ..., C_m; each k-th criterion measures the achievement of one strategic objective of the organisation. A model is built for each s-th scenario, which provides the overall evaluation of the i-th alternative under the scenario:

$$V_s(a_i) = \sum w_{s.k} v_{s.k}(a_i)$$

Where $w_{s.k}$ is the weight of the k-th criterion under the s-th scenario ($\sum w_{s.k} = 1$ for a given scenario) and $v_{s.k}$ is the value of the i-th alternative on the k-th criterion (scaled from 0 to 100). Notice that this model allows different weights for distinct scenarios, in order to reflect different future priorities.

For example, in supporting the decision of one of our clients on whether they should go ahead with a warehouse development in Italy, five strategic options were considered and two scenarios: the council would grant planning permission with a change of destination allowing the development, or it would not grant it (see details in [52]). The MCDA model, for each scenario, its weights and the performances of each strategy is shown in Figure 2.3; notice that there is no dominating option in all scenarios.

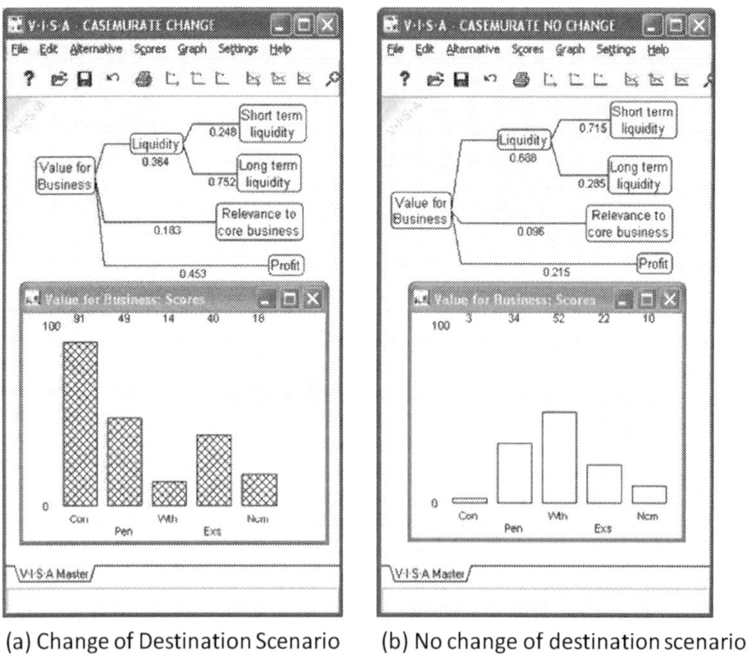

Fig. 2.3 Evaluating strategies under different scenarios with MCDA

One important change that organisations may experience, when using MCDA for strategic decisions, is the use of a value-focused framework [39] to guide the decision making process. In this case, strategies are seen as means to the achievement of the organisation's strategic objectives. This may help both in aligning the strategic vision of the organisation with its strategic objectives, and in better scoping the strategic choices it is considering (see [6]).

A key aspect in supporting strategic decisions using value-focused thinking is, therefore, the need to help the definition and structuring of these strategic objectives. As recent research has shown [10], people usually struggle to think about the fundamental objectives they need to consider in a decision. While managers have a deep understanding of their organisations, and think about what they want to achieve, our experience with management teams shows that they usually do not have a clear framework to think about decisions. Consequently, it is reasonable to argue that they invariably need support to define and negotiate the objectives considered as important and salient in a particular strategic decision context [17].

There are several tools that can be used for the structuring of objectives, such as means-ends networks [39], causal/cognitive maps [86, 7], post-it workshops with the CAUSE framework [81], affinity diagrams [54] and Soft-Systems Methodology [53]. We have used extensively cognitive maps - a network that represents ends decision makers want to achieve and means available to them, connected by links denoting perceived influence - to support the structuring of objectives and of value trees. This is a particularly useful tool, as the means-end structure permits the analyst to ladder-up the decision-makers' values and find their fundamental and strategic objectives, helping to structure a value tree (for a discussion on how they can be used for such purpose see Montibeller and Belton [50]). There are several other applications of cognitive maps for this purpose reported in the literature, for instance, Belton et al. [7], Bana e Costa et al. [5], Ensslin et al. [19], Montibeller et al. [51].

For example, in one recent application we helped a team responsible for planning and performance (PP Team) of a British city council in identifying their strategic objectives. The process was aided by a cognitive map, part of which is shown in Figure 2.4, which was interactively developed with team members using the Group Explorer networked system (www.phrontis.com) (a set of laptops connected wireless) running along the Decision Explorer mapping software (www.banxia.com). These mapping tools allow team members to input and structure their ideas and objectives in "real-time".

2.2.3 *Identifying Robust Options*

The early focus of traditional decision analysis was in providing a single solution, the one that maximises the expected value/utility. As we discussed before, this is a feasible aim, as long as the conditions required hold. In conditions of deep epistemic uncertainty, however, decision makers may be unable to define a set of exhaustive outcomes to events and/or attach realistic probabilities to them [42, 44]. Further-

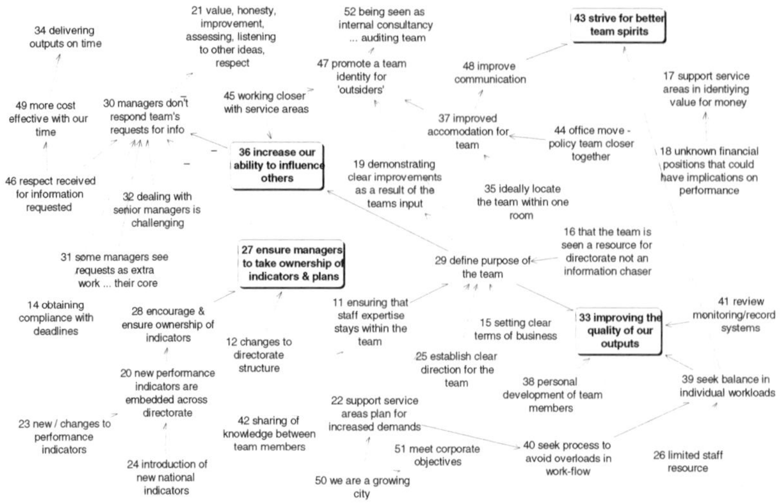

Fig. 2.4 An excerpt from a cognitive map used to identify strategic objectives (in boxes)

more, if one is using scenarios and MCDA to assess the value that each of these strategies generates for the organisation under each scenario, it is not feasible to calculate an expected overall value for each strategy as probabilities should not be attached to scenarios.

Instead of maximum expected utility, scenario planning proposers have stressed the need for finding robust strategies [72, 82], those that perform relatively well across the scenarios. This call for robustness instead of optimality has also been made within the operational research field by Rosenhead [63, 64] and Lempert [42, 44, 43]. The multi-criteria community has also made calls for a focus on robustness instead of optimisation ([4, 28, 67, 84]).

We suggested elsewhere [52] that if the analyst is using MCDA with multiple scenarios, two aspects should be of concern. The first one is the robustness of performances of a strategy across scenarios, which we denominated *inter-scenario robustness*. Thus a strategy that performs relatively well on all scenarios exhibits higher (inter-scenario) robustness than one that performs poorly on a given scenario. The second aspect is the spread of performances across scenarios, which we named *inter-scenario risk*. For example, in supporting the decision of the warehouse development in Italy, there was no dominating option, but there were more robust options as well as less risky ones, as shown in Figure 2.5.

One challenge of using the concept of robustness is that there are different ways of conceptualising it [28, 64]). A simple way, as we suggested in Montibeller et al. [52], is the use of the maximin rule. Another, suggested by Lempert et al. [44] and Lempert et al. [43] is the use of min-regret. (In both cases the analyst needs to normalise the scales under each scenario, to make them comparable - see also [61]). As it is well known, any of these rules is weaker than the expected value rule; but

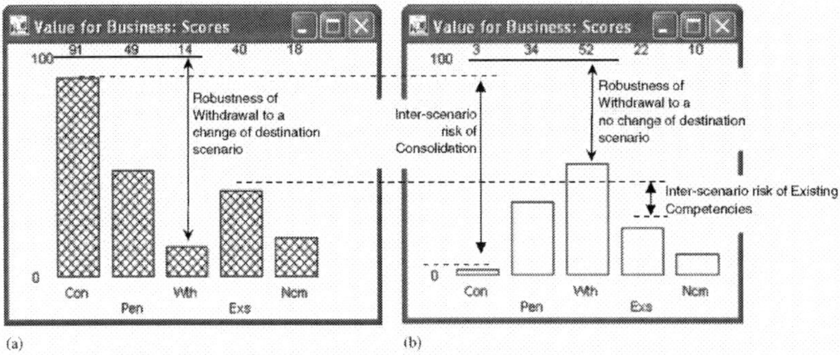

Fig. 2.5 Inter-scenario risk and inter-scenario robustness for (a) change of destination and (b) no change of destination scenarios (from [52], p. 17)

the paradox is that the expected value does require the specification of probabilities of outcomes, which is not feasible if the analyst is using scenarios. In practice, we have found that the most helpful way for supporting decision-makers' choices of strategies has been a visual inspection of the performances and spreads, with a focus on inter-scenario robustness and inter-scenario risk.

2.2.4 Designing Robust Options

Much of the focus of the MCDA literature has been on evaluating options, given a predefined set of alternatives. While this is an important aspect of many decisions, our experience is that most decisions - particularly at the strategic level - do not start with a well-defined set of options. As Keeney has emphasised [26, 39] the design of better options is a crucial aspect of a successful decision support.

In this regard, the decision analyst can help decision makers in: (1)identifying strategic options; and, (2)designing better ones. The identification of options is usually supported during the problem structuring phase. Again there are several tools that may help such as, among others, cognitive maps, the CAUSE framework [81], dialog mapping [12], and Keeney's [39] probes for generating objectives..

We have been using extensively cognitive maps for the generation of options. For example, in the project developed for the city council mentioned before, for each strategic objective shown in Figure 2.4, we asked the group members to generate a list of options, which were then input, using their laptops and shown in the cognitive map projected on a public screen. In this way we had a brainstorming focused on achieving the strategic objectives of the organisation.

The second support an analyst can provide is in the design of better options. Indeed, one main advantage of using MCDA for supporting strategic decision making stems from the specification and achievement measurement of the organisation's strategic objectives. In this way, it is easy to determine the weaknesses and strengths of each strategy and the contribution of each strategic objective to the achievement of the overall objective. The analysts can then help their clients in thinking of ways to re-design options, improving their weaknesses and assessing the marginal value of such improvements; or in creating better strategic options, by combining positive features of other alternatives. Not only inter-scenario robustness should be an aim, but also inter-scenario risk, the latter being a concern about reducing the variability of performances of a given strategic option across scenarios.

For example, in the Italian warehouse development, we managed to improve an alternative that was the best performer in the "no-change of destination" scenario but quite weak in the other one (alternative Wth in Figure 2.5), by focusing on how to improve its performance in the latter scenario and coming up with a better marketing strategy to sell the land option. In this way we managed to both increase the inter-scenario robustness of the alternative as well as reduce its inter-scenario risk.

2.2.5 Designing and Appraising Complex Strategic Options

Almost invariably, the MCDA literature has focused on alternatives that are relatively easy to describe, e.g., the choice of location for an industrial plant or the selection of the right candidate for a job. At a strategic level, however, many times decision-makers are faced with far more complex strategic choices or policies that are composed by a large set of sub-options. A challenge in this type of problem is the cognitive burden involved in appraising holistically the performance of each policy and the time burden that may be required to evaluate a large set of options.

There are some methods that deal with this situation. The strategy generation table proposed by Howard [31] is a simple way of creating strategies from the combination of option under several dimensions. Another tool is the Analysis of Interconnected Decision Areas (AIDA) technique that is part of the Strategic Choice Approach developed by Friend and Hickling [24], where the links between several "decision areas" are represented, each one with several options, and whose compatibility is explored in order to generate a list of possible option portfolios. For example, in an intervention with a major international hotel company, we used AIDA to initially shape a strategic decision concerning how to tackle "cost of sale", and produced a list of candidate interconnected strategic options, grouped in three areas (*distribution, timing launch* and *scope level*). This is shown in Figure 2.6, where the links between nodes represent incompatible combinations.

Another way of dealing with complex policies is to represent the problem as a portfolio problem, with decision areas which group options [60]. An MCDA model can then be built to assess the multiple benefits that each option generates, and software such as Equity (www.catalyze.co.uk) can then calculate the best portfolio of options, i.e. the one with the highest marginal benefit per unit of cost, given the organisation's budget.

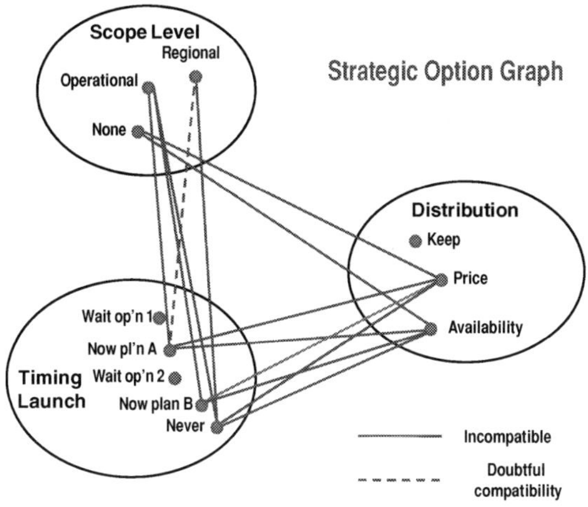

Fig. 2.6 Analysis of interconnected decision areas

For example, in the city council intervention, each strategic objective in Figure 2.4 was transformed in an area of a portfolio, as shown in the bottom row of Figure 2.7 (two of those strategic objectives were not depicted in Figure 2.4). Each of those strategic objectives had a set of options that are "stacked" vertically above the respective area name. Each option was then assessed in terms of its overall benefit and implementation cost and the rank of non-dominated portfolios was identified (for details see [51]).

2.2.6 Considering Long Term Consequences

Most of the MCDA applications reported in the literature assess single-point outcomes, which try to represent the performance of an option if it were implemented. Particularly in strategic decision-making, however, considering long-term consequences is relevant and, many times, crucial.

One relatively simple way of considering long-term consequences in these cases is by applying time discounting, as in net present value (NPV) analysis (see also [80]). A key challenge of NPV analysis is always to define a suitable discount rate. In private companies this may be relatively straightforward, as it is linked with the cost of capital. However, the same cannot be said about public decisions, where the level of discounting is debatable - a large rate can make costs in the long-term future negligible and favour short-termism [23]. Another avenue, which has been recently suggested by Santos et al. [68] is the use of system dynamics models to simulate multiple responses of a system, given some policy as input. These responses can then be employed as the policy's performances in a MCDA model.

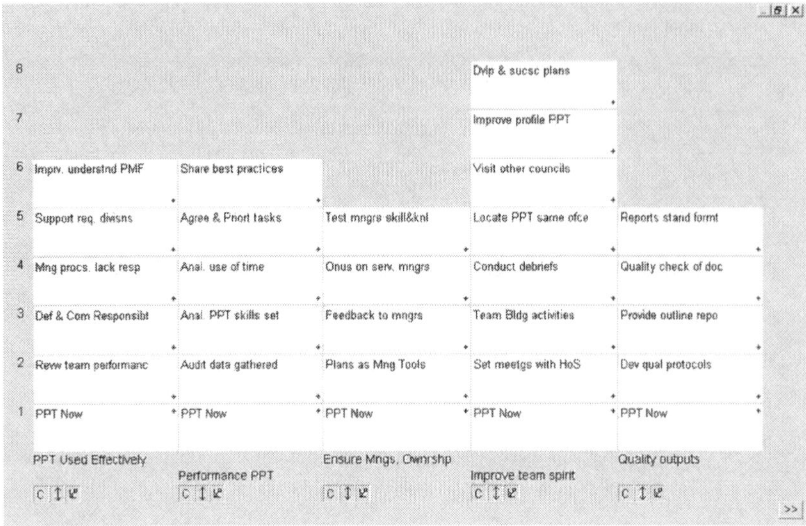

Fig. 2.7 Complex policies as a portfolio problem

A less technical way, but particularly suitable for strategy workshops, of considering long-term consequences, is the use of several MCDA models, each one concerning a particular time frame (for example, consequences after 5, 10 and 20 years). In this way the MCDA analysis can cover not only multiple-scenarios but also multiple-time frames. How to analyse the results from this kind of model is, however, still an open issue.

2.3 MCDA for Strategic Decision Making: Facilitating Process

The previous section discussed some ways to address the technical complexities associated with strategic decisions. This section will focus on designing decision support processes to tackle the social aspects associated with strategic decision making, and propose facilitated decision modelling as an effective means to provide that support. The focus of support will be at the group rather than the individual level, which is consistent with an increasing recognition of the importance of strategy workshops (e.g. [29]). Such strategy workshops, by definition, involve working with groups of diverse composition which are likely to include key organisational stakeholders.

2.3.1 Facilitated Decision Modelling

The term "facilitated decision modelling" will be used here to describe a process by which formal decision models are jointly developed with a strategy team, in real time, and with or without the assistance of computer support (Eden, 1990; Franco and Montibeller, forthcoming; Phillips 2007). We consider a decision model as "formal" if it represents a strategic decision problem either in terms of cause and effect relationships; or of relationships between decision choices and their (deterministic or uncertain) consequences. A formal decision model is amenable to analysis and manipulation, but not necessarily fully quantifiable.

A decision model produced in a facilitated manner is used by the strategy team members as a "transitional object" [16, 13]. It allows them to share their strategic concerns and increase their individual understandings of the strategic issues, appreciate the potential impact of different strategic choices, and negotiate strategic action that is politically feasible.

When members of a strategy teams participate in a facilitated modelling process, they engage in "conversations" [22] to exchange their understandings and views about the strategic decision that is being analysed. This process is a participative one, in the sense that team members are able to jointly construct the strategic decision, make sense of it, and develop and evaluate a portfolio of strategic options for the decision. This participatory process is supported by the decision analyst both as a modeller and a facilitator [2, 57].

Because interaction between the participants in the decision modelling process, and of the participants with the decision analyst, is needed to jointly build a decision model of the strategic situation, facilitated decision modelling is also an interactive process. Participants' interaction with the model reshapes the analysis, and the model analysis reshapes the group discussion. Such interactive processes continue until the situation is satisfactorily structured and analysed, so that the group feels sufficiently confident in making commitments and implementing options.

Facilitated modelling is typically organised into group work stages, which roughly correspond to: structuring the decision problem and agreeing a decision focus; developing a model of organisational objectives; creating, refining and evaluating options; and developing action plans. However, the stages of facilitated decision modelling do not have to be followed in a linear sequence; rather, it is possible for the participants to cycle between the stages.

In terms of technology, facilitated decision modelling can be a relatively unsophisticated activity, conducted in a workshop format, and one which does not necessarily require software to support it [3]. Facilitated decision modelling can also be deployed with computer support. In this case, specialised software is used to support the processes [1, 56]. This kind of software enables fast model building and real time computing [3].

Some software, such as *Group Explorer* (www.banxia.com) and *VISA Groupware* (www.simul8.com/products/visagroup.htm), also allows participants to enter their views relating to a decision problem directly and anonymously into it. The system is then operated by the facilitator/modeller who manipulates and analyses the data according to the wishes of the group. Once a decision model is built and stored in the system, several analyses can be performed "on-the-hoof".

The preceding discussion makes it clear that facilitated decision modelling is different from standard decision analysis modelling. It requires a decision analyst able to support a group model-building process that must be participatory, interactive, staged, non-linear, adaptable, and supported by appropriate technology. At the same time, the facilitative decision modeller and his/her chosen modelling approach must be responsive to the dynamics of group work and the particularities of the situation at hand [57]. The next section explores further what is required to become a facilitative decision modeller.

2.3.2 Becoming a Facilitative Decision Modeller

As already stated, facilitated decision modelling requires the decision analyst to act as a facilitator during the group decision modelling and analysis process. This means that the decision analyst should be prepared to use general facilitation skills as part of his/her modelling work. Drawing from the general facilitation literature (e.g. [38, 74]) and the work of Schuman [73] and others (e.g. [57]), we consider below three fundamental facilitation skills required in facilitated decision modelling:

- *Active listening* requires the decision analyst to be able to clarify, develop, summarise and refine participants' contributions by paraphrasing and/or mirroring what participants say; validating what they say without judging; asking them non-directive questions; gathering lists of their contributions; helping them to take turns; keeping track of the various discussion themes that may emerge simultaneously; balancing the discussion to avoid bind spots; and listening for the common ground.
- *Managing group dynamics* is perhaps one of the most fundamental skills for the facilitative modeller. Through active facilitative listening, the decision analyst must be able to sense when difficult group dynamics crop up during modelling, and treat them as group situations that must be handled supportively. A typical approach is for the decision analyst to help the group step back from the content of the ongoing discussion and talk about the process instead. This is usually achieved by, for example, encouraging more people to participate, acknowledging and handling out of-context distractions, educating participants about how to handle anxiety in the group, and helping participants deal with any unfinished business. Difficult group dynamics also require the decision analyst to know whether, how and why to intervene during the modelling process.
- *Reaching closure* is a key skill that a facilitative decision modeller uses to help the group reach agreements about the way forward. This requires the decision analyst to identify when the group has reached a point, from "playing" with the decision model, at which closure on a proposal is needed and a requisite decision model has been achieved [55]. Depending on the particular organisational context within which the decision analysts is working, those with the power and authority

to make commitments for the implementation of particular courses of action must then decide whether the proposal needs further group discussion or whether a commitment about the way forward can be made.

In Figure 2.8, we have attempted to capture the main aspects of facilitated modelling discussed in this section. The figure suggests that the decision analyst interacts simultaneously in two "spaces": as an analyst in the *decision modelling* space and as a facilitator in the *group process* space. In the former, s/he uses a particular decision methodology to inform model building and represents the decision problem as described by the group, using a decision model. Meanwhile, in the *group process* space, s/he facilitates the group (informed by the facilitation methods s/he is using), supports the group's discussion and interacts with the group's members. The group provides information that allows the facilitator to model the decision problem and, conversely, the model generates responses that support the group's discussions about the issues they are dealing with. Two types of outcomes are provided by facilitated decision modelling: in the *decision modelling* space, there are model outcomes (such as performances of options, sensitivity analyses, etc.); in the *group process* space, there are group outcomes (such as commitment to the way forward, learning, etc.). The two spaces cannot be divorced from each other [15], as there will be cross-impacts from one space to the other (for example, conceptual mistakes in the decision model may generate meaningless model results, which then impact negatively on group outcomes, such as creating low commitment to action).

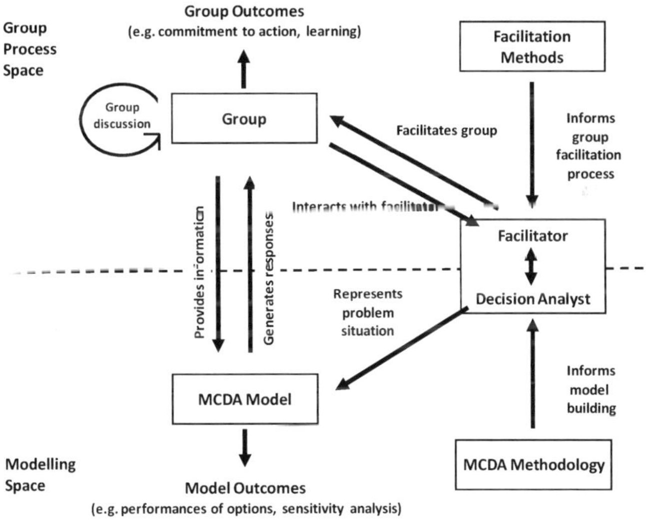

Fig. 2.8 Facilitated decision modelling (adapted from Franco and Montibeller, forthcoming)

2.4 Conclusions

Multi-Criteria Decision Analysis (MCDA) has been extensively used for supporting extremely complex decisions in both public and private organisations. We believe that there is an excellent opportunity for MCDA in supporting strategy workshops, given their prevalence nowadays. In these workshops, organisations shape their strategic vision, design strategic options and appraise strategic choices.

This chapter proposed a framework to employ Multi-Criteria Decision Analysis for supporting strategic decision making in strategy workshops. This framework comes from our practice, as decision analysts, in providing strategic decision support to a wide range of organisations, in Europe, North America and Latin America.

We suggested that there are two main aspects that have to be addressed by the decision analyst, if s/he wants to support strategic decision making processes. The first is related to content issues, in particular in dealing with epistemic uncertainty, multiple organisational objectives, complex policies and long-term consequences. We believe that the key aspect is to develop robust strategies against multiple scenarios. The second aspect concerns process issues, in particular being an active listener, dealing with group dynamics and helping the group to reach closure. Here the analyst has to facilitate the management team in making a strategic decision; and the key is to conduct, in an efficient and suitable way, facilitated decision analysis.

We recognise that further research has to be conducted on this topic, which could permit additional development of this framework. In particular, we suggest the following directions for further research:

- Robustness - more studies on robustness of strategic options under multiple scenarios is required; for example, about suitable operators and graphical displays for interacting with users/clients.
- Design of complex policies - structuring policies composed by options that are interconnected is an area almost unexplored, from a decision analysis perspective, and ideas from the field of problem structuring methods [65] may be relevant for this intent.
- Long term consequences - this is an open area for research by MCDA, and developments in other areas (e.g., cost benefit analysis) could be analysed and adapted to the context discussed here.
- Impact of facilitated decision analysis on the strategy process - there is already some systematic research about the impact of decision conferencing on group's outcomes (e.g., [69]); but given the special nature of strategy workshops, it would be interesting to assess the impacts of the framework we are suggesting on their effectiveness, as well as the overall usefulness of the framework to increase our understanding of decision analytical support at the strategic level.

Concluding the chapter, we believe that supporting strategic decision making, particularly within a strategy workshop format, represents an important - but somehow neglected - area for research in Multi-Criteria Decision Analysis. Given the importance of strategic decision making for the survival of any organisation, further developments in this field could, therefore, not only bring opportunities for research on the several challenges we highlighted here, but also have a real impact on MCDA practice.

References

1. F. Ackermann. The role of computers in group decision support. In C. Eden and J. Radford, editors, *Tackling Strategic Problems: The Role of Group Decision Support*, pages 132–141. Sage, London, 1990.
2. F. Ackermann. Participant's perceptions on the role of facilitators using group decision support systems. *Group Decision and Negotiation*, 5:93–112, 1996.
3. F. Ackermann and C. Eden. Issues in computer and non-computer supported GDSSs. *Decision Support Systems*, 12(4-5):381–390, 1994.
4. H. Aissi and B. Roy. Robustesse en aide multicritère à la decision. Cahier du Lamsade, No. 276, 2008.
5. C.A. Bana e Costa, L. Ensslin, E.C. Corrêa, and J.C. Vansnick. DSS in action: Integrated application in a multi-criteria decision aid process. *European Journal of Operational Research*, 113:315–335, 1999.
6. A. Barcus and G. Montibeller. Supporting the allocation of software development work in distributed teams with multi-criteria decision analysis. *Omega*, 36:464–475, 2008.
7. V. Belton, F. Ackermann, and I. Shepherd. Integrated support from problem structuring through to alternative evaluation using COPE and VISA. *Journal of Multi-Criteria Decision Analysis*, 6:115–130, 1997.
8. S. Benartzi and R.H. Thaler. Risk aversion or myopia? Choices in repeated gambles and retirement investments. *Management Science*, 45(3):364–381, 1999.
9. P.L. Berger and T. Luckmann. *The Social Construction of Reality: A Treause in the Sociology of Knowledge*. Penguin, Harmondsworth, 1967.
10. S.D. Bond, K.A. Carlson, and R.L. Keeney. Generating objectives: Can decision makers articulate what they want? *Management Science*, 54(1):56–70, 2008.
11. R.T. Clemen and T. Reilly. *Making Hard Decisions*. Pacific Grove, Duxbury, 2001.
12. J. Conklin. *Dialogue Mapping: Building Shared Understanding of Wicked Problems*. Wiley, Chichester, 2006.
13. A.P. de Geus. Planning as learning. *Harvard Business Review*, 6(2):70–74, 1988.
14. I. Durbach and T.J. Stewart. Integrating scenario planning and goal programming. *Journal of Multi-Criteria Decision Analysis*, 12(4-5):261–271, 2003.
15. C. Eden. The unfolding nature of group decision support: Two dimensions of skill. In C. Eden and J. Radford, editors, *Tackling Strategic Problems: The Role of Group Decision Support*, pages 48–52. Sage, London, 1990.
16. C. Eden. Strategy development as a social process. *Journal of Management Studies*, 29(6):799–811, 1992.
17. C. Eden and F. Ackermann. *Strategy Making: The Journey of Strategic Planning*. Sage, London, 1998.
18. K.M. Eisenhardt and M.J. Zbaracki. Strategic decision making. *Strategic Management Journal*, 13:17–37, 1992.
19. L. Ensslin, G. Montibeller, and M.V.A. Lima. Constructing and implementing a DSS to help evaluate perceived risk of accounts receivable. In Y.Y. Haimes and R.E. Steuer, editors, *Research and Practice in Multiple Criteria Decision Making*, pages 248–259. Springer, Berlin, 2000.
20. M. Fine. New voices in the workplace: Research directions in multicultural communication. *Journal of Business Communication*, 28:259–275, 1991.
21. A. Franco and G. Montibeller. Facilitated modelling in operational research. *European Journal of Operational Research*, 2009. Forthcoming.

22. L.A. Franco. Forms of conversation and problem structuring methods: A conceptual development. *Journal of the Operational Research Society*, 57(7):813–821, 2006.
23. S. French, T. Bedford, and E. Atherton. Supporting ALAEP decision maing lby cost benefit analysis and multi-attribute utility theory. *Journal of Risk Research*, 8(3):207–223, 2005.
24. J. Friend and A. Hickling. *Planning under Pressure: The Strategic Choice Approach*. Elsevier, Oxford, 3rd edition, 2005.
25. P. Goodwin and G. Wright. Enhancing strategy evaluation in scenario planning: A role for decision analysis. *Journal of Management Studies*, 38:1–16, 2001.
26. R. Gregory and R.L. Keeney. Creating policy alternatives using stakeholder values. *Management Science*, 40(8):1035–1048, 1994.
27. J. Hendry. Strategic decision making, discourse, and strategy as social practice. *Journal of Management Studies*, 37(7):955–978, 2000.
28. R. Hites, Y. De Smet, Y.N. Risse, M. Salazar-Neumann, and P. Vincke. About the applicability of MCDA to some robustness problems. *European Journal of Operational Research*, 174(1):322–332, 2006.
29. G.P. Hodgkinson, R. Whittington, G. Johnson, and M. Schwarz. The role of strategy workshops in strategy development processes: Formality, communication, co-ordination and inclusion. *Long Range Planning*, 39(5):479–496, 2006.
30. F.O. Hoffman and Hammonds J.S. Propagation of uncertainty in risk assessments: The need to distinguish between uncertainty due to lack of knowledge and uncertainty due to variability. *Risk Analysis*, 14(5):707–712, 1994.
31. R.A. Howard. Decision analysis: Practice and promise. *Management Science*, 34(6):679–695, 1988.
32. M. Huspek and K. Kendall. On withholding political voice: An analysis of the political vocabulary of a "non-political" speech community. *Quarterly Journal of Speech*, 77:1–19, 1991.
33. W. Isaacs. *Dialogue and the Art of Thinking Together*. Currency, New York, 1999.
34. I.L. Janis. *Groupthink: Psychological Studies of Policy Decisions and Fiascos*. Houghton Mifflin Company, Boston, MA, 2nd edition, 1982.
35. P. Jarzabkowski, J. Balogun, and D. Seidl. Strategizing: The challenges of a practice perspective. *Human Relations*, 60:5–27, 2007.
36. G. Johnson, K. Scholes, and R. Whittington. *Exploring Corporate Strategy: Text and Cases*. Prentice Hall, London, 7th edition, 2005.
37. D. Kahneman, P. Slovic, and A. Tversky. *Judgment under Uncertainty: Heuristics and Biases*. Cambridge University Press, Cambridge, 1982.
38. S. Kaner. *Facilitator's Guide to Participatory Decision Making*. Jossey Bass, San Francisco, CA, 2007.
39. R.L. Keeney. *Value-Focused Thinking: A Path to Creative Decision-making*. Harvard University Press, Cambridge, 1992.
40. R.L. Keeney and H. Raiffa. *Decisions with Multiple Objectives: Preferences and Value Trade-offs*. Cambridge University Press, Cambridge, 2nd edition, 1993.
41. H. Laroche. From decision to action in organizations: A social representation perspective. *Organization Science*, 6:62–75, 1995.
42. R.J. Lempert and M.T. Collins. Managing the risk of uncertain threshold responses: Comparison of robust, optimum, and precautionary approaches. *Risk Analysis*, 27(4):1009–1026, 2007.
43. R.J. Lempert, D.G. Groves, S.W. Popper, and S.C. Bankes. A general, analytic method for generating robust strategies and narrative scenarios. *Management Science*, 52:514–528, 2006.
44. R.J. Lempert, S.W. Popper, and S.C. Bankes. *Shaping the Next One Hundred Years: New Methods for Quantitative, Long-Term Policy Analysis*. RAND, Santa Monica, 2003.
45. L.L. Lopes. When time is of the essence: Averaging, aspiration, and the short run. *Organizational Behavior and Human Decision Processes*, 65(3):179–189, 1996.
46. R.D. Luce. Commentary on aspects of Lola Lopes' paper. *Organizational Behavior and Human Decision Processes*, 65(3):190–193, 1996.
47. H. Mintzberg. 5 Ps for strategy. *California Management Review*, 30(1):11–24, 1987.

48. H. Mintzberg, D. Raisinghani, and A. Théorêt. The structure of "unstructured" decision processes. *Administrative Science Quarterly*, 21(2):246–275, 1976.
49. H. Mintzberg and J.A. Waters. Does decision get in the way? *Organization Studies*, 111:1–6, 1990.
50. G. Montibeller and V. Belton. Causal maps and the evaluation of decision options - A review. *Journal of the Operational Research Society*, 57(7):779–771, 2006.
51. G. Montibeller, L.A. Franco, E. Lord, and A. Iglesias. Structuring resource allocation decisions: A framework for building multi-criteria portfolio models with area-grouped options. *European Journal of Operational Research*, 199(3):846–856, 2009.
52. G. Montibeller, H. Gummer, and D. Tumidei. Combining scenario planning and multi-criteria decision analysis in practice. *Journal of Multi-Criteria Decision Analysis*, 14:5–20, 2006.
53. L.P. Neves, L.C. Dias, C.H. Antunes, and A.G. Martins. Structuring an MCDA model using SSM: A case study in energy efficiency. *European Journal of Operational Research*, 199(3):834–845, 2009.
54. G.S. Parnell, H.W. Conley, J.A. Jackson, L.J. Lehmkuhl, and J.M. Andrew. Foundations 2025: A value model for evaluating future air and space forces. *Management Science*, 44(10):1336–1350, 1998.
55. L. Phillips. A theory of requisite decision models. *Acta Psychologica*, 56(1-3):29–48, 1984.
56. L. Phillips. People-centred group decision support. In G. Doukidis, F. Land, and G. Miller, editors, *Knowledge-based Management Support Systems*, pages 208–224. Ellis-Horwood, Chichester, 1989.
57. L.D. Phillips, , and M. Phillips. Facilitated work groups: Theory and practice. *Journal of Operational Research Society*, 44(6):533–549, 1993.
58. L.D. Phillips. Decision analysis and its application in the industry. In G. Mitra, editor, *Computer Assisted Decision Making*, pages 189–197. Elsevier, North-Holland, 1986.
59. L.D. Phillips. Decision conferencing. In W. Edwards, R.F. Miles Jr., and D. von Winterfeldt, editors, *Advances in Decision Analysis: From Foundations to Applications*, pages 375–399. Cambridge University Press, New York, 2007.
60. L.D. Phillips and C.A. Bana e Costa. Transparent prioritisation, budgeting, and resource allocation with multi-criteria decision analysis and decision conferencing. *Annals of Operations Research*, 154:51–68, 2007.
61. C. Ram, G. Montibeller, and A. Morton. Extending the use of scenario planning and MCDA: An application to food security in Trinidad & Tobago. Working Paper LSEOR 09.105, London School of Economics (http://www.lse.ac.uk/collections/operationalResearch/pdf/WP105 - Camelia Ram.pdf), 2009.
62. N.C. Roberts. *Transformative Power of Dialogue*. Elsevier, London, 2002.
63. J. Rosenhead. Robustness analysis: Keeping your options open. In J. Rosenhead and J. Mingers, editors, *Rational Analysis for a Problematic World Revisited: Problem Structuring Methods for Complexity, Uncertainty and Conflict*, pages 181–209. Wiley, Chichester, 2001.
64. J. Rosenhead, M. Elton, and S.K. Gupta. Robustness and optimality as criteria for strategic decisions. *Operational Research Quarterly*, 23(4):413–431, 1972.
65. J. Rosenhead and J. Mingers. *Rational Analysis for a Problematic World Revisited: Problem Structuring Methods for Complexity, Uncertainty and Conflict*. Wiley, Chichester, 2001.
66. B. Roy. *Multicriteria Methodology for Decision Aiding*. Kluwer, Dordrecht, 1996.
67. B. Roy. A missing link in or-da: Robustness analysis. *Foundations of Computing and Decision Sciences*, 23(3):141–160, 1998.
68. S. Santos, V. Belton, and S. Howick. Adding value to performance measurement by using system dynamics and multicriteria analysis. *International Journal of Operations and Production Management*, 22(11):1246–1272, 2002.
69. M.S. Schilling, N. Oeser, and C. Schaub. How effective are decision analyses? Assessing decision process and group alignment effects. *Decision Analysis*, 4(4):227–242, 2007.
70. P.J.H. Schoemaker. When and how to use scenario planning: A heuristic approach with illustration. *Journal of Forecasting*, 10(6):549–564, 1991.
71. P.J.H. Schoemaker. Multiple scenario development: Its conceptual and behavioral foundation. *Strategic Management Journal*, 14(3):193–213, 1993.

72. P.J.H. Schoemaker. Scenario planning: A tool for strategic thinking. *Sloan Management Review*, 36(2):25–40, 1995.
73. S. Schuman. *The IAF Handbook of Group Facilitation: Best Practices from the Leading Organization in Facilitation*. Jossey Bass, San Francisco, CA, 2005.
74. R. Schwarz. *The Skilled Facilitator: A Comprehensive Resource for Consultants, Facilitators, Managers, Trainers, and Coaches*. Jossey Bass, San Francisco, CA, 2002.
75. C.R. Schwenk. Strategic decision making. *Journal of Management Studies*, 21(3):471–493, 1995.
76. W.H. Starbuck. Organizations as action generators. *American Sociological Review*, 48:91–102, 1983.
77. W.H. Starbuck. Acting first and thinking later: Theory versus reality in strategic change. In J.M. Pennings, editor, *Organizational Strategy and Change*, pages 132–141. Jossey Bass, San Francisco, 1985.
78. T.J. Stewart. Scenario analysis and multicriteria decision making. In J. Climaco, editor, *Multicriteria Analysis*, pages 519–528. Springer, Berlin, 1997.
79. T.J. Stewart. Dealing with uncertainties in MCDA. In J. Figueira, S. Greco, and M. Ehrgott, editors, *Multiple Criteria Decision Analysis - State of the Art Surveys*, pages 445–470. Springer, New York, 2005.
80. T.J. Stewart and A.R. Joubert. Multicriteria support for decision-making in the ecosystems approach to fisheries management. BCLME Programme (http://www.bclme.org/projects/docs/EAF/Appendix 8a Stewart and Joubert_isa.pdf), 2006.
81. Belton V. and Stewart T. *Multiple Criteria Decision Analysis: An Integrated Approach*. Kluwer, Dordrecht, 2002.
82. K. van der Heijden. *Scenarios: The Art of Strategic Conversation*. Wiley, Chichester, 2nd edition, 2004.
83. D. Vaughan. *The Challenger Launch Decision: Risk Technology, Culture and Deviance at NASA*. University of Chicago Press, Chicago, IL, 1996.
84. P. Vincke. Robust solutions and methods in decision-aid. *Journal of Multi-Criteria Decision Analysis*, 8:181–187, 1999.
85. J. Wallenius, J.S. Dyer, P.C. Fishburn, R.E. Steuer, S. Zionts, and K. Deb. Multiple criteria decision making, multiattribute utility theory: Recent accomplishments and what lies ahead. *Management Science*, 54(7):1336–1349, 2008.
86. S.R. Watson and D.M. Buede. *Decision Synthesis*. Cambridge University Press, Cambridge, 1987.
87. K.E. Weick. *Sense Making in Organizations*. Sage, Thousand Oaks, CA, 1995.

Part II
Multiple Criteria Decision Aid Methodologies

Chapter 3
ELECTRE Methods: Main Features and Recent Developments

Jose Rui Figueira, Salvatore Greco, Bernard Roy, and Roman Słowiński

Abstract We present main characteristics of ELECTRE family methods, designed for multiple criteria decision aiding. These methods use as a preference model an outranking relation in the set of actions – it is constructed in result of concordance and non-discordance tests involving a specific input preference information. After a brief description of the constructivist conception in which the ELECTRE methods are inserted, we present the main features of these methods. We discuss such characteristic features as: the possibility of taking into account positive and negative reasons in the modeling of preferences, without any need for recoding the data; using of thresholds for taking into account the imperfect knowledge of data; the absence of systematic compensation between "gains" and "losses". The main weaknesses are also presented. Then, some aspects related to new developments are outlined. These are related to some new methodological developments, new procedures, axiomatic analysis, software tools, and several other aspects. The chapter ends with conclusions.

Jose Rui Figueira
CEG-IST, Center for Management Studies, Instituto Superior Técnico, Technical University of Lisbon, Tagus Park, Av. Cavaco Silva, 2744-016, Porto Salvo, Portugal e-mail: figueira@ist.utl.pt

Salvatore Greco
Faculty of Economics, The University of Catania, Corso Italia, 55, 95 129 Catania, Italy
e-mail: salgreco@unict.it

Bernard Roy
LAMSADE, Université Paris-Dauphine, Place du Maréchal De Lattre de Tassigny, F-75 775 Paris Cedex 16, France e-mail: roy@lamsade.dauphine.fr

Roman Słowiński
Institute of Computing Science, Poznań University of Technology, Street Piotrowo 2, 60-965 Poznań, Poland, and Institute for Systems Research, Polish Academy of Sciences, 01-447 Warsaw, Poland e-mail: roman.slowinski@cs.put.poznan.pl

C. Zopounidis and P.M. Pardalos (eds.), *Handbook of Multicriteria Analysis*, Applied Optimization 103, DOI 10.1007/978-3-540-92828-7_3,

3.1 Introduction

Since their conception, which started in the sixties of the last century, ELECTRE methods have been widely used for Multiple Criteria Decision Aiding (MCDA) in many real-world decision problems, ranging from agriculture to environment and water management, from finance to project selection, from personnel recruiting to transportation, and many others. The theoretical research on the foundations of ELECTRE methods has also been intensive all this time. In this chapter, we present the main features of ELECTRE methods, we characterize their strengths, as well as their weaknesses, paying a particular attention to the developments of the last decade.

In the next sub-section, we explain what we mean by constructivist or "European" conception of MCDA. The term "European" does not mean, however, that this conception was only developed and followed by Europeans. A large number of researchers all over the Globe are being working in this area and are applying the techniques in real-world problems, for example, in Canada, Tunisia, Poland, Switzerland, Italy, Spain, Portugal, Germany, New Zealand, and many other countries.

In the sub-section 3.2.1, the basic notation are introduced.

3.1.1 The Constructivist Conception of MCDA

This sub-section is based on the speech of Roy [58], delivered on the 30th of January 2009, on receiving a honorary doctoral degree from the *Università degli Studi di Catania*, Italy.

Before introducing the constructivist or "European" conception of an MCDA methodology, we should present the meaning of a decision aiding situation and its key elements, and three fundamental pillars that support such a conception. In what follows, the term "decision aiding", rather than "decision support", "decision making", or "decision analysis", will be adopted for escaping from simplistic assimilations.

3.1.1.1 Decision Aiding Situation/Decision Aiding Process

Consider a company or a public institution, where a manager and/or a group of people are confronted with a decision situation or "problem" that requires them to make a decision. They call on an in-house operational research service or an outside consultant or even a university research team to get help in making "the best possible" decision. This aspect allows to characterize a *decision aiding situation*, where two key actors are relevant for co-interaction that will lead to build and make evolve the *decision aiding process*; a process that comprises several phases, see [57]. The two key actors will be designated in what follows as the *analyst*, who is appointed to give this decision aiding, and as the *decision maker*, in whose name or for whom this decision aiding is to be given.

3.1.1.2 Three Fundamental Pillars

In the operational research and decision aiding community, to which we belong, the *decision-aiding activity* (which is meant to be scientific) is founded on three pillars:

1. The *actions* (formal definition of the possible actions or alternatives),
2. The *consequences* (aspects, attributes, characteristics, ... of the actions, that allow to compare one action to another), and
3. The *modeling of one or several preference systems* (it consist of an implicit or explicit process, that for each pair of actions envisioned, assigns one and only one of the three possibilities (see sub-section 3.2.1): *indifference*, *preference*, or *incomparability*).

The last pillar needs further explanation. When given two possible actions, any individual, whoever he/she may be, based on the actions' consequences, and his/her value system, can state: "I prefer the first to the second" or *vice-versa*, "I am indifferent between the two", or "I am unable to compare these two actions". *Modeling a preference system* means to specify a process that will provide this type of results based on a pre-established model of the action consequences. These consequences are most often complex and inadequately known. They can be modeled in quantitative or qualitative terms, in a deterministic or stochastic manner, with a part of arbitrariness or ill determination. We will designate by $C(a)$ the model of the consequences of action a.

3.1.1.3 The "European" Conception of MCDA

According to the "European" conception, the analyst must seek for obtaining *a coherent and structured set of results*. These results should be sought in order to guide the decision aiding process and facilitate communication about the decisions. To do so, the analyst must use an approach that aims at *producing knowledge from working hypotheses*, taking into account the objectives and the value systems involved in a particular decision context. This approach should be based on models that are, at least partially, *co-constructed through interaction* with the decision maker. This co-construction first concerns the way the considered actions are taken into account, as well as the consequences on which these actions will be judged. Secondly, the *co-construction process* concerns the way that certain characteristics (notably the values attributed to the different parameters) of the preference model were judged the most appropriate given the specificities of the decision context and the working hypotheses retained. In this conception, it is no longer necessary to assume that there exists, in the mind of the decision maker, a stable procedure capable of defining the decision maker's preference system completely, before even beginning the decision aiding process.

To elaborate results likely to make things more clear to the decision maker (e.g., if…, then… results), in the "European" conception, the analyst must propose working hypotheses which will allow the co-construction of the preference model to play an appropriate role in the decision aiding process. The co-constructed model must be a tool for looking more thoroughly into the subject, by exploring, interpreting, debating and even arguing. To guide this process of co-construction, the analyst must also interact with the decision maker assuming that he/she understands the questions that are asked. Nevertheless, in the "European" conception, it is not necessary to assume that the given responses are produced through a stable pre-existing process, but only that these responses are made up through interaction with the decision maker's value system, which is *rarely free of ambiguity or even contradiction*. In particular, the analyst must make sure that the person who responds to the questions is able to place these questions in the context of the current study. The analyst must also admit that these questions can bring the person thus questioned to revise certain pre-existing preferences momentarily and locally.

According to the "European" conception, the knowledge produced does not aim to help the decision maker to discover a good approximation of a decision which would objectively be one of the best, taking into account his/her own value system, but rather more humbly to provide the decision maker with a set of results derived from the reasoning modes and working hypotheses. The decision maker will better understand the results produced and will appropriate them (and potentially share with others) if the analyst makes sure that understanding of the underlying reasoning modes and working hypotheses is integrated into the model co-construction process.

In this "European" conception, the analyst does not need to accept either of the following two postulates [58]:

- *Postulate of the decision maker's optimum*. In the decision context studied, there exists at least one optimal decision, or, in other words, there exists one decision for which it is possible (if sufficient time and means are available) to establish objectively that there are no strictly better decisions with respect to the decision maker's preference system.
- *Postulate of the decision context reality*. The principal aspects of the reality on which the decision aiding is based (particularly the decision maker's preferences) are related to objects of knowledge that can be seen as data (i.e., existing outside of the way they are modeled); these objects can also be seen as sufficiently stable over time and for the questions asked, such that it is possible to refer to the exact state or the exact value (deterministic or stochastic) of given characteristics judged to accurately portray an aspect of that reality.

He/she may find these postulates as totally unrealistic, or may even have good reasons for accepting the existence of *incomparabilities* in the preference models used.

3.1.2 Notation

For a suitable description of the main features and recent developments of ELECTRE methods (sections 3.2 and 3.3, respectively) it is necessary to introduce a few notation related to the basic data.

The basic data needed for any MCDA problem can be represented as follows:

- $A = \{a_1, a_2, \ldots, a_i, \ldots, a_m\}$ is the set of m *potential actions*; this set is, possibly, only partially known *a priori*, which is common in sorting problems (see 3.2.3),
- $F = \{g_1, g_2, \ldots, g_j, \ldots, g_n\}$ is a *coherent family of criteria* with $n \geqslant 3$,
- $g_j(a_i)$ is the *performance* of action a_i on criterion g_j, for all $a_i \in A$ and $g_j \in F$; an $m \times n$ *performance matrix* M can thus be built, with $g_j(a_i)$ in row i and column j ($i = 1, \ldots, m$; $j = 1, \ldots, n$).

In the following, we assume without loss of generality that the higher the performance $g_j(a)$ is, the better it is for the decision makers (*increasing direction of preference*).

Since the recent appearance of new ELECTRE methods for sorting problems, in what follows we rename the well-known ELECTRE TRI method by ELECTRE TRI-B, where B means "bounds".

3.2 Main Features

The distinctive features of ELECTRE methods, to which analysts should pay special attention on, when dealing with real-world decision aiding situations, are presented in this section. These are: the four preference situations handled by ELECTRE methods, the preference modeling through outranking relations, the concepts of concordance and non-discordance, the structure of the methods, the main strengths as well as the weaknesses of ELECTRE methods.

3.2.1 Modeling Four Main Preference Situations

The ELECTRE methods are based on the following four preference situations concerning the comparison of two actions [57]:

I (*Indifference*): it corresponds to a situation where there are clear and positive reasons that justify an equivalence between the two actions (it leads to a reflexive and symmetric binary relation);

P (*Strict Preference*): it corresponds to a situation where there are clear and positive reasons in favor of one (identified) of the two actions (it leads to a nonreflexive and asymmetric binary relation);

Q (*Weak Preference*): it corresponds to a situation where there are clear and positive reasons that invalidate strict preference in favor of one (identified) of the two actions, but they are insufficient to deduce either the strict preference in favor of the other action or indifference between both actions, thereby not allowing either of the two preceding situations to be distinguished as appropriate (it leads to a nonreflexive and asymmetric binary relation);

R (*Incomparability*): it corresponds to an absence of clear and positive reasons that would justify any of the three preceding relations (it leads to a nonreflexive and symmetric binary relation).

3.2.2 Preference Modeling Through Outranking Relations

This sub-section presents four fundamental aspects of preference modeling in ELECTRE methods: modeling hesitation (partial and comprehensive) situations, modeling comprehensive incomparabilities, the concept of concordance and the relative importance of criteria, and the concept of non-discordance and the veto thresholds.

Let us observe that aggregation procedures used in ELECTRE methods are better adapted to situations where decision models include more than five criteria (up to twelve or thirteen).

3.2.2.1 The Concept of Pseudo-criterion

Definition 3.1 (pseudo-criterion). A pseudo-criterion is a function g_j associated with two threshold functions, $q_j(\cdot)$ and $p_j(\cdot)$, satisfying the following condition: for all ordered pairs of actions $(a,a') \in A \times A$, such that $g_j(a) \geqslant g_j(a')$, $g_j(a) + p_j(g_j(a'))$ and $g_j(a) + q_j(g_j(a'))$ are non-decreasing monotone functions of $g_j(a')$, such that $p_j(g_j(a')) \geqslant q_j(g_j(a')) \geqslant 0$ for all $a \in A$.

For more details about the concept of pseudo-criterion see [56] and [61]. Here we consider the thresholds as variables, but they can also be defined as constant values. Moreover, not necessarily all the criteria are subject to a definition of these discriminating thresholds. It should also be noted, that the way a pseudo-criterion was defined above, takes into account only *direct thresholds*, since the arguments of the threshold functions are the worst of the two performances $g_j(a)$ and $g_j(a')$. When the thresholds are expressed as a function of more preferred of the two values, we call them *inverse thresholds*. In the case of constant thresholds, there is no distinction between direct and inverse thresholds.

According to the above definition,

- $q_j(g_j(a'))$ is the greatest performance difference for which the situation of indifference holds on criterion g_j between two actions a and a', where $q_j(g_j(a')) = g_j(a) - g_j(a')$,

- $p_j(g_j(a'))$ is the smallest performance difference for which the situation of preference occurs on criterion g_j between two actions a and a', where $p_j(g_j(a')) = g_j(a) - g_j(a')$.

The reader can find more details about the discrimination thresholds in [55, 57].

3.2.2.2 The Definition of the Partial Binary Relations

Consider an ordered pair of actions $(a,a') \in A \times A$, and the two thresholds associated with the pseudo-criterion $g_j \in F$, which is used to model the following situations (note that no assumption is made here about which one of the two actions is better on criterion g_j):

1) $g_j(a) - g_j(a') > p_j(g_j(a'))$ $\Leftrightarrow$ aP_ja',
2) $q_j(g_j(a')) < g_j(a) - g_j(a') \leqslant p_j(g_j(a'))$ $\Leftrightarrow$ aQ_ja' (hesitation between aI_ja' and aP_ja'),
3) $-q_j(g_j(a)) \leqslant g_j(a) - g_j(a') \leqslant q_j(g_j(a'))$ $\Leftrightarrow$ aI_ja'.

The above three binary relations can be grouped into one partial outranking relation S_j comprising the three corresponding situations. $S_j = P_j \cup Q_j \cup I_j$, where aS_ja' means that "a is at least as good as a'" on criterion g_j. When aS_ja', the voting power of criterion g_j, denoted by w_j (assume w.l.g. that $\sum_{\{j \mid g_j \in F\}} w_j = 1$), is taken in total. Figure 3.1 illustrates the different zones of the partial binary relations previously defined, i.e. the situations aP_ja', aQ_ja', aI_ja', $a'Q_ja$, and $a'P_ja$, as well as the fraction φ_j of the voting power associated with each one of these situations.

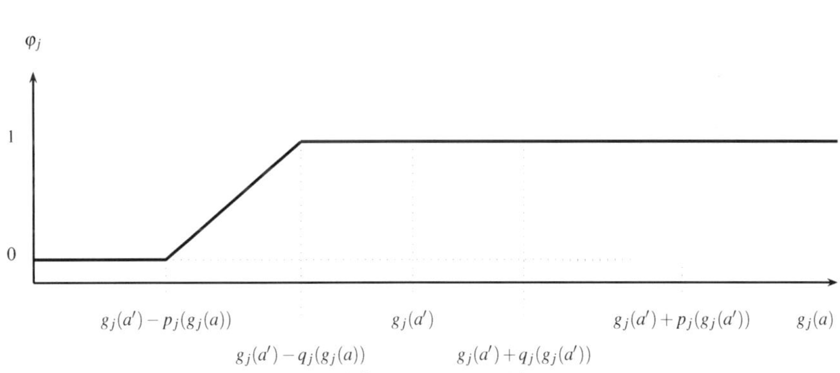

Fig. 3.1 Variation of φ_j for a given $g_j(a')$ and variable $g_j(a)$

From the definition of the partial binary relations and from Figure 3.1 it is easy to see that the two types of thresholds, direct and inverse, have to be taken into account.

3.2.2.3 The Comprehensive Outranking Relation

Preferences in ELECTRE methods are modeled by the comprehensive binary outranking relation S, whose meaning is *"at least as good as"*; in general, $S = P \cup Q \cup I$. Consider two actions $(a, a') \in A \times A$. Modeling comprehensive preference information leads to the four cases:

1. aSa' and not($a'Sa$), i.e., aPa' (a is *strictly preferred* to a');
2. $a'Sa$ and not(aSa'), i.e., $a'Pa$ (a' is *strictly preferred* to a);
3. aSa' and $a'Sa$, i.e., aIa' (a is *indifferent* to a');
4. not(aSa') and not($a'Sa$), i.e., aRa' (a is incomparable to a').

3.2.3 *The Concepts of Concordance and Discordance*

All outranking based methods rely on the concepts of concordance and discordance which represent, in a certain sense, the raisons *for* and *against* an outranking situation.

3.2.3.1 Concordance

Definition 3.2. To validate an outranking aSa', *a sufficient majority of criteria in favor* of this assertion must occur.

The comprehensive concordance index

The *comprehensive concordance index* can be defined as follows:

$$c(a,a') = \sum_{\{j \mid g_j \in \mathscr{C}(aSa')\}} w_j + \sum_{\{j \mid g_j \in \mathscr{C}(a'Qa)\}} w_j \varphi_j,$$

where

$$\varphi_j = \frac{g_j(a) - g_j(a') + p_j(g_j(a))}{p_j(g_j(a)) - q_j(g_j(a))},$$

and $\mathscr{C}(a\{Rel\}a')$ is the coalition of criteria in which relations $\{Rel\}$ hold for a, a'.

The concordance index is used to model the concept of concordance. It permits to build an $m \times m$ concordance matrix C composed of elements $c(a,a')$, for all $a, a' \in A$.

One can see that the criteria can be classified in three groups:

1. Those that are in favor of the assertion aSa' with no reservation,
2. Those that hesitate between the indifference and the opposition,
3. Those that are in the opposition.

The element $c(a,a')$ is the comprehensive concordance index with the assertion aSa'. This index results from a summation of the voting power of criteria from the first group, and of the fraction φ_j of the voting power of criteria from the second group (see also 3.2.4.1).

3.2.3.2 Discordance

Definition 3.3. The assertion aSa' cannot be validated if *a minority of criteria is strongly against* this assertion.

The concept of veto threshold

When criterion g_j opposes strongly to the assertion aSa', g_j puts its veto to the assertion aSa'. This occurs if $g_j(a') - g_j(a) > v_j(g_j(a))$. The value $v_j(g_j(a))$ is called the veto threshold of g_j.

For each criterion $g_j \in F$, an opposition power is determined, for each $(a,a') \in A \times A$.

Partial discordance indices

$$d_j(a,a') = \begin{cases} 1 & \text{if } g_j(a) - g_j(a') < -v_j(g_j(a)), \\ \frac{g_j(a)-g_j(a')+p_j(g_j(a))}{p_j(g_j(a)) - v_j(g_j(a))} & \text{if } -v_j(g_j(a)) \leqslant g_j(a) - g_j(a') < -p_j(g_j(a)), \\ 0 & \text{if } g_j(a) - g_j(a') \geqslant -p_j(g_j(a)). \end{cases}$$

where $d_j(a,a')$ is the partial discordance index of criterion g_j.

It permits to build an $m \times m$ discordance matrix D_j composed of elements $d_j(a,a')$, for all $(a,a') \in A \times A$ and for each criterion $g_j \in F$.

3.2.4 *Structure*

Each ELECTRE method comprises two main procedures: an aggregation procedure and an exploitation procedure.

3.2.4.1 Multiple Criteria Aggregation Procedures (MCAPs)

By definition, MCAP is a procedure that builds one or possibly several outranking relations on the basis of the performances of each action on each criterion, which leads to assign to each ordered pair of actions one and only one of the four situations presented in sub-section 3.2.1. Let us notice that the decision maker does not make any pairwise comparison; all the comparisons are done by the procedure itself.

The MCAP has to take into account the role played by the criteria: some of them can play a "very important" role, while others can play a "totally secondary" role. For this purpose, ELECTRE methods make use of intrinsic weights and possible veto thresholds [29, chap. 4]. The intrinsic weights can be interpreted as the voting power of each criterion. The higher the intrinsic weight, the more important the criterion is. Note that the voting power neither depends on the range of the criterion scale nor on the encoding chosen (in particular the unit selected) to express the evaluation (score) on this scale. In ELECTRE methods it is not assumed that the weights, as well as the veto thresholds, have a real existence in the mind of the decision maker. They do not have a "true value". Such parameters are artifacts, co-constructed abstract "objects" [58].

For example, in ELECTRE III, its MCAP associates for each ordered pair of action $(a,a') \in A \times A$ a credibility index of the assertion aSa'. This credibility index is a fuzzy measure denoted by $\sigma(a,a') \in [0,1]$, which combines $c(a,a')$ and $d_j(a,a')$:

$$\sigma(a,a') = c(a,a') \prod_{j \in \mathscr{J}(a,a')} \frac{1-d_j(a,a')}{1-c(a,a')},$$

where $j \in \mathscr{J}(a,a')$ if and only if $d_j(a,a') \geqslant c(a,a')$.

3.2.4.2 Exploitation Procedures (EPs)

By definition, EP is a procedure used to obtain *adequate results* from which *recommendations* can be derived.

3.2.4.3 The Nature of the Results

The *nature of the results* depends on the specific *problematic*. The three major problematics in MCDA can be stated as follows:

Choosing

Selecting a restricted number of the most interesting potential actions, as small as as possible, which will justify to eliminate all others.

Sorting

Assigning each potential action to one of the categories among a family previously defined; the categories are ordered, in general, from the worst to the best one. An example of a family of categories suitable for assignment procedures is given below:

C_1: actions whose implementation is not advised
C_2: actions whose implementation could only be advised after significant modifications,
C_3: actions whose implementation could only be advised after slight modifications,
C_4: actions whose implementation is always advised without any reservation.

Ranking

Ranking of actions from the best to the worst, with the possibility of ties (*ex æquo*) and incomparabilities.

The way in which MCDA is envisaged (the problematic) will constrain the form of the results to be obtained.

Remark

1. In *sorting problematic* the result depends on *absolute evaluation* of actions: the assignment of an action takes into account, *only its intrinsic evaluation on all the criteria*, and it *neither depends on nor influences* the category to be selected for the assignment of another action.
2. As in the *remaining problematics* the actions are compared against each other, the result depends in these cases on *relative evaluation* instead of absolute one as in the previous case.

3.2.4.4 Software

The family of ELECTRE methods includes several variants of methods designed for the three main problematics defined above. Below, we list these variants, together with an associated software available at LAMSADE.

1. Choosing: ELECTRE I, ELECTRE Iv, and ELECTRE IS. ELECTRE IS is a generalization of ELECTRE Iv. ELECTRE IS software runs on an IBM compatible PC with Windows 98 or higher.
2. Ranking: ELECTRE II, ELECTRE III, and ELECTRE IV. ELECTRE III-IV software runs on a PC with Windows 3.1, 95, 98, 2000, Millennium and XP.
3. Sorting: ELECTRE TRI-B, ELECTRE TRI-V, and ELECTRE TRI-NC. ELECTRE TRI-B software runs on Microsoft Windows 3.1, 95, 98, Me, 2000, XP, and NT. It incorporates an Assistant tool which enables the user to define the weights and the λ-cutting level indirectly, from an elicitation process based on a set of assignment examples. The weights are thus inferred through a certain form of

regression. Due to the indirect elicitation, ELECTRE TRI-B Assistant reduces the cognitive effort required from the decision makers.

3.2.5 *Strong Features*

This sub-section goes through major strong features of ELECTRE family methods. They include the possibility of dealing with the qualitative as well as the quantitative nature of criteria. The heterogeneity of scales and the non-relevance of compensatory effects are also discussed here. The imperfect knowledge of data and some arbitrariness when building criteria can be taken into account in ELECTRE methods, and, finally, they can deal with the reasons for and against an outranking.

3.2.5.1 The Qualitative Nature of Some Criteria

ELECTRE family methods have the provision for taking into account the *qualitative nature of some criteria*. They allow to consider original data, without the need of recoding them. In fact, all the criteria are processed as qualitative criteria, even if some are quantitative by their very nature.

3.2.5.2 The Heterogeneity of Scales

The ELECTRE family methods can deal with *heterogeneous* scales to model such diversified notions as noise, delay, aesthetics, cost, ... Whatever the nature of scales, every procedure can run with preserved original performances of the actions on the criteria, without the need of recoding them, for example, by using a normalization technique or the assessment of the corresponding evaluations through the use of a utility or a value function.

3.2.5.3 The Non-relevance of Compensatory Effects

The MCAP of ELECTRE methods are not based on the notion of compensation. Such MCAPs where conceived such that they do not allow for compensation of performances among criteria, i.e. the degradation of performances on certain criteria cannot be compensated by improvements of performances on other criteria.

The weights of criteria do not mean substitution rates as it is the case in many other methods. The limited possibility of compensation can be brought into light through the concordance and discordance indices:

- Concerning the concordance index, when comparing action a to action a', with the exclusion of the ambiguity zone, only the fact that a outranks or does not outrank a' with respect to criteria from F is relevant, while it is not relevant how much the performance of a is better or worse than the performance of a' on criteria from F;

- The existence of veto thresholds strengthening the non-compensatoiriness effect is yet another reason of the possibility of non compensation in ELECTRE methods. For example, when $d_j(a,a') = 1$, no improvement of the performance of a and no deterioration of the performance of a', with respect to the other criteria than g_j, can compensate this veto effect.

Consider the following example with 4 criteria and only 2 actions (scales: [0,10]). The performance matrix for this example is given in Table 3.1. Suppose that the weighted-sum model was chosen, i.e, $V(a) = w_1g_1(a) + \ldots + w_jg_j(a) + \ldots + w_ng_n(a)$. In the considered example, the weights w_j are equal for all criteria ($w_j = 0.250$, for all $j = 1,\ldots,4$):

Table 3.1 Performance matrix

	g_1	g_2	g_3	g_4
a_1	9.500	9.500	8.100	5.400
a_2	8.300	8.300	7.300	8.500

$V(a_1) = 8.125 > V(a_2) = 8.100$ (notice that $V(a_1) - V(a_2) = 0.025$), and so a_1Pa_2 (a_1 is strictly preferred to a_2). The difference between the performances of the two actions is small on the first 3 criteria, while this difference on the fourth criterion (3.100) is very big in favor of a_2. The compensatory effect led a_1 to be strictly preferred to a_2. This example shows, in an obvious way, the possibility that a big preference difference not favorable to a_1 on one of the criteria (g_4) can be compensated by 3 differences of small amplitude on the remaining criteria, in such a way that a_1 becomes finally strictly preferred to a_2.

In ELECTRE methods the type of compensatory effect shown in the above example, does not occur in a systematic way (see 3.2.5.4 and 3.2.5.5). Thus, contrarily to many other methods, for example, Choquet integral, Sugeno integral, and MACBETH, there is no need in ELECTRE methods to use identical and commensurable scales, which by their very nature, do not have, in general, the property called commensurability.

3.2.5.4 Taking into Account the Imperfect Knowledge of Data and Some Arbitrariness when Building Criteria

ELECTRE methods are adequate to take into account the *imperfect knowledge* of the data and the *arbitrariness* related to the construction of the family of criteria. This is modeled through the indifference and preference thresholds (discrimination thresholds). Consider the same example with the (constant) discrimination thresholds, given in Table 3.2.

Table 3.2 Performances and discrimination thresholds

	g_1	g_2	g_3	g_4
a_1	9.500	9.500	8.100	5.400
a_2	8.300	8.300	7.300	8.500
q_j	1.000	1.000	1.000	1.000
p_j	2.000	2.000	2.000	2.000

On the one hand, it should be noticed that any small variation of some performance will not affect in a significant way the preference difference resulting from the MCAP used in ELECTRE methods, but it will modify the weighted-sum value. For example, if on criterion g_3 we would change the performance of action a_2 from 7.300 to 7.100, then the weighted-sum score $V(a_2)$ would move from 8.100 to 8.050 ($V(a_1) - V(a_2) = 0.050$). Consequently, there would be a reinforcement of the preference in favor of a_1.

On the other hand, in ELECTRE methods, $c(a_1,a_2)$ and $c(a_2,a_1)$ remain unchanged as it will be shown hereafter. Since the weighted-sum based models do not allow for the inclusion of thresholds, a_1 is still better than a_2. Now, if we consider 7.500 instead of 7.300, then $V(a_2) = 8.150$, and consequently a_2Pa_1. This slight variation is really too small to invert the preference between a_1 and a_2, but since the weighted-sum based models do not allow for the inclusion of thresholds, a_2 became preferred to a_1. This phenomenon shows the sensitivity of the weighted-sum with respect to non significant variations of the performances, due to the compensatory character of the model.

The performances of the actions can be affected by the imperfect knowledge coming from different sources. At the same time, the way the criteria are built or conceived contains some part of arbitrariness. These are the two major reasons that led to define the discrimination thresholds in ELECTRE methods. When considering the discrimination thresholds and using ELECTRE methods, $c(a_1,a_2) = 0.250 + 0.250 + 0.250 = 0.750$ and $c(a_2,a_1) = 0.200 + 0.200 + 0.250 + 0.250 = 0.800$ (see 3.2.3.1).

3.2.5.5 Reasons *for* and Reasons *against* an Outranking

The ELECTRE methods are based, in a certain sense, on the *reasons for* (concordance) and the *reasons against* (discordance) of an outranking between two actions. Consider the same example and a veto threshold $v_j = 3$, for all $j = 1,\ldots,4$ (see Table 3.3).

If the concordance threshold $s = 0.800$, then a_2Sa_1 and $not(a_1Sa_2)$. But, if $s = 0.700$, then a_2Sa_1 and a_1Sa_2, i.e., a_2Ia_1.

Table 3.3 Performances, discrimination and veto thresholds

	g_1	g_2	g_3	g_4
a_1	9.500	9.500	8.100	5.400
a_2	8.300	8.300	7.300	8.500
q_j	1.000	1.000	1.000	1.000
p_j	2.000	2.000	2.000	2.000
v_j	3.000	3.000	3.000	3.000

The discordance index (see 3.2.3.2) of g_4, $d_4(a_2, a_1) = 1$, and whatever the value of concordance threshold s, we get $not(a_1 S a_2)$. This means that g_4 imposes its veto power on the assertion $a_1 S a_2$. Weighted-sum based models do not allow for the inclusion of veto effects.

The above shows, moreover, that the consideration of a veto threshold reinforce the non-compensatory character of the ELECTRE methods.

3.2.6 *Weaknesses*

This sub-section gives account of the main drawbacks or weaknesses of ELECTRE methods, notably when the quantitative nature of the family of criteria requires the use of a different method, when a score should be assigned to each action, when the independence with respect to irrelevant actions and the possible instability of results is required, or the possible and frequent occurrence of intransitivities would make a problem.

3.2.6.1 Scoring Actions

In certain contexts it is required to assign a *score* to each action. When the decision makers require that each action should get a score, the ELECTRE methods are not adequate for such a purpose and the scoring methods should be applied instead. The decision makers should be aware, however, that using a score method they cannot provide information that leads, for example, to intransitivities or to incomparabilities between some pairs of actions. Moreover, the score is very fragile.

For the time being, there is no outranking-like method allowing to assign a score to different actions in a convincing manner. This seems, however, a very difficult task to accomplish, because it is assumed to take into account a measure of preference difference (or intensity of preference). In PROMETHEE [19, chap. 5] there was an attempt to define a measure of preference differences, but the way in which it was presented seems to contain matter for some criticism (see 6.4.1 in [59]).

3.2.6.2 The Quantitative Nature of the Family of Criteria

When all the *criteria are quantitative* it is "better" to use some other methods. But, if we would like to take into account a completely or even a partially non-compensatory method, as well as the reasons for and against, then, even if the criteria would be all quantitative, we should use ELECTRE methods. Assume that all the criteria are quantitative and defined on the same scale with the same unit. Also, then, if we are dealing with imperfect knowledge with respect to at least one criterion, ELECTRE methods are suitable.

3.2.6.3 The Independence with Respect to Irrelevant Actions

Except ELECTRE TRI-B, ELECTRE TRI-C, and ELECTRE TRI-NC, the remaining ELECTRE methods do not fulfill the property of *independence with respect to irrelevant actions*, which says that when comparing two actions, the preference relation should not depend on the presence or absence of other actions. Roy [54] shows that rank reversal may occur and, consequently, the property of independence w.r.t. irrelevant actions can be violated when dealing with outranking relations. Notice that rank reversal may occur only when the set of potential actions is subject to evolve, which is quite a natural assumption, however, it is not present in many real-world decision aiding processes where the number of actions is rather small and easily identified. Roy [54] presents an example illustrating that such phenomena can be interpreted quite naturally and the author also suggests that allowing the independence property is not realistic in many real-world decision aiding situations. Other works devoted to the same kind of concern include, for example, Perny [51], Roy and Bouyssou [59], Simpson [63], and Wang and Triantaphyllou [72].

In fact, the *instability* of the results in ELECTRE methods was recently re-analyzed by Wang and Triantaphyllou [72] with respect to ELECTRE II and III. When the decision makers feel more comfortable and confident with an evaluation model that provides a stable result, they might be a little bit surprised by the results provided by ELECTRE methods in certain circumstances. In our perspective, a stable result is not necessarily the evidence of an adequate processing of data because some aggregation procedures assume that the data have a meaning, but very often they do not really have it. For example, this is often the case of the weighed-sum based methods, where the results may be stable but not necessarily meaningful [43]. Moreover, if one uses different normalization procedures (as is the case when one deals with multiple units of measurement) such methods may alter the derived results [71]. What the ELECTRE methods show is related to the poorly determined margins on the results, very often related to the poor quality of data since the scales are processed as ordinal ones.

Regarding the rank reversal it is important to underline the following aspects:

1. It is quite natural that MCAPs based on pairwise comparisons violate the principle of independence with respect to irrelevant actions. The possibility of what is called rank reversal is a consequence of this violation.

2. In ELECTRE methods when there exists a phenomenon of rank reversal between action a and action a', this shed some light on the fact that the way a and a' are compared is not robust. This is due to the following two reasons:

   a) the existence of discriminating thresholds and the values that should have been assigned to them,

   b) the fact that such a comparison is conditioned by the way the actions a and a' are compared to the remaining actions [72, 28].

3.2.6.4 Intransitivities

Intransitivities may also occur in ELECTRE methods [54]. It is also well-known that methods using outranking relations (not only the ELECTRE methods) do not need to satisfy the transitivity property. This aspect represents a weakness only if we impose *a priori* that preferences should be transitive. There are, however, some reasons for which the transitivity should not be imposed:

1. It is quite natural that the binary relation of indifference should be considered intransitive (see an example illustrating this phenomenon in [41]); there is also no reason to avoid defining indifference thresholds for certain criteria.
2. It is also possible to have insensitivities with respect to the binary relation of preference; we would say that it is possible and rather frequent to have a majority of the criteria in favor of a over b, and majority of the criteria of b over c, without necessarily implying that there is a majority of the criteria in favor of a over c; we can also have a majority of criteria in favor of c over a; this is the well-known Condorcet Paradox, described, e.g., in [11]. In fact, Gerhlein [30] proved that for 25 voters and 11 candidates, the probability that the Condorcet Paradox occurs is 50%.

Let us notice that there is no such intransitivity phenomena in ELECTRE TRI-B and ELECTRE TRI-C methods.

3.2.7 A Discussion of the Weak and the Strong Points of ELECTRE *Methods*

A *discussion on the weak and the strong points* of ELECTRE methods can be found in [28]. The objective of this discussion was also to draw the attention of the readers to a particular philosophy of interpreting the results of ELECTRE methods, which is different from philosophies of interpreting the results obtained by other methods, notably the ones adopted in the paper by Wang and Triantaphyllou [72]. In [28], as well as in [57], the authors try to show, that the objective of decision aiding is not to discover an absolute truth or, a pre-existing "real" best action, ranking, or assignment. The modifications that may occur when adding or removing an action, emphasizing the limitations of the conclusions that can be derived by using ELECTRE methods, without any robustness concern in mind. Clearly, this is also what decision aiding is designed to do: to show how the conclusions can be tenuous and do not reveal a pre-existing truth.

3.3 Recent Developments

The recent developments presented in this section are mainly methodological. They concern new approaches, an axiomatic analysis of ELECTRE methods, as well as some aspects related to the meaningfulness of the methods.

3.3.1 Methodological Developments

This section is devoted to the presentation of the inference based approaches and some related issues, the inference-robustness based approaches, the pseudo robustness based approaches, and the new concepts of robustness that can be applied to ELECTRE methods.

3.3.1.1 Pure-inference Based Approaches

Mousseau and Słowiński [48] proposed the first general algorithm for inferring the values of the model parameters of ELECTRE TRI-B method from assignment examples given by the decision maker, i.e., from holistic judgments. The assignment examples serve to build a set of mathematical constraints and the inference of the model parameters consists in solving a mathematical programming problem. This approach represents the paradigm of disaggregation-aggregation of preferences [39], which aims at extracting implicit information contained in holistic statements given by a decision maker. In this case, the statements to be disaggregated are assignment examples. Such an indirect elicitation of preferences requires from the decision maker a much smaller cognitive effort than direct elicitation of the model parameters.

The proposed interactive disaggregation-aggregation procedure finds values of the model parameters that best restore the assignment examples provided by the decision maker. Finding values of all the model parameters at once, i.e., the weights, all thresholds, category bounds and the cutting level λ used in ELECTRE TRI-B, requires, however, solving a hard non-linear programming problem. In order to overcome this difficulty, one can decompose the inference procedure into a series of linear programs specialized in finding values of subsets of these parameters. A computer implementation of this inference method with respect to weights and the $\lambda-$cutting level gave birth to a software tool called ELECTRE TRI ASSISTANT [49].

The tool is also able to identify "inconsistent judgements" in the assignment examples.

Let us notice that in all inference procedures concerning the ELECTRE TRI-B method, the "pessimistic" version of the assignment procedure was considered only (the "optimistic" version is even more difficult to model in terms of mathematical programming because it requires binary variables).

The *inference-based approaches* proposed after the work by Mousseau and Słowiński [48] are the following:

1. Inferring the *weights* and the $\lambda-$cutting level of ELECTRE TRI-B by linear programming (the discrimination and the veto thresholds as well as the category bounds being fixed) [46]. In this work, the authors consider the linear programming model of Mousseau et al. [49], and perform several numerical experiments related to checking the behavior of this inference disaggregation tool. These experiments show that $2n$ (n being the number of criteria) assignment examples are sufficient to infer adequately the weights and the $\lambda-$cutting level.
2. Inferring the *bounds of categories* [50]. This work deals with the possibility of inferring the the bounds of categories of ELECTRE TRI-B. After making a slightly simplifying assumptions, the authors developed linear programming and 0-1 linear programming models to infer the bounds.
3. Inferring *veto thresholds* [20]. This work is a complement of the previous ones. The authors proposed mathematical programming models to assess veto thresholds for the original outranking relation and its two other variants, which may be used in ELECTRE methods, including ELECTRE III. In this case, the inference tools make use of linear programming, 0-1 linear programming, or separable programming.
4. Some *manageable disaggregation procedures* for valued outranking relations were proposed [44]. The authors used a modified definition of the valued outranking relation, preserving the original discordance concept. This modification makes easier to solve inference problem via mathematical programming. These procedures can be used within ELECTRE III and ELECTRE TRI methods.
5. For some decision examples given by decision makers, there may be no feasible values of model parameters which would permit the model to represent these examples. We then say that the preference information is *inconsistent with respect to the model*. Resolving inconsistency is a problem of utmost importance, as shown in [47, 45]. The authors proposed algorithms for resolving inconsistency, where the decision makers must choose between different options of withdraw ing or relaxing inconsistent examples. It should be noted, however, that unless inconsistency does not come from violation of dominance, it is not a fault of the decision maker but a deficiency of the preference model to restore the decision examples. Thus, instead of withdrawing or relaxing inconsistent examples, one should also consider the possibility of using a more adequate preference model [26].

3.3.1.2 Inference-robustness Based Approaches

The *disaggregation-aggregation* approach for *inferring weights* and deriving *robust conclusions* in sorting problems was proposed in [21]. This work presents a new interactive approach that combines two different approaches, the inference based approach with the robustness based approach. It is also applied to ELECTRE TRI-B. The first approach was described in the previous sub-section. The second approach considers a set of constraints with respect to the parameter values (weights and $\lambda-$cutting level), used to model the imperfect character of the information provided by the decision maker. Then, for each action, the best and worst categories compatible with the constraints are determined. This type of results allows to derive some robust conclusions about the assignments. The robustness analysis is used in this study to guide the decision maker through an interactive inference of weights and $\lambda-$cutting level.

3.3.1.3 Pseudo-robustness Based Approaches

Stability analysis or pseudo-robust conclusions based on Monte Carlo simulation methods, mainly for ranking and sorting problems [70]. The authors proposed a new method SMAA-TRI based on stochastic multiple criteria acceptability analysis (SMAA), for analyzing the stability of some parameters of the ELECTRE TRI-B method. The method consists of analyzing finite spaces of arbitrarily distributed parameter values. Then a Monte Carlo simulation is applied in these spaces for describing each action in terms of the share of parameter values that have been assigned to different categories. This is a kind of stability analysis that can be used to derive pseudo-robust conclusions. For each action, the result obtained is the share of parameter values for each category (in terms of percentage).

3.3.1.4 New Concepts for Robustness Concerns

Although having a more general range of applicability, the works that will be described below should be able to bring answers to the robustness concerns, when applied to decision aiding using ELECTRE methods.

In [1, section 3.4], the authors propose a measure of robustness, which is applied to ranking of potential actions $a \in A$ obtained when using ELECTRE III or ELECTRE IV, in the case where it is necessary to take into account a family $\hat{S}$ of scenarios (or of "variable settings"). Let P_s denote a (partial or complete) order provided by ELECTRE with scenario $s \in \hat{S}$, and let $P = \{P_s \mid s \in \hat{S}\}$. First, the authors consider the following measure of robustness:

$$r_\alpha(a) = \text{Proportion of pre-orders } P_s \in \hat{S},$$

in which a occupies a position in the ranking at least equal to α; where α denotes an *a priori* fixed position. Under such basis we can judge that action "a is at least as robust as action a'", when $r_\alpha(a) \geqslant r_\alpha(a')$. Then, the authors proposed to improve this measure by taking into account another position in the ranking β (also defined *a priori*) in order to penalize the actions with a very bad position in certain scenarios. Thus, they propose the following robustness measure:

$$r_{\alpha\beta}(a) = r_\alpha(a) - \text{ Proportion of } P_s \in \hat{S},$$

in which action a occupies a position in the ranking greater than or equal to β.

The results obtained with this robustness measure (possibly supplemented by a sensitivity analysis with respect to the reference positions α and β) must be able to be synthesized in the form of robust conclusions (concept with which the authors deal in section 5) easily understandable by the decision maker (for more details on this subject see section 6 in Chapter 1 of the present book).

Still in [1, section 5.3], the authors propose two frameworks intended to generalize an approach that was successfully used in two concrete cases by one of the authors. In these formal frameworks (using different ELECTRE methods), the approach allows to work out some conclusions and then recommendations answering to certain robustness concerns. The approach mainly aims at restricting the number of combinations of the options to be explored. This restriction is supported by making in clear positions those combinations of options, which appear to have the most significant effect for answering robustness concerns.

In Chapter 1 of this book, B. Roy introduce in section 5, various suggestions and proposals for answering to certain robustness concerns by weakening the role of the worst case. These suggestions and proposals do not concern in particular the ELECTRE methods but, at least for some of them, they can be useful.

3.3.2 *Improvements and New Approaches*

This section presents the main novelties of ELECTRE-like methods, such as a concept of bi-polar outranking relations implemented in the RUBIS method, the modeling of three different types of interaction among criteria, the research done to modify the credibility index through the use of the reinforced preference thresholds and the counter-veto thresholds, the ELECTRE TRI-C and ELECTRE TRI-NC methods, and the ELECTRE^{GMS} method.

3.3.2.1 Bi-polar Outranking Based Procedures

The concept of *bi-polar outranking relations* was proposed by Bisdorff et al. [7] and implemented in the RUBIS software. The RUBIS method is a progressive multiple criteria decision aiding method for choice problems. It is also an outranking based method. It is, however, based on a new concept of bi-polar outranking relation.

The bipolar outranking index $\widetilde{S} : A \times A \rightarrow [-1, 1]$ is defined as follows: for $(a, a') \in A \times A$,

$$\widetilde{S}(a,a') = \min\left\{\widetilde{C}(a,a'), -V_1(a,a'), \ldots, -V_n(a,a')\right\}$$

where

$$\widetilde{C}(a,a') = \sum_{\{j \mid g_j \in \mathscr{C}(a\{P,Q,I\}a')\}} w_j - \sum_{\{j \mid g_j \in \mathscr{C}(a'Pa)\}} w_j$$

and for all $g_j \in F$,

$$V_j(a,a') = \begin{cases} 1 & \text{if } g_j(a) - g_j(a') \leqslant -v_j(g_j(a)), \\ -1 & \text{if } g_j(a) - g_j(a') > -wv_j(g_j(a)), \\ 0 & \text{otherwise} \end{cases}$$

where $wv_j(g_j(a))$ and $p_j(g_j(a)) \leqslant wv_j(g_j(a)) \leqslant v_j(g_j(a))$ is a weak veto threshold.

The maximum value +1 of the bipolar outranking index is reached in the case of unanimous concordance, whereas the minimum value –1 is obtained either in the case of unanimous discordance, or if there exists a strong veto situation on at least one criterion. The median situation 0 represents a case of indetermination: either the arguments in favor of an outranking are compensated by those against it, or a positive concordance in favor of the outranking is outbalanced by a potential (weak) veto situation.

The semantics linked to this bipolar outranking index is the following:

- $\widetilde{S}(a,a') = +1$ means that assertion "aSa'" is *clearly validated*,
- $\widetilde{S}(a,a') > 0$ means that assertion "aSa'" is *more validated than non-validated*,
- $\widetilde{S}(a,a') = 0$ means that assertion "aSa'" is *undetermined*,
- $\widetilde{S}(a,a') < 0$ means that assertion "aSa'" is *more non-validated than validated*,
- $\widetilde{S}(a,a') = -1$ means that assertion "aSa'" is *clearly non-validated*.

On the basis of the bipolar outranking index, a recommendation for choice problems is given by a procedure based on five pragmatic principles ($\mathscr{P}_1$: non-retainment for well motivated reasons; $\mathscr{P}_2$: minimal size; $\mathscr{P}_3$: efficient and informative refinement; $\mathscr{P}_4$: effective recommendation; $\mathscr{P}_5$: maximal credibility) and the theoretical concepts of hyper-kernel and augmented cordless circuits in a digraph.

3.3.2.2 Taking into Account the Interaction Between Criteria

The interaction between criteria is modeled through the *weights of the interaction coefficients* and the modifications in the concordance index [27]. This work presents an extension of the comprehensive (overall) concordance index of ELECTRE methods, which takes the interaction among criteria into account. Three types of interactions have been considered: mutual strengthening, mutual weakening, and antagonism. The new concordance index correctly takes into account these three types of interactions by imposing such conditions as boundary, monotonicity, and continuity. The following types of interactions were considered (let us notice that the cases

$a-b$ are mutually exclusive, but cases $a-c$ and $b-c$ are not). Let $\bar{\mathscr{C}}(a'Pa)$ denote the coalition of criteria that strongly opposes to the assertion "a outranks a':

a) *Mutual strengthening effect*
If both criteria g_i and g_j strongly, or even weakly, support the assertion aSa' (more precisely, $g_i, g_j \in \bar{\mathscr{C}}(a'Pa)$), we consider that their contribution to the concordance index must be greater than the sum of $k_i + k_j$, because these two weights represent the contribution of each of the two criteria to the concordance index when the other criterion does not support aSa'. We suppose that the effect of the combined presence of both g_i and g_j among the criteria supporting the assertion aSa' can be modeled by a mutual strengthening coefficient $k_{ij} > 0$, which intervenes algebraically in $c(a,b)$. Note that $k_{ij} = k_{ji}$.

b) *Mutual weakening effect*
If both criteria g_i and g_j strongly, or even weakly, support the assertion aSa' (more precisely, $g_i, g_j \in \bar{\mathscr{C}}(a'Pa)$), we consider that their contribution to the concordance index must be smaller than the sum of $k_i + k_j$, because these two weights represent the contribution of each of the two criteria to the concordance index when the other criterion does not support aSa'. We suppose that this effect can be modeled using a mutual weakening coefficient $k_{ij} < 0$, which intervenes algebraically in $c(a,a')$. Note that $k_{ij} = k_{ji}$.

c) *Antagonistic effect*
If criterion g_i strongly, or weakly, supports the assertion aSa', and criterion g_h strongly opposes to this assertion, we consider that the contribution of criterion g_i to the concordance index must be smaller than the weight k_i that was considered in the cases in which g_h does not belong to $\mathscr{C}(a'Pa)$. We suppose that this effect can be modeled by introducing an antagonism coefficient $k'_{ih} > 0$, which intervenes negatively in $c(a,a')$. Note that the presence of an antagonism coefficient $k'_{ih} > 0$ is compatible with both the absence of antagonism in the reverse direction ($k'_{hi} = 0$) and the presence of a reverse antagonism ($k'_{hi} > 0$).
The antagonistic effect does not double the influence of the veto effect; in fact, they are quite different. If criterion g_h has a veto power, it will always be considered, regardless of whether g_i belongs to the concordant coalition or not. The same is not true for the antagonistic effect, which occurs only when criterion g_i belongs to the concordant coalition. Let us notice that a veto threshold expresses the power attributed to a given criterion g_j to be against the assertion "a outranks a'", when the difference between performances $g_j(a')$ and $g_j(a)$ is greater than this threshold.

The authors demonstrated that the generalized index is able to take satisfactorily into account the three types of interactions or dependencies among criteria, and they also examined the links between the new concordance index and the Choquet integral. Nevertheless, this extension is appropriate only when the number of pairs of interacting criteria is rather small. Otherwise, we consider that the family of criteria should be rebuilt, since it contains too many interactions and (possibly) incoherencies.

3.3.2.3 The Reinforced Preference and the Counter-veto Effects

The credibility index $\sigma(a,a')$ of the outranking relation aSa' (see sub-section 3.2.2) involves preference scales which are purely ordinal. For this reason, as soon as on criterion g_j, the difference of performances $g_j(a) - g_j(a')$ becomes greater than the preference threshold, the value of this difference does not influence the credibility of outranking of action a over action a'. If one would judge that a very large value of this difference gets the meaning of "very strong preference", then one could wish to take this judgment into account in the definition of the credibility of outranking of a over a'. To satisfy such a wish, Roy and Słowiński [60] proposed two complementary ways:

- The first one involves a new threshold called reinforced preference threshold: it corresponds to the value of the difference of performances $g_j(a) - g_j(a')$ which is "judged meaningful" for considering criterion g_j as more important in the concordant coalition (by increasing its weight), comparing to the situation where (all things equal elsewhere) the difference of performances is smaller than this threshold (however, not smaller than the preference threshold);
- The second one involves another threshold called counter-veto threshold (it is not necessarily equal to the previous one, as it has a different meaning and it plays a different role): it corresponds to the value of the difference of performances $g_j(a) - g_j(a')$ which is "judged meaningful" for weakening the mechanism of veto against the credibility of outranking (from the side of discordant criteria), comparing to the situation where (all things equal elsewhere) the difference of performances is smaller than this threshold (however, not smaller than the preference threshold).

After defining some principles and requirements for the new formula of the credibility index $\sigma(a,a')$ giving account of the two ways above, Roy and Słowiński gave the following proposal which satisfies these requirements.

Let $rp_j(g_j(a))$ denote the reinforced preference threshold for criterion g_j. When this threshold is crossed, the importance coefficient w_j in the formula for concordance index $c(a,a')$ should be replaced by $\omega_j w_j$, where $\omega_j > 1$ is called reinforcement factor. Let $\mathscr{C}(aRPa')$ denote the set of criteria for which $g_j(a) > g_j(a') + rp_j(g_j(a))$. The new concordance index in then defined as follows,

$$\hat{c}(a,a') = \frac{\sum_{\{j \mid g_j \in \mathscr{C}(aRPa')\}} \omega_j w_j + \sum_{\{j \mid g_j \in \mathscr{C}(aSa') \setminus \mathscr{C}(aRPa')\}} w_j + \sum_{\{j \mid g_j \in \mathscr{C}(a'Qa)\}} w_j \varphi_j}{\sum_{\{j \mid g_j \in \mathscr{C}(aRPa')\}} \omega_j w_j + \sum_{\{j \mid g_j \in F \setminus \mathscr{C}(aRPa')\}} w_j}$$

Let $cv_j(g_j(a))$ denote the counter-veto threshold for criterion g_j, and k the number of criteria for which this threshold has been crossed.

In order to give account of the reinforced preference and the counter-veto effects, the credibility index $\sigma(a,a')$ of the assertion aSa' has to be adequately adapted. For example, the the credibility index $\sigma(a,a')$ defined in point 3.2.4.1, takes the following form:

$$\hat{\sigma}(a,a') = c(a,a') \left[\prod_{j \in \mathscr{J}(a,a')} \frac{1 - d_j(a,a')}{1 - c(a,a')} \right]^{(1-k/n)},$$

where $j \in \mathscr{J}(a,a')$ if and only if $d_j(a,a') \geqslant c(a,a')$. Again, $\hat{\sigma}(a,a') \in [0,1]$.

For any criterion g_j, $g_j \in F$, the two thresholds $rp_j(g_j(a))$ and $cv_j(g_j(a))$ can be chosen equal, and, moreover, one may wish to consider only one of the two effects; deleting an effect means giving to the corresponding threshold an infinite or very large value. Consequently, no order relation is imposed between these two thresholds.

The new formula for the index of the credibility of outranking $\hat{\sigma}(a,a')$ can be substituted to similar formulae used in original versions of ELECTRE III, ELECTRE TRI-B, ELECTRE TRI-C, and ELECTRE TRI-NC.

The assignment of values to the new thresholds $rp_j(g_j(a))$ and $cv_j(g_j(a))$ can be done in a constructive way of thinking about the model of decision problem at hand. One can use for this some protocols of inquiry similar to those proposed for assigning appropriate values to indifference and preference thresholds [59], or to the weights [25]. These protocols involve few easy questions which do not require from the addressee to speculate about completely unrealistic situations. Another way could be to proceed via disaggregation-aggregation approach, so as to get thresholds $rp_j(g_j(a))$ and $cv_j(g_j(a))$ as compatible as possible with some exemplary pairwise comparisons of few real actions [48].

The way of introducing the two new effects is consistent with the handling of purely ordinal preference scales. Each of the two new thresholds is like a frontier representing a qualifier without any reference to a notion of quantity. The weights remain intrinsic weights, and do not become substitution rates, the indifference and preference thresholds play exactly the same role as before.

The new formula could also be used outside ELECTRE methods, for example, as similarity or closeness index [65, 66, 67], or as a filtering operator [52].

3.3.2.4 The ELECTRE TRI-C and ELECTRE TRI-NC methods for sorting problems

ELECTRE TRI-C [2] is a new method for sorting problems designed for dealing with decision aiding situations where each category from a completely ordered set is defined by a single characteristic reference action. The characteristic reference actions are co-constructed through an interactive process involving the analyst and the decision maker. ELECTRE TRI-C has been also conceived to verify a set of natural structural requirements (conformity, homogeneity, monotonicity, and stability). The method makes use of two joint assignment rules, where the result is a range of categories for each action to be assigned.

Set A of the considered actions is either completely known *a priori* or may appear progressively during the decision aiding process. The objective is to assign these actions to a set of completely ordered categories, denoted by $C_1, \ldots, C_h, \ldots, C_q$ with

$q \geqslant 2$. The two joint rules, called descending rule and ascending rule, can be presented as follows:

Descending rule
Choose a credibility level $\lambda \in [0.5, 1]$. Decrease h from $(q+1)$ until the first value t, such that $\sigma(a, b_t) \geqslant \lambda$:

$a)$ For $t = q$, select C_q as a possible category to assign action a.
$b)$ For $0 < t < q$, if $\rho(a, b_t) > \rho(a, b_{t+1})$, then select C_t as a possible category to assign a; otherwise, select C_{t+1}.
$c)$ For $t = 0$, select C_1 as a possible category to assign a.

Ascending rule
Choose a credibility level $\lambda \in [0.5, 1]$. Increase h from 0 until the first value k, such that $\sigma(b_k, a) \geqslant \lambda$:

$a)$ For $k = 1$, select C_1 as a possible category to assign action a.
$b)$ For $1 < k < (q+1)$, if $\rho(a, b_k) > \rho(a, b_{k-1})$, then select C_k as a possible category to assign a; otherwise, select C_{k-1}.
$c)$ For $k = (q+1)$, select C_q as a possible category to assign a.

Each one of the two joint rules requires the selecting function $\rho(a, b_h)$, which allows to choose between the two consecutive categories where an action a can be assigned to. The results appear in one of the following forms, and the decision maker may choose:

1. A single category, when the two selected categories are the same;
2. One of the two selected categories, when such categories are consecutive;
3. One of the two selected categories or one of the intermediate categories, when such categories are not consecutive.

In [3], ELECTRE TRI-C method was generalized to ELECTRE TRI-NC method where each category is defined by a set of several reference characteristic actions, rather than one. This aspect is enriching the definition of each category and allows to obtain more narrow ranges of categories to which an action can be assigned to, than the ELECTRE TRI-C method. The joint assignments rules are similar to the previous ones.

3.3.2.5 The Possible and the Necessary Approach: ELECTRE^{GMS} method

In [36], the authors considered the problem of robustness for ELECTRE methods within the *robust ordinal regression* (ROR) approach [38]. When the preference parameters of a decision model are inferred from some holistic preference comparisons of some reference actions made by the decision maker, there are, in general, many sets of values of preference model parameters representing this preference information, however usually, only one specific set is selected and used to work out a recommendation. Since the selection of one among many sets of parameter values compatible with the preference information provided by the decision maker is rather arbitrary, the ROR approach proposes taking into account all these sets in order to give a recommendation in terms of necessary and possible consequences of applying all the compatible preference models on the considered set of actions [35, 24]. With respect to ELECTRE methods, the ROR approach was applied in the method ELECTRE^{GMS}, where the possible and the necessary outranking are calculated as follows. Given an ordered pair of actions $(a,a') \in A \times A$, *a necessarily* outranks a', which is denoted by $aS^N a'$, if for all compatible sets of parameter values, a outranks a', while *a possibly* outranks a', denoted by $aS^P a'$, if for at least one compatible set of parameter values, a outranks a'. The ELECTRE^{GMS} method has been adapted also to the case of group decision making, and called ELECTRE^{GMS}-GROUP method [36]. In this case, several decision makers cooperate in a decision problem to make a collective decision. Decision makers share the same "description" of the decision problem (the same set of actions, evaluation criteria, and performance matrix). Each decision maker provides his/her own preference information, composed of pairwise comparisons of some reference actions. The collective preference model accounts for preferences expressed by each decision maker.

Let us denote the set of decision makers by $\mathscr{D}=\{d_1,\ldots,d_p\}$. For each decision maker $d_r \in D' \subseteq D$, we consider all compatible outranking models. Four situations are interesting for an ordered pair $(a,a') \in A \times A$:

- $a\ S^{N.N}(D')\ a'$: $aS^N a'$ for all $d_r \in D'$, $a\ S^{N.P}(D')\ a'$: $aS^N a'$ for at least one $d_r \in D'$,
- $a\ S^{P.N}(D')\ a'$: $aS^P a'$ for all $d_r \in D'$, $a\ S^{P.P}(D')\ a'$: $aS^P a'$ for at least one $d_r \in D'$.

3.3.3 Axiomatic and Meaningfulness Analysis

This section is devoted to theoretical foundations of ELECTRE methods, concerning their axiomatization and the meaningfulness of statements they provide with respect to different types of scales of considered criteria.

3.3.3.1 Axiomatic Analysis

Concerns about axiomatic basis of ELECTRE methods have been described in a long series of papers started in the last millennium [8, 18, 53, 17]. The works on this topic were continued in this millennium [23, 12]. We will not review in detail all the works on the axiomatic analysis of ELECTRE methods, but we will concentrate our attention on contributions related to conjoint measurement analysis of ELECTRE methods done by Bouyssou and Pirlot on one hand, and Greco, Matarazzo and Słowiński, on the other hand.

Greco et al. [32] introduced the first conjoint measurement model of an ELECTRE method, namely, ELECTRE IV. Let $X = X_1 \times X_2 \times \ldots \times X_n$ be a product space, where X_j is the value set of criterion $j = 1,2,\ldots,n$. Let (x_j, z_{-j}), $x_j \in X_j$ and $z_{-j} \in X_{-j} =$

$\prod_{i=1, i \neq j}^{n} X_i$, denote an element of X equal to z except for its jth coordinate being equal to x_j. Analogously, let $(x_{\bar{A}}, z_{-\bar{A}})$, $x_{\bar{A}} \in X_{\bar{A}} = \prod_{j \in \bar{A}}$ and $z_{-\bar{A}} \in X_{-\bar{A}} = \prod_{j \notin \bar{A}} X_j$, $\bar{A} \subseteq \{1, 2, \ldots, n\}$, denote an element of X equal to $x_{\bar{A}}$ for coordinates $j \in \bar{A}$ and to $z_{-\bar{A}}$ for coordinates $j \notin \bar{A}$. A comprehensive outranking relation $\gtrsim$ is defined on X such that $x \gtrsim y$ means that "x is at least as good as y". The symmetric part of $\gtrsim$ is the indifference relation denoted by $\sim$, while the asymmetric part of $\gtrsim$ is the preference relation denoted by $\succ$. The only minimal requirement imposed on $\gtrsim$ is its reflexivity. In the following, for each $j = 1, \ldots, n$, we consider a marginal outranking relation $\gtrsim_j$, such that $x_j \gtrsim_j y_j$ means "criterion j is in favor of the comprehensive outranking of x over y".

For each ordered pair $(x, y) \in X$, let $S(x, y) = \{j \,|x_j \gtrsim_j y_j\}$.

We say that a comprehensive outranking relation $\gtrsim$ on X and the marginal outranking relations $\gtrsim_j$, $j = 1, \ldots, n$, constitute a concordance structure if and only if for all $x, y, w, z \in X$:

$$[S(x, y) \supseteq S(w, z)] \Rightarrow [w \gtrsim z \Rightarrow x \gtrsim y].$$

Greco et al. [32] proposed the following result.

Theorem 3.1. *[32] The three following propositions are equivalent:*

1) for each $x_j, y_j, u_j, v_j, w_j, z_j \in X_j, a_{-j}, b_{-j}, c_{-j}, d_{-j} \in X_{-i}, j = 1 \ldots, n$,

$$(\mathbf{A}) \quad [(x_j, a_{-j}) \gtrsim (y_j, b_{-j}) \text{ and } (u_j, c_{-j}) \gtrsim (v_j, d_{-j})]$$

$$\Rightarrow$$

$$[(x_j, c_{-j}) \gtrsim (y_j, d_{-j}) \text{ or } (w_j, a_{-j}) \gtrsim (z_j, b_{-j})],$$

and

$$(\mathbf{B}) \quad (x_j, a_{-j}) \gtrsim (y_j, b_{-j}) \Rightarrow (x_j, a_{-j}) \gtrsim (x_j, b_{-j});$$

2) there exists a marginal outranking relation $\gtrsim_j$ *for each criterion* $j = 1, \ldots, n$, *such that the comprehensive outranking relation* $\gtrsim$ *on* X *is a concordance structure;*

3) there exists

- *a marginal outranking relation* $\gtrsim_j$ *for each criterion* $j = 1, \ldots, n$,
- *a set function (capacity)* $v : 2^{\{1, \ldots, n\}} \to [0, 1]$, *such that* $v(\emptyset) = 0$, $v(\{1, \ldots, n\})$ *and for each* $\bar{A} \subseteq \bar{B} \subseteq \{1, \ldots, n\}$, $v(\bar{A}) \leqslant v(\bar{B})$, *and*
- *a threshold* $t \in]0, 1[$ *such that* $v(S(x, y)) \geqslant t \Leftrightarrow x \gtrsim y$.

ELECTRE methods are based not only on the concordance relation but also on the discordance relation. For each criterion $j = 1, \ldots, n$, there is defined a veto relation V_j, such that for each $x_j, y_j \in X_j$, $x_j V_j y_j$ means that "the preference of y_j over x_j is so strong that, for all $a_{-j}, b_{-j} \in X_{-j}$, it is not true that $(x_j, a_{-j}) \gtrsim (y_j, b_{-j})$", i.e. (x_j, a_{-j}) cannot be as good as (y_j, b_{-j}).

We say that a comprehensive outranking relation $\gtrsim$ on X is a concordance structure with veto if and only if for all $x, y, w, z \in X$:

$$[S(x,y) \supseteq S(w,z) \text{ and } non(x_j V_j y_j) \text{ for all } j=1,\ldots,n] \Rightarrow [w \succsim z \Rightarrow x \succsim y].$$

Greco et al. [32] proposed the following result.

Theorem 3.2. *[32] The three following propositions are equivalent:*

1) for each $x_j, y_j, u_j, v_j, w_j, z_j \in X_j, a_{-j}, b_{-j}, c_{-j}, d_{-j}, e_{-j}, f_{-j} \in X_{-j}, j=1,\ldots,n,$

$$(\mathbf{A}') \quad [(x_j,a_{-j}) \succsim (y_j,b_{-j}) \text{ and } (u_j,c_{-j}) \succsim (v_j,d_{-j}) \text{ and } (w_j,e_{-j}) \succsim (z_j,f_{-j})]$$

$$\Rightarrow$$

$$[(x_j,c_{-j}) \succsim (y_j,d_{-j}) \text{ or } (w_j,a_{-j}) \succsim (z_j,b_{-j})],$$

and above condition (**B**) *holds;*

2) there exists a marginal outranking relation $\succsim_j$ *and a veto relation* V_j *for each criterion* $j=1,\ldots,n$*, such that the comprehensive outranking relation* $\succsim$ *on* X *is a concordance structure with the veto relation;*

3) there exists,

- *a marginal outranking relation* $\succsim_j$ *for each criterion* $j=1,\ldots,n$,
- *a set function (capacity)* $v: 2^{\{1,\ldots,n\}} \rightarrow [0,1]$*, such that* $v(\emptyset)=0$, $v(\{1,\ldots,n\})$ *and for each* $\bar{A} \subseteq \bar{B} \subseteq \{1,\ldots,n\}$, $v(\bar{A}) \leqslant v(\bar{B})$ *and*
- *a threshold* $t \in]0,1[$ *such that,*

$$v(S(x,y)) \geqslant t \text{ and } V(x,y)=\emptyset \Leftrightarrow x \succsim y.$$

Bouyssou and Pirlot [13] introduced another axiomatic analysis of ELECTRE I that proposed a certain number of results aiming at presenting the ELECTRE I method as a special case of their non-additive non-transitive model.

Theorem 3.3. *[14] The above Theorem 3.1 holds also when Proposition 1) is replaced by the following one:*

1') for each $x_j, y_j, w_j, z_j \in X_j, a_{-j}, b_{-j}, c_{-j}, d_{-j} \in X_{-j}, j=1\ldots,n,$

$$(\mathbf{RC2}) \quad [(x_j,a_{-j}) \succsim (y_j,b_{-j}) \text{ and } (y_j,c_{-j}) \succsim (x_j,d_{-j})]$$

$$\Rightarrow$$

$$[(z_j,a_{-j}) \succsim (w_j,b_{-j}) \text{ or } (w_j,c_{-j}) \succsim (z_j,d_{-j})],$$

$$(\mathbf{UC}) \quad [(x_j,a_{-j}) \succsim (y_j,b_{-}) \text{ and } (z_j,c_{-j}) \succsim (w_j,d_{-j})]$$

$$\Rightarrow$$

$$[(y_j,a_{-j}) \succsim (x_j,b_{-j}) \text{ or } (x_j,c_{-j}) \succsim (y_j,d_{-j})],$$

$$(\mathbf{LC}) \quad [(x_j,a_{-j}) \succsim (y_j,b_{-j}) \text{ and } (y_j,c_{-j}) \succsim (x_j,d_{-j})]$$

$$\Rightarrow$$

$$[(y_j,a_{-j}) \succsim (x_j,b_{-j}) \text{ or } (z_j,c_{-j}) \succsim (w_j,d_{-j})].$$

The axioms of the first result, however, interact with the axioms of their non-additive and non-transitive model [12], and, therefore, they produced another result.

Theorem 3.4. *[15] The above Theorem 3.1 holds also when Proposition 1′) is replaced by the following one:*

1″) for each $x_j, y_j, w_j, z_j \in X_j, a_{-j}, b_{-j}, c_{-j}, d_{-j} \in X_{-j}, j = 1, \ldots, n,$

$$(\mathbf{RC1}) \quad [(x_j, a_{-j}) \succsim (y_j, b_{-j}) \text{ and } (z_j, c_{-j}) \succsim (w_j, d_{-j})] \Rightarrow [(x_j, c_{-j}) \succsim (y_j, d_{-j}) \text{ or } (z_j, a_{-j}) \succsim (w_j, b_{-j})],$$

$$(\mathbf{M1}) \quad [(x_j, a_{-j}) \succsim (y_j, b_{-j}) \text{ and } (z_j, c_{-j}) \succsim (w_j, d_{-j})] \Rightarrow [(y_j, a_{-j}) \succsim (x_j, b_{-j}) \text{ or } (w_j, a_{-j}) \succsim (z_j, b_{-j}) \text{ or } (x_j, c_{-j}) \succsim (y_j, d_{-j})],$$

$$(\mathbf{M2}) \quad [(x_j, a_{-j}) \succsim (y_j, b_{-j}) \text{ and } (y_j, c_{-j}) \succsim (x_j, d_{-j})] \Rightarrow [(y_j, a_{-j}) \succsim (x_j, b_{-j}) \text{ or } (z_j, a_{-j}) \succsim (w_j, b_{-j}) \text{ or } (z_j, c_{-j}) \succsim (w_j, d_{-j})],$$

*and above condition (***RC2***) holds.*

Finally, Bouyssou and Pirlot [16] considered also the veto condition, proposing the following result.

Theorem 3.5. *[16] The above Theorem 3.2 holds also when Proposition 1) is replaced by the following one:*

for each $x_j, y_j, w_j, z_j \in X_j, a_{-j}, b_{-j}, c_{-j}, d_{-j}, e_{-j}, f_{-j} \in X_{-j}, j = 1, \ldots, n,$

$$(\mathbf{M3}) \quad [(x_j, a_{-j}) \succsim (y_j, b_{-j}) \text{ and } (y_j, c_{-j}) \succsim (x_j, d_{-j}) \text{ and } (z_j, e_{-j}) \succsim (w_j, f_{-j})] \Rightarrow [(y_j, a_{-j}) \succsim (x_j, b_{-j}) \text{ or } (z_j, a_{-j}) \succsim (w_j, b_{-j}) \text{ or } (z_j, c_{-j}) \succsim (w_j, d_{-j})],$$

*and above conditions (***RC1***), (***RC2***), and (***M1***) hold.*

The approach of Bouyssou and Pirlot [16] has the merit of putting the axiomatic basis of ELECTRE methods in the larger context of their general non-additive and non-transitive model. However, their conditions are more numerous and complex than the conditions proposed by Greco et al. [32].

3.3.3.2 Representing Preferences by Decision Rules

In Greco et al. [34], an equivalence of preference representation by conjoint measure and decision rules induced using the Dominance-based Rough Set Approach

(DRSA) [33] was demonstrated for choice and ranking problems. One of the most important conclusions in this context is that ELECTRE IV method can be represented in terms of DRSA. In this case, for all $a \in A$ and for all $g_j \in F$, $q_j(g_j(a)) = p_j(g_j(a))$, such that $Q_j = \emptyset$, and $d_j(a,a') \in \{0,1\}$. Then, the set of decision rules describing the aggregation procedure of ELECTRE IV has the following form:

$$\text{if } aS_{j_1}a' \text{ and} \ldots aS_{j_p}a' \text{ and} \ldots aV^c_{j_{p+1}}a' \text{ and} \ldots aV^c_{j_n}a', \text{ then } aSa'$$

where $aV^c_j a'$ means that $d_j(a,a') = 0$ (i.e., there is no veto with respect to criterion $g_j \in F$) and

$$w_{j_1} + \ldots + w_{j_p} \geqslant s$$

with s being a specific concordance threshold. Not all the above decision rules are necessary to obtain a representation of the outranking relation S on A, because it is enough to consider only those decision rules that involve subsets $\widetilde{F} = \{g_{j_1}, \ldots, g_{j_p}\} \subseteq F$ including no $g_i \in \widetilde{F}$ for which

$$w_{j_1} + \ldots + w_{j_p} - w_i \geqslant s.$$

Using this result, Greco et al. [37] proposed a methodology to infer preference parameters (weights and veto thresholds) of ELECTRE methods on the basis of a set of decision rules obtained by DRSA.

3.3.3.3 A Conjoint Measurement Analysis of a Simplified Version of ELECTRE TRI-B

An axiomatic analysis of a simplified variant of ELECTRE TRI-B has been proposed in [9] and in [10], in the framework of conjoint measurement theory. This variant only takes into account the "pessimistic" assignment rule, and does not make use of veto thresholds; preference and indifference thresholds are considered equal.

From a technical point of view, the authors make use of conjoint measurement techniques to work with partitions, instead of binary relations. This aspect of dealing with the problem was first proposed by Goldstein [31] and after generalized by Greco et al. [32]. Based, moreover, on the concepts of conjoint measurement theory, these authors analyze a certain type of "non-compensatory sorting methods" close to the "pessimistic" version of ELECTRE TRI-B, and make a comparison with other sorting methods. They proved that the simplified version of ELECTRE TRI-B is non-compensatory. This result does not hold, however, for the "optimistic" version of ELECTRE TRI-B with the same simplifications.

Some hints to elicit parameters from assignment examples within the framework of the studied version of ELECTRE TRI-B were also provided in their work.

To give an axiomatic basis to ELECTRE TRI-B, they considered the following simplified model. Consider a twofold partition $\langle \mathscr{A}, \mathscr{U} \rangle$ of X, which means that the two sets $\mathscr{A}$ and $\mathscr{U}$ are non-empty and disjoint, and that their union makes the entire set X. For the sake of simplicity, one can imagine $\mathscr{A}$ as a set of all good actions, and $\mathscr{U}$ as a set of all bad actions. In ELECTRE TRI-B, the sorting of action $x \in X$ is based on comparison of x with profile p separating the categories, using outranking relation S. Then, in the "pessimistic" version of ELECTRE TRI-B, for all $x \in X$,

$$x \in \mathscr{A} \Leftrightarrow xSp,$$

while in the "optimistic" version of ELECTRE TRI-B,

$$x \in \mathscr{A} \Leftrightarrow not(pPx),$$

where P is the asymmetric part of S, i.e. xSp and $not(pSx)$. A partition $\langle \mathscr{A}, \mathscr{U} \rangle$ has a representation in the non-compensatory sorting model if:

- for all $j = 1, \ldots, n$, there is a set $\mathscr{A}_j \subseteq X_j$,
- there is a subset $\mathscr{F}$ of 2^N, such that, for all $I, J \in 2^N$, $N = \{1, \ldots, n\}$,

$$[I \in \mathscr{F} \text{ and } I \subseteq J] \Rightarrow J \in \mathscr{F},$$

such that, for all $x \in X$,

$$x \in \mathscr{A} \Leftrightarrow \{j \in N \mid x_j \in \mathscr{A}_j\} \in \mathscr{F}.$$

Bouyssou and Marchant [9] proposed the following result.

Theorem 3.6. *[9] A partition $\langle \mathscr{A}, \mathscr{U} \rangle$ has a representation in the non-compensatory sorting model if and only if*

for each $x_j, y_j \in X_j$ and all $a_{-j}, b_{-j} \in X_{-j}$, $j = 1 \ldots, n$,

(Linear) $[(x_j, a_{-j}) \in \mathscr{A} \text{ and } (y_j, b_{-j}) \in \mathscr{A}] \Rightarrow [(y_j, a_{-j}) \in \mathscr{A} \text{ or } (x_j, b_{-j}) \in \mathscr{A}]$,

$$(2-graded)\ [(x_j, a_{-j}) \in \mathscr{A} \text{ and } (y_j, a_{-j}) \in \mathscr{A} \text{ and } (y_j, b_{-j}) \in \mathscr{A}]$$
$$\Rightarrow$$
$$[(x_j, b_{-j}) \in \mathscr{A} \text{ and } (z_j, a_{-j}) \in \mathscr{A}].$$

It is interesting to note that the same axioms have been given by Słowiński et al. [64] as an axiomatic basis to the sorting procedure based on the Sugeno integral [68]. Therefore, the non-compensatory sorting model is equivalent to the sorting model based on the Sugeno integral.

Bouyssou and Marchant [9] considered also a non-compensatory sorting model with veto, that augment the above non-compensatory sorting model by consideration of sets $\mathscr{V}_j \subseteq X_j, j = 1, \ldots, n$, such that for all $x \in X$,

$$x \in \mathscr{A} \Leftrightarrow [\{j \in N \mid x_j \in \mathscr{V}_j\} \in \mathscr{F} \text{ and } \{j \in N \mid x_j \in \mathscr{V}_j\} = \emptyset].$$

Indeed, Bouyssou and Marchant [9] proposed the following result.

Theorem 3.7. *[9] A partition $\langle \mathscr{A}, \mathscr{U} \rangle$ has a representation in the non-compensatory sorting model with veto if only if for each $x_j, y_j \in X_j$ and all $a_{-j}, b_{-j} \in X_{-j}, j = 1, \ldots, n$,*

$$(3v\text{-}graded)\ [(x_j,a_{-j})\in\mathscr{A}\ and\ (y_j,a_{-j})\in\mathscr{A}\ and\ (y_j,b_{-j})\in\mathscr{A}\ and\ (z_j,c_{-j})\in\mathscr{A}]$$

$$\Rightarrow$$

$$[(x_j,b_{-j})\in\mathscr{A}\ and\ (z_j,a_{-j})\in\mathscr{A}],$$

*and above condition (***Linear***) holds.*

In [10], this approach has been extended to give an axiomatic basis to the non-compensatory sorting in the case of more than two classes.

3.3.3.4 The Meaningfulness of ELECTRE Methods

In [43], the authors analyze the meaningfulness of the assertions of the type "a outranks a' for such and such method", in particular, for ELECTRE methods.

The notion of meaningfulness [69] comes from the measurement theory. This theory [42] deals with the way one can represent certain information (in particular, information of qualitative nature) coming from a given category of phenomena through a set of numerical values, in such a way that this representation must adequately reflect certain properties of the considered category of phenomena.

In order to obtain a meaningful assertion (with respect to a considered category of phenomena) based on the computations that make use of the numerical representation, it is necessary that its validity or non-validity will not be affected when one uses another adequate measure or way of representing the phenomena. Indeed, meaningfulness in MCDA is related to invariance of results with respect to some admissible transformation of the criterion scales. The ELECTRE methods are meaningful up to interval scales. In general ELECTRE methods make use of purely ordinal scales. With such scales, a lack of meaningfulness may only appear locally, due to existence of the ambiguity zones.

3.3.4 Other Aspects

This section is devoted to other aspects related to ELECTRE methods, that do not fit the previous sections, but, nevertheless are important for several reasons.

3.3.4.1 The Relative Importance of Criteria

The metaphor of weight is very often a source of misunderstanding [58]. Knowing the weight of different objects allows to line them up from the heaviest to the lightest one. Similarly, the talk about the (relative) weight of two criteria assumes implicitly that the assertion "this criterion is more important than the other one" makes a sense. It leads to suppose that the weight of a criterion has an intrinsic character, that is to say that it depends only on the point of view reflected by it, and does not depend on the manner in which it is modeled (the nature of the scales, the range of the scales, the possible unit, ...). Very often researchers and practitioners had the opportunity to notice that it is in such a way that a decision maker uses (even before talking to him/her) the expression "weight of a criterion". This parameter holds different names, according to the type of model in which it intervenes. It is, nevertheless, the term *weight* which is the most often used.

It is, in general, the notion of more or less big importance between two criteria that makes naturally sense in the head of the decision makers. Simos [62] proposed a procedure that was further revised by Figueira and Roy [25]. These authors proposed a method, called SRF, for assessing the importance coefficients of criteria having exactly the above meaning. They also stressed the fact that SRF must not be used for the coefficients (called inappropriately weights) of a weighted-sum, and that it must be reserved for intrinsic weights (independent on the very nature of the scales) corresponding to the number of voices which could be allocated to every criterion in a voting process. It should be noted that SRF first exploits the ordinal character of the criteria scales, which means that the units and the range of the scales play no role in the assessment of the importance coefficients (to be more rigorous, a very local and minimal role). As mentioned before, the decision makers, who express themselves spontaneously about the notion of importance of criteria, make, in general, no link between this notion and the nature of the scales. The MCAP used to aggregate this information must reflect such a fact adequately.

3.3.4.2 Concordant Outranking with Criteria of Ordinal Significance

In [6], a new contribution to robustness concerns in MCDA was proposed. More precisely, a complete preorder π on the family of criteria F is considered, which is a ranking of significance of criteria, to be taken into account in the construction of the comprehensive outranking relation S. The weights are π-compatible if for all $g_j, g_{j'} \in F$, $w_j = w_{j'}$ if g_j and $g_{j'}$ have the same rank of significance in π, and $w_j > w_{j'}$ if g_j has a higher rank of significance than $g_{j'}$ in π. If for $(a,a') \in A \times A$ the concordance index $c(a,a') > 0.5$, for every π-compatible set of weights, there is an ordinal concordance of a over a', which is denoted by $aC_{\pi}a'$.

3.3.4.3 Evolutionary Approaches

Evolutionary algorithms are starting to be used in order to deal with large scale problems, as well as, to mitigate the complexity of some computations in ELECTRE methods, mainly due to some non-linearities existing in the formulas used in these methods.

In [22], an evolutionary approach was proposed to deal with construction of outranking relations in the context of ELECTRE TRI-B.

In [40], a new MCDA method was proposed for ranking problems. It makes use of the ELECTRE III method to build a fuzzy outranking relation and exploit it through the application of a multi-objective genetic algorithm.

3.3.4.4 The EPISSURE Method for the Assessment of Non-financial Performance

A new approach making use of the ELECTRE TRI-B method is presented in this book (see Chapter 13). This new approach, called EPISSURE (splice, which is a nautical term meaning a joint made by splicing) has been designed by André [4] for evaluating non-financial performances of companies.

Because of the fierce competition in markets among companies and institutions, and because of a strong pressure by international entities to take into account other kinds of performance criteria than financial ones, there was a need of a new approach to the evaluation of non-financial performance of the companies. EPISSURE responds to this need.

Two normative principles were laid down *ex-ante* to ground the approach:

Principle 1: The approach must be hierarchical, i.e., classified into successive levels, wherein the levels match a hierarchy of responsibilities *vis-à-vis* the successive aggregates of performance that contribute to the performance summary.

Principle 2: At each hierarchical level (except perhaps for some at the lowest levels), the evaluations rely on ordinal verbal scales. The number of degrees on the scales must be adjusted to its matching levels; the number of degrees must be high enough to mirror evolutions and be understandable by the stakeholders operating at the said level.

A consultation process, called a framed consultation process, is an integral part of the EPISSURE approach. As any other consultation approach, the objective is that the different stakeholders involved in the evaluation reach a common outlook.

The EPISSURE approach was tested and set up within several companies for the purpose of evaluating sponsorship projects and deciding on their follow-up. The results seem to indicate that this approach is decidedly appropriate for evaluating non-financial performance. Another application concerning evaluation of the environmental performance of the Company *Total* is described in [5].

3.4 Concluding Remarks

ELECTRE methods have a long history of successful real-world applications with considerable impact on human decisions. Several application areas can be pointed out (see [29]): agriculture and forest management, energy, environment and water management, finance, military, project selection (call for tenders), transportation, medicine, nano-technologies, ... As every MCDA method, also ELECTRE methods have their theoretical limitations. This is why, when applying these methods, analysts should first check if their theoretical characteristics respond to the characteristics of the context in which they will be used.

In this chapter, we tried to show that research on ELECTRE methods is not a dead field. Rather the opposite, it is still evolving and gains acceptance thanks to new application areas, new methodological and theoretical developments, as well as user-friendly software implementations.

Acknowledgements The first and third authors acknowledge the financial support from the Luso-French bilateral cooperation agreement between LAMSADE and CEG-IST, under "Programa CNRS/FCT". The fourth author wishes to acknowledge financial support from the Ministry of Science and Higher Education, grant N N519 314435. This research has also partially benefited from the COST Action grant IC0602 on "Algorithmic Decision Theory".

References

1. H. Aissi and B. Roy. Robustness in multi-criteria decision aiding. In M. Ehrgott, J.R. Figueira, and S. Greco, editors, *Trends in Multiple Criteria Decision Analysis*, pages 101–139. Springer Science + Business Media, Inc., New York, U.S.A., 2010.
2. J. Almeida-Dias, J.R. Figueira, and B. Roy. ELECTRE TRI-C: A multiple criteria sorting method based on characteristic reference actions. *European Journal of Operational Research*, 2009. [doi:10.1016/j.ejor.2009.10.018].
3. J. Almeida-Dias, J.R. Figueira, and B. Roy. A multiple criteria sorting method defining the categories by several characteristic reference actions: The ELECTRE TRI-NC methods. *European Journal of Operational Research*, 2010. Submitted.
4. S. André. *Evaluation de la Performance non Financière des Entreprises : Apport des Méthodes Multicritère dAide à la Décision*. Thèse de Doctorat, LAMSADE, Université Paris-Dauphine, Paris, France, 2009.
5. S. André and B. Roy. Conception et mise en place d'un outil d'évaluation de la performance environnementale : Le cas des raffineries de la société TOTAL. *Journal of Decision Systems*, 16:335–367, 2007.
6. R. Bisdorff. Concordant outranking with multiple criteria of ordinal significance. *4OR*, 2:293–308, 2004.
7. R. Bisdorff, P. Meyer, and M. Roubens. RUBIS: A bipolar-valued outranking method for the choice problem. *4OR*, 6:143–165, 2008.
8. D. Bouyssou. Some remarks on the notion of compensation in MCDM. *European Journal of Operational Research*, 26:150–160, 1986.
9. D. Bouyssou and T. Marchant. An axiomatic approach to noncompensatory sorting methods in MCDM I: The case of two categories. *European Journal of Operational Research*, 178:217-245, 2007.
10. D. Bouyssou and T. Marchant. An axiomatic approach to noncompensatory sorting methods in MCDM II: More than two categories. *European Journal of Operational Research*, 178:246276, 2007.
11. D. Bouyssou, T Marchant, M. Pirlot, P. Perny, A. Tsoukiàs, and Ph. Vincke. *Evaluation and Decision Models: A Critical Perspective*. Kluwer Academic Publishers, Dordrecht, Holland, 2000.
12. D. Bouyssou and M. Pirlot. A characterization of strict concordance relations. In D. Bouyssou, E. Jacquet-Lagrèze, P. Perny, R. Słowiński, D. Vanderpooten, and Ph. Vincke, editors, *Aiding Decisions with Multiple Criteria, Essays in Honour of Bernard Roy*, pages 121–145. Kluwer Academic Publisher, Boston, U.S.A., 2002.

13. D. Bouyssou and M. Pirlot. Non transitive decomposable conjoint measurement. *Journal of Mathematical Psychology*, 46:677703, 2002.
14. D. Bouyssou and M. Pirlot. A characterization of concordance relations. *European Journal of Operational Research*, 167:427–443, 2005.
15. D. Bouyssou and M. Pirlot. Further results on concordance relations. *European Journal of Operational Research*, 181:505–514, 2007.
16. D. Bouyssou and M. Pirlot. An axiomatic analysis of concordance-discordance relations. *European Journal of Operational Research*, 199:468477, 2009.
17. D. Bouyssou, M. Pirlot, and Ph. Vincke. A general model of preference aggregation. In M. Karwan, J. Spronk, and J. Wallenius, editors, *Essays in Decision Making*, pages 120–134. Springer-Verlag, Berlin, Germany, 1997.
18. D. Bouyssou and J-C. Vansnick. Noncompensatory and generalized noncompensatory preference structures. *Theory and Decision*, 21:251266, 1986.
19. J-P. Brans and Mareschal. PROMETHEE methods. In J.R. Figueira, S. Greco, and M. Ehrgott, editors, *Multiple Criteria Decision Analysis: State of the Art Surveys*, pages 163–195. Springer Science + Business Media, Inc., New York, U.S.A., 2005.
20. L. Dias and V. Mousseau. Inferring ELECTRE's veto-related parameters from outranking examples. *European Journal of Operational Research*, 170:172–191, 2006.
21. L. Dias, V. Mousseau, J. Figueira, and J. Clímaco. An aggregation/disaggregation approach to obtain robust conclusions with ELECTRE TRI. *European Journal of Operational Research*, 138:332–348, 2002.
22. M. Doumpos, Y. Marinakis, M. Marinaki, and C. Zopounidis. En evolutionary aproach to construction of outranking models for multicriteria classification: The case of the ELECTRE TRI method. *European Journal of Operational Research*, 199:496–505, 2009.
23. D. Dubois, H. Fargier, P. Perny, and H. Prade. A characterization of generalized concordance rules in multicriteria decision-making. *International Journal of Intelligent Systems*, 18:751–774, 2003.
24. J. Figueira, S. Greco, and R. Słowiński. Building a set of additive value functions representing a reference preorder and intensities of preference: GRIP method. *European Journal of Operational Research*, 195:460–486, 2009.
25. J. Figueira and B. Roy. Determining the weights of criteria in the ELECTRE type methods with a revised Simos' procedure. *European Journal of Operational Research*, 139:317–326, 2002.
26. J.R. Figueira. *La Prise en Compte de Critères Multiples en Recherche Opérationnelle et Aide à la Décision*. Habilitation à Diriger des Recherches (HDR), LAMSADE, Université Paris-Dauphine, Paris, France, 2009.
27. J.R. Figueira, S. Greco, and B. Roy. ELECTRE methods with interaction between criteria: An extension of the concordance index. *European Journal of Operational Research*, 199(2):496–505, 2009.
28. J.R. Figueira and B. Roy. A note on the paper,"ranking irregularities when evaluating alternatives by using some ELECTRE methods, by Wang and Triantaphyllou, omega (2008)". *OMEGA*, 37:731–733, 2009.
29. J.R. Figueira, B. Roy, and V. Mousseau. ELECTRE methods. In J.R. Figueira, S. Greco, and M. Ehrgott, editors, *Multiple Criteria Decision Analysis: State of the Art Surveys*, pages 133–162. Springer Science + Business Media, Inc., New York, U.S.A., 2005.
30. W. Gerhlein. Condorcets paradox. *Theory and Decision*, 15:161–169, 1983.
31. W. Goldstein. Decomposable threshold models. *Journal of Mathematical Psychology*, 35:64-79, 1991.
32. S. Greco, B. Matarazzo, and R. Słowiński. Axiomatic basis of noncompensatory preferences. Communication at *FUR X*, 30 May-2 June, Torino, Italy, 2001.
33. S. Greco, B. Matarazzo, and R. Słowiński. Rough sets theory for multicriteria decision analysis. *European Journal of Operational Research*, 129:1–47, 2001.
34. S. Greco, B. Matarazzo, and R. Słowiński. Preference representation by means of conjoint measurement and decision rule model. In D. Bouyssou, E. Jacquet-Lagrèze, P. Perny, R. Słowiński, D. Vanderpooten, and Ph. Vincke, editors, *Aiding Decisions with Multiple Criteria, Essays in Honour of Bernard Roy*, pages 263–313. Kluwer Academic Publisher, Boston, U.S.A., 2002.

35. S. Greco, V. Mousseau, and R. Słowiński. Ordinal regression revisited: multiple criteria ranking with a set of additive value functions. *European Journal of Operational Research*, 191:415–435, 2008.
36. S. Greco, V. Mousseau, and R. Słowiński. The possible and the necessary for multiple criteria group decision. In F. Rossi and A. Tsoukiàs, editors, *Algorithmic Decision Theory*, volume 5783 of *Lectures Notes on Computer Science*, pages 203–214. Springer-Verlag, Berlin, Germany, 2009.
37. S. Greco, B. Prędki, and R. Słowiński. Searching for an equivalence between decision rules and concordance-discordance preference model in multicriteria choice problems. *Control and Cybernetics*, 31:921–935, 2002.
38. S. Greco, R. Słowiński, J.R. Figueira, and V. Mousseau. Robust ordinal regression. In M. Ehrgott, J.R. Figueira, and S. Greco, editors, *Trends in Multiple Criteria Decision Analysis*, pages 273–320. Springer Science + Business Media, Inc., New York, U.S.A., 2010.
39. E. Jacquet-Lagrèze and Y. Siskos. Assessing a set of additive utility functions for multicriteria decision making: The UTA method. *European Journal of Operational Research*, 10:151–164, 1982.
40. J.C. Leyva-López, L. Sánchez, and M. Contreras. A multicriteria decision support system with an evolutionary algorithm for deriving final ranking from a fuzzy outranking relation. *Operations Research: An International Journal*, 8:4762, 2008.
41. R. Luce. Semiorders and the theory of utility discrimination. *Econometrica*, 54:178–191, 1956.
42. R. Luce, D. Krantz, P. Suppes, and A. Tversky. *Foundations of Measurement: Representation, Axiomatization, and Invariance*, volume Vol. 3. Academic Press, New York, U.S.A., 1990.
43. J-M. Martel and B. Roy. Analyse de la signifiance de diverses procédures dagrégation multicritère. *INFOR*, 44:119–214, 2006.
44. V. Mousseau and L. Dias. Valued outranking relations in ELECTRE providing manageable disaggregation procedures. *European Journal of Operational Research*, 156:467482, 2006.
45. V. Mousseau, L. Dias, and J.R. Figueira. Dealing with inconsistent judgments in multiple criteria sorting models. *4OR*, 4:145158, 2006.
46. V. Mousseau, J. Figueira, and J. Naux. Unsing assignment examples to infer weights for ELECTRE TRI method: Some experimental results. *European Journal of Operational Research*, 130:263–275, 2001.
47. V. Mousseau, J.R. Figueira, L. Dias, C. Gomes da Silva, and J. Clímaco. Resolving inconsistencies among constraints on the parameters of an MCDA model. *European Journal of Operational Research*, 147:7293, 2003.
48. V. Mousseau and R. Słowiński. Inferring an ELECTRE TRI model from assignment examples. *Journal of Global Optimization*, 12:157–174, 1998.
49. V. Mousseau, R. Słowiński, and P. Zielniewicz. A user-oriented implementation of the ELECTRE TRI method integrating preference elicitation support. *Computers & Operations Research*, 27:757–777, 2000.
50. A. Ngo The and V. Mousseau. Using assignment examples to infer category limits for the ELECTRE TRI method. *Journal of Multi-Criteria Decision Analysis*, 11:29–43, 2002.
51. P. Perny. Sur le non-respect de laxiome dindépendance dans les méthodes de type ELECTRE. *Cahiers du CERO*, 34:211–232, 1992.
52. P. Perny. Multicriteria filtering methods based on concordance/non-discordance principles. *Annals of Operations Research*, 80:137–167, 1998.
53. M. Pirlot. A common framework for describing some outranking procedures. *Journal of Multi-Criteria Decision Analysis*, 6:86–92, 1997.
54. B. Roy. How outranking relation helps multiple criteria decision making. In J. Cochrane and Zeleny M., editors, *Multiple Criteria Decision Making*, page 179201. University of South Carolina Press, Columbia, SC, U.S.A., 1973.
55. B. Roy. *Méthodologie Multicritère d'aide à la Décision*. Economica, Paris, France, 1985.
56. B. Roy. The outranking approach and the foundations of ELECTRE methods. *Theory and Decision*, 31(1):49–73, 1991.

57. B. Roy. *Multicriteria Methodology for Decision Aiding*. Kluwer Academic Publishers, Dordrecht, Holland, 1996.
58. B. Roy. Two concepts of decision aiding. *International Journal of Multiple Criteria Decision-Making*, 1, 2010. To appear.
59. B. Roy and D. Bouyssou. *Aide Multicritère à la Décision : Méthodes et Cas*. Production et Techniques Quantitatives Appliquées à la Gestion. Economica, Paris, France, 1993.
60. B. Roy and R. Słowiński. Handling effects of reinforced preference and counter-veto in credibility of outranking. *European Journal of Operational Research*, 188(1):185–190, 2008.
61. B. Roy and Ph. Vincke. Relational systems of preference with one or more pseudo-criteria: Some new concepts and results. *Management Science*, 30(11):1323–1335, 1984.
62. J. Simos. *Gestion des Déchets Solides Urbains Genevois: Les Faits, le Traitement, l'Analyse*. Presses Polytechniques et Universitaires Romandes, Lausanne, Switzerland, 1990.
63. L. Simpson. Do decision makers know what they prefer?: MAVT and ELECTRE II. *Journal of the Operational Research Society*, 47:919–929, 1996.
64. R. Słowiński, S. Greco, and B. Matarazzo. Axiomatization of utility, outranking and decision-rule preference models for multiple-criteria classification problem under partial inconsistency with the dominance principle. *Control and Cybernetics*, 31:1005–1035, 2002.
65. R. Słowiński and J. Stefanowski. Rough classification with valued closeness relation. In E. Diday, Y. Lechevallier, M. Schrader, P. Bertrand, and B. Burtschy, editors, *New Approaches in Classification and Data Analysis*, pages 482–489. Springer-Verlag, Berlin, Germany, 1994.
66. R. Słowiński and D. Vanderpooten. Similarity relation as a basis for rough approximations. In P.P. Wang, editor, *Advances in Machine Intelligence & Soft-Computing*, volume Vol. IV, pages 17–33. Duke University Press, Durham, NC, U.S.A., 1997.
67. R. Słowiński and D. Vanderpooten. A generalized definition of rough approximations based on similarity. *IEEE Transactions on Data and Knowledge Engineering*, 12:331–336, 2000.
68. M. Sugeno. *Theory of Fuzzy Integrals and its Applications*. Thèse de Doctorat, Tokyo Institute of Technology, Tokyo, Japan, 1974.
69. P. Suppes. Measurement empirical meaningfulness and three-valued logic. In Churchman and Ratoosh, editors, *Measurement: Definition and Theories*, pages 129–143. John Wiley & Sons, New York, U.S.A., 1959.
70. T. Tervonen, J. R. Figueira, R. Lahdelma, J. Almeida Dias, and P. Salminen. A stochastic method for robustness analysis in sorting problems. *European Journal of Operational Research*, 192:236–242, 2009.
71. E. Triantaphyllou. *Multi-Criteria Decision Making Methods: A Comparative Study*. Kluwer Academic Publishers, Dordrecht, Holland, 2000.
72. X. Wang and E. Triantaphyllou. Ranking irregularities when evaluating alternatives by using some ELECTRE methods. *OMEGA*, 36:45–63, 2008.

Chapter 4
The Analytic Hierarchy and Analytic Network Measurement Processes: The Measurement of Intangibles
Decision Making under Benefits, Opportunities, Costs and Risks

Thomas L. Saaty and Mariya Sodenkamp

Abstract Multicriteria thinking demonstrates that in order to make a best choice in a decision, discussion and cause-effect reasoning are inadequate to learn what the best overall outcome is. The Analytic Hierarchy Process (AHP) and its generalization to dependence and feedback, the Analytic Network Process (ANP), provide a comprehensive structure and mathematics to incorporate measurements for tangible criteria and derive priorities for intangible criteria to enable one to choose a best alternative for a decision. It overcomes so-called bounded rationality that is based on the assumption of transitivity by including in its structures and calculations, the sensitivity and depth of feelings associated with understanding and the imagination and awareness needed to address all the concerns. The AHP can cope with the inherent subjectivity in all decision making, and make it explicit to the stakeholders through relative quantitative priorities. It also provides the means to validate outcomes when measurements are available to show that it does not do number crunching without meaningful justification. It can deal with the benefits, opportunities, costs and risks separately and bring them together to determine the best overall outcome. One can also perform dynamic sensitivity analysis of changes in judgments to ensure that the best outcome is stable.
In an award from the Institute for Operations Research and the Management Sciences (INFORMS) given to the author in October 2008 it is written: "The AHP has revolutionized how we resolve complex decision problems ... the AHP has been applied worldwide to help decision makers in every conceivable decision context across both the public and private sectors, with literally thousands of reported applications."

Thomas L. Saaty
University of Pittsburgh, USA e-mail: saaty@katz.pitt.edu (this chapter which originally appeared in Int. J. Applied Decision Sciences, Vol. 1, No. 1, 2008 [7] (Interscience retains the copyright), is based on material provided by this author and on his research)

Mariya Sodenkamp
University of Paderborn, Germany e-mail: msodenk@mail.uni-paderborn.de (visiting student who assisted in preparation of paper)

C. Zopounidis and P.M. Pardalos (eds.), *Handbook of Multicriteria Analysis*, Applied Optimization 103, DOI 10.1007/978-3-540-92828-7_4, © Springer-Verlag Berlin Heidelberg 2010

4.1 Introduction

The Analytic Hierarchy Process (AHP) and its generalization to dependence and feedback, the Analytic Network Process (ANP) are psychophysical theories of measurement. This means that they make the assumption that judgments about subjective feelings and understanding are essentially not very different than and depend on judgments about the physical world in which we acquire our experience and understanding. In the physical world we respond to intensities of occurrence, such as the varying intensities of sight, sound and smell. These intensities fall in different threshold intervals of just noticeable differences because we are unable to detect change in intensity until a stimulus is increased by a noticeable amount. Judgments must reflect not only knowledge about influences, but also the strengths with which these influences occur. These strengths are expressed by us, and especially by experts who have experienced the complexity with which we are concerned, through judgments from which priorities are derived in relative form that reflect numerical intensities that can be validated in those cases where we have measurement in order to improve our confidence in the applicability of our quantified judgments in those cases where measurements are not available. Measurements in science are made on scales with arbitrary units and need interpretation through judgment to indicate the degree to which they serve our value systems. Occasionally the measurements are used directly in normalized form as priorities that reflect our values if indeed we think they do. In decision making we have to trade-off different kinds of measurement by filtering them through our judgments from which priorities are derived thereby reducing a multidimensional scaling problem to a one-dimensional scale of priorities of the importance of influences on which our actions are based. All this tells us that it is not enough to advocate the use of a theory with numbers as a justifiable way to make decisions because judgments are subjective any way. There has to be validation of the process through a variety of examples to make it a science based on reason, quantity and mathematics, not a religion based on the strength of authority, belief and lots of statistics devoid of understanding, tradeoffs and interpretation.

To make complex risky decisions we need not only judgments but also structures that represent our best understanding of the flow of influences. The basic structure in doing this is a hierarchy for the AHP and an influence network of clusters and nodes contained within the clusters for the ANP. Priorities are established in the AHP and ANP using pairwise comparisons and judgment. Many decision problems cannot be structured hierarchically because they involve the interaction and dependence of higher-level elements such as objectives and criteria in a hierarchy on lower-level elements. Not only does the importance of the criteria determine the importance of the alternatives as in a hierarchy, but also the importance of the alternatives themselves determines the importance of the criteria as in a network Two bridges, both strong, but the stronger is also uglier, would lead one to choose the strong but ugly one unless the criteria themselves are evaluated in terms of the bridges, and strength receives a smaller value and appearance a larger value because both bridges are strong. Feedback enables us to factor the future into the present to determine what we have to do to attain a desired future.

The feedback structure does not have the top-to-bottom form of a hierarchy but looks more like a network, with cycles connecting its components of elements, which we can no longer call levels, and with loops that connect a component to itself. It also has sources and sinks. A **source** node is an origin of paths of influence (importance) and never a destination of such paths. A **sink** node is a destination of paths of influence and never an origin of such paths. A full network can include source nodes; intermediate nodes that fall on paths from source nodes, lie on cycles, or fall on paths to sink nodes; and finally sink nodes. Some networks can contain only source and sink nodes. Still others can include only source and cycle nodes or cycle and sink nodes or only cycle nodes. A decision problem involving feedback arises frequently in practice. It can take on the form of any of the networks just described. The challenge is to determine the priorities of the elements in the network and in particular the alternatives of the decision and even more to justify the validity of the outcome. Because feedback involves cycles, and cycling is an infinite process, the operations needed to derive the priorities become more demanding than is with hierarchies.

4.2 Paired Comparisons, The Fundamental Scale, Eigenvectors, Consistency, Homogeneity

How to measure intangibles is the main concern of the mathematics of the AHP. In the end we must fit our entire world experience into our system of priorities if we need to understand it in both its details and its general workings. As we said above, the AHP reduces a multidimensional problem into a one dimensional one. Decisions are determined by a single number for the best outcome or by a vector of priorities that gives an ordering of the different possible outcomes. We can also combine our judgments or our final choices obtained from a group when we wish to cooperate to agree on a single outcome.

4.2.1 Paired Comparisons and the Fundamental Scale

To make tradeoffs among the many objectives and criteria, the judgments that are usually made in qualitative terms are expressed numerically. To do this, rather than simply assigning a seemingly arbitrary score out of a person's memory that appears reasonable, one must make reciprocal pairwise comparisons in a carefully designed scientific way. In paired comparisons the smaller or lesser element is used as the unit, and the larger or greater element is estimated as a multiple of that unit with respect to the common property or criterion for which the comparisons are made. In this sense measurement with judgments is made more scientifically than by assigning numbers more or less arbitrarily. Because human beings are limited in size and the firings of their neurons are limited in intensity, it is clear that there is a limit on their ability to compare the very small with the very large. It is precisely for this reason that pairwise comparisons are made on elements or alternatives that are close or homogeneous and the more separated they are, the more need there is to put them in different groups and link these groups with a common element from one group to an adjacent group of slightly greater or slightly smaller elements. In this way one can gradually compare grains of sand of varying sizes increasing to small pebbles and larger stones. When done properly, the largest element in one group is used as the smallest one in the next group, and in the end each group is compared separately and the measurement combined.

From all the paired comparisons, one derives a scale of relative values for the priorities. As we shall see below, due to inevitable inconsistency among the judgments and more importantly because of the need for the invariance of priorities, it is mathematically necessary to derive the priorities in the form of the principal eigenvector of the matrix of paired comparisons.

We learn from making paired comparisons in the AHP that if A is 5 times larger than B and B is 3 times larger than C, then A is 15 times larger than C and A dominates C 15 times. That is different from A having 5 dollars more than B and B having 3 dollars more than C implies that A has 8 dollars more than C. Defining intensity along the arcs of a graph and raising the matrix to powers measures the first kind of dominance precisely and never the second. It has definite meaning and as we shall see below, in the limit it is measured uniquely by the principal eigenvector. There is a useful connection between what we do with dominance priorities in the AHP and what is done with transition probabilities both of which use matrix algebra to find their answers. Probabilities of transitions between states are multiplied and added. To compose the priorities for the alternatives of a decision with respect to different criteria, it is also necessary that the priorities of the alternatives with respect to each criterion be multiplied by the priority of that criterion and then added over all the criteria.

The Fundamental Scale used for the judgments applied to compare homogeneous (close) elements is given in Table 4.1. Judgments are first given verbally as indicated in the scale and then a corresponding number is associated with that judgment.

Judgments that represent dominance belong to an absolute scale of numbers which unlike interval and ratio scales that can be transformed to other interval or ratio scales respectively and yield different numbers that mean the same thing, an absolute scale is invariant under the identity transformation that is its numbers cannot be changed to other numbers and mean the same thing. From such numbers priorities can be derived which also belong to an absolute scale of relative numbers whose total sum is equal to one.

Table 4.2 exhibits an example in which the scale is used to compare the relative consumption of drinks in the United States (done by an audience many years ago). One compares a drink indicated on the left with another indicated at the top and answers the question: How many times more, or how strongly more is that drink consumed in the US than the one at the top? More simply which drink of a pair

Table 4.1 The fundamental scale of absolute numbers

Intensity of Importance	Definition	Explanation
1	Equal importance	Two activities contribute equally to the objective
2	Weak or slight	
3	Moderate importance	Experience and judgment slightly favor one activity over another
4	Moderate plus	
5	Strong importance	Experience and judgment strongly favor one activity over another
6	Strong plus	
7	Very strong or demonstrated importance	An activity is favored very strongly over another; its dominance demonstrated in practice
8	Very, very strong	
9	Extreme importance	The evidence favoring one activity over another is of the highest possible order of affirmation
1.1–1.9	When activities are very close a decimal is added to 1 to show their difference as appropriate	Perhaps a better way than assigning the small decimals is to compare two close activities with other widely contrasting ones, favoring the larger one a little over the smaller one when using the 1–9 values.
Reciprocals of above	If activity i has one of the above nonzero numbers assigned to it when compared with activity j, then j has the reciprocal value when compared with i	A logical assumption

dominates the other and how strongly? In general, one uses the verbal explanation to develop a judgment and then enters its numerical value: for example enter 9 in the (coffee, wine) position meaning that coffee consumption is extremely more than wine consumption. It is automatic that $1/9$ is what one needs to use in the (wine, coffee) position. Note that water is consumed a little more than coffee, so one enters 2 in the (water, coffee) position, and $1/2$ in the (coffee, water) position. One always enters the whole number in its appropriate position and automatically enters its reciprocal in the transpose position.

The priorities, (obtained in exact form by raising the matrix to large powers and summing each row and dividing each by the total sum of all the rows, or approximately by adding each row of the matrix and dividing by their total and taking the average of the resulting columns) are shown at the bottom of the table along with the true values expressed in relative form by dividing the consumption of each drink (volume) by the sum of the consumption of all drinks. The information about actual consumption was obtained from the US Statistical Abstracts. We see the answers are very close and pair-wise comparison judgments of someone who knows can lead to accurate results of drink consumption. There are numerous examples of this kind of validation.

Table 4.2 Which drink is consumed more in the U.S.?

An example of estimation using judgments:
Drink consumption in the U.S.

	Coffee	Wine	Tea	Beer	Sodas	Milk	Water
Coffee	1	9	5	2	1	1	1/2
Wine	1/9	1	1/3	1/9	1/9	1/9	1/9
Tea	1/5	2	1	1/3	1/4	1/3	1/9
Beer	1/2	9	3	1	1/2	1	1/3
Sodas	1	9	4	2	1	2	1/2
Milk	1	9	3	1	1/2	1	1/3
Water	2	9	9	3	2	3	1

Very early in the history of the subject, T.L. Saaty and M. Khouja did the following exercise on an airplane in 1973. They simply used their common knowledge about the relative influence and standing of these countries in the world and without referring to any specific economic data related to GNP values. The two results are close and demonstrate that the general understanding an interested person has about a problem can be used to advantage to make fairly good estimates through paired comparisons.

Table 4.3 gives the judgments using the AHP 1–9 scale and Table 4.4 provides the derived priorities, the actual and relative GNP values.

Table 4.3 Paired comparisons of the relative dominance in wealth of seven nations

	U.S	U.S.S.R	China	France	U.K	Japan	W. Germany
U.S	1	4	9	6	6	5	5
U.S.S.R	1/4	1	7	5	5	3	4
China	1/9	1/7	1	1/5	1/5	1/7	1/5
France	1/6	1/5	5	1	1	1/3	1/3
U.K	1/6	1/5	5	1	1	1/3	1/3
Japan	1/5	1/3	7	3	3	1	2
W. Germany	1/5	1/4	5	3	3	1/2	1

The reader may now want to know how the foregoing integer-valued scale of response used in making paired comparison judgments can be derived mathematically from the well-known psychophysical logarithmic response function of Weber-Fechner. For a given value of the stimulus, the magnitude of response remains the same until the value of the stimulus is increased sufficiently large in proportion to the value of the stimulus, thus preserving the proportionality of relative increase in stimulus for it to be detectable for a new response. This suggests the idea of just noticeable differences (jnd), well known in psychology.

Table 4.4 Outcome of estimated relative wealth and the actual and relative values

	Normalized eigenvector	Actual GNP (1972)	Normalized GNP values
U.S	0.427	1,167	0.413
U.S.S.R	0.230	635	0.225
China	0.021	120	0.043
France	0.052	196	0.069
U.K	0.052	154	0.055
Japan	0.123	294	0.104
W. Germany	0.094	257	0.091

To derive the values in the scale starting with a stimulus s_0 successive magnitudes of the new stimuli take the form:

$$s_1 = s_0 + \Delta s_0 = s_0 + \frac{\Delta s_0}{s_0} s_0 = s_0(1+r) \equiv s_0 \alpha$$

$$s_2 = s_1 + \Delta s_1 = s_1(1+r) = s_0(1+r)^2 = s_0 \alpha^2$$

$$\vdots$$

$$s_n = s_{n-1}\alpha = s_0 \alpha^n \quad (n = 0, 1, 2, \ldots)$$

We consider the responses to these stimuli to be measured on a ratio scale $(b = 0)$. A typical response has the form $M_i = a \log \alpha^i$, $i = 1, \ldots, n$, or one after another they have the form:

$$M_1 = a \log \alpha, \quad M_2 = 2a \log \alpha, \ldots, \quad M_n = na \log \alpha$$

We take the ratios M_i/M_1 $i = 1, \ldots, n$ of these responses in which the first is the smallest and serves as the unit of comparison, thus obtaining the integer values $1, 2, \ldots, n$ of the fundamental scale of the AHP. It appears that numbers are intrinsic to our ability to make comparisons, and that they were not an invention by our primitive ancestors. We must be grateful to them for the discovery of the symbolism. In a less mathematical vein, we note that we are able to distinguish ordinally between high, medium and low at one level and for each of them in a second level below that also distinguish between high, medium and low giving us nine different categories. We assign the value one to (low, low) which is the smallest and the value nine to (high, high) which is the highest, thus covering the spectrum of possibilities between two levels, and giving the value nine for the top of the paired comparisons scale as compared with the lowest value on the scale. Because of increase in inconsistency when we compare more than about 7 elements, we don't need to keep in mind more than 7 ± 2 elements. This was first conjectured by the psychologist George Miller in the 1950's. Finally, we note that the scale just derived is attached to the importance we assign to judgments. If we have an exact measurement such as 2.375 and want to use it as it is for our judgment without attaching significance to it, we can use its entire value without approximation.

In the judgment matrix A, instead of assigning two numbers w_i and w_j and forming the ratio w_i/w_j we assign a single number drawn from the Fundamental Scale of absolute numbers to represent the ratio $(w_i/w_j)/1$. It is a nearest integer approximation to the ratio w_i/w_j. The derived scale will reveal what w_i and w_j are. This is a central fact about the relative measurement approach. It needs a fundamental scale to express numerically the relative dominance relationship. The general eigenvalue formulation is obtained by perturbation of the following consistent formulation:

$$Aw = \begin{bmatrix} \frac{w_1}{w_1} & \cdots & \frac{w_1}{w_n} \\ \vdots & & \vdots \\ \frac{w_n}{w_1} & \cdots & \frac{w_n}{w_n} \end{bmatrix} \begin{bmatrix} w_1 \\ \vdots \\ w_n \end{bmatrix} = n \begin{bmatrix} w_1 \\ \vdots \\ w_n \end{bmatrix} = nw$$

where A has been multiplied on the right by the transpose of the vector of weights $w = (w_1, \ldots, w_n)$. The result of this multiplication is nw. Thus, to recover the scale from the matrix of ratios, one must solve the problem $Aw = nw$ or $(A - nI)w = 0$. This is a system of homogeneous linear equations. It has a nontrivial solution if and only if the determinant of $A - nI$ vanishes, that is, n is an eigenvalue of A. Now A has unit rank since every row is a constant multiple of the first row. Thus all its eigenvalues except one are zero. The sum of the eigenvalues of a matrix is equal to its trace, that is, the sum of its diagonal elements. In this case the trace of A is equal to n. Thus n is an eigenvalue of A, and one has a nontrivial solution. The solution consists of positive entries and is unique to within a multiplicative constant.

The foregoing matrix of ratios of measurements is consistent. Its entries satisfy the relationship $a_{ij}a_{jk} = a_{ik}$ for all i, j, k. Note that the ratio of two readings from a ratio scale is an absolute (dimensionless) number. If we were to use judgment we would estimate this absolute number by using the Fundamental Scale of Table 4.1. When we use judgment we no longer can ensure consistency. It becomes important for us to know how inconsistent we are and which are the most inconsistent judgments and how they can be changed to improve the consistency. But our knowledge may not be adequate to correct our inconsistency as needed. If the inconsistency remains very high despite the changes we make that are compatible with our understanding, we cannot make a decision. The priority weights are obtained directly by adding and normalizing to one the sum of the rows of the matrix, or any of its columns. The intransitivity of influences (how much A dominates B and how much B dominates C and then how much C dominates A) cannot occur when the judgments are consistent. However, when the judgments are inconsistent, such dominance may happen along with the fact that $a_{ij}a_{jk} = a_{ik}$ for all i, j, k no longer holds. It is known that the different order transitivity of influences can be measured by raising the matrix to different powers. Each power of the matrix yields a set of priorities obtained as the normalized sum of its rows. It is not difficult to show that the average priority of the all these priority vectors is their Cesaro sum that leads to taking the limiting power of the matrix. Perron's theory about positive matrices tells us that this limit is the principal eigenvector of the matrix thus requiring us to solve the principal eigenvalue problem for our positive matrix. This shows that the principal eigenvector is a necessary condition for deriving priorities from inconsistent judgments.

Associated with the weights is an inconsistency index. The consistency index of a matrix is given by $C.I. = \frac{\lambda_{max}-n}{n-1} \equiv \mu$. The consistency ratio (C.R.) is obtained by forming the ratio of C.I. and the appropriate one of the following set of numbers shown in Table 4.5, each of which is an average random consistency index computed for $n \leq 10$ for very large samples. They create randomly generated reciprocal matrices using the scale $1/9, 1/8, \ldots, 1/2, 1, 2, \ldots, 8, 9$ and calculate the average of their eigenvalues. This average is used to form the Random Consistency Index *R.I.* Table 4.5 shows the values obtained from one set of such simulations and also their first order differences, for matrices of size $1, 2, \ldots, 15$. Of course we do not recommend comparing more than 7 items in any single matrix.

Table 4.5 Random index

Order	1	2	3	4	5	6	7	8	9	10	11	12	13	14	15
R.I.	0	0	0.52	0.89	1.11	1.25	1.35	1.40	1.45	1.49	1.52	1.54	1.56	1.58	1.59
First order differences		0	0.52	0.37	0.22	0.14	0.10	0.05	0.05	0.04	0.03	0.02	0.02	0.02	0.01

Figure 4.1 below is a plot of the first two rows of Table 4.5. It shows the asymptotic nature of random inconsistency.

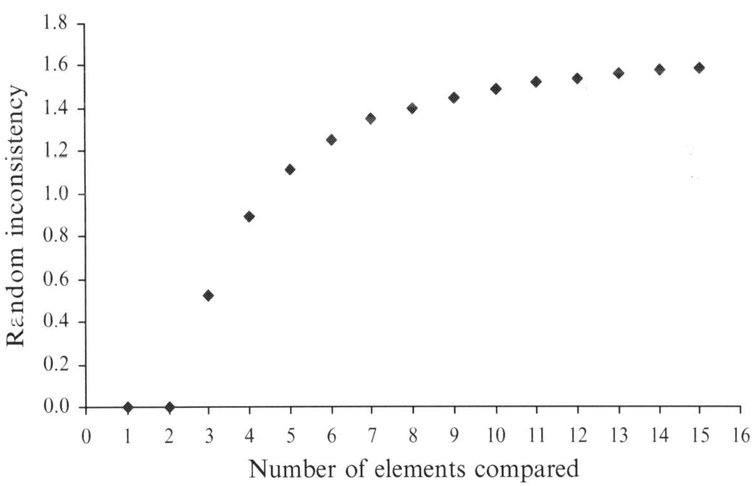

Fig. 4.1 Plot of random inconsistency

Since it would be pointless to try to discern any priority ranking from a set of random comparison judgments, we should probably be uncomfortable about proceeding unless the consistency index of a pairwise comparison matrix is very much smaller than the corresponding random index value in Table 4.5. The *consistency ratio* (C.R.) of a pairwise comparison matrix is the ratio of its consistency index $\sim$ to the corresponding random index value in Table 4.5. The notion of order of magnitude is essential in any mathematical consideration of changes in measurement. When one has a numerical value say between 1 and 10 for some measurement and one wishes to determine whether change in this value is significant or not, one reasons as follows: A change of a whole integer value is critical because it changes the magnitude and identity of the original number significantly. If the change or perturbation in value is of the order of a percent or less, it would be so small (by two orders of magnitude) and would be considered negligible. However if this perturbation is a decimal (one order of magnitude smaller) we are likely to pay attention to modify the original value by this decimal without losing the significance and identity of the original number as we first understood it to be. Thus in synthesizing near consistent judgment values, changes that are too large can cause dramatic change in our understanding, and values that are too small cause no change in our understanding. We are left with only values of one order of magnitude smaller that we can deal with incrementally to change our understanding. It follows that our allowable consistency ratio should be not more than about 0.10 for a matrix larger than 5 by 5, 8 for a 4 by 4 matrix and 5 for a 3 by 3 matrix. This requirement cannot be made smaller such as 1 or 0.1 without trivializing the impact of inconsistency. But inconsistency itself is important because without it, new knowledge that changes preference cannot be admitted. Assuming that all knowledge should be consistent contradicts experience that requires continued revision of understanding.

If the *C.R.* is larger than desired, we do three things: 1) Find the most inconsistent judgment in the matrix (for example, that judgment for which $\varepsilon_{ij} = \frac{a_{ij}w_j}{w_i}$ is largest), 2) Determine the range of values to which that judgment can be changed corresponding to which the inconsistency would be improved, 3) Ask the judge to consider, if he can, change his judgment to a plausible value in that range. If he is unwilling, we try with the second most inconsistent judgment and so on. If no judgment is changed the decision is postponed until better understanding of the stimuli is obtained. Judges who understand the theory are always willing to revise their judgments often not the full value but partially and then examine the second most inconsistent judgment and so on. It can happen that a judge's knowledge does not permit the improvement of consistency and more information is required to improve that consistency.

Before proceeding further, the following observations may be useful for a better understanding of the importance of the concept of a limit on our ability to process information and also change in information. The quality of response to stimuli is determined by three factors. Accuracy or validity, consistency, and efficiency or amount of information generated. Our judgment is much more sensitive and responsive to large perturbations. When we speak of perturbation, we have in mind numerical change from consistent ratios obtained from priorities. The larger the inconsistency and hence also the larger the perturbations in priorities, the greater is our sensitivity to make changes in the numerical values assigned. Conversely, the smaller the inconsistency, the more difficult it is for us to know where the best changes should be made to produce not only better consistency but also better validity of the outcome. Once near consistency is attained, it becomes uncertain which coefficients should be perturbed by small amounts to transform a near consistent matrix to a consistent one. If such perturbations were forced, they could be arbitrary and thus distort the validity of the derived priority vector in representing the underlying decision.

The third row of Table 4.5 gives the differences between successive numbers in the second row. Figure 4.2 is a plot of these differences and shows the importance of the number seven as a cutoff point beyond which the differences are less than 0.10 where we are not sufficiently sensitive to make accurate changes in judgment on several elements simultaneously. A similar argument and plot can be made by using the ratios of the numbers in the third row of Table 4.5 for $n \geq 3$.

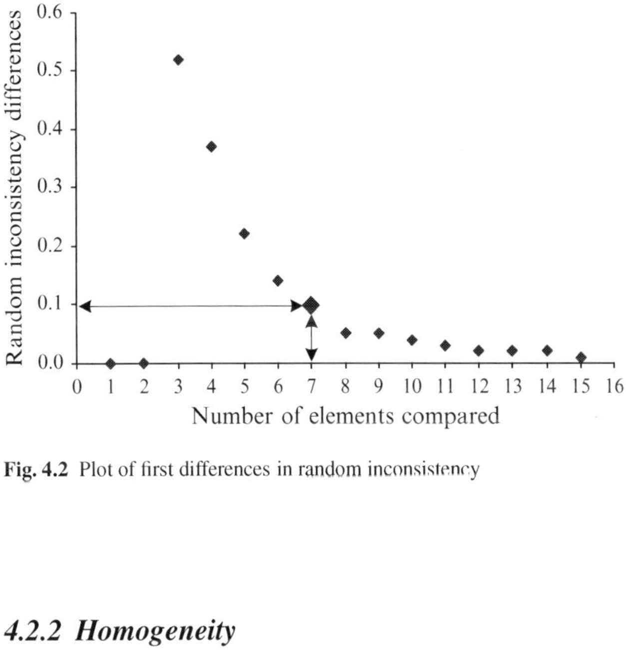

Fig. 4.2 Plot of first differences in random inconsistency

4.2.2 *Homogeneity*

Homogeneity as an important concept to ensure consistency in the paired comparisons requires the elements to be of the same order of magnitude which means that our perceptions in comparing them should be of nearly the same order of magnitude. It is a fact that people are unable to directly compare widely disparate objects such as a ping-pong ball and a basketball according to volume. To do that, we need a range greater than the 1–9 scale. To resolve this problem, we can use a method in which we cluster different elements so we can rate them within a cluster and then rate them across the clusters. We need to add other objects to make the comparison possible and then form groups of comparable elements. A common element, the pivot, could be the largest in one cluster and the smallest in the next cluster of the next higher order of magnitude. The weights of the elements in the second group are divided by the priority of the pivot in that group and then multiplied by the priority of the same pivot element from the first group, making them comparable with the first group. The process is then continued.

Table 4.6 shows how this process works in comparing a cherry tomato with a water melon, which appears to be two orders of magnitude bigger in size, by introducing intermediate objects in stages.

Table 4.6 Comparisons according to volume

 0.07 Cherry tomato	 0.28 Small green tomato	 0.65 Lime
 0.08 Lime $0.08/0.08 = 1$ $0.65 \times 1 = 0.65$	 0.22 Grapefruit $0.22/0.08 = 2.75$ $0.75 \times 2.75 = 1.79$	 0.70 Honeydew $0.70/0.08 = 8.75$ $0.65 \times 8.75 = 5.69$
 0.10 Honeydew $0.10/0.10 = 1$ $5.69 \times 1 = 5.69$	 0.30 Sugar baby watermelon $0.30/0.10 = 3$ $5.69 \times 3 = 17.07$	 0.60 Oblong watermelon $0.60/0.10 = 6$ $5.69 \times 6 = 34.14$

For a given positive reciprocal matrix $A = [a_{ij}]$ and a given pair of distinct indices $k > l$, define $A(t) = [a_{ij}(t)]$ by $a_{kl}(t) \equiv a_{kl} + t$, $a_{lk}(t) \equiv (a_{lk} + t)^{-}1$, and $a_{ij}(t) \equiv a_{ij}$ for all $i \neq k$, $j \neq l$, so $A(0) = A$. Let $\lambda_{max}(t)$ denote the Perron eigenvalue of $A(t)$ for all t in a neighborhood of $t = 0$ that is small enough to ensure that all entries of the reciprocal matrix $A(t)$ are positive there. Finally, let $v = [v_i]$ be the unique positive eigenvector of the positive matrix $A^\top$ that is normalized so that $v^\top w = 1$. Then a classical perturbation formula tells us that

$$\left.\frac{d\lambda_{\max}(t)}{dt}\right|_{t=0} = \frac{v^\top A'(0)w}{v^\top w} = v^\top A'(0)w = v_k w_l - \frac{1}{a_{kl}^2} v_l w_k.$$

We conclude that:

$$\frac{\partial \lambda_{\max}}{\partial a_{ij}} = v_i w_j - a_{ji}{}^2 v_j w_i, \quad \text{for all } i, j = 1, \ldots, n$$

Because we are operating within the set of positive reciprocal matrices, $\frac{\partial \lambda max}{\partial a_{ij}} = -\frac{\partial \lambda_{max}}{\partial a_{ij}}$ for all i and j. Thus, to identify an entry of A whose adjustment within the class of reciprocal matrices would result in the largest rate of change in λ_{max} we should examine the $n(n-1)/2$ values $\{v_i w_j - a_{ji}^2 v_j w_i\}$, $i > j$ and select (any) one of largest absolute value.

4.3 Additive Composition is Necessary

Sometimes people have assigned criteria different weights when they are measured in the same unit. Others have used different ways of synthesis than multiplying and adding. An example should clarify what we must do. Synthesis in the AHP involves weighting the priorities of elements compared with respect to an element in the next higher level, called a parent element, by the priority of that element and adding over all such parents for each element in the lower level. Consider the example of two criteria C_1 and C_2 and three alternatives A_1, A_2 and A_3 measured in the same scale such as dollars. If the criteria are each assigned the value 1, then the weighting and adding process produces the correct dollar value as in Table 4.7.

Table 4.7 Calculating returns arithmetically

Alternatives	Criterion C_1 Unnormalized weight = 1.0	Criterion C_2 Unnormalized weight = 1.0	Weighted sum Unnormalized	Normalized or relative values
A_1	200	150	350	$350/1300 = 0.269$
A_2	300	50	350	$350/1300 = 0.269$
A_3	500	100	600	$600/1300 = 0.462$
Column totals	1000	300	1300	1

However, it does not give the correct outcome if the weights of the criteria are normalized, with each criterion having a weight of 0.5. Once the criteria are given in relative terms, so must the alternatives also be given in relative terms. A criterion that measures values in pennies cannot be as important as another measured in thousands of dollars. In this case, the only meaningful importance of a criterion is the ratio of the total money for the alternatives under it to the total money for the alternatives under both criteria. By using these weights for the criteria, rather than 0.5 and 0.5, one obtains the correct final relative values for the alternatives.

What is the relative importance of each criterion? Normalization indicates relative importance. Relative values require that criteria be examined as to their relative importance with respect to each other. What is the relative importance of a criterion, or what numbers should the criteria be assigned that reflect their relative importance? Weighting each criterion by the proportion of the resource under it, as shown below in Table 4.8, and multiplying and adding as in the additive synthesis of the

AHP, we get the same correct answer. For criterion C_1 we have

$$(200+300+500)/[(200+300+500)+(150+50+100)] = 1000/1300$$

and for criterion C_2 we have

$$(150+50+100)/[(200+300+500)+(150+50+100)] = 300/1300.$$

Here the weights of the criteria are automatically given in normalized form, and their weights sum to one. We see that when the criteria are normalized, the alternatives must also be normalized to get the right answer. For example, if we look in Table 4.7 we have $350/1300$ for the priority of alternative A_1. Now if we simply weight and add the values for alternative A_1 in Table 4.8 we get for its final value $(200/1000)(1000/1300)+(150/300)(300/1300) = 350/1300$ which is the same as in Table 4.7. It is clear that if the priorities of the alternatives are not normalized one does not get the desired outcome.

Table 4.8 Normalized criteria weights and normalized alternative weights from measurements in the same scale (additive synthesis

Alternatives	Criterion C_1 Normalized weight= $1000/1300 = 0.7692$	Criterion C_2 Normalized weight= $300/1300 = 0.2308$	Weighted Sum
A_1	$200/1000$	$150/300$	$350/1300 = 0.2692$
A_2	$300/1000$	$50/300$	$350/1300 = 0.2692$
A_3	$500/1000$	$100/300$	$600/1300 = 0.4615$

We have seen in this example that in order to obtain the correct final relative values for the alternatives when measurements on a measurement scale are given, it is essential that the priorities of the criteria be derived from the priorities of the alternatives. Thus when the criteria depend on the alternatives we need to normalize the values of the alternatives to obtain the final result. This procedure is known as the distributive mode of the AHP. It is also used in case of functional (real life not paired comparison) dependence of the alternatives on the alternatives and of the criteria on the alternatives. The AHP is a special case of the Analytic Network Process. The dominant mode of synthesis in the ANP with all its interdependencies is the distributive mode. The ANP automatically assigns the criteria the correct weights, if one only uses the normalized values of the alternatives under each criterion and also the normalized values for each alternative under all the criteria without any special attention to weighting the criteria.

4.3.1 *Benefits, Opportunities, Costs and Risks*

The process of decision-making requires us to analyze a decision according to Benefits (B), the good things that would result from taking the decision; Opportunities (O), the potentially good things that can result in the future from taking the decision; Costs (C), the pains and disappointments that would result from taking the decision; and Risks (R), the potential pains and disappointments that can result from taking the decision. We then create control criteria and subcriteria or even a network of criteria under each and develop a subnet and its connection for each control criterion. Next we determine the best outcome for each control criterion and combine the alternatives in what is known as the ideal form for all the control criteria under each of the BOCR merits. Then we take the best alternative under B and use it to think of benefits and the best one under O, which may be different than the one under C, and use it to think of opportunities and so on for costs and risks. Finally we must rate these four with respect to the strategic criteria (criteria that underlie the evaluations of the merits all the decisions we make) using the ratings mode of the AHP to obtain priority ratings for B, O, C, and R. We then normalize (not mandatory but recommended) and use these weights to combine the four vectors of outcomes for each alternative under BOCR to obtain the overall priorities. We can form the ratio BO/CR which does not need the BO/CR ratings to obtain marginal overall outcomes. Alternatively, and better, we can use the ratings to weight and subtract the costs and risks from the sum of the weighted benefits and opportunities.

4.4 Hierarchies

Structuring a complex decision is perhaps the most important task along with the process of prioritization. Experience has shown that one can prescribe guidelines for structuring a hierarchy. Here are some suggestions for an elaborate design of a hierarchy:

1. Identify the overall goal. What are you trying to accomplish? What is the main question?
2. Identify the subgoals of the overall goal. If relevant, identify time horizons that affect the decision.
3. Identify criteria that must be satisfied to fulfill the subgoals of the overall goal.
4. Identify subcriteria under each criterion. Note that criteria or subcriteria may be specified in terms of ranges of values of parameters or in terms of verbal intensities such as high, medium, low.
5. Identify the actors involved.
6. Identify the actors' goals.
7. Identify the actors' policies.
8. Identify the people affected by the decision.
9. Identify the objectives of these people.

10. Identify options or outcomes to take that serve people's objectives best.
11. For restricted yes-no decisions, take the most preferred outcome and compare the benefits and costs of making the decision with those of not making it.
12. Do a benefit/cost analysis using marginal values and total priority values. Because we deal with dominance hierarchies, ask which alternative yields the greatest benefit; for costs: which alternative costs the most, and for risks, which alternative is more risky. Rate the top ranked alternative for each of the BOCR with respect to strategic criteria and subcriteria that are compared to obtain their priorities. Normalize the derived weights and use them to weight the BOCR. In the marginal case for each alternative divide the weighted benefit multiplied by the weighted opportunity and divide by the weighted cost times the weighted risk. Note whether the outcome is more or less than one. It is generally better to have the result more than one for the best alternative but may not always be necessary as is the case of having a major airline in a poor country. To obtain the total priority values, for each alternative add the weighted benefits and opportunities and subtract the weighted costs and risks. The outcome in this case may be negative. The final outcome from the marginal and the total results may not be the same. In general the total is what is needed, but the marginal is also valuable for actions with immediate rather than long term results.
13. Perform sensitivity analysis on the outcome to determine its stability to changes in the judgments. If desired, include a criterion in each hierarchy called "other" or "the unknown" for which appropriate priority values may be derived from paired comparisons. Sensitivity testing with respect to such a criterion can determine the impact of the unknown on the outcome to the best of an experienced person's understanding. It must be understood that such a factor cannot be included to cover up for total ignorance about a decision. Only the wise should use it.

4.5 An Example: The Hospice Problem

Westmoreland County Hospital in Western Pennsylvania, like hospitals in many other counties around the nation, has been concerned with the costs of the facilities and manpower involved in taking care of terminally ill patients. Normally these patients do not need as much medical attention as do other patients. Those who best utilize the limited resources in a hospital are patients who require the medical attention of its specialists and advanced technology equipment, whose utilization depends on the demand of patients admitted into the hospital. The terminally ill need medical attention only episodically. Most of the time such patients need psychological support. Such support is best given by the patient's family, whose members are able to supply the love and care the patients most need. For the mental health of the patient, home therapy is a benefit. Most patients need the help of medical professionals only during a crisis. Some will also need equipment and surgery.

The planning association of the hospital wanted to develop alternatives and to choose the best one considering various criteria from the standpoint of the patient, the hospital, the community, and society at large.

In this problem, we need to consider the costs and benefits of the decision. Cost includes economic costs and all sorts of intangibles, such as inconvenience and pain. Such disbenefits are not directly related to benefits as their mathematical inverses, because patients infinitely prefer the benefits of good health to these intangible disbenefits. To study the problem, one needs to deal with benefits and with costs separately.

To keep matters simple we give an example of a decision made by considering benefits and costs only. No opportunities and risks were included as one usually must do in a more complex decision. The first author met with representatives of the planning association for several hours to decide on the best alternative. To make a decision by considering benefits and costs, one must first answer the question: In this problem, do the benefits justify the costs? If they do, then either the benefits are so much more important than the costs that the decision is based simply on benefits, or the two are so close in value that both the benefits and the costs should be considered. Then we use two hierarchies for the purpose and make the choice by forming the ratio from them of the (benefits priority/cost priority) for each alternative. One asks which is most beneficial in the benefits hierarchy of Figure 4.3 and which is most costly in the costs hierarchy of Figure 4.4.

If the benefits do not justify the costs, the costs alone determine the best alternative, that which is the least costly. In this example, we decided that both benefits and costs had to be considered in separate hierarchies. In a risk problem, a third hierarchy is used to determine the most desired alternative with respect to all three: benefits, costs, and risks. In this problem, we assumed risk to be the same for all contingencies. Whereas for most decisions one uses only a single hierarchy, we constructed two hierarchies for the hospice problem, one for benefits or gains (which model of hospice care yields the greater benefit) and one for costs or pains (which model costs more).

The planning association thought the concepts of benefits and costs were too general to enable it to make a decision. Thus, the planners further subdivided each (benefits and costs) into detailed subcriteria to enable the group to develop alternatives and to evaluate the finer distinctions the members perceived between the three alternatives. The alternatives were to care for terminally ill patients at the hospital, at home, or partly at the hospital and partly at home.

For each of the two hierarchies, benefits and costs, the goal clearly had to be choosing the best hospice. We placed this goal at the top of each hierarchy. Then the group discussed and identified overall criteria for each hierarchy; these criteria need not be the same for the benefits as for the costs.

The two hierarchies are fairly clear and straightforward in their description. They descend from the more general criteria in the second level to secondary subcriteria in the third level and then to tertiary subcriteria in the fourth level on to the alternatives at the bottom or fifth level.

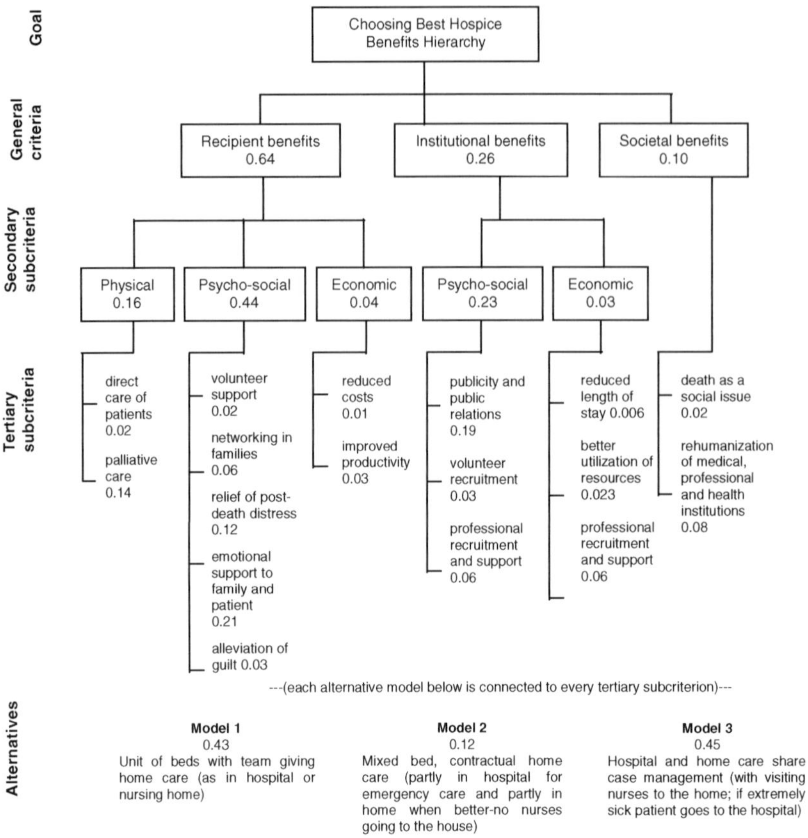

Fig. 4.3 Benefits hierarchy to choose the best hospice plan

At the general criteria level, each of the hierarchies, benefits or costs, involved three major interests. The decision should benefit the recipient, the institution, and society as a whole and their relative importance is the prime determinant as to which outcome is more likely to be preferred. We located these three elements on the second level of the benefits hierarchy. As the decision would benefit each party differently and the importance of the benefits to each recipient affects the outcome, the group thought that it was important to specify the types of benefit for the recipient and the institution. Recipients want physical, psycho-social and economic benefits, while the institution wants only psychosocial and economic benefits. We located these benefits in the third level of the hierarchy. Each of these in turn needed further decomposition into specific items in terms of which the decision alternatives could be evaluated. For example, while the recipient measures economic benefits in terms of reduced costs and improved productivity, the institution needed the more specific measurements of reduced length of stay, better utilization of resources, and increased financial support from the community. There was no reason to decompose

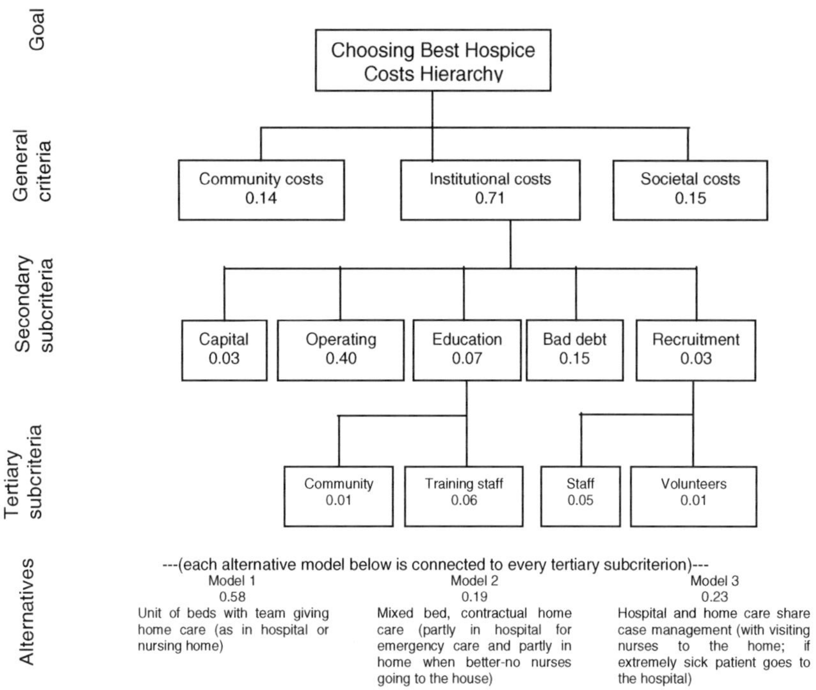

Fig. 4.4 Costs hierarchy to choose the best hospice plan

the societal benefits into third level subcriteria and hence societal benefits connects directly to the fourth level. The group considered three models for the decision alternatives, and they are at the bottom (or fifth level in this case) of the hierarchy: in Model 1, the hospital provided full care to the patients; in Model 2, the family cares for the patient at home, and the hospital provides only emergency treatment (no nurses go to the house); and in Model 3, the hospital and the home share patient care (with visiting nurses going to the home).

In the costs hierarchy there were also three major interests in the second level that would incur costs or pains: community, institution, and society. In this decision the costs incurred by the patient were not included as a separate factor. Patient and family could be thought of as part of the community. We thought decomposition was necessary only for institutional costs. We included five such costs in the third level: capital costs, operating costs, education costs, bad debt costs, and recruitment costs. Educational costs apply to educating the community and training the staff. Recruitment costs apply to staff and volunteers. Since both the costs hierarchy and the benefits hierarchy concern the same decision, they both have the same alternatives in their bottom levels, even though the costs hierarchy has fewer levels.

The question now is how to use pairwise comparison judgments and derive priorities and synthesize them to obtain the overall benefits and costs of each of the three alternatives, and then again for the costs and combine the two outcomes into a single overall outcome. To do that we need to first explain in simple terms how the process of prioritization is carried out.

4.5.1 Judgments and Comparisons

A judgment is an expression of an opinion. A comparison is an expression of an opinion about the dominance (importance, preference or likelihood) of one thing over another. Dominance represents the intensity of strength. It is done every day through verbal expression that has some quantitative significance that we need to use to combine the many dominance judgments involved in a decision. The set of all such judgments in making comparisons with respect to a single property or goal can be represented in a square matrix in which the set of elements is compared with itself. It is a way of organizing all the judgments with respect to that property to be processed and synthesized along with other matrices of comparison judgments involved in that decision. Each judgment represents the dominance of an element in the column on the left of the matrix over an element in the row on top. It reflects the answers to two questions: which of the two elements is more important with respect to a higher level criterion, and how strongly.

As usual with the AHP, in both the benefits and costs models, we compared the criteria and subcriteria according to their relative importance with respect to the parent element in the adjacent upper level. For example, the entries in the matrix shown in Table 4.9 are responses to the question: which general criterion is more important with respect to choosing the best hospice alternative and how strongly? Here recipient benefits are moderately more important than institutional benefits and are assigned the absolute number 3 in the (1,2) or first-row second-column position. Three signifies three times more. The reciprocal value is automatically entered in the (2,1) position, where institutional benefits on the left are compared with recipient benefits at the top. Similarly a 5, corresponding to strong dominance or importance, is assigned to recipient benefits over social benefits in the (1,3) position, and a 3, corresponding to moderate dominance, is assigned to institutional benefits over social benefits in the (2,3) position with corresponding reciprocals in the transpose positions of the matrix.

Judgments in a matrix may not be consistent. In eliciting judgments, one makes redundant comparisons to improve the validity of the answer, given that respondents may be uncertain or may make poor judgments in comparing some of the elements. Redundancy gives rise to multiple comparisons of an element with other elements and hence to numerical inconsistencies.

For example, where we compare recipient benefits with institutional benefits and with societal benefits, we have the respective judgments 3 and 5. Now if $x = 3y$ and $x = 5z$ then $3y = 5z$ or $y = 5/3z$. If the judges were consistent, institutional benefits would be assigned the value $5/3$ instead of the 3 given in the matrix. Thus

Table 4.9 Judgment matrix for the criteria of the benefits hierarchy

Choosing best hospice	Recipient benefits	Institutional benefits	Social benefits	Priorities
Recipient benefits	1	3	5	0.64
Institutional benefits	1/3	1	3	0.26
Social benefits	1/5	1/3	1	0.11

C.R.= 0.033

the judgments are inconsistent. In fact, we are not sure which judgments are the accurate ones and which are the cause of the inconsistency.

The process is repeated in all the matrices by asking the appropriate dominance or importance question. For example, the entries in the judgment matrix shown in Table 4.10 are responses to the question: which subcriterion yields the greater benefit with respect to institutional benefits and how strongly?

Here psycho-social benefits are regarded as very strongly more important than economic benefits, and 7 is entered in the (1, 2) position and 1/7 in the (2, 1) position.

Table 4.10 Judgment matrix of subcriteria with respect to institutional benefits

Institutional benefits	Psycho-social	Economic	Priorities
Psycho-social	1	7	0.875
Economic	1/7	1	0.125

In comparing the three models for patient care, we asked members of the planning association which model they preferred with respect to each of the covering or parent secondary criterion in level 3 or with respect to the tertiary criteria in level 4. For example, for the subcriterion direct care (located on the left-most branch in the benefits hierarchy), we obtained a matrix of paired comparisons in Table 4.11 in which Model 1 is preferred over Models 2 and 3 by 5 and 3 respectively and Model 3 is preferred by 3 over Model 2. The group first made all the comparisons using semantic terms for the fundamental scale and then translated them to the corresponding numbers.

Table 4.11 Relative benefits of the models with respect to direct care of patients

Direct care of patient	Model I	Model II	Model III	Priorities
Model I: Unit team	1	5	3	0.64
Model II: Mixed/home care	1/5	1	1/3	0.10
Model III: Case management	1/3	3	1	0.26

C.R.= 0.003

For the costs hierarchy, we again illustrate with three matrices. First the group compared the three major cost criteria and provided judgments in response to the question: which criterion is a more important determinant of the cost of a hospice model? Table 4.12 shows the judgments obtained.

Table 4.12 Judgment matrix for the criteria of the costs hierarchy

Choosing best hospice (costs)	Community	Institutional	Societal	Priorities
Community costs	1	1/5	1	0.14
Institutional costs	5	1	5	0.71
Societal costs	1	1/5	1	0.14

C.R.= 0.000

The group then compared the subcriteria under institutional costs and obtained the importance matrix shown in Table 4.13. The entries are responses to the question: which criterion incurs greater institutional costs and how strongly? Finally we compared the three models to find out which incurs the highest cost for each criterion or subcriterion. Table 4.14 shows the results of comparing them with respect to the costs of recruiting staff.

Table 4.13 Judgment matrix of subcriteria under institutional costs

Institutional costs	Capital	Operating	Education	Bad debt	Recruitment	Priorities
Capital	1	1/7	1/4	1/7	1	0.05
Operating	7	1	9	4	5	0.57
Education	4	1/9	1	1/2	1	0.01
Bad debt	7	1/4	2	1	3	0.21
Recruitment	1	1/5	1	1/3	1	0.07

C.R.= 0.000

Table 4.14 Relative costs of the models with respect to recruiting staff

Institutional costs for recruiting staff	Model I	Model II	Model III	Priorities
Model I: Unit team	1	5	3	0.64
Model II: Mixed/home care	1/5	1	1/3	0.1
Model III: Case management	1/3	3	1	0.26

C.R.= 0.08

As shown in Table 4.15, we divided the benefits priorities by the costs priorities for each alternative to obtain the best alternative, model 3, the one with the largest value for the ratio.

Table 4.15 shows two ways or modes of synthesizing the local priorities of the alternatives using the global priorities of their parent criteria: The distributive mode and the ideal mode. In the distributive mode, the weights of the alternatives sum to one. It is used when there is dependence among the alternatives and a unit priority is distributed among them. The ideal mode is used to obtain the single best alternative regardless of what other alternatives there are. In the ideal mode, the local priorities of the alternatives are divided by the largest value among them. This is done for each criterion; for each criterion one alternative becomes an ideal with value one. In both modes, the local priorities are weighted by the global priorities of the parent criteria and synthesized and the benefit-to-cost ratios formed. In this case, both modes lead to the same outcome for hospice, which is model 3. As we shall see below, we need both modes to deal with the effect of adding (or deleting) alternatives on an already ranked set.

Model 3 has the largest ratio of benefits to costs in both the distributive and ideal modes, and the hospital selected it for treating terminal patients. This need not always be the case. In this case, there is dependence of the personnel resources allocated to the three models because some of these resources would be shifted based on the decision. Therefore the distributive mode is the appropriate method of synthesis. If the alternatives were sufficiently distinct with no dependence in their definition, the ideal mode would be the way to synthesize.

We also performed marginal analysis to determine where the hospital should allocate additional resources for the greatest marginal return. To perform marginal analysis, we first ordered the alternatives by increasing cost priorities and then formed the benefit-to-cost ratios corresponding to the smallest cost, followed by the ratios of the differences of successive benefits to costs. If this difference in benefits is negative, the new alternative is dropped from consideration and the process continued. The alternative with the largest marginal ratio is then chosen. For the costs and corresponding benefits from the synthesis rows in Table 4.15 we obtained:

Costs:	0.20	0.21	0.59
Benefits:	0.12	0.45	0.43

From these values we compute the marginal ratios as the final priorities:

$$\frac{0.12}{0.20} = 0.60 \quad \frac{0.45 - 0.12}{0.21 - 0.20} = 33 \quad \frac{0.43 - 0.45}{0.59 - 0.21} = -0.051$$

The third alternative is not a contender for resources because its marginal return is negative. The second alternative is best. In fact, in addition to adopting the third model, the hospital management chose the second model of hospice care for further development.

Table 4.15 Global and ideal modes of synthesizing the local priorities of the alternatives

		Distributive mode			Ideal mode		
		Models					
Benefits	Priorities	1	2	3	1	2	3
Direct care of patient	0.02	0.64	0.10	0.26	1.00	0.16	0.41
Palliative care	0.14	0.64	0.10	0.26	1.00	0.16	0.41
Volunteer support	0.02	0.09	0.17	0.74	0.12	0.23	1.00
Networking in families	0.06	0.46	0.22	0.32	1.00	0.48	0.70
Relief of post death stress	0.12	0.30	0.08	0.62	0.48	0.13	1.00
Emotional support of family and patient	0.21	0.30	0.08	0.62	0.48	0.13	1.00
Alleviation of guilt	0.03	0.30	0.08	0.62	0.48	0.13	1.00
Reduced economic costs for patient	0.01	0.12	0.65	0.23	0.19	1.00	0.35
Improved productivity	0.03	0.12	0.27	0.61	0.20	0.44	1.00
Publicity and public relations	0.19	0.63	0.08	0.29	1.00	0.13	0.46
Volunteer recruitment	0.03	0.64	0.10	0.26	1.00	0.16	0.41
Professional recruitment and support	0.06	0.65	0.23	0.12	1.00	0.35	0.19
Reduced length of stay	0.01	0.26	0.10	0.64	0.41	0.41	1.00
Better utilization of resources	0.02	0.09	0.22	0.69	0.13	0.13	1.00
Increased monetary support	0.00	0.73	0.08	0.19	1.00	1.00	0.26
Death as a social issue	0.02	0.20	0.20	0.60	0.33	0.33	1.00
Rehumanization of institutions	0.08	0.24	0.14	0.62	0.39	0.23	1.00
Synthesis		0.43	0.12	0.45	0.42	0.12	0.45
Costs							
Community costs	0.14	0.33	0.33	0.33	1.00	1.00	1.00
Institutional capital costs	0.03	0.76	0.09	0.15	1.00	0.12	0.20
Institutional operating costs	0.40	0.73	0.08	0.19	1.00	0.11	0.26
Institutional costs for educating the community	0.01	0.65	0.24	0.11	1.00	0.37	0.17
Institutional costs for training staff	0.06	0.56	0.32	0.12	1.00	0.57	0.21
Institutional bad debt	0.15	0.60	0.20	0.20	1.00	0.33	0.33
Institutional costs of recruiting staff	0.05	0.66	0.17	0.17	1.00	0.26	0.26
Institutional costs of recruiting volunteers	0.01	0.60	0.20	0.20	1.00	0.33	0.33
Societal costs	0.15	0.33	0.33	0.33	1.00	1.00	1.00
Synthesis		0.58	0.19	0.22	0.52	0.23	0.25
Benefit/cost ratio		0.73	0.63	2.01	0.81	0.54	1.82

4.6 Absolute Measurement - Rating Alternatives One at a Time

People are able to make two kinds of comparisons - absolute and relative. In absolute comparisons, people compare alternatives with a standard in their memory that they have developed through experience. In relative comparisons, they compared alternatives in pairs according to a common attribute, as we did throughout the hospice example.

People use absolute measurement (sometimes also called rating) to rank independent alternatives one at a time in terms of rating intensities for each of the criteria. An intensity is a range of variation of a criterion that enables one to distinguish the quality of an alternative for that criterion. An intensity may be expressed as a numerical range of values if the criterion is measurable or defined in qualitative terms.

For example, if ranking students is the objective and one of the criteria on which they are to be ranked is performance in mathematics, the mathematics ratings might be: excellent, good, average, below average, poor; or, using the usual school terminology, A, B, C, D, and F. Relative comparisons are first used to set priorities on the ratings themselves. If desired, one can fit a continuous curve through the derived intensities. This concept may go against our socialization. However, it is perfectly reasonable to ask how much an A is preferred to a B or to a C. The judgment of how much an A is preferred to a B might be different under different criteria. Perhaps for mathematics an A is very strongly preferred to a B, while for physical education an A is only moderately preferred to a B. So the end result might be that the ratings are scaled differently. For example one could have the scale values for the ratings as shown in Table 4.16:

Table 4.16 Examples of scale values for ratings

	Math	Physical Education
A	0.50	0.30
B	0.30	0.30
C	0.15	0.20
D	0.04	0.10
E	0.01	0.10

The alternatives are then rated or ticked off one at a time using the intensities. We will illustrate absolute measurement with an example. A firm evaluates its employees for raises. The criteria are dependability, education, experience, and quality. Each criterion is subdivided into intensities, standards, or subcriteria (Figure 4.5). The managers set priorities for the criteria by comparing them in pairs. They then pairwise compare the intensities according to priority with respect to their parent criterion (as in Table 4.17) or with respect to a subcriterion if they are using a deeper hierarchy. The priorities of the intensities are divided by the largest intensity for each criterion (second column of priorities in Figure 4.5).

Table 4.17 shows a paired comparison matrix of intensities with respect to dependability. The managers answer the question: which intensity is more important and by how much with respect to dependability. The priorities of the intensities for each criterion are divided by the largest one and multiplied by the priority of the criterion. Finally the managers rate each individual (Table 4.18) by assigning the intensity rating that applies to him or her under each criterion. The scores of these intensities are each weighted by the priority of its criterion and summed to derive a total ratio scale score for the individual (shown on the right of Table 4.18). These numbers belong to an absolute scale, and the managers can give salary increases precisely in proportion to the ratios of these numbers. Adams gets the highest score and Kesselman the lowest. This approach can be used whenever it is possible to

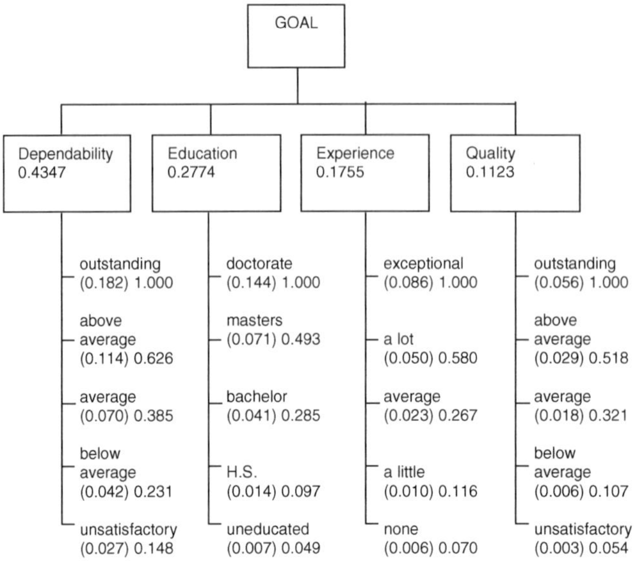

Fig. 4.5 Hierarchy with absolute measurement

set priorities for intensities of criteria; people can usually do this when they have sufficient experience with a given operation. This normative mode requires that alternatives be rated one by one without regard to how many there may be and how high or low any of them rates on prior standards. Some corporations have insisted that they no longer trust the normative standards of their experts and that they prefer to make paired comparisons of their alternatives. Still, when there is wide agreement on standards, the absolute mode saves time in rating a large number of alternatives.

Table 4.17 Ranking intensities: Which intensity is preferred most with respect to dependability and how strongly?

Dependability	Outstanding	Above average	Average	Below average	Unsatisfactory	Priorities	Idealized priorities
Outstanding	1	2	3	4	5	0.419	1.000
Above average	1/2	1	2	3	4	0.263	0.628
Average	1/3	1/2	1	2	3	0.160	0.382
Below average	1/4	1/3	1/2	1	2	0.097	0.232
Unsatisfactory	1/5	1/4	1/3	1/2	1	0.062	0.148

C.R. = 0.015

Table 4.18 Rating alternatives

Employees	Dependability 0.4347	Education 0.2774	Experience 0.1775	Quality 0.1123	Total
1. Adams, V	Outstanding	Bachelor	A little	Outstanding	0.646
2. Becker, L	Average	Bachelor	A little	Outstanding	0.379
3. Hayat, F	Average	Masters	A lot	Below avg.	0.418
4. Kessel, S	Above	H.S.	None	Above avg.	0.369
5. O'Shea, K	Average	Doctorate	A lot	Above	0.605
6. Peters, T	Average	Doctorate	A lot		0.583
7. Tobias, K	Above	Bachelor	Averag		0.456

4.7 On the Admission of China to the World Trade Organization (WTO) [4]

This section was taken from an analysis done in 2000 carried out before the US Congress acted favorably on China joining the WTO and was hand-delivered to many of the members of the committee including its Chairperson. Since 1986, China had been attempting to join the multilateral trade system, the General Agreement on Tariffs and Trade (GATT) and, its successor, the World Trade Organization (WTO). According to the rules of the 135-member nations of WTO, a candidate member must reach a trade agreement with any existing member country that wishes to trade with it. By the time this analysis was done, China signed bilateral agreements with 30 countries - including the US (November 1999) - out of 37 members that had requested a trade deal with it [8].

As part of its negotiation deal with the US, China asked the US to remove its annual review of China's Normal Trade Relations (NTR) status, until 1998 called Most Favored Nation (MFN) status. In March 2000, President Clinton sent a bill to Congress requesting a Permanent Normal Trade Relations (PNTR) status for China. The analysis was done and copies sent to leaders and some members in both houses of Congress before the House of Representatives voted on the bill, May 24, 2000. The decision by the US Congress on China's trade-relations status will have an influence on US interests, in both direct and indirect ways. Direct impacts include changes in economic, security and political relations between the two countries as the trade deal is actualized. Indirect impacts will occur when China becomes a WTO member and adheres to WTO rules and principles. China has said that it would join the WTO only if the US gives it Permanent Normal Trade Relations status.

It is likely that Congress will consider four options the least likely is that the US will deny China both PNTR and annual extension of NTR status. The other three options are:

1. **Passage of a clean PNTR bill**: Congress grants China Permanent Normal Trade Relations status with no conditions attached. This option would allow implementation of the November 1999 WTO trade deal between China and the Clinton

administration. China would also carry out other WTO principles and trade conditions.

2. **Amendment of the current NTR status bill**: This option would give China the same trade position as other countries and disassociate trade from other issues. As a supplement, a separate bill may be enacted to address other matters, such as human rights, labor rights, and environmental issues.
3. **Annual Extension of NTR status**: Congress extends China's Normal Trade Relations status for one more year, and, thus, maintains the status quo.

The conclusion of the study is that the best alternative is granting China PNTR status. China now has that status.

Our analysis involves four steps. First, we prioritize the criteria in each of the benefits, costs, opportunities and risks hierarchies with respect to the goal. Figure 4.6 shows the resulting prioritization of these criteria. The alternatives and their priorities are shown under each criterion both in the distributive and in the ideal modes. The ideal priorities of the alternatives were used appropriately to synthesize their final values beneath each hierarchy.

The priorities shown in Figure 4.6 were derived from judgments that compared the elements involved in pairs. For readers to estimate the original pairwise judgments (not shown here) one forms the ratio of the corresponding two priorities shown, leave them as they are, or take the closest whole number, or its reciprocal if it is less than 1.0.

The idealized values are shown in parentheses after the original distributive priorities obtained from the eigenvector. The ideal values are obtained by dividing each of the distributive priorities by the largest one among them. For the Costs and Risks structures, the question is framed as to which is the *most* costly or risky alternative. That is, the most costly alternative ends up with the highest priority.

It is likely that, in a particular decision, the benefits, costs, opportunities and risks (BOCR) are not equally important, so we must also prioritize them. This is shown in Table 4.19. The priorities for the economic, security and political factors themselves were established as shown in Figure 4.7 and used to rate the importance of the top ideal alternative for each of the benefits, costs, opportunities and risks from Table 4.19. Finally, we used the priorities of the latter to combine the synthesized priorities of the alternatives in the four hierarchies, using the normalized reciprocal - priorities of the alternatives under costs and risks, to obtain their final ranking, as shown in Table 4.20.

How to derive the priority shown next to the goal of each of the four hierarchies shown in Figure 4.7 is outlined in Table 4.19. We rated each of the four merits: benefits, opportunities, costs and risks of the dominant PNTR alternative, as it happens to be in this case, in terms of intensities for each assessment criterion. The intensities, Very High, High, Medium, Low, and Very Low were themselves prioritized in the usual pairwise comparison matrix to determine their priorities. We then assigned the appropriate intensity for each merit on all assessment criteria. The outcome is as found in the bottom row of Table 4.19.

We are now able to obtain the overall priorities of the three major decision alternatives listed earlier, given as columns in Table 4.20 which gives two ways of

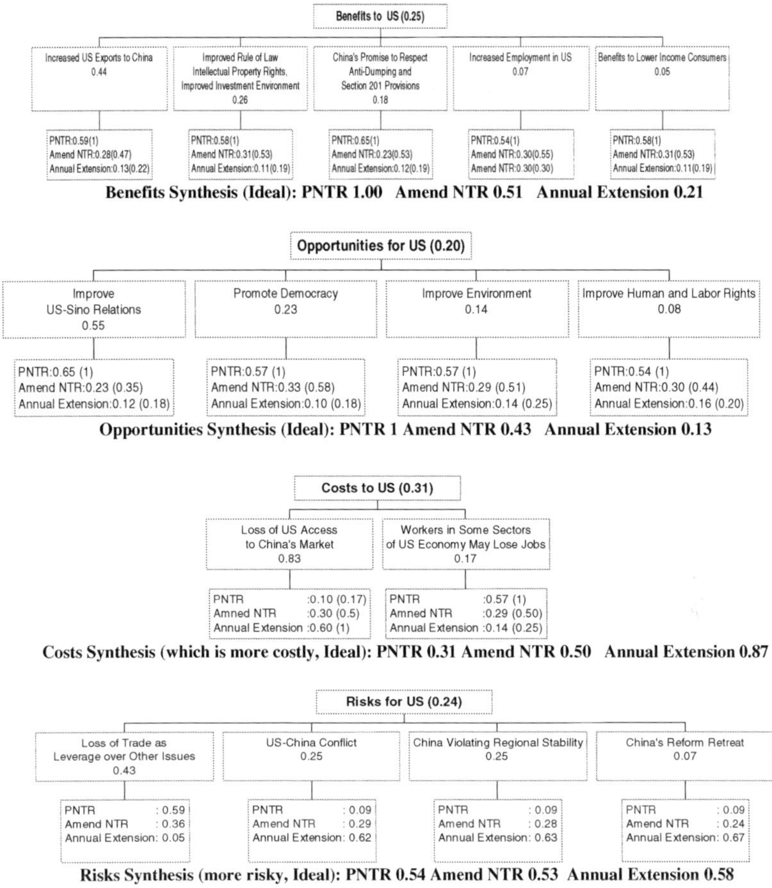

Fig. 4.6 Hierarchies for rating benefits, costs, opportunities, and risks

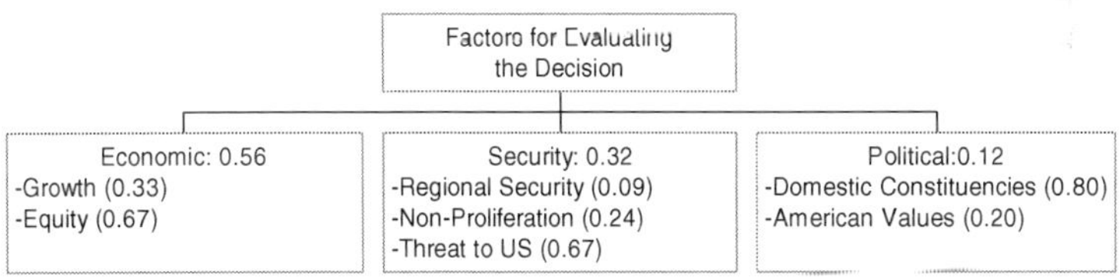

Fig. 4.7 Prioritizing the strategic criteria to be used in rating the BOCR

Table 4.19 Priority ratings for the merits: Benefits, costs, opportunities, and risks

		Benefits	Opportunities	Costs	Risks
Economic	Growth (0.19)	High	Medium	Very low	Very low
(0.56)	Equity (0.37)	Medium	Low	High	Low
	Regional (0.03)	Low	Medium	Medium	High
Security	Non-proliferation (0.08)	Medium	High	Medium	High
(0.32)	Threat to US (0.21)	High	High	Very high	Very high
Political	Constituencies (0.1)	High	Medium	Very high	High
(0.12)	American values (0.02)	Very low	Low	Low	Medium
Priorities		0.25	0.2	0.31	0.24

Intensities: Very high (0.42), High (0.26), Medium (0.16), Low (0.1), Very low (0.06)

synthesis for the ideal mode. We see in bold that PNTR is the dominant alternative any way we synthesize as in the last four columns.

Table 4.20 Two methods of synthesizing BOCR using the ideal mode

Alternatives	Benefits (0.25)	Opportunities (0.2)	Costs (0.31)	Risks (0.24)	BO/CR	bB+oO−cC− rR
PNTR	1.00	1.0	0.31	0.54	4.00	0.22
Amend NTR	0.51	0.43	0.50	0.53	0.51	−0.07
Annual Exten.	0.21	0.13	0.87	0.58	0.05	−0.31

In general one can weight the B, O, C and R values by the corresponding b,o,c and r in the BO/CR formula to determine if it is advantageous to implement the best alternative with a value greater or less than one.

We have laid the basic foundation with hierarchies for what we need to deal with networks involving interdependencies. Let us now turn to that subject.

4.8 Networks, Dependence and Feedback [2, 5]

In Figure 4.8, we exhibit a hierarchy and a network. A hierarchy is comprised of a goal, levels of elements and connections between the elements. These connections are oriented only to elements in lower levels. A hierarchy is authoritarian. It passes the word down from higher up. It describes our commitments, what is important to us and what we prefer even if we imagine it all. A hierarchy is a special case of a network. In a hierarchy connections go only in one direction. In the view of a hierarchy such as that shown in Figure 4.8 the levels correspond to clusters in a network. A network has clusters of elements, with the elements in one cluster being connected to elements in another cluster (outer dependence) or the same cluster (inner dependence). A network is concerned with all the influences from people and from nature that can affect an outcome. It is a model of continual change because everything affects everything else and what we do now can change the importance of the criteria that control the evolution of the outcome.

There are two kinds of influence: outer and inner. In the first one compares the influence of elements in a cluster on elements in another cluster with respect to a control criterion. In inner influence one compares the influence of elements in a group on each one. For example if one takes a family of father mother and child, and then take them one at a time say the child first, one asks who contributes more to the child's survival, its father or its mother, itself or its father, itself or its mother. In this case the child is not so important in contributing to its survival as its parents are. But if we take the mother and ask the same question on who contributes to her survival more, herself or her husband, herself would be higher, or herself and the child, again herself. Another example of inner dependence is making electricity. To make electricity you need steel to make turbines, and you need fuel. So we have the electric industry, the steel industry and the fuel industry. What does the electric industry depend on more to make electricity, itself or the steel industry? Steel is more important; itself or the fuel industry? The fuel industry is much more important; the steel or fuel industry? Fuel is more important. The electric industry does not need its own electricity to make electricity. It needs fuel. Its electricity is only used to light the rooms, which it may not even need.

If we think about it carefully everything can be seen to influence everything else including itself according to many criteria. The world is more interdependent than we know how to deal with using our ways of thinking and taking action. The ANP is our logical way to deal with dependence.

The priorities derived from pairwise comparison matrices are entered as parts of the columns of a supermatrix. The supermatrix represents the influence priority of an element on the left of the matrix on an element at the top of the matrix with respect to a particular control criterion. A supermatrix along with an example of one of its general entry matrices is shown in Figure 4.9. The component C_1 in the supermatrix includes all the priority vectors derived for nodes that are "parent" nodes in the C_1 cluster. Figure 4.10 gives the supermatrix of a hierarchy and Figure 4.11 shows the kth power of that supermatrix which is the same as hierarchic composition in the $(k,1)$ position.

The $(n,1)$ entry of the limit supermatrix of a hierarchy as shown in Figure 4.11 above gives the hierarchic composition principle.

In the ANP we look for steady state priorities from a limit super matrix. To obtain the limit we must raise the matrix to powers. Each power of the matrix captures all transitivities of an order that is equal to that power. The limit of these powers, according to Cesaro Summability, is equal to the limit of the average sum of all the powers of the matrix. All order transitivities are captured by this series of powers of the matrix. The outcome of the ANP is nonlinear and rather complex. The limit may not converge unless the matrix is column stochastic that is each of its columns sums to one. If the columns sum to one then from the fact that the principal eigenvalue of a matrix lies between its largest and smallest column sums, we know that the principal eigenvalue of a stochastic matrix is equal to one.

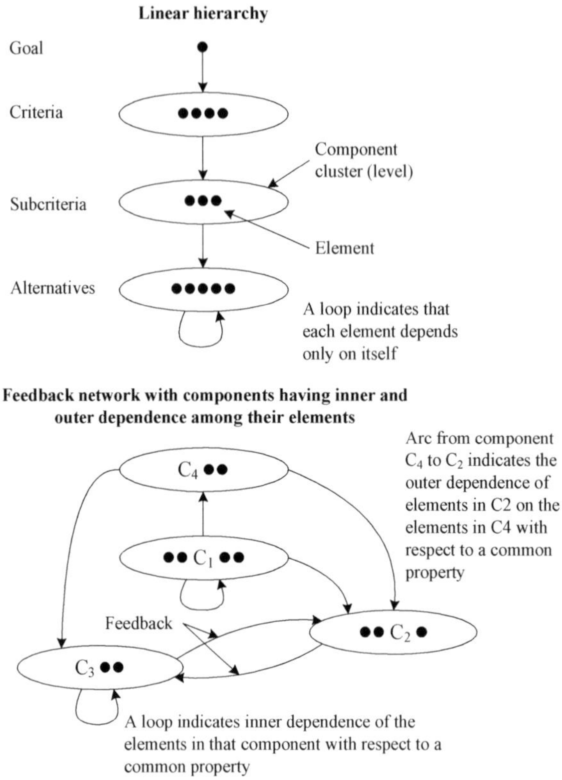

Fig. 4.8 How a hierarchy compares to a network

But for the supermatrix we already know that $\lambda_{max}(T) = 1$ which follows from:

$$\max \sum_{j=1}^{n} a_{ij} \geq \sum_{j=1}^{n} a_{ij} \frac{w_j}{w_i} = \lambda_{\max} \quad \text{for max } w_i$$

$$\min \sum_{j=1}^{n} a_{ij} \leq \sum_{j=1}^{n} a_{ij} \frac{w_j}{w_i} = \lambda_{\max} \quad \text{for min } w_i$$

Thus for a row stochastic matrix we have

$$1 = \min \sum_{j=1}^{n} a_{ij} \leq \lambda_{\max} \leq \max \sum_{j=1}^{n} a_{ij} = 1$$

The same type of argument applies to a matrix that is column stochastic.

Now we know, for example, from a theorem due to J.J. Sylvester [2] that when the eigenvalues of a matrix W are distinct that an entire function $f(x)$ (power series expansion of $f(x)$ converges for all finite values of x) with x replaced by W, is given

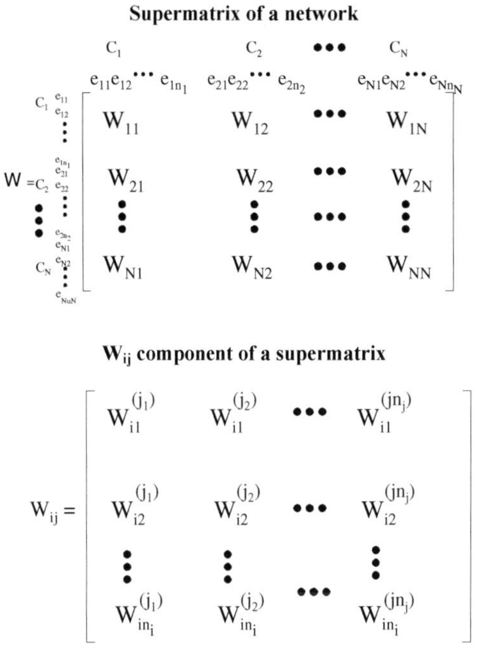

Fig. 4.9 The supermatrix of a network and detail of a component in it

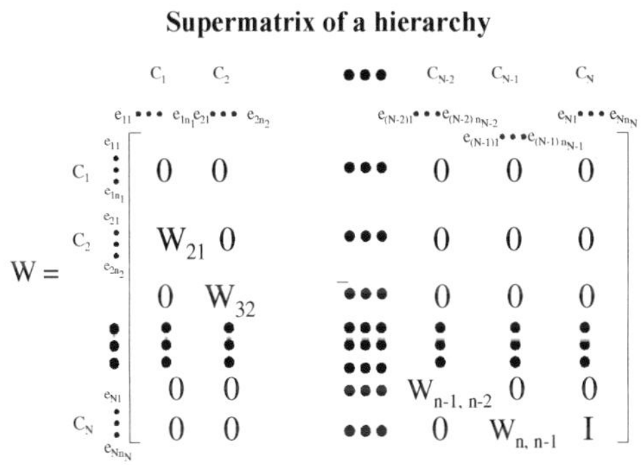

Fig. 4.10 The supermatrix of hierarchy

$$W^k = \begin{bmatrix} 0 & 0 & \cdots & 0 & 0 & 0 \\ 0 & 0 & \cdots & 0 & 0 & 0 \\ \vdots & \vdots & \vdots & \vdots & \vdots & \vdots \\ 0 & 0 & \cdots & 0 & 0 & 0 \\ W_{n,n-1}W_{n-1,n-2}\cdots W_{32}W_{21} & W_{n,n-1}W_{n-1,n-2}\cdots W_{32} & \cdots W_{n,n-1}W_{n-1,n-2} & W_{n,n-1} & I \end{bmatrix}$$

Fig. 4.11 The limit supermatrix of a hierarchy (corresponds to hierarchical composition)

by

$$f(W) = \sum_{i=1}^{n} f(\lambda_i) Z(\lambda_i)$$

$$Z(\lambda_i) = \frac{\prod_{j \neq i} (\lambda_j I - A)}{\prod_{j \neq i} (\lambda_j - \lambda_i)}$$

$$\sum_{i=1}^{n} Z(\lambda_i) = I$$

$$Z(\lambda_i) Z(\lambda_j) = 0$$

$$Z^2(\lambda_i) = Z(\lambda_i)$$

where I and 0 are the identity and the null matrices respectively.

A similar expression is also available when some or all of the eigenvalues have multiplicities. We can see that if, as we need in our case, $f(W) = W^k$, then $f(\lambda_i) = \lambda_i^k$ and as $k \to \infty$ the only terms that give a finite nonzero value are those for which the modulus of λ_i is equal to one. The fact that W is stochastic ensures this because its largest eigenvalue is equal to one. The priorities of the alternatives (or any set of elements in a component) are obtained by normalizing the corresponding values in the appropriate columns of the limit matrix. When W has zeros and is reducible (its graph is not strongly connected so there is no path from some point to some other point) the limit can cycle and a Cesaro average over the different limits of the cycle is taken. For a more complete treatment, see the book by Saaty on the ANP.

4.8.1 ANP Formulation of the Classic AHP School Example

We show in Figure 4.12 below the hierarchy for choosing a best school, and in the corresponding supermatrix, and its limit supermatrix in Figure 4.13 the priorities of three schools involved in a decision to choose the best one. They are precisely what one obtains by hierarchic composition using the AHP. The priorities of the criteria with respect to the goal and those of the alternatives with respect to each criterion are clearly discernible in the supermatrix itself. Note that there is an identity submatrix for the alternatives with respect to the alternatives in the lower right hand part of the matrix. The level of alternatives in a hierarchy is a sink cluster of nodes that absorbs priorities but does not pass them on. This calls for using an identity submatrix for them in the supermatrix.

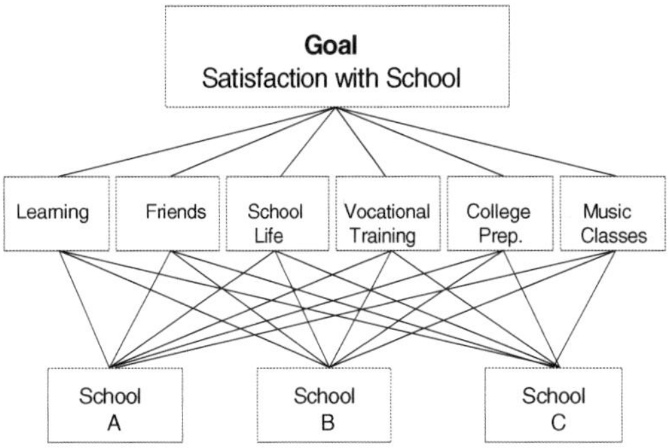

Fig. 4.12 The school choice hierarcy

The school hierarchy as a supermatrix

	Goal	Learning	Friends	School life	Vocational training	College preparation	Music classes	A	B	C
Goal	0	0	0	0	0	0	0	0	0	0
Learning	0.32	0	0	0	0	0	0	0	0	0
Friends	0.14	0	0	0	0	0	0	0	0	0
School life	0.03	0	0	0	0	0	0	0	0	0
Vocational training	0.13	0	0	0	0	0	0	0	0	0
College preparation	0.24	0	0	0	0	0	0	0	0	0
Music classes	0.14	0	0	0	0	0	0	0	0	0
Alternative A	0	0.16	0.33	0.45	0.77	0.25	0.69	1	0	0
Alternative B	0	0.59	0.33	0.09	0.06	0.5	0.09	0	1	0
Alternative C	0	0.25	0.34	0.46	0.17	0.25	0.22	0	0	1

Limiting supermatrix & hierarchic composition

	Goal	Learning	Friends	School life	Vocational training	College preparation	Music classes	A	B	C
Goal	0	0	0	0	0	0	0	0	0	0
Learning	0	0	0	0	0	0	0	0	0	0
Friends	0	0	0	0	0	0	0	0	0	0
School life	0	0	0	0	0	0	0	0	0	0
Vocational training	0	0	0	0	0	0	0	0	0	0
College preparation	0	0	0	0	0	0	0	0	0	0
Music classes	0	0	0	0	0	0	0	0	0	0
Alternative A	0.3676	0.16	0.33	0.45	0.77	0.25	0.69	1	0	0
Alternative B	0.3781	0.59	0.33	0.09	0.06	0.5	0.09	0	1	0
Alternative C	0.2543	0.25	0.34	0.46	0.17	0.25	0.22	0	0	1

Fig. 4.13 The limit supermatrix of the school choice hierarchy shows same result as hierarchic composition

4.9 Market Share Examples Mainly to Justify with Existing Measurements Subjective Judgment that does not Refer to any Numerical Data

4.9.1 An ANP Network with a Single Control Criterion - Market Share

A market share estimation model is structured as a network of clusters and nodes. The object is to try to determine the relative market share of competitors in a particular business, or endeavor, by considering what affects market share in that business and introduce them as clusters, nodes and influence links in a network. The decision alternatives are the competitors and the synthesized results are their relative dominance. The relative dominance results can then be compared against some outside measure such as dollars. If dollar income is the measure being used, the incomes of the competitors must be normalized to get it in terms of relative market share.

The clusters might include customers, service, economics, advertising, and the quality of goods. The customers' cluster might then include nodes for the age groups of the people that buy from the business: teenagers, 20–33 year olds, 34–55 year olds, 55–70 year olds, and over 70. The advertising cluster might include newspapers, TV, Radio, and Fliers. After all the nodes are created one starts by picking a node and linking it to the other nodes in the model that influence it. The "children" nodes will then be pairwise compared with respect to that node as a "parent" node. An arrow will automatically appear going from the cluster the parent node cluster to the cluster with its children nodes. When a node is linked to nodes in its own cluster, the arrow becomes a loop on that cluster and we say there is inner dependence.

The linked nodes in a given cluster are pairwise compared for their influence on the node they are linked from (the parent node) to determine the priority of their influence on the parent node. Comparisons are made as to which is more important to the parent node in capturing "market share". These priorities are then entered in the supermatrix for the network.

The clusters are also pairwise compared to establish their importance with respect to each cluster they are linked from, and the resulting matrix of numbers is used to weight the corresponding blocks of the original unweighted supermatrix to obtain the weighted supermatrix. This matrix is then raised to powers until it converges to yield the limit supermatrix. The relative values for the companies are obtained from the columns of the limit supermatrix that in this case are all the same because the matrix is irreducible. Normalizing these numbers yields the relative market share.

If comparison data in terms of sales in dollars, or number of members, or some other known measures are available, one can use these relative values to validate the outcome. The AHP/ANP has a compatibility index to determine how close the ANP result is to the known measure. It involves taking the Hadamard product of the matrix of ratios of the ANP outcome and the transpose of the matrix of ratios of the actual outcome summing all the coefficients and dividing by n^2. The requirement is that the value should be close to 1.

4.9.2 Compatibility Index

Let us show first that the priority vector $w = (w_1, \ldots, w_n)$ is completely compatible with itself. Thus we form the matrix of all possible ratios $W = (w_{ij}) = (w_i/w_j)$ from this vector. This matrix is reciprocal, that is $w_{ji} = 1/w_{ij}$. The Hadamard product of a reciprocal matrix W and its transpose W^T is given by:

$$W \circ W^T = \begin{pmatrix} \frac{w_1}{w_1} & \cdots & \frac{w_1}{w_n} \\ \vdots & & \vdots \\ \frac{w_n}{w_1} & \cdots & \frac{w_n}{w_n} \end{pmatrix} \circ \begin{pmatrix} \frac{w_1}{w_1} & \cdots & \frac{w_1}{w_n} \\ \vdots & & \vdots \\ \frac{w_n}{w_1} & \cdots & \frac{w_n}{w_n} \end{pmatrix} = \begin{pmatrix} 1 & \cdots & 1 \\ \vdots & & \vdots \\ 1 & \cdots & 1 \end{pmatrix} = \begin{pmatrix} 1 \\ \vdots \\ 1 \end{pmatrix} (1 \ldots 1) \equiv ee^{\top}$$

The sum of the elements of a matrix A can be written as $e^\top Ae$. In particular we have $e^\top A \circ A^\top e = n^2$ for the sum of the elements of the Hadamard product of a matrix and its transpose. The index of compatibility is the sum resulting from the Hadamard product divided by n^2. Thus a vector is completely compatible with itself as $n^2/n^2 = 1$. Now we have an idea of how to define a measure of compatibility for two matrices A and B. It is given by $\frac{1}{n^2}e^\top A \circ B^\top e$. Note that a reciprocal matrix of judgments that is inconsistent is not itself a matrix of ratios from a given vector. However, such a matrix has a principal eigenvector and thus we speak of the compatibility of the matrix of judgments and the matrix formed from ratios of the principal eigenvector. We have the following theorem for a reciprocal matrix of judgments and the matrix W of the ratios of its principal eigenvector:

Theorem 4.1.

$$\frac{1}{n^2}e^\top A \circ W^\top e = \frac{\lambda_{max}}{n}$$

Proof. From $Aw = \lambda_{max}w$ we have

$$\sum_{j=1}^{n} a_{ij}w_j = \lambda_{max}w_i$$

and

$$\frac{1}{n^2}e^\top A \circ W^\top e = \frac{1}{n^2}\sum_{i=1}^{n}\sum_{j=1}^{n} a_{ij}\frac{w_j}{w_i} = \frac{\lambda_{max}}{n}$$

We want this ratio to be close to one or in general not much more than 1.01 and be less than this value for small size matrices. It is in accord with the idea that a 10% deviation is at the upper end of acceptability.

We will give two examples of market share estimation showing details of the process in the first example and showing only the models and results in the second example.

4.9.3 Example 1: Estimating the Relative Market Share of Walmart, Kmart and Target

The network for the ANP model shown in Figure 4.14 describes quite well the influences that determine the market share of these companies. We will not use space in this chapter to describe the clusters and their nodes in greater detail.

4.9.3.1 The Unweighted Supermatrix

The unweighted supermatrix is constructed from the priorities derived from the different pairwise comparisons. The nodes, grouped by the clusters they belong to, are the labels of the rows and columns of the supermatrix. The column for a node a

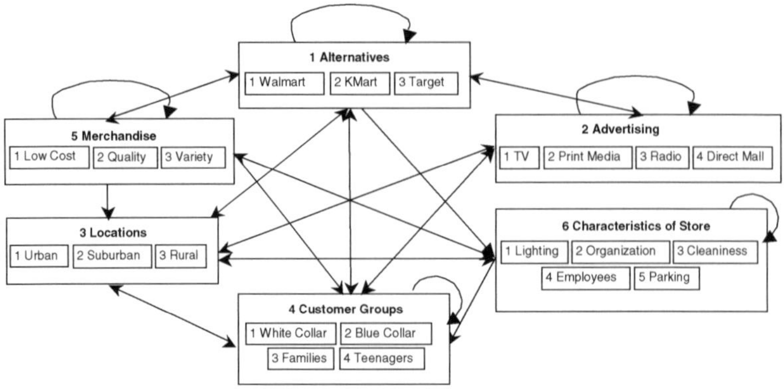

Fig. 4.14 The clusters and nodes of a model to estimate the relative market share of Walmart, Kmart and Target

contains the priorities of the nodes that have been pairwise compared with respect to a. The supermatrix for the network in Figure 4.14 is shown in Table 4.21.

4.9.3.2 The Cluster Matrix

The cluster themselves must be compared to establish their relative importance and use it to weight the supermatrix to make it column stochastic. A cluster impacts another cluster when it is linked from it, that is, when at least one node in the source cluster is linked to nodes in the target cluster. The clusters linked from the source cluster are pairwise compared for the importance of their impact on it with respect to market share, resulting in the column of priorities for that cluster in the cluster matrix. The process is repeated for each cluster in the network to obtain the matrix shown in Table 4.22. An interpretation of the priorities in the first column is that Merchandise (0.442) and Locations (0.276) have the most impact on Alternatives, the three competitors.

4.9.3.3 The Weighted Supermatrix

The weighted supermatrix shown in Table 4.23 is obtained by multiplying each entry in a block of the component at the top of the supermatrix by the priority of influence of the component on the left from the cluster matrix in Table 4.22. For example, the first entry, 0.137, in Table 4.22 is used to multiply each of the nine entries in the block (Alternatives, Alternatives) in the unweighted supermatrix shown in Table 4.21. This gives the entries for the (Alternatives, Alternatives) component in the weighted supermatrix of Table 4.23. Each column in the weighted supermatrix has a sum of 1, and thus the matrix is stochastic and thus converges or is periodic.

Table 4.21 The unweighted supermatrix

		1 Alternatives			2 Advertising				3 Locations			4 Customer groups				5 Merchandise			6 Characteristics of store				
		Alt1	Alt2	Alt3	Adv1	Adv2	Adv3	Adv4	Loc1	Loc2	Loc3	Cg1	Cg2	Cg3	Cg4	Mer1	Mer2	Mer3	Ch1	Ch2	Ch3	Ch4	Ch5
1 Alternatives	Alt1	0.000	0.833	0.833	0.687	0.540	0.634	0.661	0.614	0.652	0.683	0.637	0.661	0.630	0.691	0.661	0.614	0.648	0.667	0.655	0.570	0.644	0.558
	Alt2	0.750	0.000	0.167	0.186	0.297	0.174	0.208	0.268	0.235	0.200	0.105	0.208	0.218	0.149	0.208	0.117	0.122	0.111	0.095	0.097	0.085	0.122
	Alt3	0.250	0.167	0.000	0.127	0.163	0.192	0.131	0.117	0.113	0.117	0.258	0.131	0.151	0.160	0.131	0.268	0.230	0.222	0.250	0.333	0.271	0.320
2 Advertising	Adv1	0.553	0.176	0.188	0.000	0.000	0.000	0.000	0.288	0.543	0.558	0.323	0.510	0.508	0.634	0.000	0.000	0.000	0.000	0.000	0.000	0.000	0.000
	Adv2	0.202	0.349	0.428	0.750	0.000	0.800	0.000	0.381	0.231	0.175	0.214	0.221	0.270	0.170	0.000	0.000	0.000	0.000	0.000	0.000	0.000	0.000
	Adv3	0.062	0.056	0.055	0.000	0.000	0.000	0.000	0.059	0.053	0.048	0.059	0.063	0.049	0.096	0.000	0.000	0.000	0.000	0.000	0.000	0.000	0.000
	Adv4	0.183	0.420	0.330	0.250	0.000	0.200	0.000	0.273	0.173	0.219	0.404	0.206	0.173	0.100	0.000	0.000	0.000	0.000	0.000	0.000	0.000	0.000
3 Locations	Loc1	0.114	0.084	0.086	0.443	0.126	0.080	0.099	0.000	0.000	0.000	0.167	0.094	0.096	0.109	0.268	0.105	0.094	0.100	0.091	0.091	0.111	0.067
	Loc2	0.405	0.444	0.628	0.387	0.416	0.609	0.537	0.000	0.000	0.000	0.833	0.280	0.308	0.309	0.117	0.605	0.627	0.433	0.455	0.455	0.444	0.293
	Loc3	0.481	0.472	0.285	0.169	0.458	0.311	0.364	0.000	0.000	0.000	0.000	0.627	0.596	0.582	0.614	0.291	0.280	0.466	0.455	0.455	0.444	0.641
4 Customer groups	Cg1	0.141	0.114	0.208	0.165	0.155	0.116	0.120	0.078	0.198	0.092	0.000	0.000	0.279	0.085	0.051	0.222	0.165	0.383	0.187	0.242	0.165	0.000
	Cg2	0.217	0.214	0.117	0.165	0.155	0.198	0.203	0.223	0.116	0.224	0.000	0.000	0.649	0.177	0.112	0.159	0.165	0.383	0.187	0.208	0.165	0.000
	Cg3	0.579	0.623	0.620	0.621	0.646	0.641	0.635	0.656	0.641	0.645	0.857	0.857	0.000	0.737	0.618	0.566	0.621	0.185	0.583	0.494	0.621	0.000
	Cg4	0.063	0.049	0.055	0.048	0.043	0.045	0.041	0.043	0.045	0.038	0.143	0.143	0.072	0.000	0.219	0.053	0.048	0.048	0.043	0.056	0.048	0.000
5 Merchandise	Mer1	0.362	0.333	0.168	0.000	0.000	0.000	0.000	0.000	0.000	0.000	0.000	0.000	0.000	0.000	0.000	0.800	0.800	0.000	0.000	0.000	0.000	0.000
	Mer2	0.261	0.140	0.484	0.000	0.000	0.000	0.000	0.000	0.000	0.000	0.000	0.000	0.000	0.000	0.750	0.000	0.200	0.000	0.000	0.000	0.000	0.000
	Mer3	0.377	0.528	0.349	0.000	0.000	0.000	0.000	0.000	0.000	0.000	0.000	0.000	0.000	0.000	0.250	0.200	0.000	0.000	1.000	0.000	0.000	0.000
6 Characteristics	Ch1	0.000	0.000	0.000	0.000	0.000	0.000	0.000	0.000	0.000	0.000	0.000	0.000	0.000	0.000	0.000	0.000	0.000	0.000	0.169	0.121	0.000	0.250
	Ch2	0.000	0.000	0.000	0.000	0.000	0.000	0.000	0.000	0.000	0.000	0.000	0.000	0.000	0.000	0.000	0.000	0.000	0.251	0.000	0.575	0.200	0.750
	Ch3	0.000	0.000	0.000	0.000	0.000	0.000	0.000	0.000	0.000	0.000	0.000	0.000	0.000	0.000	0.000	0.000	0.000	0.673	0.469	0.000	0.800	0.000
	Ch4	0.000	0.000	0.000	0.000	0.000	0.000	0.000	0.000	0.000	0.000	0.000	0.000	0.000	0.000	0.000	0.000	0.000	0.000	0.308	0.304	0.000	0.000
	Ch5	0.000	0.000	0.000	0.000	0.000	0.000	0.000	0.000	0.000	0.000	0.000	0.000	0.000	0.000	0.000	0.000	0.000	0.075	0.055	0.000	0.000	0.000

Alt1: Walmart, Alt2: KMart, Alt3: Target, Adv1: TV, Adv2: Print media, Adv3: Radio, Adv4: Direct mail, Loc1: Urban, Loc2: Suburban, Loc3: Rural, Cg1: White collar, Cg2: Blue collar, Cg3: Families, Cg4: Teenagers, Mer1: Low cost, Mer2: Quality, Mer3: Variety, Ch1: Lighting, Ch2: Organization, Ch3: Clean, Ch4: Employees, Ch5: Parking

Table 4.22 The cluster matrix

	1 Alts.	2 Advert.	3 Loc.	4 Cust. grps.	5 Merchand.	6 Charact.
1 Alternatives	0.137	0.174	0.094	0.057	0.049	0.037
2 Advertising	0.091	0.220	0.280	0.234	0.000	0.000
3 Locations	0.276	0.176	0.000	0.169	0.102	0.112
4 Customer groups	0.054	0.429	0.627	0.540	0.252	0.441
5 Merchandise	0.442	0.000	0.000	0.000	0.596	0.316
6 Characteristics of store	0.000	0.000	0.000	0.000	0.000	0.094

The limit supermatrix is not shown here to save space. It is obtained from the weighted supermatrix by raising it to powers until it converges so that all columns are identical. From the top part of the first column of the limit supermatrix we get the priorities we seek and normalize. These results are compared with the actual values shown in Table 4.24.

4.9.3.4 Synthesized Results from the Limit Supermatrix

The relative market shares of the alternatives Walmart, Kmart and Target from the limit supermatrix are: 0.057, 0.024 and 0.015. When normalized they are 0.599, 0.248 and 0.154.

The relative market share values obtained from the model were compared with the actual sales values by computing the compatibility index. The compatibility Index enables is used to determine how close two sets of numbers from a ratio scale or an absolute scale are to each other. In this example the result is equal to 1.016 and falls below 1.1 and therefore is an acceptable outcome.

4.9.3.5 Actual Relative Market Share Based on Sales

The object was to estimate the market share of Walmart, Kmart, and Target. The normalized results from the model were compared with sales shown in Table 4.11 as reported in the Discount Store News of July 13, 1998, p.77, of 58, 27.5 and 20.3 billions of dollars respectively. Normalizing the dollar amounts shows their actual relative market shares to be 54.8, 25.9 and 19.2. The relative market share from the model was compared with the sales values by constructing a pairwise matrix from the results vector in column 1 below and a pairwise matrix from results vector in column 3 and computing the compatibility index using the Hadamard multiplication method. The index is equal to 1.016. As that is about 1.01 the ANP results may be said to be close to the actual relative market share.

Table 4.23 The weighted supermatrix

		1 Alternatives			2 Advertising				3 Locations			4 Customer groups				5 Merchandise			6 Characteristics of store				
		Alt1	Alt2	Alt3	Adv1	Adv2	Adv3	Adv4	Loc1	Loc2	Loc3	Cg1	Cg2	Cg3	Cg4	Mer1	Mer2	Mer3	Ch1	Ch2	Ch3	Ch4	Ch5
1 Alternatives	Alt1	0.000	0.114	0.114	0.120	0.121	0.110	0.148	0.058	0.061	0.064	0.036	0.038	0.036	0.040	0.033	0.030	0.032	0.036	0.024	0.031	0.035	0.086
	Alt2	0.103	0.000	0.023	0.033	0.066	0.030	0.047	0.025	0.022	0.019	0.006	0.012	0.012	0.009	0.010	0.006	0.006	0.006	0.004	0.005	0.005	0.019
	Alt3	0.034	0.023	0.000	0.022	0.037	0.033	0.029	0.011	0.011	0.011	0.015	0.007	0.009	0.009	0.006	0.013	0.011	0.012	0.009	0.018	0.015	0.049
2 Advertising	Adv1	0.050	0.016	0.017	0.000	0.000	0.000	0.000	0.080	0.152	0.156	0.076	0.119	0.119	0.148	0.000	0.000	0.000	0.000	0.000	0.000	0.000	0.000
	Adv2	0.018	0.032	0.039	0.165	0.000	0.176	0.000	0.106	0.064	0.049	0.050	0.052	0.063	0.040	0.000	0.000	0.000	0.000	0.000	0.000	0.000	0.000
	Adv3	0.006	0.005	0.005	0.000	0.000	0.000	0.000	0.016	0.015	0.014	0.014	0.015	0.012	0.023	0.000	0.000	0.000	0.000	0.000	0.000	0.000	0.000
	Adv4	0.017	0.038	0.030	0.055	0.000	0.044	0.000	0.076	0.048	0.061	0.095	0.048	0.040	0.023	0.000	0.000	0.000	0.000	0.000	0.000	0.000	0.000
3 Locations	Loc1	0.031	0.023	0.024	0.078	0.028	0.014	0.022	0.000	0.000	0.000	0.028	0.016	0.016	0.018	0.027	0.011	0.010	0.016	0.010	0.015	0.018	0.031
	Loc2	0.112	0.123	0.174	0.068	0.094	0.107	0.121	0.000	0.000	0.000	0.141	0.047	0.052	0.052	0.012	0.062	0.064	0.071	0.051	0.074	0.073	0.135
	Loc3	0.133	0.130	0.079	0.030	0.103	0.055	0.082	0.000	0.000	0.000	0.000	0.106	0.101	0.098	0.063	0.030	0.029	0.076	0.051	0.074	0.073	0.295
4 Customer groups	Cg1	0.008	0.006	0.011	0.071	0.086	0.050	0.066	0.049	0.124	0.058	0.000	0.000	0.151	0.046	0.013	0.056	0.042	0.247	0.082	0.156	0.107	0.000
	Cg2	0.012	0.011	0.006	0.071	0.086	0.085	0.112	0.140	0.073	0.141	0.000	0.000	0.350	0.096	0.028	0.040	0.042	0.247	0.082	0.134	0.107	0.000
	Cg3	0.031	0.033	0.033	0.267	0.356	0.275	0.350	0.411	0.402	0.404	0.463	0.463	0.000	0.398	0.156	0.143	0.157	0.119	0.257	0.318	0.400	0.000
	Cg4	0.003	0.003	0.003	0.021	0.024	0.019	0.023	0.027	0.028	0.024	0.077	0.077	0.039	0.000	0.055	0.013	0.012	0.031	0.019	0.036	0.031	0.000
5 Merchandise	Mer1	0.160	0.147	0.074	0.000	0.000	0.000	0.000	0.000	0.000	0.000	0.000	0.000	0.000	0.000	0.000	0.477	0.477	0.000	0.000	0.000	0.000	0.000
	Mer2	0.115	0.062	0.214	0.000	0.000	0.000	0.000	0.000	0.000	0.000	0.000	0.000	0.000	0.000	0.447	0.000	0.119	0.000	0.000	0.000	0.000	0.000
	Mer3	0.166	0.233	0.154	0.000	0.000	0.000	0.000	0.000	0.000	0.000	0.000	0.000	0.000	0.000	0.149	0.119	0.000	0.000	0.316	0.000	0.000	0.000
6Characteristics	Ch1	0.000	0.000	0.000	0.000	0.000	0.000	0.000	0.000	0.000	0.000	0.000	0.000	0.000	0.000	0.000	0.000	0.000	0.000	0.016	0.017	0.000	0.097
	Ch2	0.000	0.000	0.000	0.000	0.000	0.000	0.000	0.000	0.000	0.000	0.000	0.000	0.000	0.000	0.000	0.000	0.000	0.035	0.000	0.079	0.027	0.290
	Ch3	0.000	0.000	0.000	0.000	0.000	0.000	0.000	0.000	0.000	0.000	0.000	0.000	0.000	0.000	0.000	0.000	0.000	0.092	0.044	0.000	0.110	0.000
	Ch4	0.000	0.000	0.000	0.000	0.000	0.000	0.000	0.000	0.000	0.000	0.000	0.000	0.000	0.000	0.000	0.000	0.000	0.000	0.029	0.042	0.000	0.000
	Ch5	0.000	0.000	0.000	0.000	0.000	0.000	0.000	0.000	0.000	0.000	0.000	0.000	0.000	0.000	0.000	0.000	0.000	0.010	0.005	0.000	0.000	0.000

Alt1: Walmart, Alt2: KMart, Alt3: Target, Adv1: TV, Adv2: Print media, Adv3: Radio, Adv4: Direct mail, Loc1: Urban, Loc2: Suburban, Loc3: Rural, Cg1: White collar, Cg2: Blue collar, Cg3: Families, Cg4: Teenagers, Mer1: Low cost, Mer2: Quality, Mer3: Variety, Ch1: Lighting, Ch2: Organization, Ch3: Clean, Ch4: Employees, Ch5: Parking

Table 4.24 The synthesized results for the alternatives

Alternatives	Values from limit supermatrix	Actual values 13-Jul-98	Normalized values from supermatrix	Actual market share as dollar sales normalized
Walmart	0.057	58 billion $	0.599	54.8
KMart	0.024	27.5 billion $	0.248	25.9
Target	0.015	20.3 billion $	0.254	19.2

4.9.4 Example 2: Estimating Relative Market Share of Airlines (close outcome)

An ANP model to estimate the relative market share of eight American Airlines is shown in Figure 4.15. The results from the model and the comparison with the relative actual market share are shown in Table 4.25.

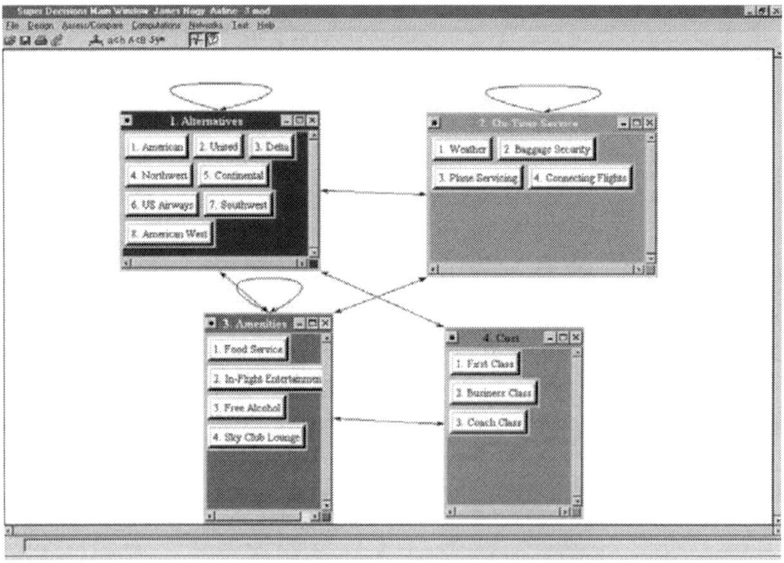

Fig. 4.15 ANP network to estimate relative market share of eight US airlines

We summarize by giving the reader a list of the steps we have followed in applying the ANP.

Table 4.25 Comparing model results with actual market share data

	Model results	Actual market share (yr 2000)
American	23.9	24.0
United	18.7	19.7
Delta	18.0	18.0
Northwest	11.4	12.4
Continental	9.3	10.0
US Airways	7.5	7.1
Southwest	5.9	6.4
American West	4.4	2.9

Compatibility Index: 1.0247

4.10 Outline of Steps of the ANP

1. Describe the decision problem in detail including its objectives, criteria and subcriteria, actors and their objectives and the possible outcomes of that decision. Give details of influences that determine how that decision may come out.
2. Determine the control criteria and subcriteria in the four control hierarchies one each for the benefits, opportunities, costs and risks of that decision and obtain their priorities from paired comparisons matrices. If a control criterion or subcriterion has a global priority of 3 or less, you may consider carefully eliminating it from further consideration. The software automatically deals only with those criteria or subcriteria that have subnets under them. For benefits and opportunities, ask what gives the most benefits or presents the greatest opportunity to influence fulfillment of that control criterion. For costs and risks, ask what incurs the most cost or faces the greatest risk. Sometimes (very rarely), the comparisons are made simply in terms of benefits, opportunities, costs, and risks in the aggregate without using control criteria and subcriteria.
3. Determine the most general network of clusters (or components) and their elements that apply to all the control criteria. To better organize the development of the model as well as you can, number and arrange the clusters and their elements in a convenient way (perhaps in a column). Use the identical label to represent the same cluster and the same elements for all the control criteria.
4. For each control criterion or subcriterion, determine the clusters of the general feedback system with their elements and connect them according to their outer and inner dependence influences. An arrow is drawn from a cluster to any cluster whose elements influence it.
5. Determine the approach you want to follow in the analysis of each cluster or element, influencing (the preferred approach) other clusters and elements with respect to a criterion, or being influenced by other clusters and elements. The sense (being influenced or influencing) must apply to all the criteria for the four control hierarchies for the entire decision.
6. For each control criterion, construct the supermatrix by laying out the clusters in the order they are numbered and all the elements in each cluster both vertically on

the left and horizontally at the top. Enter in the appropriate position the priorities derived from the paired comparisons as subcolumns of the corresponding column of the supermatrix.

7. Perform paired comparisons on the elements within the clusters themselves according to their influence on each element in another cluster they are connected to (outer dependence) or on elements in their own cluster (inner dependence). In making comparisons, you must always have a criterion in mind. Comparisons of elements according to which element influences a given element more and how strongly more than another element it is compared with are made with a control criterion or subcriterion of the control hierarchy in mind.
8. Perform paired comparisons on the clusters as they influence each cluster to which they are connected with respect to the given control criterion. The derived weights are used to weight the elements of the corresponding column blocks of the supermatrix. Assign a zero when there is no influence. Thus obtain the weighted column stochastic supermatrix.
9. Compute the limit priorities of the stochastic supermatrix according to whether it is irreducible (primitive or imprimitive [cyclic]) or it is reducible with one being a simple or a multiple root and whether the system is cyclic or not. Two kinds of outcomes are possible. In the first all the columns of the matrix are identical and each gives the relative priorities of the elements from which the priorities of the elements in each cluster are normalized to one. In the second the limit cycles in blocks and the different limits are summed and averaged and again normalized to one for each cluster. Although the priority vectors are entered in the supermatrix in normalized form, the limit priorities are put in idealized form because the control criteria do not depend on the alternatives.
10. Synthesize the limiting priorities by weighting each idealized limit vector by the weight of its control criterion and adding the resulting vectors for each of the four merits: Benefits (B), Opportunities (O), Costs (C) and Risks (R). There are now four vectors, one for each of the four merits. An answer involving marginal values of the merits is obtained by forming the ratio BO/CR for each alternative from the four vectors. The alternative with the largest ratio is chosen for some decisions. Companies and individuals with limited resources often prefer this type of synthesis.
11. Governments prefer this type of outcome. Determine strategic criteria and their priorities to rate the four merits one at a time. Normalize the four ratings thus obtained and use them to calculate the overall synthesis of the four vectors. For each alternative, subtract the costs and risks from the sum of the benefits and opportunities. At other times one may subtract the costs from one and risks from one and then weight and add them to the weighted benefits and opportunities. This is useful for predicting numerical outcomes like how many people voted for an alternative and how many voted against it. In all, we have three different formulas for synthesis.
12. Perform sensitivity analysis on the final outcome and interpret the results of sensitivity observing how large or small these ratios are. Can another outcome that is close also serve as a best outcome? Why? By noting how stable this outcome

is. Compare it with the other outcomes by taking ratios. Can another outcome that is close also serve as a best outcome? Why?

The next section includes real ANP applications of many different areas from business to public policy. We intentionally included not only simple examples that have a single network such as market share examples but also more complicated decision problems. The second group includes BOCR merit evaluations using strategic criteria, with control criteria (and perhaps subcriteria) under them for each of the BOCR and their related decision networks.

In the light of the above explanations, we cover a rather wide spectrum in the variety of examples. There are examples that do not have BOCR merit evaluation and/or sensitivity analysis while some of them analyze the problem with all the details. In order to start with a clear understanding, we preferred to give the example below that has all the possible analyses from the BOCR merit evaluation to the control criteria and subcriteria, decision networks for each of these and sensitivity analysis with regard to the BOCR.

4.11 A Complete BOCR Example

Disney Decision: A New Theme Park in Greater China (By Amber Ling-Hui, Lin SzuLun Peng)

4.11.1 Introduction / Background

In order to enhance operations in foreign market, Disney is constantly searching for areas where it can expand into new markets. According to the projected number of foreign visitors, Walt Disney World expects to increase the current level from 20 percent foreign visitors in domestic parks to 50 percent as well as to expand its theme park business outside the U.S. To achieve these projected numbers Disney needs to make an aggressive attempt to expand its presence in foreign markets, especially Greater China. However, considering the diverse social and economic backgrounds within this area, Disney needs to carefully evaluate the possible benefits as well as the costs and potential risks. In this model, we narrow down the alternatives to Hong Kong, Shanghai, Taiwan and no investment in Greater China. In fact, an awakening and growing middle class in these three areas is exactly the prime target audience for a Disney theme park.

4.11.2 Ultimate Goal for Disney

Disney's intention is to make a minimal equity investment in any operating entity and generate most of its returns through royalty, licensing, and fee income streams.

4.11.3 Main Model

4.11.3.1 BOCR Networks and Cluster Definitions

Under the benefits, opportunities, costs, and risks (BOCR) models, different clusters define interactions with respect to the control hierarchy established. The benefits networks indicate the alternatives that yield the most benefit and the opportunities networks indicate the alternative that offers the most opportunities, whereas the costs and risks networks indicate the alternatives that are the most costly or pose the most risk on each alternative.

The flow of the decision process is to first build the networks and sub-networks for each of the BOCR models (Figure 4.16), make the judgments and evaluate which is the best alternative in each case for this particular decision. The importance of the BOCR must then be determined by rating them with respect to the strategic criteria of the organization or decision maker.

4.11.3.2 Control Criteria and Subnets of the BOCR

Each of the BOCR has control criteria whose priorities are established through pairwise comparison. The control criteria in turn have associated network sub-models that contain the alternatives of the decision and clusters of elements. Thus priorities for the alternatives are determined in each of the subnets. These are weighted by their control criterion, and these results are multiplied by the BOCR weights from the rating model and combined to give the final results. The alternatives appear in a cluster in every decision subnet, so we define them only once here. There are three locations being considered for the first Disney theme park in Greater China plus the alternative of not building at all.

4.11.3.3 Alternatives (in Every Subnet)

- Don't invest in Greater China
- Hong Kong
- Shanghai
- Taiwan

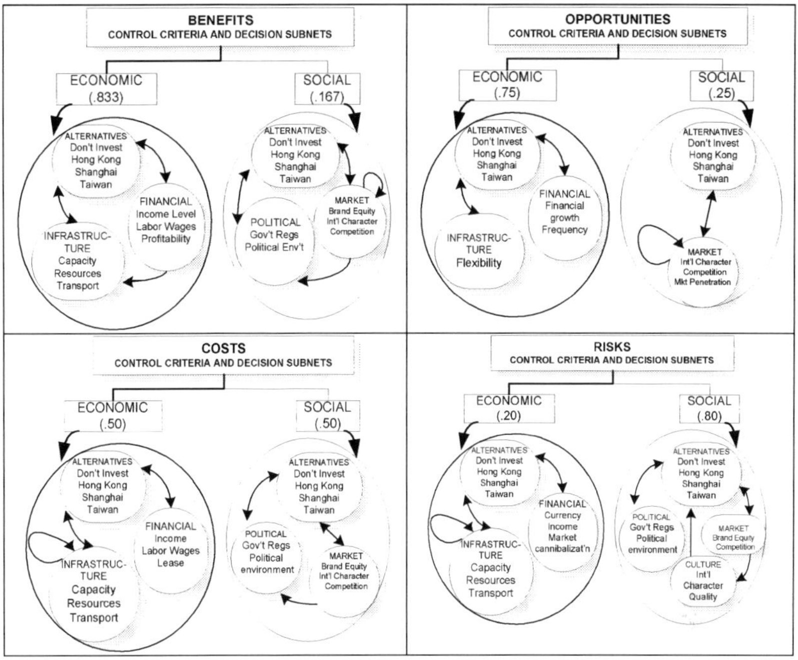

Fig. 4.16 Decision sub-networks with clusters and nodes for each of the BOCR

Moving on to the first subnet, under the Social control criterion for Benefits we show the clusters in that network below:

4.11.3.4 Clusters in Benefits/Social Subnet

- Alternatives
- Market

  Brand Equity: For the brand equity, we consider it as an intangible asset to Walt Disney. Brand equity represents Disney's reputation and image in the market. Within this subnet, we will examine how much benefit each alternative can bring to Disney in terms of increasing their brand equity.

  International Character: International character refers to having a diversified visitor base. The higher the diversification of the visitor base, the more it benefits Disney.

  Market Competition: Market competition refers to the number of competitors with **comparable scale** in one market. Within the benefit cluster, we will discuss the level that Disney can benefit from the competition in the market under each alternative.
- Political Factors

Government Regulation: We believe a favorable local government regulation on the theme park business will definitely benefit Disney's operation in that area and vise versa.

Political Environment: We believe a stable political environment will create a promising investment environment. Thus, the benefits will be measured base on the current political stability and potential political instability of each alternative.

4.11.3.5 Interactions between Clusters in the Benefits/Social subnet

In this subnet, we can see the interactions among clusters as well as interactions within clusters.

Market Factors: First of all, since the government regulations and political environment will affect the international character and the market competition in a market, we can see an interaction between market cluster and political factors cluster. Besides, different choices that Disney makes will affect the company itself in terms of brand equity, international character and competition in the market. Finally, the competitive ability of the company and the international character of the market may also affect Disney's brand equity at the end. Thus, we can see another interaction within the market cluster itself.

Political Factors: Besides the interaction with the market cluster, the political factors cluster also interacts with the alternative cluster because the political factors are also affected by different alternatives.

Alternatives: While each alternative affects factors in the market and political clusters, those factors also have effect on Disney's decision among alternatives in return. Thus, there are also backward interactions between the alternatives cluster and the other two clusters.

4.11.3.6 Nodes in the Benefits/Economic Subnet Clusters

- Alternatives
- Financial Factors

  Gross and disposable income level: Under this factor, only the current gross and disposable income level of the area's citizens will be considered. We assume that a higher income level in the local area will bring more business to the Disney facility and further increase Disney's revenue.

  Labor Wage: Labor refers to the current level of local labor wage. A lower labor wage will benefit Disney from reducing operating overheads.

  Profitability: Profitability refers to the forecasted profits based on the current market situation.
- Infrastructure

  Accommodation Capacity: This refers to the current hotel accommodation capacity of that area.

Resources: The resources factor refers to the current construction quality and efficiency of the area.

Transportation: Transportation here means the current development of local railroads, airports, tunnels, etc. If the area is already well developed, Disney can benefit from an instant resource of transportation system for customers.

Table 4.26 Alternative rankings from the benefits/economic subnet

	Alternatives	Total from supermatrix	Total Normalized	Total Idealized	Ranking
	Don't invest in Greater China	0.0273	0.0579	0.1242	4
	Hong Kong	0.2201	0.4662	1.0000	1
	Shanghai	0.1379	0.2922	0.6267	2
	Taiwan	0.0867	0.1837	0.3940	3

Table 4.27 Alternative rankings from the benefits/social subnet

Graphic	Alternatives	Total	Normalized	Idealized	Ranking
	Don't invest in Greater China	0.0045	0.0099	0.0219	4
	Hong Kong	0.2059	0.4521	1.0000	1
	Shanghai	0.1556	0.3417	0.7558	2
	Taiwan	0.0894	0.1963	0.4342	3

Combining the outcomes from the social and economic decision subnets for the benefits model produces the results shown below. The normalized values (in bold) show that Hong Kong offers the most benefits, and by a significant amount, at 46.4.

In the opportunities, costs and risks models, the decision subnets are built based on the same logic as that of the benefits subnets. The details of their clusters and nodes are similar to that of benefits and will not be shown here. A general idea of what they are can be obtained from the figure above showing the decision subnetworks. The results for each of the control criteria for opportunities, costs and risks are given below.

We show only the final synthesized results for opportunities, costs, and risks

Table 4.28 Synthesized result for the benefits model

Graphic	Alternatives	Total	Normal	Ideal	Ranking
	Don't invest in Greater China	0.107	**0.050**	0.107	4
	Hong Kong	1.000	**0.464**	1.000	1
	Shanghai	0.648	**0.301**	0.648	2
	Taiwan	0.401	**0.186**	0.401	3

Table 4.29 Synthesized results for the opportunities model

Graphic	Alternatives	Total	Normal	Ideal	Ranking
	Don't invest in Greater China	0.019	**0.010**	0.019	4
	Hong Kong	0.428	**0.224**	0.428	3
	Shanghai	1.000	**0.524**	1.000	1
	Taiwan	0.462	**0.242**	0.462	2

Table 4.30 Synthesized results for the costs model

Graphic	Alternatives	Total	Normal	Ideal	Ranking
	Don't invest in Greater China	0.104	**0.040**	0.105	4
	Hong Kong	0.610	**0.233**	0.617	3
	Shanghai	0.989	**0.378**	1.000	1
	Taiwan	0.912	**0.349**	0.922	2

Table 4.31 Synthesized results for the risks model

Graphic	Alternatives	Total	Normal	Ideal	Ranking
	Don't invest in Greater China	0.116	**0.051**	0.118	4
	Hong Kong	0.425	**0.188**	0.434	3
	Shanghai	0.981	**0.434**	1.000	1
	Taiwan	0.736	**0.326**	0.751	2

4.11.3.7 Decision Model for Rating Strategic Criteria

The final step in the decision is to determine the strategic criteria that are more or less the same for the organization or individual in making any decision and use them to rate the BOCR with respect to competition, income level, infrastructure, international character and political support as shown in Table 4.32. We thought the five strategic criteria below pretty well captured Disney's main corporate concerns about their theme parks.

To prepare to rate the strategic criteria one first pairwise compares them for importance in a hierarchy resulting in the priorities shown underneath their names in Table 4.32. Then one establishes intensities to indicate the degree of fulfillment (in the case of benefits and opportunities) or impact (in the case of costs and risks). The intensities and their priorities (in the ideal form) are Very Strong (1.000), Strong (0.627), Medium (0.382), Moderate (0.232) and Weak (0.148). Priorities are determined for them by pairwise comparing. In this case the same intensities and priorities are used for each strategic criterion, although they could be different.

4.11.3.8 Strategic Criteria Definitions

The strategic criteria are defined below and pairwise compared for importance with respect to Disney's goal. Ratings are then established for each of these criteria and pairwise compared to establish their priorities in turn. These ratings are then used to determine the priority or importance of Benefits, Opportunities, Costs and Risks and these values are used to weight the results in the submodels attached to them.

Competition: Successful theme parks in the area of the Disney Facility may be viewed both positively and negatively. Other theme parks already in the areas represent competition for Disney; however, competitors may also bring more people to the area to visit both facilities at the same time.

Income Level: Gross and disposable income levels of the area's citizens may also affect the success of the park. Consider Tokyo Disney Land for example. Approximately 95 of its visitors are local Japanese; thus, the high average income level of Japanese does appear to contribute to the tremendous success of Disney in Japan.

Infrastructure: Infrastructure in the area of the park and the regional support are also important. Visitors should be able to access the park easily. The transportation system should be well established or enhanced while the park is being constructed. A good area should have the infrastructure to support a park efficiently. Besides, the region should also contribute to extending the time visitors are able to spend at the Disney facilities. For example, a stock of hotel rooms to support park visitors is important and rooms at a variety of price levels, from economy all the way to luxury, should be available when the park opens.

International Character: Disney is looking for "international character" for any theme park it builds in Greater China. A diversified visitor base will reduce the risks of problems in one country having an adverse effect on the flow of international visitors.

Political Support: In all Disney's international operations, support from local government is critical to the Disney Company. This support ranges from providing a good location to build the theme park to insuring sufficient capital flow.

4.11.3.9 Rating the Benefits, Opportunities, Costs and Risks

To select the ratings in Table 4.32 for the Benefits, for example, one must keep in mind the top alternative in the synthesized results for the benefits model given in Table 4.28 that has the highest priority, Hong Kong. For example, Hong Kong's benefits to fulfill the Competition strategic criterion or objective is thought to be strong. For fulfilling benefits for Income Level, Hong Kong would be very strong as people there have high disposable income, and so on for all the Strategic Criteria.

When making ratings for Costs and Risks, keep in mind that the highest priority alternative is the most costly or most risky. To select the ratings for Risks keep in mind Shanghai. Shanghai has very strong risks so far as Competition is concerned, and strong risks for Income Level as people have less disposable income there, and medium risks for Political Support which means the risk is not too great for Disney in Shanghai as they believe they would have the support of the Chinese Government.

The overall priorities for each of the BOCR are computed by multiplying and adding across each row and normalizing the final result shown in the last column of Table 4.32. The priorities show that the most important merit is Benefits at 31.9 followed by Opportunities at 26.4. This means that the priorities of the alternatives under benefits are weighted more heavily. Benefits at 31.9 drive the decision more than the Risks at 19.3.

Table 4.32 BOCR ratings and priorities

	Competition (0.127)	Income level (0.19)	Infrastructure (0.147)	Internat'l char. (0.323)	Political support (0.214)	**Priorities**
Benefits	Strong	Very strong	Strong	Very strong	Very strong	**0.32**
Opportunities	Very strong	Strong	Strong	Very strong	Medium	**0.26**
Costs	Very strong	Medium	Strong	Strong	Strong	**0.22**
Risks	Very strong	Strong	Strong	Medium	Medium	**0.19**

Very Strong (1.000), Strong (0.627), Medium (0.382), Moderate (0.232) and Weak (0.148)

The final results shown in Table 4.33 are obtained using the formula $b\text{B} + o\text{O} - c\text{C} - r\text{R}$ where b, o, c and r are the priorities for Benefits, Opportunities, Costs and Risks just obtained from rating the BOCR with respect to the strategic criteria in Table 4.32. This formula is applied to the alternatives using the priority vectors from the synthesized results (the B, O, C, and R of the formula) in the previous tables. Since this formula involves negatives, the overall synthesized results in Table 4.33 may be negative, meaning that the alternative is undesirable. Sometimes all results are negative, and one is forced to take the least undesirable one. In Table

4.33 positive results are labeled in blue and negative ones in red. Here Hong Kong is the best with the highest positive value and Taiwan is the worst with the highest negative value.

Table 4.33 BOCR model: Overall synthesized results

Graphic	Alternatives	Total	Normal	Ideal	Ranking
(red)	Don't invest in Greater China	−0.006	−0.017	−0.030	3
(blue)	Hong Kong	0.214	0.567	1.000	1
(blue)	Shanghai	0.061	0.161	0.284	2
(red)	Taiwan	−0.096	−0.255	−0.449	4

As we can see, from the overall synthesized results in Table 4.33, Disney's best option is to build their new theme park in **Hong Kong**.

4.11.3.10 Sensitivity Analysis Graphs

Sensitivity analysis in Figure 4.17 shows that when the importance of benefits is greater than 0.05, investing in Hong Kong is the best choice. The dotted vertical line indicates the priority of Benefits, for example. At a priority of less than about 0.35 for opportunities, Hong Kong is the best choice, but above that the choice shifts to Shanghai. One might interpret this as meaning that there are great opportunities in Shanghai, but it is also risky as can be seen from the risks sensitivity graph. As the priority of costs increases beyond about 0.38, the best choice shifts from investing in Hong Kong to not investing at all. As the importance of risk increases the preferred alternative is to not to invest as all in Greater China, but since the priority is negative, below the x-axis, this is not a particularly good alternative, though it is the least negative. When risk is less than about 0.50, the preferred alternative is to invest in Hong Kong.

The vertical dotted line represents the priority of Benefits and Opportunities. To see what happens as the importance of Benefits increases, move the vertical line to the right. Above a Benefits priority of about 40 the least preferred alternative changes from Taiwan to Don't Invest in Greater China. The line immediately under Hong Kong at down is Shanghai. One might interpret this as indicating that investing in China somewhere is imperative in terms of benefits.

As the importance of Opportunities increases past about 35, the top line would be Shanghai and the bottom line Taiwan. This can be interpreted to mean that the greatest opportunities lie in Shanghai. Similar conclusions can be made from the costs and risks graphs in Figure 4.18.

BENEFITS
At Benefits = 32.9%, Hong Kong (top line) is best; Shanghai second and Taiwan (bottom line) is worst

OPPORTUNITIES
At Opportunities = 26.4%, Hong Kong (top line) is best; Shanghai is second and Taiwan (bottom line) is worst

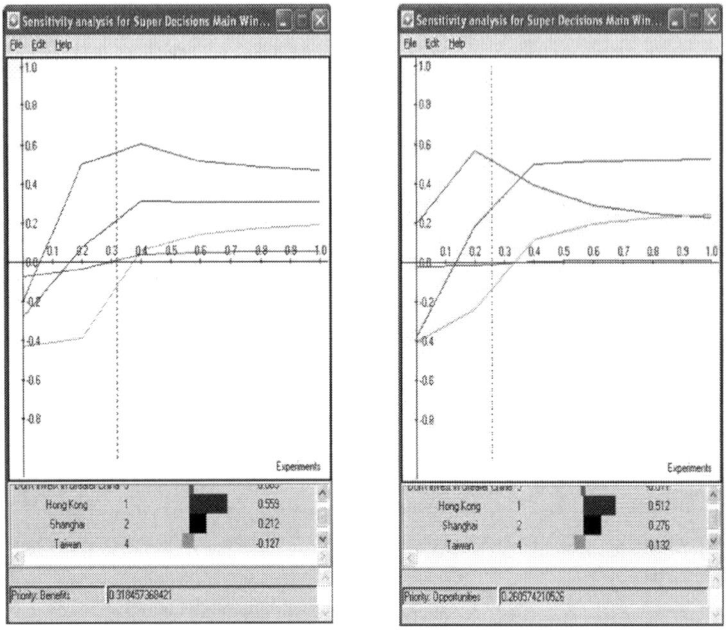

Fig. 4.17 Sensitivity graphs for benefits and opportunities

In Figure 4.18 (left part) at an importance of 22.3 for Costs, Hong Kong (the top line) is most costly and Taiwan (the bottom line) is least costly, perhaps because of the political uncertainty and lack of supporting infrastructure in Hong Kong, and as Costs increases, not investing in China is the top line (after about 40). So it is extremely risky to not invest in China at all.

In Figure 4.18 (right part) at an importance of 19.3 for Risks the top line is Hong Kong, so it is most risky and the bottom line is Taiwan, meaning least risky.

In sum, the greatest benefits and opportunities lie in mainland China, along with the greatest costs and risks, but netting it out, Hong Kong is best overall.

4.12 Decision on National Missile Defense (NMD) - An Application of the ANP with Strong Risks (Analysis done in 2000, decision made to implement in December 2002)

Not long ago, the United States government faced the crucial decision of whether or not to commit itself to the deployment of a National Missile Defense (NMD)

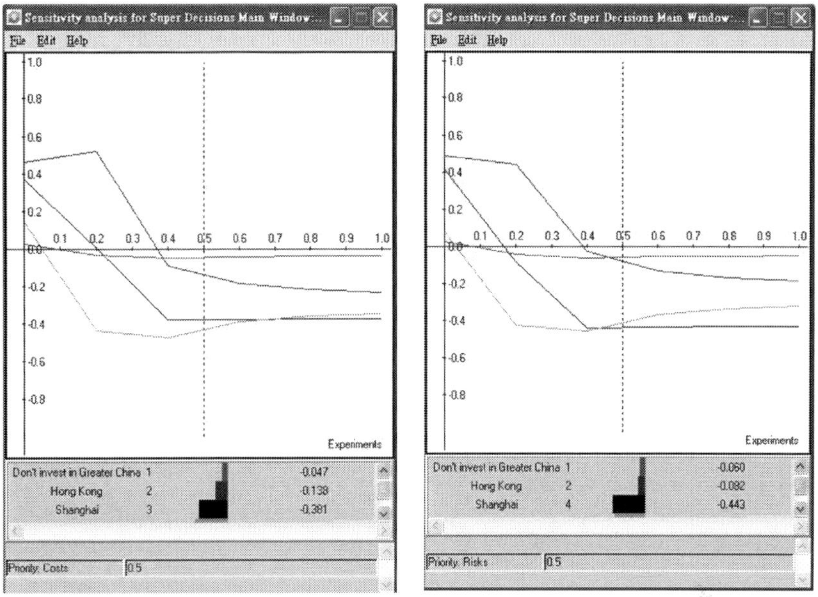

Fig. 4.18 Sensitivity graphs for costs and risks

system. Many experts in politics, the military, and academia had expressed different views regarding this decision. The most important rationale behind supporters of the NMD system was protecting the U.S. from potential threats said to come from countries such as North Korea, Iran and Iraq. According to the Central Intelligence Agency, North Korea's Taepo Dong long-range missile tests were successful, and it has been developing a second generation capable of reaching the U.S. Iran also tested its medium-range missile Shahab-3 in July 2000. Opponents expressed doubts about the technical feasibility, high costs (estimated at 60 billion), political damage, possible arms race, and the exacerbation of foreign relations.

The idea for the deployment of a ballistic missile defense system has been around since the late 1960s but the current plan for NMD originated with President Reagan's Strategic Defense Initiative (SDI) in the 1980s. SDI investigated technologies for destroying incoming missiles. The controversies surrounding the project were intensified with the National Missile Defense Act of 1996, introduced by Senator Sam Nunn (D-GA) in June 25, 1996. The bill required Congress to make a decision on whether the U.S. should deploy the NMD system by 2000. The bill also targeted the end of 2003 as the time for the U.S. to be capable of deploying NMD. The idea explored in this project is to develop and illustrate the three phases with a timely example, the intricate and very costly decision regarding a National Missile Defense (NMD) system. Because of the possibility of dependence and feedback, we use the Analytic Network Process (ANP) and its software Super-Decisions with its sensitivity analysis option to examine the NMD decision. On February 21, 2002 this author gave a half-day presentation on the subject to National Defense University in Washington. In December 2002, President George W. Bush and his advisors decided to build the NMD. This study may have had no influence on the decision but still two years earlier (September 2000) it had arrived at the same outcome. The alternatives we considered for this analysis are: Deploy NMD, Global defense, R D, Termination of the NMD program.

4.12.1 Criteria and Decision Networks [5]

The second column of Table 4.34 shows the criteria of each BOCR. For example, there are four benefits criteria: Economic (0.157), Political (0.074), Security (0.481) and Technology (0.288). The priorities attached to each are obtained through pairwise comparisons. Each criterion under benefits has subcriteria such as Local Economy and Defense Industry under Economic. Again, the priorities of the two subcriteria are obtained from pairwise comparisons and similarly for the remaining criteria and subcriteria under opportunities, costs and risks. Opportunities and risks have no subcriteria. The total number of criteria and subcriteria used as control criteria for the comparisons made in the networks is 2. The global priorities of these criteria (subcriteria) shown in the last column of Table 4.34 are obtained by weighting their priorities by those of their parent criterion if there is one. For example, for local economy we have $0.157 \times 0.141 = 0.022$. We will see later, after the BOCR merits are weighted, that the priorities of nine of these (shown in boldface), Military Capability, Technological Advancement, Arms Sales, Spin-Off, Security Threat, Sunk Cost, Further Investment, Arms Race, and Technical Failure account for approximately 0.760 of the total. To economize effort, we used these nine as control criteria each with its decision network to do the analysis. Actually we simply chose the top ones under each merit without being fully consistent about the cutoff point. For example we left out U.S. Reputation under Risks. All economic cost factors were included. We proceeded as if these nine criteria and subcriteria, called covering criteria for the alternatives, were the only criteria to drive the outcome. Their decision networks and connections are shown in Figures 4.19 to 4.27. A more thorough analysis might include a few more criteria or subcriteria.

4.12.2 Full Development of the Analysis with Respect to a Single Criterion

We explain in outline form our thinking about the network under one of the criteria. We have chosen Military Capability, one of the main control subcriteria, to elaborate the details of its decision network. There are five main parties involved in the decision making process of NMD: Congress, President/Military, Foreign Countries, Technical Experts and the Defense Industry. The latter two influence Congress and

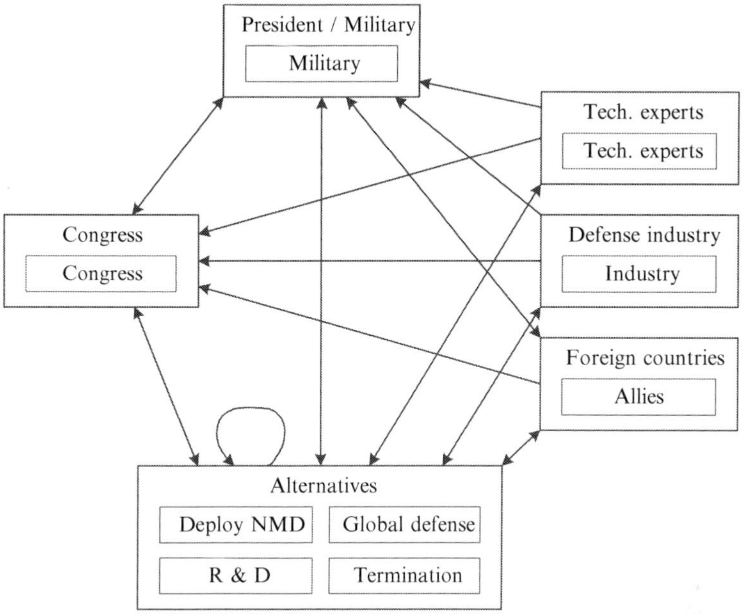

Fig. 4.19 Decision network under the military capability

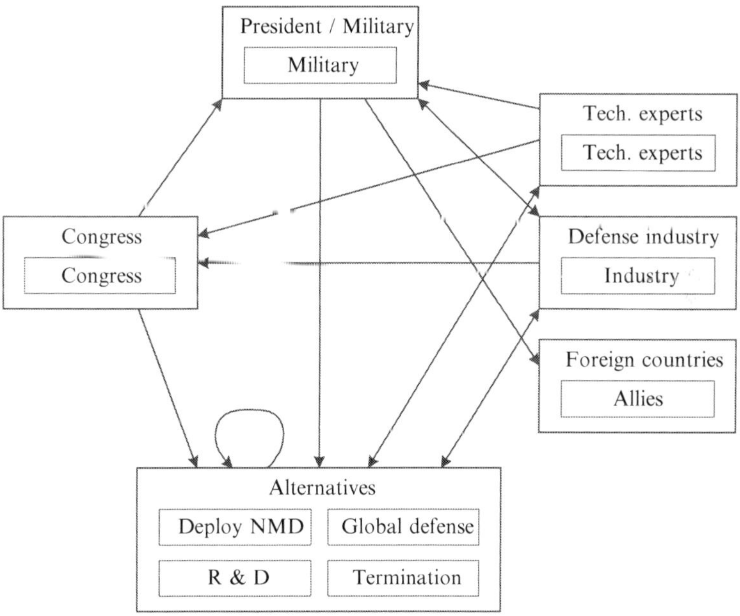

Fig. 4.20 Decision network under the technological advancement

Table 4.34 Criteria and their priorities

Merits	Criteria	Sub-criteria	Global Priorities (Normalized)
Benefits	Economic (0.157)	Local Economy (0.141)	0.022
		Defense Industry (0.859)	0.135
	Political (0.074)	Bargaining Power (0.859)	0.064
		U.S. Military Leadership (0.141)	0.010
	Security (0.481)	Deterrence (0.267)	0.128
		Military Capability (0.590)	**0.284**
		Anti-terrorism (0.143)	0.069
	Technology (0.288)	Tech. Advancement (0.834)	0.240
		Tech. Leadership (0.166)	0.048
Opportunities	Arms Sales (0.520)		**0.520**
	Spin- off (0.326)		**0.326**
	Space Development (0.051)		0.051
	Protection of Allies (0.103)		0.103
Costs	Security Threat: Vulnerability to the security threat (0.687)		**0.687**
	Economic (0.228)	Sunk Cost (0.539)	**0.120**
		Further Investment (0.461)	**0.105**
	Political (0.085)	ABM Treaty (0.589)	0.050
		Foreign Relations (0.411)	0.035
Risks	Technical Failure (0.430)		**0.430**
	Arms Race (0.268)		**0.268**
	Increased Terrorism (0.052)		0.052
	Environmental Damage (0.080)		0.080
	U.S. Reputation (0.170)		0.170

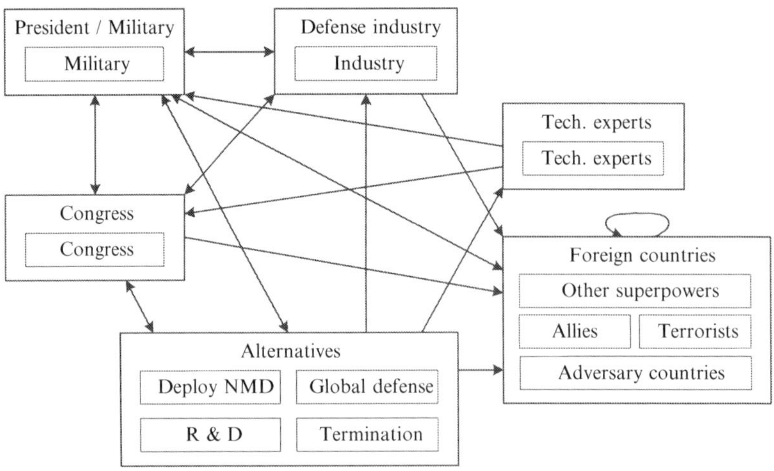

Fig. 4.21 Decision network under the arms sales

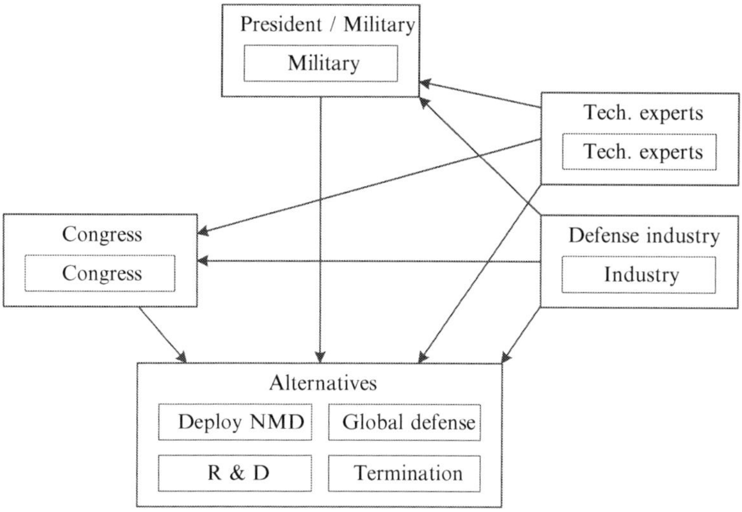

Fig. 4.22 Decision network under the spin-off

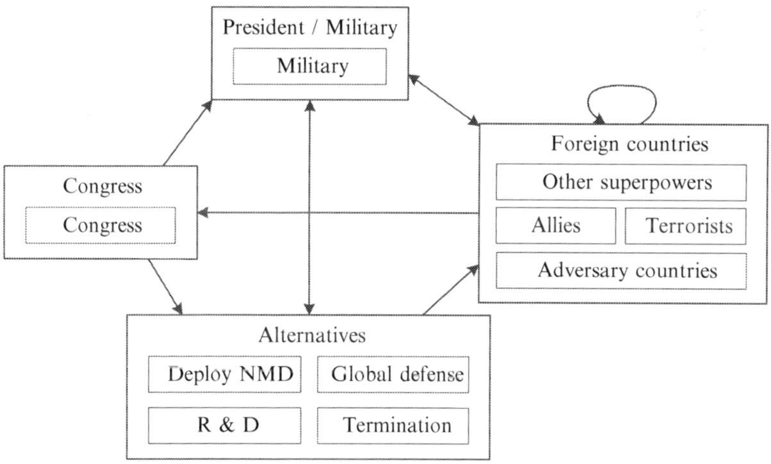

Fig. 4.23 Decision network under the security threat control

President/Military by providing their professional expertise and technical information. Allies among Foreign Countries can have a partial influence on Global Defense among the four alternatives through economic and technological cooperation.

The first block of four rows and four columns in Table 4.35, The Unweighted Supermatrix, indicates that Deploy NMD (NMD) and RD (RD) are influenced by Global Defense (Glob.) with priorities of 0.5760 and 0.4240 respectively. The next five columns and first four rows of Table 4.35, The Unweighted Supermatrix, summarize the different views of actors on the contribution of each of the four alter-

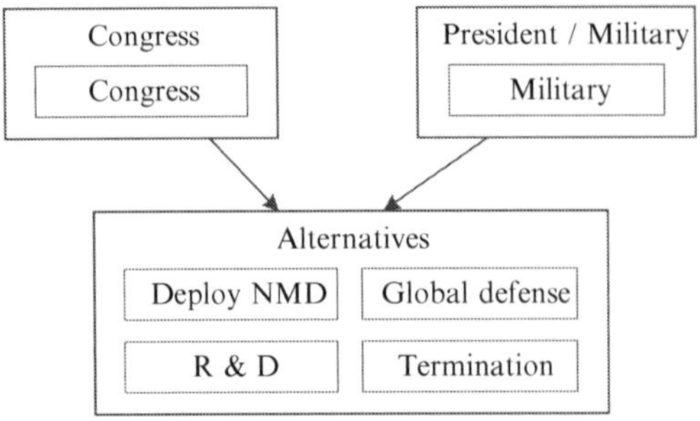

Fig. 4.24 Decision network under the sunk cost

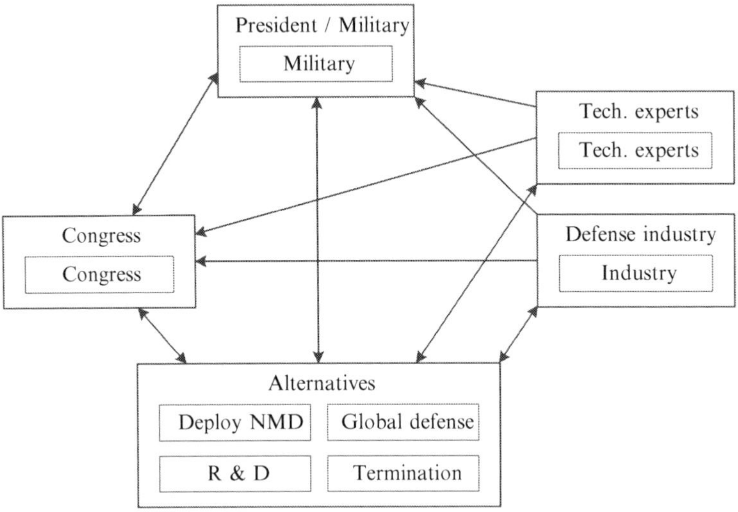

Fig. 4.25 Decision network under the further investment

natives to U.S. military capability. Congress, President/Military, Defense Industry, and Technical Experts all have a say as to what extent the decision contributes to the Military Capability of the U.S. All domestic actors think that Deploy NMD will increase military capability followed by Global Defense, RD and Termination (Term) but to different degrees. Deploy NMD (0.5587) was given the highest priority by Defense Industry, followed by the priority given by President/Military (0.5158), and Congress (0.5060). The lowest priority given to NMD is by Technical Experts (0.2878). It reflects the opinion of scientists who think Deploy NMD is technically infeasible and would not contribute to the enhancement of U.S. military capability. Only Global Defense is influenced by Allies and thus the priority of Global Defense is equal to 1.0000.

The rows from the fifth to the last of Table 4.35 show connections among components (clusters) each consisting of a single element except for the component of

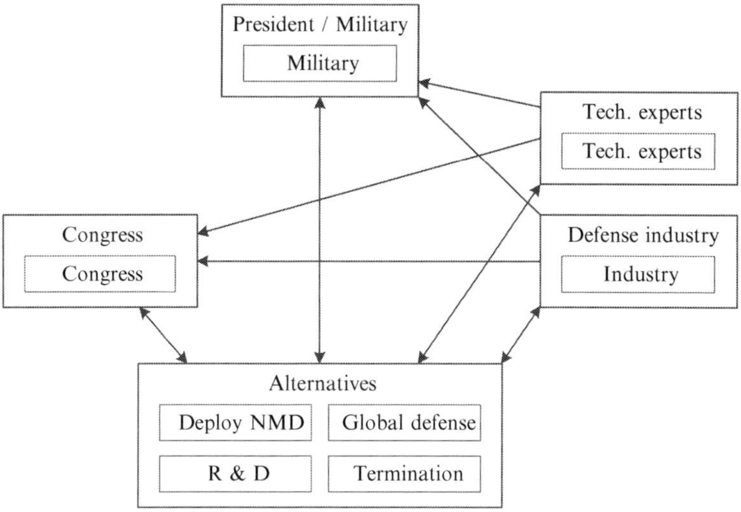

Fig. 4.26 Decision networks under the technical feasibility

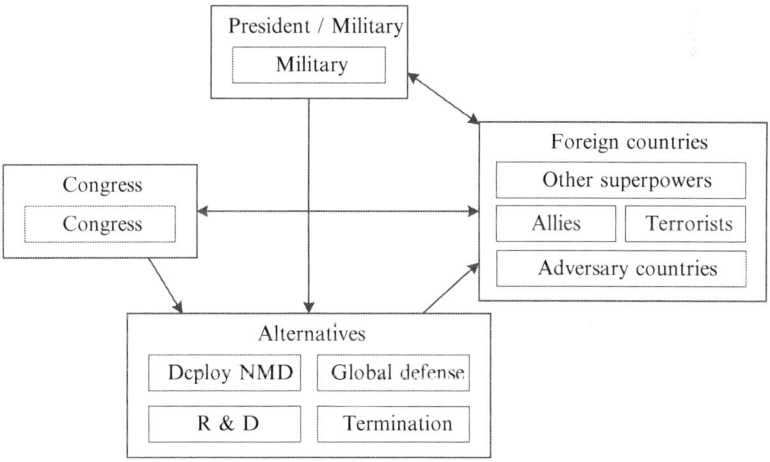

Fig. 4.27 Decision network under the arms race

Alternatives that has four elements. The priorities of the entries in these rows must be either 1.0000 or 0.0000 depending on whether there is influence among them. For example, the fifth to the ninth entries of column one have unit entries obtained from answering the question "Is the component of Congress influenced by Deploy NMD?", "Is the component of Defense Industry influenced by Deploy NMD?" and similarly for the other three alternatives. All actors are influenced by the three alternatives of Deploy NMD, Global Defense and RD. Note that an entire column under Termination in the Unweighted Supermatrix of Table 4.34 consists of zeros because nothing is influenced by Termination and that leads to dropping the entire matter of missile defense. It is worth noting that under the Security Threat criterion of Costs (not shown here), the column under Termination in the Unweighted Supermatrix consists of non-zero values because security threat to the U.S. would continue particularly if Termination is chosen as it accentuates vulnerability of U.S. security.

Table 4.35 The unweighted supermatrix

MilCap		Altern.				Cong.	Def. ind.	For.	Pre/Mil.	Tech.
Unweighted		NMD	Glob.	R & D	Term.	Cong.	Industry	Allies	Military	Tech.
Altern.	NMD	0.0000	0.5760	0.0000	0.0000	0.5060	0.5587	0.0000	0.5158	0.2878
	Glob.	0.0000	0.0000	0.0000	0.0000	0.2890	0.2574	1.0000	0.2929	0.2620
	R & D	0.0000	0.4240	0.0000	0.0000	0.1307	0.1382	0.0000	0.1367	0.2690
	Term.	0.0000	0.0000	0.0000	0.0000	0.0744	0.0457	0.0000	0.0546	0.2130
Cong.	Cong.	1.0000	1.0000	1.0000	0.0000	0.0000	1.0000	1.0000	1.0000	1.0000
Defense ind.	Industry	1.0000	1.0000	1.0000	0.0000	0.0000	0.0000	0.0000	0.0000	0.0000
For.	Allies	1.0000	1.0000	1.0000	0.0000	0.0000	0.0000	0.0000	1.0000	0.0000
Pre/Mil.	Military	1.0000	1.0000	1.0000	0.0000	1.0000	1.0000	1.0000	0.0000	1.0000
Tech.	Tech.	1.0000	1.0000	1.0000	0.0000	0.0000	0.0000	0.0000	0.0000	0.0000

Table 4.36 shows the pairwise comparisons of the components. The judgments were obtained by answering the question "Which of two components is influenced more by a third component with respect to military capability?" The eigenvectors of the pairwise comparisons of the components in the matrices of Table 4.36 are exhibited in Table 4.37, augmented by zeros in those positions where the components on the left are not influenced by the component on top of the column. The Weighted Supermatrix of Table 4.38 illustrates the weighting of the blocks of the supermatrix by the priorities from the corresponding eigenvector of comparisons of the components in Table 4.36. Table 4.39, The Limit Supermatrix, yields the stable priorities of all the elements. From it, the priorities of the four alternatives are extracted and normalized. We obtain for (Deploy NMD, Global Defense, R&D, and Termination) the corresponding values (0.1532, 0.0968, 0.0438, 0.0201) which when normalized by dividing by their sum yields the priority vector (0.488, 0.308, 0.140, and 0.064. Similar computations are done for the remaining eight high priority criteria. An entry in each subcolumn of the supermatrix indicates the relative priority within the block to which that subcolumn belongs that an element on the left is influence by the element on top of the column with respect to Military Capability. Each subcolumn is an eigenvector imported from a corresponding pairwise comparisons matrix not shown here because its elements can be approximately formed from the ratios of the corresponding priority vector. A subcolumn of zeros indicates no influence and therefore no comparisons matrix is needed.

Table 4.36 Pairwise comparisons matrices and priorities of components

Pairwise comparing components with respect to the alternatives component
Q: Which of a pair of components is influenced more by the alternatives component with respect to military capability?

	Altern.	Cong.	Def. ind.	For.	Pres.	Tech.	Prior.
Altern.	1.0000	0.1667	0.2500	1.3300	0.1429	0.5556	**0.0486**
Cong.	5.9999	1.0000	2.2000	6.2000	0.7407	3.2000	**0.2889**
Def. ind.	4.0000	0.4546	1.0000	4.0000	0.4115	2.2600	**0.1653**
For.	0.7519	0.1613	0.2500	1.0000	0.1250	0.5263	**0.0425**
Pres.	7.0000	1.3500	2.4300	8.0000	1.0000	5.1000	**0.3742**
Tech.	1.8000	0.3125	0.4425	1.9000	0.1961	1.0000	**0.0805**

Pairwise comparing components with respect to the Congress component
Q: Which of a pair of components is influenced more by the Congress component with respect to Military Capability?

	Altern.	Pres.	Prior.
Altern.	1.0000	0.5638	**0.3605**
Pres.	1.7736	1.0000	**0.6395**

Pairwise comparing components with respect to the Defense Industry component
Q: Which of a pair of components is influenced more by the Defense Industry component with respect to Military Capability?

	Altern.	Cong.	Pres.	Prior.
Altern.	1.0000	0.6769	0.5388	**0.2292**
Congr.	1.4773	1.0000	0.6600	**0.3181**
Pres.	1.8561	1.5152	1.0000	**0.4528**

Pairwise comparing components with respect to the Foreign Countries component
Q: Which of a pair of components is influenced more by the Foreign Countries component with respect to Military Capability?

	Altern.	Cong.	Pres.	Prior.
Altern.	1.0000	0.5556	0.3259	**0.1671**
Congr.	1.8000	1.0000	0.4632	**0.2781**
Pres.	3.0682	2.1591	1.0000	**0.5548**

Pairwise comparing components with respect to the Presidnet/Military component
Q: Which of a pair of components is influenced more by the President/ Military component with respect to Military Capability?

	Altern.	Cong.	For.	Prior.
Altern.	1.0000	2.1887	3.6604	**0.5735**
Congr.	0.4569	1.0000	2.0377	**0.2799**
For.	0.2732	0.4907	1.0000	**0.1467**

Pairwise comparing components with respect to the Technical Experts component
Q: Which of a pair of components is influenced more by the Technical Experts component with respect to Military Capability?

	Altern.	Cong.	Pres.	Prior.
Altern.	1.0000	2.5379	2.5379	**0.5593**
Congr.	0.3940	1.0000	1.0000	**0.2204**
Pres.	0.3940	1.0000	1.0000	**0.2204**

Table 4.37 Priorities matrix of eigenvectors (how much components are influenced by each component; imported from the matrices of Table 4.36 above)

Clusters	Altern.	Cong.	Def. ind.	For.	Pres.	Tech.
Altern.	0.0486	0.3605	0.2292	0.1671	0.5735	0.5593
Cong.	0.2889	0.0000	0.3181	0.2781	0.2799	0.2204
Def. ind.	0.1653	0.0000	0.0000	0.0000	0.0000	0.0000
For.	0.0425	0.0000	0.0000	0.0000	0.1467	0.0000
Pres.	0.3742	0.6395	0.4528	0.5548	0.0000	0.2204
Tech.	0.0805	0.0000	0.0000	0.0000	0.0000	0.0000

Table 4.38 The weighted supermatrix (Priorities from Table 4.37 are used to weight corresponding blocks of unweighted supermatrix of Table 4.35)

MilCap		Altern.				Cong.	Def. ind.	For.	Pre/Mil.	Tech.
Weighted		NMD	Glob.	R & D	Term.	Cong.	Industry	Allies	Military	Tech.
Altern.	NMD	0.0000	0.0280	0.0000	0.0000	0.1824	0.1280	0.0000	0.2958	0.1610
	Glob.	0.0000	0.0000	0.0000	0.0000	0.1042	0.0590	0.1671	0.1680	0.1467
	R & D	0.0000	0.0206	0.0000	0.0000	0.0471	0.0317	0.0000	0.0784	0.1325
	Term.	0.0000	0.0000	0.0000	0.0000	0.0268	0.0105	0.0000	0.0313	0.1191
Cong.	Cong.	0.3037	0.2889	0.3037	0.0000	0.0000	0.3181	0.2780	0.2799	0.2204
Defense ind.	Industry	0.1737	0.1653	0.1737	0.0000	0.0000	0.0000	0.0000	0.0000	0.0000
For.	Allies	0.0446	0.0425	0.0446	0.0000	0.0000	0.0000	0.0000	0.1467	0.0000
Pre/Mil.	Military	0.3933	0.3742	0.3933	0.0000	0.6395	0.4528	0.5548	0.0000	0.2204
Tech.	Tech.	0.0846	0.0805	0.0846	0.0000	0.0000	0.0000	0.0000	0.0000	0.0000

Table 4.39 The limit supermatrix (The weighted supermatrix raised to sufficiently large powers to stabilize within rounded off four place decimals)

MilCap		Altern.				Cong.	Def. ind.	For.	Pre/Mil.	Tech.
Limited		NMD	Glob.	R & D	Term.	Cong.	Industry	Allies	Military	Tech.
Altern.	NMD	0.1532	0.1532	0.1532	0.0000	0.1532	0.1532	0.1532	0.1532	0.1532
	Glob.	0.0968	0.0968	0.0968	0.0000	0.0968	0.0968	0.0968	0.0968	0.0968
	R & D	0.0438	0.0438	0.0438	0.0000	0.0438	0.0438	0.0438	0.0438	0.0438
	Term.	0.0201	0.0201	0.0201	0.0000	0.0201	0.0201	0.0201	0.0201	0.0201
Cong.	Cong.	0.2224	0.2224	0.2224	0.0000	0.2224	0.2224	0.2224	0.2224	0.2224
Defense ind.	Industry	0.0513	0.0513	0.0513	0.0000	0.0513	0.0513	0.0513	0.0513	0.0513
For.	Allies	0.0619	0.0619	0.0619	0.0000	0.0619	0.0619	0.0619	0.0619	0.0619
Pre/Mil.	Military	0.3255	0.3255	0.3255	0.0000	0.3255	0.3255	0.3255	0.3255	0.3255
Tech.	Tech.	0.0250	0.0250	0.0250	0.0000	0.0250	0.0250	0.0250	0.0250	0.0250

4.12.3 BOCR Weight Development

The judgments used in this analysis were our interpretation of what experts thought about the various issues obtained from the vast reading of the literature we examined and from following the news closely for a period of more than six months. We also consulted some knowledgeable people on the subject in the area. We quickly realized there is no single expert in all the criteria we considered. Sensitivity analysis given later would essentially vary these judgments widely to determine the stability of the outcome. The assessment criteria used to determine the priorities of the BOCR merits are shown in Figure 4.28. These are World Peace, Human Well-being, and International Politics. All these criteria have subcriteria under them. The three subcriteria, Adversary Countries, Security Dilemma and Terrorism cover all the causes disturbing or stabilizing peace in the world. The first subcriterion, Adversary Countries, concerns the potential threats by adversary countries. The second criterion, Security Dilemma, means that increasing one country's security inevitably decreases other countries' security. Terrorism indicates any possibility of the rise or decline of terrorism in the world. Human Well-being includes Technological Advancement and Market Creation. Technological Advancement driven by the NMD research and development process can ultimately benefit all people, particularly in providing possible space exploration that can lead to the creation of new markets. Moreover, the 21st century is characterized as a post-industrialization era. Service industries in communication and transportation will benefit not only businesses associated with these industries, but also consumers who can enjoy the products from the new market. The last criterion is International Politics. It is composed of two subcriteria, Military Relations and Diplomatic Relations. Military Relations refer to the impact of NMD on relations with U.S. allies for better or for worse. Also, the impact of NMD on diplomatic relations among all countries should be considered. The priorities shown next to the criteria and subcriteria in Figure 4.28 were obtained through the usual pairwise comparison process of the AHP according to their importance with respect to their higher-level goal or parent criterion.

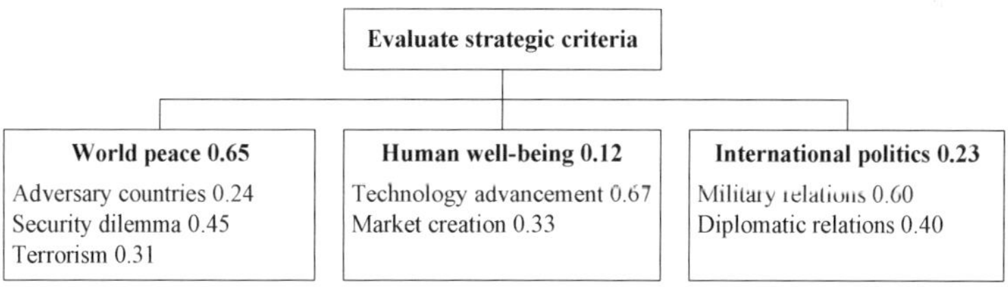

Fig. 4.28 Strategic criteria for BOCR ratings

The four merits of BOCR were rated according to the five intensities listed in Table 4.40. The priorities of the intensities shown in the table were derived from pairwise comparisons. The weighted values of the subcriteria shown in Table 4.40, are multiplied times the value of the assigned intensity and summed for each column to yield the outcome, the BOCR values at the bottom of the table.

Table 4.40 Priority ratings for the merits: Benefits, opportunities, costs, and risks

Criteria	Subcriteria	Benefits	Opportunities	Costs	Risks
World peace	Adversary countries (0.156)	Very high	Medium	High	Very low
	Security dilemma (0.293)	Very low	Very low	Very high	Very low
	Terrorism (0.202)	Medium	Very low	High	High
Human well-being	Technological advancement (0.080)	High	High	Low	Very low
	Market creation (0.04)	Medium	High	Very low	Very low
International politics	Military relations (0.138)	High	High	Medium	Very low
	Diplomatic relations (0.092)	Low	Low	Low	Very high
Priorities		0.264	0.185	0.361	0.19

Very high (0.419), High (0.263), Medium (0.160), Low (0.097), Very low (0.061)

Note that BOCR are rated one at a time and are not obtained from paired comparisons. They are obtained using the rating approach of the AHP.

As we said earlier if we weight the priorities derived in Table 4.34 shown in boldface by the corresponding priorities of the merits just derived in Table 4.40 above and then add we get:

$$0.264 \times \mathbf{0.284} + 0.264 \times \mathbf{0.240} + 0.185 \times \mathbf{0.520} + 0.185 \times \mathbf{0.326} + 0.361 \times \mathbf{0.687} \\ + 0.361 \times \mathbf{0.12} + 0.361 \times \mathbf{0.105} + 0.190 \times \mathbf{0.430} + 0.190 \times \mathbf{0.268} \approx 0.76$$

In most of our studies we attempt to use enough control criteria factors under any given merit so that they total not less than 70%.

Table 4.41 shows the idealized priorities of the four alternatives (developed from their values in the corresponding limit supermatrix) with respect to the control criteria of the merits and also the synthesis with respect to the criteria of the alternatives for each merit to obtain the final outcome for that merit.

The merit priorities are represented in the formulas in Table 4.42 by **b**, **o**, **c** and **r** respectively and imported from Table 4.40, while the priorities of the alternatives imported from the right column of Table 4.41 are represented by **B**, **O**, **C** and **R**.

Here we see that all three formulas give the same outcome to deploy as the best alternative. The conclusion of this analysis is that pursuing the deployment of NMD is the best alternative. But we must now examine how realistic this outcome is.

One might have different judgments in comparing the importance of BOCR or of the nine control criteria. To ensure the stability of the outcome of our analysis, we conducted sensitivity analysis.

Table 4.41 Synthesis of the alternatives for each BOCR merit

Merit	Alternatives	Idealized results under the control criteria of each merit (the values are obtained by normalizing the selected control criteria in the goal column in the limiting matrix)			Multiply idealized vectors in previous column by CC priorities and sum to obtain composite vector
		Military cap. (0.542)		Tech advanc. (0.458)	
Benefits	A1	1.000		0.928	0.967
	A2	0.620		1.000	0.796
	A3	0.282		0.448	0.358
	A4	0.129		0.085	0.109
		Arms sales (0.614)		Spinoff (0.386)	
Opportunities	A1	1.000		1.000	1.000
	A2	0.674		0.520	0.615
	A3	0.341		0.288	0.321
	A4	0.190		0.166	0.181
		Security (0.687)	Sunk costs (0.12)	Further inv. (0.105)	
Costs	A1	0.183	1.000	1.000	0.386
	A2	0.344	0.574	0.496	0.393
	A3	0.579	0.332	0.279	0.512
	A4	1.000	0.193	0.147	0.794
		Technical failure (0.43)		Arms race (0.268)	
Risks	A1	1.000		1.000	1.000
	A2	0.621		0.693	0.648
	A3	0.375		0.441	0.401
	A4	0.262		0.302	0.277

Table 4.42 Overall synthesis of the alternatives

Alternatives	BO / CR		bB + oO − cC − rR	
	From unweighted col's in table above	Normalized	From weighted col's in table above	Unitized*
Deploy	**2.504**	**0.493**	**0.111**	**1.891**
Global	1.921	0.379	0.059	**1.000**
R&D	0.560	0.110	−0.108	**−1.830**
Terminate	0.090	0.018	−0.278	**−4.736**

* Unitized means to divide each number in the column by the number with the smallest absolute value

4.12.3.1 Sensitivity Analysis at the BOCR Level

First, we increased and decreased one of the four merits of BOCR keeping the others proportionally the same. For example, if benefits were to be increased from its original priority 0.264 to 0.500, the sum of the other three merits would comprise the other 0.500 and the proportion among them would remain the same as before and their new priorities would be: opportunities, 0.124, costs, 0.246, and risks, 0.130. We found that no matter how much we increased or decreased the priorities of benefits, opportunities and costs the overall ranks of the final outcome were preserved although these experiments changed the magnitude of the superiority of the best alternative, Deploy NMD (for example, from 0.301 to 0.431 for benefits as Figure 4.29 shows). Only changing the priority of risks reversed the ranks of the four alternatives as shown in Figure 4.29. This occurred only when the priority of the risks was 0.375 or more. Then, Termination gradually became third then second and finally the best alternative as the priority of risks was increased more and more.

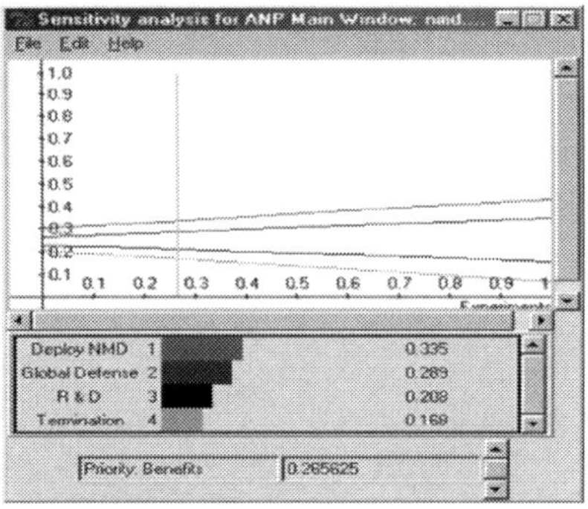

Sensitivity analysis for benefits

The rank remains the same regardless of the priorities of benefits

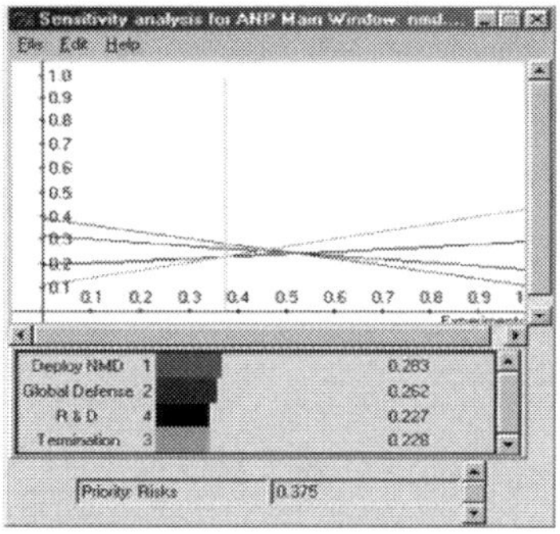

Sensitivity analysis for risks

Termination becomes the more preferred alternative as the priority of risks increases

Fig. 4.29 Sensitivity analysis for benefits and for risks

4.12.3.2 Sensitivity analysis at the Control Criteria Level

We did similar tests for the nine criteria that have decision networks. We found that the outcome was very stable and did not change the overall ranks except for changes in the importance of the three criteria: Security Threat, Sunk Cost and Further Investment all under costs. When the priority of Security Threat decreased to about 0.172 from 0.21 or the priority of Sunk Cost increased to 0.753 or the priority of Further Investment increased to 0.734 Figures 4.30 and 4.31 Termination gradually began to move to third, second and finally to first rank position.

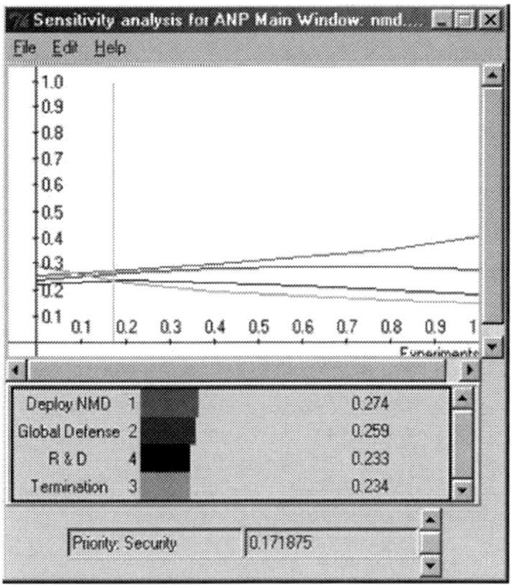

Fig. 4.30 Sensitivity analysis for security threat

If the priority of Security Threat becomes less than about 0.172, Termination becomes the more preferred alternative.

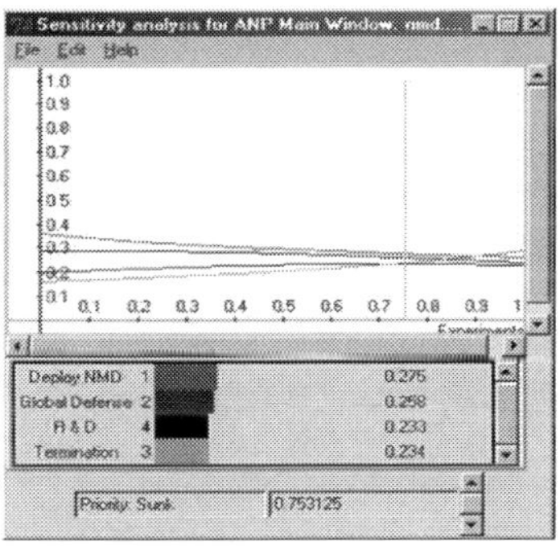

Sensitivity analysis for sunk cost

If the priority of Sunk Cost becomes larger than 0.753, Termination becomes the more preferred alternative.

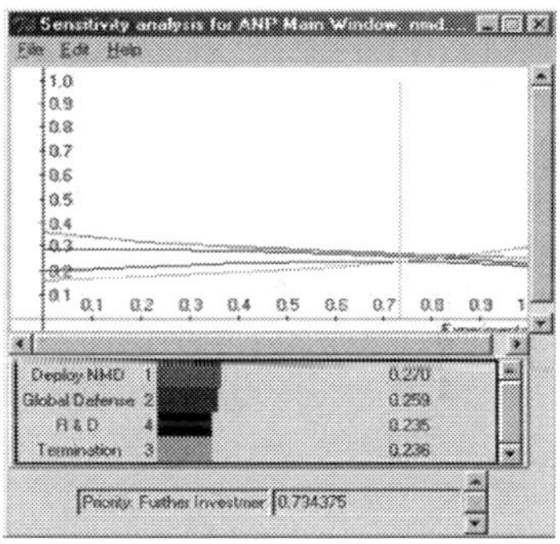

Sensitivity analysis for further investment

If the priority of Further Investment becomes larger than 0.734 Termination becomes the more preferred alternative.

Fig. 4.31 Sensitivity analysis for sunk cost and further investment

Some are highly concerned with risks associated with NMD, such as Technical Failure and Arms Race. We did another test using larger priorities for risks to see if it would change the outcome. In that case, the control criterion, U.S. Reputation, under risks replaced the control criterion, Further Investment, under costs. Interestingly enough, the ranks of the alternatives were the same as in Table 4.42 with a slightly higher priority for Deploy NMD.

Our sensitivity analysis indicates that the final ranks of the alternatives might change, but such change requires making extreme assumptions on the priorities of BOCR and of their corresponding control criteria. The outcome in Table 4.42 is very stable and the United States should choose Deploy NMD as the best alternative for the decision. I lectured on this on February 21, 2002 at National Defense University and from the comments I received in writing which I still have, it was very well received. It is not very surprising that the United States did make the decision to Deploy in December 2002. I am not sure what effect this analysis had on that decision if any. But the U.S. has excellent think tanks who can do a good job of analyzing a decision. It is certain that it took us a lot less time to do this analysis and no money at all than it takes a think tank in time and resources.

4.13 Group Decision Making [3, 6, 9]

Here we consider two issues in group decision making. The first is how to aggregate individual judgments, and the second is how to construct a group choice from individual choices. In reality group decisions should not go by consensus because not all people feel the same about things. A minority can have very strong commitments to a cause and can give rise to disruptions that the majority feels lukewarm about. There is no hiding from this issue in the real world. The reciprocal property plays an important role in combining the judgments of several individuals to obtain a judgment for a group. Judgments must be combined so that the reciprocal of the synthesized judgments must be equal to the syntheses of the reciprocals of these judgments. It has been proved that the geometric mean is the unique way to do that. If the individuals are experts, they my not wish to combine their judgments but only their final outcome from a hierarchy. In that case one takes the geometric mean of the final outcomes. If the individuals have different priorities of importance their judgments (final outcomes) are raised to the power of their priorities and then the geometric mean is formed.

4.13.1 How to Aggregate Individual Judgments

Let the function $f(x_1,\ldots,x_n)$ for synthesizing the judgments given by n judges, satisfy the

1. *Separability condition* (S): $f(x_1,\ldots,x_n) = g(x_1)\cdots g(x_n)$, for all $x_1,\ldots,x_n$ in an interval P of positive numbers, where g is a function mapping P onto a proper interval J and is a continuous, associative and cancellative operation. (S) means that the influences of the individual judgments can be separated as above.

2. *Unanimity condition* (U):$f(x_1,...,x_n) = x$ for all x in P. (U) means that if all individuals give the same judgment x, that judgment should also be the synthesized judgment.
3. *Homogeneity condition* (H): $f(ux_1,...,ux_n) = uf(x_1,...,x_n)$ where $u > 0$ and x_k, ux_k $(k = 1,2,...,n)$ are all in P. For ratio judgments (H) means that if all individuals judge a ratio u times as large as another ratio, then the synthesized judgment should also be u times as large.
4. *Power conditions* (P_p): $f(x_1^p,...,x_n^p) = f^p(x_1,...,x_n)$. ($P_2$) for example means that if the kth individual judges the length of a side of a square to be x_k, the synthesized judgment on the area of that square will be given by the square of the synthesized judgment on the length of its side.
   Special case ($R = P_1$):

$$f\left(\frac{1}{x_1},...,\frac{1}{x_n}\right) = \frac{1}{f(x_1,...,x_n)}$$

   (R) is of particular importance in ratio judgments. It means that the synthesized value of the reciprocal of the individual judgments should be the reciprocal of the synthesized value of the original judgments. Aczel and Saaty proved the following theorem [3]:

Theorem 4.2. *The general separable (S) synthesizing functions satisfying the unanimity (U) and homogeneity (H) conditions are the geometric mean and the root-mean-power. If moreover the reciprocal property (R) is assumed even for a single n-tuple $(x_1,...,x_n)$ of the judgments of n individuals, where not all x_k are equal, then only the geometric mean satisfies all the above conditions.*

In any rational consensus, those who know more should, accordingly, influence the consensus more strongly than those who are less knowledgeable. Some people are clearly wiser and more sensible in such matters than others, others may be more powerful and their opinions should be given appropriately greater weight. For such unequal importance of voters not all g's in (S) are the same function. In place of (S), the weighted separability property (WS) is now: $f(x_1,...,x_n) = g_1(x_1)\cdots g_n(x_n)$ (WS) implies that not all judging individuals have the same weight when the judgments are synthesized and the different influences are reflected in the different functions $g_1,...,g_n$.

In this situation, Aczel and Alsina proved the following theorem [3]:

Theorem 4.3. *The general weighted-separable (WS) synthesizing functions with the unanimity (U) and homogeneity (H) properties are the weighted geometric mean $f(x_1,...,x_n) = x_1^{q_1}x_2^{q_2}\cdots x_n^{q_n}$ and the weighted root-mean-powers $f(x_1,...,x_n) = (q_1x_1^\gamma + q_2x_2^\gamma + \cdots + q_nx_n^\gamma)^{1/\gamma}$, where $q_1 + \cdots + q_n = 1$, $q_k > 0$, $k = 1,...,n$, $\gamma > 0$, but otherwise $q_1,\ldots,q_n,\gamma$ are arbitrary constants.*

If f also has the reciprocal property (R) and for a single set of entries $(x_1,...,x_n)$ of judgments of n individuals, where not all x_k are equal, then *only the weighted geometric mean* applies. We give the following theorem which is an explicit statement of the synthesis problem that follows from the previous results, and applies to the second and third cases of the deterministic approach.

Theorem 4.4. *If $x_1^{(i)},\ldots,x_n^{(i)}$, $i=1,\ldots,m$ are rankings of n alternatives by m independent judges and if a_i is the importance of judge i developed from α hierarchy for evaluating the judges, and hence*

$$\sum_{i=1}^{m} a_i = 1, \text{ then } \left(\prod_{i=1}^{m} x_1^{a_i}\right),\ldots,\left(\prod_{i=1}^{m} x_n^{a_i}\right)$$

are the combined ranks of the alternatives for the m judges.

The power or priority of judge i is simply a replication of the judgment of that judge (as if there are as many other judges as indicated by his/her power a_i), which implies multiplying his/her ratio by itself a_i times, and the result follows.

The first requires knowledge of the functions which the particular alternative performs and how well it compares with a standard or benchmark. The second requires comparison with the other alternatives to determine its importance.

4.13.2 On the Construction of Group Choice from Individual Choices

Given a group of individuals, a set of alternatives (with cardinality greater than 2), and individual ordinal preferences for the alternatives, Arrow proved with his Impossibility Theorem that it is impossible to derive a rational group choice (construct a social choice function that aggregates individual preferences) from ordinal preferences of the individuals that satisfy the following four conditions, i.e., at least one of them is violated:

Decisiveness: the aggregation procedure must generally produce a group order.

Unanimity: if all individuals prefer alternative A to alternative B, then the aggregation procedure must produce a group order indicating that the group prefers A to B.

Independence of irrelevant alternatives: given two sets of alternatives which both include A and B, if all individuals prefer A to B in both sets, then the aggregation procedure must produce a group order indicating that the group, given any of the two sets of alternatives, prefers A to B.

No dictator: no single individual preferences determine the group order.

Using the absolute scale approach of the AHP, it can be shown that because now the individual preferences are cardinal rather than ordinal, it is possible to derive a rational group choice satisfying the above four conditions. It is possible because: a) Individual priority scales can always be derived from a set of pairwise cardinal preference judgments as long as they form at least a minimal spanning tree in the completely connected graph of the elements being compared; and b) The cardinal preference judgments associated with group choice belong to an absolute scale that represents the relative intensity of the group preferences [9].

4.14 Rank Preservation and Reversal

This section deals with the old question: Having made a ranking of alternatives that are assumed to be completely independent of one another, what happens to the order in that ranking when a new alternative is added or an existing one deleted? We say there is rank preservation if the order previously determined among the old alternatives is maintained when new alternatives are added (or deleted), and there is rank reversal otherwise. If the alternatives are dependent among themselves then anything can happen to rank. Traditionally people assigned the alternatives a number from a scale one at a time. Consequently, rank is always preserved when a new alternative is added or an old one deleted because the number assigned is unaffected by the numbers assigned to other alternatives. Of course if new criteria are added (or old ones deleted) or judgments are changed, then the rank may also change. As a result of this number assigning approach people came to believe that in reality, barring change in criteria or in judgment, a previously determined rank should not change on introducing or deleting alternatives. Some, blinded by the use of the technique of assigning numbers one at a time, went on to demand it from any other method of measurement even if that method, like the Analytic Hierarchy Process, derived its priorities and ranks by comparing the alternatives thus making them dependent on one another.

The question is whether rank can in fact reverse in practice without adding new criteria or changing judgments on the old alternatives. The answer to this is in the affirmative as numerous examples given by a diversity of people show. There are still people from the old school who attempt to rationalize why the examples are not correct.

Ranking or ordering things according to preference is a purely human activity. On the other hand, ranking according to importance or likelihood is a more scientific or objective activity in which one attempts to project what can happen in the natural world. Nature has no predetermined rank for the preference of alternatives on specially chosen criteria of its own. It is people who establish the criteria and make their ranking on these criteria.

Ranking alternatives on a single criterion involves use of the senses and elementary reasoning and scientific measurement. To rank alternatives on several criteria which they have in common, we need to not only evaluate the alternatives with respect to each criterion but also evaluate the criteria themselves with respect to higher criteria or directly with respect to a goal and then use their derived weights to synthesize the resulting individual rankings of the alternatives to produce an overall ranking. The ranks of the criteria are not usually known in advance, nor are they somehow hidden in our intuitive understanding. We must reason analytically in order to surface and create their priorities. By making comparisons we are concerned with the strength or multiplicity of dominance of one element over another (the smaller or lesser one) taken as the unit with respect to a give property or criterion. It is necessary that the elements be close or homogeneous in order for judgments to be accurate. Otherwise, they are put into homogeneous clusters of successive orders of magnitude with a common pivot from one cluster to an adjacent cluster. The cognitive psychologist Arthur Blumenthal describes in his book that making comparisons is an intrinsic capability of the mind [3]. It is not an invention by man like all scales of measurement which of course each has an origin and an arbitrarily chosen unit based on which the measurements must be interpreted with "subjective judgment" as to their importance in each application..

Ranking belongs to the field of order topology in mathematics in contrast to metric topology used in science, engineering and economics measurement that is concerned with how close differences in the numbers are. In ranking and deriving priorities, closeness depends more on the ratios of the numbers (how much more important is one criterion than another) and not on their numerical differences. It turns out that to derive numbers for which ratios are meaningful does not require the use of scales with arbitrarily chosen units and with an origin. The numbers can simply be relative in a more general way than forming the ratio of numbers obtained from a scale with a unit. Such ratios are dimensionless and belong to an absolute scale that is invariant under the identity transformation.

There are two major types of orientations needed in ranking alternatives: One is philosophical (also ethical and hence normative) which says that rank must be preserved for moral, economic or political reasons. It is concerned with what is good, a subjective concern. The other is technical. It is concerned with what is *right* or reasonable or true and is descriptive based more on reason than on value and is objective allowing rank naturally after making comparisons.

It turns out that there is no single categorical answer to when to preserve rank and when not to. Rank is mostly preserved not because of a law of nature or a result of mathematical dicta, but because it causes less trouble with people due to its apparent fairness. People erroneously believed in the past that rank has to be preserved mostly because of the limitations of the techniques they used to create rank order. It is safe to say that to almost any rule one may wish to put down for preserving rank there is a counter example which violates that rule. The techniques we use to establish rank order are created to serve our needs and values and not the other way around. There is a good reason for that. We believe that ranking not only depends on our criteria but also on the way our brains and memory work to evaluate alternatives even when we have some natural metric to use in measurement. When we deal with intangibles, it is difficult to declare that two alternatives are independent of one another in their ranks, because they are both in our memory and when we think of one we are also aware of the other and have difficulty assigning it a value uninfluenced by our knowledge of the other. Alternatives are related in the judgment process and cannot be arbitrarily assumed to be independent. When alternatives are slightly dependent and they are compared, their ranks can only be preserved by force.

What we believe are the important factors that have bearing on rank and its preservation and reversal have to do with the measurement of tangibles and intangibles in relative or absolute form, by assigning numerical values to alternatives one at a time or through comparisons, and how assumptions of dependence and independence affect the outcome. As we said above, much of the literature about ranking alternatives assumes independence among the alternatives and of the criteria from other criteria and of course also from the alternatives. An exception is the Analytic Network Process (ANP) with its very broad approach to the importance, preference and likelihood (probability) of influences. It is concerned with the dependence of criteria among themselves and on alternatives and of alternatives on other alternatives.

Thus for our purpose the quest for causes of rank reversal narrow down to two basic factors that influence the ranking of alternatives: the method of measurement and the assumption of independence. If alternatives depend on each other or if the criteria depend on the alternatives then it is known that anything can happen to the rank of the alternatives. It appears that treating alternatives as if they are independent is a questionable matter and at best is an assumption that one has had to live with for convenience and not because it can be logically justified in for all applications.

4.15 Conclusions

The ANP is a useful way to deal with complex decisions that involve dependence and feedback analyzed in the context of benefits, opportunities, costs and risks. It has been applied literally to hundreds of examples both real and hypothetical. What is important in decision making is to produce answers that are valid in practice. The ANP has also been validated in several examples. People often argue that judgment is subjective and that one should not expect the outcome to correspond to objective data. But that puts one in the framework of garbage in garbage out without the assurance of the long term validity of the outcome. In addition, most other approaches to decision making are normative. They say, "If you are rational you do as I say". But what people imagine is best to do and what conditions their decisions face after they are made can be very far apart in the real world. That is why the framework of the ANP is descriptive as in science rather than normative and prescriptive. It produces outcomes that are best not simple according to the decision maker's values, but also to the risks and hazards faced by the decision.

It is unfortunate that there are people who use fuzzy sets without proof to alter the AHP when it is known that fuzzy applications to decision making have been ranked as the worst among all methods. Buede and Maxwell [1] write about their findings, "*These experiments demonstrated that the MAVT and AHP techniques, when provided with the same decision outcome data, very often identify the same alternatives as 'best'. The other techniques are noticeably less consistent with MAVT, the Fuzzy algorithm being the least consistent*". The fundamental scale used in the AHP/ANP to represent judgments is already fuzzy. To fuzzify it further does not improve the outcome as we have shown through numerous examples (Buede and Maxwell, 1995). The intention of fuzzy seems to be to perturb the judgments in the AHP. It is already known in mathematics that perturbing the entries of a matrix perturbs the eigenvector by a small amount but not necessarily in a more valid direction. We invite the reader to examine my latest book on the matter. It is regrettable that air post often costs more than the price of the paperback book itself.

The SuperDecisions software used to analyze complex decisions is named after the supermatrix. It can be downloaded free from creativedecisions.com. and is available free on the internet along with a manual together with numerous applications to enable the reader to apply it to his/her decision. Alternatively, go to http://www.superdecisions.com/ saaty and download the SuperDecisions software. The installation file is the .exe file in the software folder. The serial number is located in the .doc file that is in the same folder. The important thing may be not the software but the models which are in a separate folder called models. The military are constantly involved in making complex decisions and appear to like using the ANP and investing in its development. Why do we do all this with so much effort? Because we believe strongly in the creativity of the human race and hope that our world will become increasingly more rational in making its decisions and resolving its conflicts.

References

1. D. Buede and D.T. Maxwell. Rank disagreement: A comparison of multi-criteria methodologies. *Journal of Multi-Criteria Decision Analysis*, 5:1–21, 1995.
2. T.L. Saaty. *Theory and Applications of the Analytic Network Process*. RWS Publications, Pittsburgh, PA, 2005.
3. T.L. Saaty. *Fundamentals of the Analytic Hierarchy Process*. RWS Publications, Pittsburgh, PA, 2006.
4. T.L. Saaty and Y. Cho. The decision by the US congress on China's trade status: A multicriteria analysis. *Socio-Economic Planning Sciences*, 35:243–252, 2001.
5. T.L. Saaty and M. Ozdemir. *The Encyclicon*. RWS Publications, Pittsburgh, PA, 2005.
6. T.L. Saaty and K. Peniwati. *Group Decision Making*. RWS Publications, Pittsburgh, PA, 2008.
7. T.L. Saaty and M. Sodenkamp. Making decisions in hierarchic and network systems. *International Journal of Applied Decision Sciences*, 1(1):24–79, 2008.
8. T.L. Saaty and L.T. Tran. On the invalidity of fuzzifying numerical judgments in the analytic hierarchy process. *Mathematical and Computer Modelling*, 46(7-8):962–975, 2007.
9. T.L. Saaty and L.G. Vargas. The possibility of group welfare functions. *Information Technology and Decision Making*, 4(2):167–176, 2005.

Chapter 5
Preference Programming – Multicriteria Weighting Models under Incomplete Information

Ahti Salo and Raimo P. Hämäläinen

Abstract Useful decision recommendations can often be provided even if the model parameters are not exactly specified. The recognition of this fact has spurred the development of multicriteria methods which are capable of admitting and synthesizing incomplete preference information in hierarchical weighting models. These methods share similarities in that they (i) accommodate incomplete preference information through set inclusion, (ii) offer decision recommendations based on dominance concepts and decision rules, and (iii) support the iterative exploration of the decision maker's preferences. In this Chapter, we review these methods which are jointly referred to by the term 'preference programming'. Specifically, we discuss the potential benefits of using them, and provide tentative guidelines for their deployment.

5.1 Introduction

Hierarchical weighting methods–such as value trees [30] and the Analytic Hierarchy Process [54]–are widely used in the analysis of decision problems that are characterized by incommensurate objectives, competing alternatives and conflicting stakeholder interests (see, e.g., [21, 28]). In these methods, the decision maker (DM) is engaged in a process where the decision objectives are structured as a hierarchy of attributes. This phase is often one of the most instructive phases of problem solving due to the insights that it may give [8, 29]. In effect, the resulting hierarchical problem representation provides a framework for synthesizing information about

Ahti Salo
Helsinki University of Technology, Systems Analysis Laboratory, P.O. Box 1100, FIN-02015 TKK, Finland e-mail: ahti.salo@tkk.fi

Raimo P. Hämäläinen
Helsinki University of Technology, Systems Analysis Laboratory, P.O. Box 1100, FIN-02015 TKK, Finland e-mail: raimo@tkk.fi

C. Zopounidis and P.M. Pardalos (eds.), *Handbook of Multicriteria Analysis*, Applied Optimization 103, DOI 10.1007/978-3-540-92828-7_5, © Springer-Verlag Berlin Heidelberg 2010

(i) how the alternatives perform on the attributes (=scores) and (ii) how important the attributes are (=weights). Based on these two forms of statements, an overall performance measure can be attached to each alternative.

Yet, most multicriteria methods are based on the assumption that complete information about the model parameters (scores, attribute weights) need to be elicited as 'exact' point estimates. This assumption can be questioned on many grounds. For example, it may be impossible to obtain exact *ex ante* estimates what impacts the alternatives will have on the attributes; and even if such information can be obtained, it may come at a prohibitively high cost, which makes it of interest to examine what tentative conclusions are supported by the information that can be acquired at an affordable cost [33, 71]. This is the motivation for many of the rank-based methods such as SMARTER [17] which converts the DMs ordinal statements into centroid-weights; however, in doing so, an initial incomplete (ordinal) preference specification is essentially transformed into 'exact' cardinal weights which no longer reflect this incompleteness (see [6, 23]) . Furthermore, if the DMs are not confident with providing a complete specification of their preferences, they may regard the results with untrustworthy if they are not confident with the inputs. From the viewpoint of sensitivity analysis, too, it is advisable to examine how recommendations will change when the model parameters vary within plausible limits ([46, 53]).

The above concerns have motivated the development of methods which accommodate incomplete preference information in hierarchical weighting models (see, e.g., [1, 18, 26, 33, 34, 38, 42, 43, 50, 49, 52, 59, 60, 61, 65, 71, 73]). Even though these methods differ in their details, they share many similarities: in particular, they (i) model incomplete preference information through set inclusion, (ii) apply dominance structures and decision rules to derive decision recommendations and, in many cases, (iii) guide the DM during the iterative phases of preference elicitation. In view of these similarities, we therefore employ 'preference programming'–a term that was coined by Arbel [5]–as a general term for all methods which fulfil at least the two first of the above conditions. This term seems pertinent also because these methods engage the DM in an interactive exploration of preferences and offer intermediate results by solving mathematical programming problems.

Apart from numerous incremental methodological contributions, there is a growing number of papers that describe promising real-life applications of preference programming methods (see, e.g., [19, 22, 25, 48]. Yet, not much work has been done to synthesize 'lessons learned' from this applied work. Nor has it been examined in what decision contexts preference programming methods work best, or how they should be best employed in such contexts. We therefore give a structured review of these methods and identify conditions which suggest that the modeling of incomplete information can be particularly helpful. We also argue that, in some conditions, preference programming methods may outperform 'conventional' approaches, particularly if the costs of preference elicitation are high, or if there is a need to focus the analysis on the few most preferred alternatives.

This review Chapter is structured as follows. Section 5.2 describes the essential features of preference programming. Section 5.3 reviews selected applications and presents relevant software tools. Section 5.4 considers what problem characteristics may call for the deployment of preference programming methods. Section 5.5 provides some tentative guidelines for preference elicitation.

5.2 Key Characteristics of Preference Programming

Among the different hierarchical weighting methods, the multiattribute value theory (MAVT) [30] has a strong axiomatic foundation in measurement theory. Specifically, if the DM's preference relation satisfies axioms that characterize rational decision making, this relation has a *value function* representation with the help of which the overall value of an alternative can be expressed as the attribute-weighted sum of its attribute-specific values (i.e., scores).

In terms of the aggregation of the overall performance measure, MAVT shares similarities with the Analytic Hierarchy Process [54] where the overall priority weight of an alternative is expressed as the weighted sum of its local priorities with regard to the attributes at the lowest level of the hierarchy of objectives. In view of these similarities, we therefore provide the following generic formulation of additive preference models, in the understanding that this formulation can be interpreted in the context of MAVT and AHP models. A more detailed comparative analysis of the two methodologies can be found in [62].

5.2.1 Additive Preference Representation

We assume there are n attributes at the lowest level of the hierarchical representation of decision objectives. The importance of the i-th attribute is indicated by a non-negative *weight* $w_i \in [0,1]$. These attribute weights are normalized so that they add up to one, i.e., $\sum_{i=1}^{n} w_i = 1$.

There are m alternatives $x^1,\ldots,x^m$. The achievement level of the j-th alternative on the i-th attribute is denoted by x_i^j (for instance, this could be the fuel consumption of a car). The single-attribute value associated with this achievement level is called the *score* $v_i(x_i^j) = v_i^j \in [0,1]$. These scores map the actual achievement levels onto a possibly non-linear scale of subjective value. The overall value of alternative x^j is expressed by the sum $V(x^j) = \sum_{i=1}^{n} w_i v_i^j$ which is based on the model parameters (i.e., weights $w = (w_1,\ldots,w_n) \in W = \{w \mid w_i \geq 0, \sum_{i=1}^{n} w_i = 1\}$ and scores $v^j = (v_1^j,\ldots,v_n^j), j = 1,\ldots,m$). In preference programming, incomplete preference information is typically modeled through set inclusion. Specifically, the DM's preference statements are transformed into constraints on the model parameters (i.e., attribute weights and score vectors). For instance, as shown in Figure 5.1, the DM could state that the score of an alternative like Job B is between 50 % and 70 % of the maximum score of 1; or that the weight of attribute B is at least half of the weight of attribute A, but at most twice as high as the weight of attribute A.

These kinds of constraints define sets of feasible weights and score vectors S_w, S_{v^j} (where $w \in S_w \subset W$ and $v^j = (v_1^j, \ldots, v_n^j) \in S_j \subset [0,1]^n, j = 1, \ldots, m$). For any consistent set of DM's preference statements, the resulting *feasible sets* will be non-empty. If the DM's preference statements are not very informative, these feasible sets will be large because the corresponding constraints will be satisfied by many parameters.

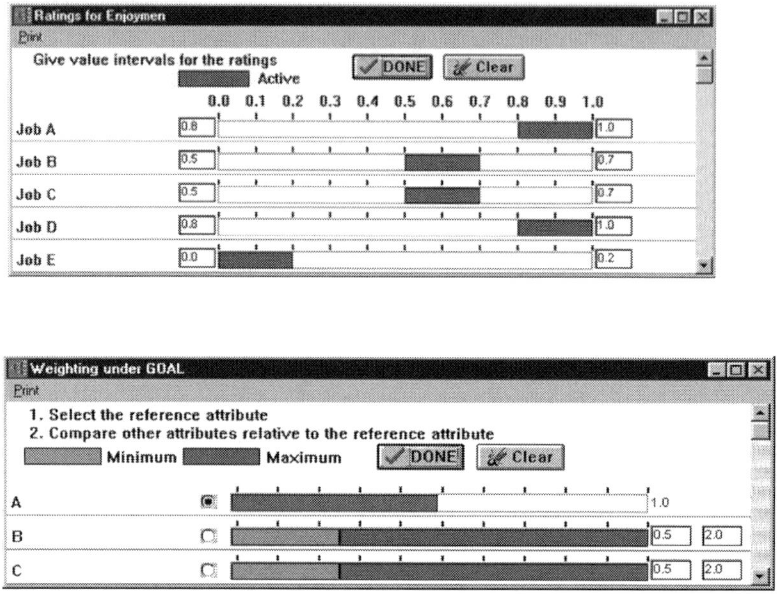

Fig. 5.1 Elicitation screens for the specification of incomplete information about scores and attribute weights in WinPRE

Table 5.1 gives an overview of selected preference programming methods, with particular attention to preference elicitation and the specific characteristcs of the methods. The evolution of methods has progressed from the mere incorporation of incomplete information towards the delivery of supplementary information–such as consistency bounds, measures of incompleteness, and decision rules–that help the DM to decide whether or not the elicitation phase should be continued and, if so, how the continuation of the process should be organized. Another trend is the increasing availability of decision support tools for interactive decision support processes.

It is worth noting that even other terms–such as 'imprecise information' (e.g., [18, 60] or 'partial information' (e.g., [9])–have been employed when referring to set inclusion. Yet we feel that 'incompleteness' is more adequate as a term, because it stresses that while the available preference information–as captured by the sets

Table 5.1 Selected examples of preference programming methods

Method	Score elicitation	Weight elicitation	Remarks
ARIADNE, White et al. [73]	Upper and lower bounds on scores	Linear constraints on attribute weights	Eliminates inconsistencies through linear programming
Hazen [27]	Completely specified score information	Linear constraints on attribute weights	Gives an extensive mathematical treatment of optimality conditions
HOPIE, Weber [70]	Derived indirectly from holistic comparisons among alternatives	Derived indirectly from holistic comparisons among alternatives	Offers recommendations from the consideration of probability distributions over the alternatives' values
PAIRS [22]	Lower and upper bounds on score information	Interval-valued statements about ratios of attribute weights	Computes dominance structures through hierarhically structured linear optimization problems
Preference Programming [61]	Interval-valued ratio statements using AHP-style pairwise comparisons	Interval-valued ratio statements using AHP-style pairwise comparisons	Offers an ambiguity index for measuring the amount of incompleteness in the preference specification
Ahn et al. [2]	Linear constraints on alternatives' scores	Linear constraints on attribute weights	Suggests the use of aggregated net preference as a decision rule
PRIME [59]	Upper or lower bounds on scores	Interval-valued statements about ratios of value differences	Introduces several decision rules and examines their computational properties
Eum et al. [18]	Both complete and incomplete score information considered	Several kinds of preference statements that correspond to linear constraints on weights	Offers a taxonomy of several forms of incomplete information
RICH, Salo and Punkka [65]	Lower and upper bounds on attribute-specific scores	Incomplete ordinal preference information about the relative importance of attributes	Introduces an incompleteness measure for ordinal preference information
Interval SMART/SWING [47]	Score intervals about the alternatives	Interval-valued ratio statements in SMART/SWING	Extends the SMART/SWING method by allowing the choice of reference attributes
Smart Swaps [44]	Complete score information	Dominance statements among alternatives	Supports the Even Swaps process by using preference programming to identify practically dominated alternatives and candidate attributes for the next swap
RPM, Liesiö et al. [40]	Lower and upper bounds on attribute-specific scores	Incomplete ordinal preference information and also other forms based on set inclusion	Extends dominance concepts to multicriteria portfolio selection problems

S_w and S_j–may not imply a full ranking of alternatives, such information could, in principle, be completed by eliciting further preference statements. 'Incompleteness' also appears better on the grounds that the constraints on feasible weights and scores are not 'imprecise': for instance, the lower and upper bounds of interval-valued ratio statements in weight elicitation are crisp numbers with no associated uncertainties.

Although we do not consider approaches based on probabilistic modelling, fuzzy sets (e.g., [9, 10, 57]) or outranking relationships (e.g., [66]), there are computational and other parallels to these other approaches. The consideration of incompletely specified probabilities in Bayesian updating, for instance, leads to related problem formulations [58]. Thus, many of our observations apply in other settings as well.

5.2.2 *Preference Elicitation*

The elicitation techniques of preference programming methods often extend those employed by the more conventional decision analytic methods. For example, many preference programming methods allow the DM to provide interval-valued estimates instead of exact crisp numerical estimates. The popular ratio-based techniques –such as the AHP [54] and SMART [16]– have also been extended to methods where the DM may provide interval-valued statements about the relative importance of attributes [5, 47, 60, 61].

Park et al. [51] present the following taxonomy which illustrates different approaches to the elicitation of incomplete information. Specifically, in the context of attribute weights, they consider both interval-valued and ordinal preference statements defined by: (1) a weak ranking (i.e., $w_i \geq w_j$), (2) a strict ranking ($w_i - w_j \geq \alpha_{ij}$) (3) a ranking with multiples ($w_i \geq \alpha_{ij} w_j$), (4) an interval form ($\alpha_i \leq w_i \leq \alpha_i + \varepsilon_i$), (5) a ranking of differences ($w_i - w_j \geq w_k - w_l$ for $j \neq k \neq l$,); here $\alpha_{ij}, \varepsilon_i \; \forall \; i,k$. All of these statements correspond to linear constraints on attribute weights.

In an extension of these elicitation techniques, Salo and Punkka [65] develop the *Rank Inclusion in Criteria Hierarchies* (RICH) approach which allows the DM to provide incomplete ordinal information about the relative importance of attributes (e.g., 'cost is among the three most important attributes' or 'the most important attribute is either cost or quality'). Such statements correspond to constraints that define possibly non-convex sets of feasible attribute weights. The resulting sets can be readily examined to obtain decision recommendations based on the application of dominance concepts and decision rules.

5.2.3 Dominance Structures

Once the incomplete preference specification has been elicited (as characterized by feasible weights S_w and scores S_j), it is of interest to examine what, if any, inferences can be made about what alternatives are 'better' than the others.

These inferences can be based on the concept of (pairwise) *dominance*. In particular, alternative x^k dominates x^l if the overall value of x^k is higher than that of x^k for all feasible model parameters and strictly higher for some parameters. The case of dominance can be checked by considering whether or not the inequality

$$V(x^k) = \sum_{i=1}^{n} w_i v_i^k \geq \sum_{i=1}^{n} w_i v_i^l = V(x^l), \tag{5.1}$$

holds for all combinations of feasible weights $w \in S_w$ and scores $v^j \in S_j$. This definition establishes a transitive and asymmetric binary relation among the alternatives. Moreover, if (5.1) holds, the value of alternative x^k will be at least as high as that of x^l, even if additional preference statements were to be acquired until the sets of feasible weights and scores become singletons.

5.2.4 Decision Rules

If there are several non-dominated alternatives, it is not possible to derive conclusive statements about which alternative is the 'best' one. This is because for any alternative x^k, there exists a combination of feasible weights and scores such that the overall value of some other alternative x^l will be higher than that of x^k. In consequence, other principles–called *decision rules*–can be applied to derive a decision recommendation. Several such decision rules have been proposed:

1. **Choice of representative parameters:** Based on 'representative' parameters from feasible regions, the recommendation can be based on the comparison of alternatives' overall values for some representative parameters. For instance, the PRIME method uses, as one of several possibilities, *central weights* that are near the center of the feasible weight set [63]. Even the approaches to the computation of rank based weights (e.g., rank sum, rank reciprocal, rank exponent, rank order centroid; see, [6, 17, 67]) can be viewed as ways of converting ordinal preference information into representative weight vectors.
2. **Alternatives' value ranges:** Recommendations can be based on an analysis of the ranges of values that alternatives may take. Examples of such rules include the *maximax rule* (i.e., choose the alternative which has the highest possible overall value), *maximin* (i.e., choose the alternative for which the smallest possible overall value is the highest among alternatives) and *central values* (i.e., choose the alternative for which the mid-point of the value interval is highest) (see [63]).

The advantage of these rules is that they can be readily computed and communicated.

3. **Pairwise value differences between alternatives:** Decision rules can be based on measures on how well alternatives perform relative to each other. One such measure is the *maximum loss of value* which indicates how much *more* value the DM could at most acquire in comparison with x^i, if she would choose some other alternative [63]). The corresponding *minimax regret* decision rule recommends the alternative which has the smallest maximum loss value. This rule is appealing because it allows the DM to take an informed decision on whether or not the possible loss of value is small enough so that elicitation efforts can be stopped. Even measures of preference strength (see, e.g., [2])–which are computed by aggregating value differences across several alternatives–belong to this class of decision rules.
4. **Maximization of expected value:** If there are grounds for making plausible assumptions about how probable the feasible parameters are, it is possible to recommend the alternative with the highest expected overall value (see, e.g., [70]). Although this decision rule is conceptually appealing, it is not necessarily easy to apply because the elicitation of required probability distributions is likely to be a major effort. Also computational difficulties may be encountered.
5. **Likelihood maximization of potentially optimal alternatives:** If probability distributions on the feasible regions can be elicited, the alternative which has the highest probability of receiving the largest overall value can be offered as the decision recommendation. This approach is, in effect, the SMAA method [35] which in its basic formulation recommends potentially optimal alternatives. Subsequently, this method has been extended so that it considers not only non-dominated alternatives, but considers the alternatives' relative rankings and, based, on an analysis of these, may recommend alternatives that are not necessarily potentially optimal for any combination of feasible parameters (see, e.g., [37, 36].

Although several decision rules have been proposed, the literature does not offer conclusive guidance as to what decision rules should be applied in specific decision contexts. Simulation studies suggest that decision rules based on the use of central values tend to outperform others in terms of minimizing the expected loss of value [59]. But even this tentative conclusion depends on context-specific assumptions (e.g., absence of correlations among alternatives). It therefore appears that further computational and empirical studies are needed.

It is also possible, particularly in group decision making, that an examination of different recommendations based on different decision rules may provoke discussions about which decision rules are 'better'. Such discussions may be driven by strategically motivated arguments if the DMs defend certain decision rules on the grounds that these favor their own favorite alternatives. But rather than focusing on the comparative merits of decision rules, it may be instructive to examine several decision rules in parallel, or to agree what decision rules will be applied before the phases of preference elicitation and synthesis are started.

5.2.5 Management of Inconsistencies

The derivation of decision recommendations from an incomplete preference specification presumes that the DM's preference statements are consistent and thus define non-empty sets of feasible weights and scores. Yet, without adequate decision support, the DM may be inclined to provide preference statements that are not consistent with the previous statement, in which case the set of feasible parameters would become empty. Two main approaches (which are also supported by software tools, see Section 5.3) have been proposed to avoid this possibility:

- **Consistency restoration:** Taking the set of conflicting constraints as a point of departure, the DM can be requested to modify or withdraw earlier statements until the remaining, possibly revised constraints are not in conflict with each other any more (see, e.g., [33, 73]).
- **Consistency preservation:** Before the elicitation of each new preference statement, full information about the implications of earlier preference statements can be computed and presented to the DM, to ensure that the new statement is not in conflict with the earlier ones (see, e.g., by [56, 60, 63]).

Consistency restoration may be problematic if the DM is not able or willing to revisit earlier statements. Furthermore, the withdrawal of earlier statements may undermine the credibility of the analysis, because it insinuates that there are 'errors' in some inputs without guaranteeing that the other inputs are less 'erroneous'. Also, although automated procedures can be used to identify the least number of constraints that should be removed to re-establish consistency, such procedures are computational interventions with little interaction on the part of the DM (see, e.g., [55, 73]). At worst, this approach may thus lead to the removal of statements that the DM feels most confident about. Computationally, however, consistency restoration can be applied in a batch mode so that possible problems with inconsistencies–if they do arise–can be addressed after the preference elicitation phase.

Consistency preservation requires that the implications of earlier preference statements are presented to the DM whenever new preference statements are elicited. These implications can be presented through the *consistency bounds* which convey the smallest and upper bounds that previously entered ratio-statements imply for the ratio statement that is to be elicited next [60]. For instance, in Figure 5.2, the two statements 'neither attribute A nor attribute B is more than two times more important than the other' and 'attribute C is twice as important as attribute B' logically imply that 'attribute C is more important than attribute A, but no more than twice as important'. If the decision maker is willing to provide a new statement that is within these consistency bounds, the new augmented constraint set will be consistent, too; however, if she wishes to enter a statement that is *not* within these bounds, some of the earlier statements would have be to revisited and revised. As a result, the management of inconsistencies has broader implications for preference elicitation: should the DM be encouraged to provide relatively 'narrow' statements (which tend to support more conclusive dominance results, but are more prone to inconsistencies) or 'broad' statements (which entail a lower risk of inconsistencies, but are likely to produce less conclusive dominance results)? (see also [47]). Related issues of consistency preservation arise also in group decision contexts: for example, when synthesizing individual statements, the correct interpretation of criterion weights may have to be ensured through the explicit consideration of trade-offs [22, 25].

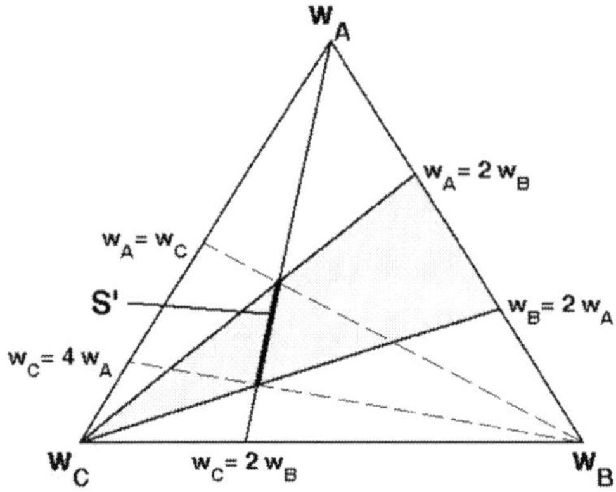

Fig. 5.2 Consistency bounds implied by two ratio-based comparisons of attributes (A and B, B and C) for the third pairwise comparison (A and C)

5.2.6 Dominance Structures, Decision Rules and Rank Reversals

A much contested topic in the literature on hierarchical weighting models is the phenomenon of *rank reversals* which, in brief terms, means that the introduction or removal of an alternative changes the relative ranks of *other* alternatives. Such changes would suggest that the DM's preferences for the alternatives that are being compared depend not only on these alternatives, but also on what other alternatives may be included in or excluded from the analysis. Rank reversals have been a source of considerable controversy, and many researchers have regarded them as a major flaw in decision support methodologies such as the Analytic Hierarchy Process (see, e.g., [7, 11, 15, 62]).

It is therefore pertinent to ask which decision rules may exhibit rank reversals. To begin with, we may conclude that rank reversal cannot occur in decision rules which attach a single performance measure to an alternative, based on the weighted aggregation of numerical parameters that are not affected by the other alternatives, for instance through the use of normalization rules (this is not the case in the AHP where the introduction of an additional alternative typically affects the local priorities of *all* alternatives due to normalization [7, 15, 62]). Clearly, no decision rules based on such *unitary* performance measures will exhibit rank reversals, even if additional alternatives are introduced or existing alternatives are removed. This observation implies that several of the decision rules discussed above (e.g., maximax, maximin, central values, central weights) are immune to rank reversals.

However, if the scores can be impacted by other alternatives, or if the decision rule is based on the comparison of several alternatives, rank reversals may be possible. This is the case, for instance when using the minimization of maximum loss of value decision rule. To demonstrate this, assume that there are two attributes and two alternatives with scores $v^1 = (v_1^1, v_2^1) = (0.4, 0.6)$ and $v^2 = (v_1^2, v_2^2) = (0.6, 0.3)$, and that no weight information is available (i.e., the set of feasible weights is $W = \{(w_1, w_2) \mid w_1 + w_2 = 1, w_i \geq 0, i = 1, 2\}$). Then, the first of these alternatives x^1 has the smaller maximum loss value, because $\max_{w \in W}[(6-4)w_1 + 0w_2] = 2$ while the corresponding maximum for x^2 is 3. Now, if a third alternative with scores $v^3 = (v_1^3, v_2^3) = (0.8, 0.1)$ is added, the maximum value losses for the alternatives become 4,3 and 5, respectively, indicating that x^2 now becomes the recommended alternative, although x^1 was the recommended alternative before the introduction of x^3. Proceeding in much the same way, one can show that even other measures based on the comparison of value differences among two or more alternatives (e.g, aggregated net intensity; [32, 2]) may exhibit rank reversals.

Another example is the Stochastic Multiobjective Acceptability Analysis (SMAA; Lahdelma et al. [35]) where the decision recommendation is based on the comparison of the relative sizes over which a given alternative is optimal. Technically, this means that the set $W(x^j) = \{w \in W \mid \sum_{i=1}^n w_i v_i^j \geq \sum_{i=1}^n w_i v_i^k \ \forall \ x_k \neq x_j\}$ (where $W = \{(w_1, \ldots, w_n) \mid \sum_{i=1}^n w_i = 1, w_i \geq 0, i = 1, \ldots, n\}$) contains those attribute weights for which alternative x^j will have the highest aggregate overall value. This weight set is used to establish a performance measure–called the *acceptability index* $AI(x^j)$–which is defined as the ratio between the volumes of $W(x^j)$ and W. The larger the acceptability index, the more support it will receive relative to the other alternatives.

To demonstrate that SMAA, too, exhibits rank reversals, assume that there are two attributes and two alternatives x^1, x^2 characterized by the score information $v^1 = (v_1^1, v_2^1) = (0.7, 0)$ and $(v_1^2, v_2^2) = (0, 0.6)$. Because the corresponding weight sets are $W(x^1) = \{w \in W \mid w_1 \geq \frac{6}{13}\}$ and $W(x^2) = \{w \in W \mid w_1 \leq \frac{6}{13}\}$, alternative x^1 has the larger acceptability index in SMAA and is therefore the recommended alternative. Next, assume that a third alternative x^3 with scores $v^3 = (v_1^3, v_2^3) = (0.6, 0.225)$ is introduced. The weight sets then become $W(x^1) = \{w \in W \mid w_1 \geq \frac{9}{13}\}, W(x^3) = \{w \in W \mid \frac{5}{13} \leq w_1 \leq \frac{9}{13}\}$ and $W(x^2) = \{w \in W \mid w_1 \leq \frac{5}{13}\}$. Thus, the second alternative x^2 now obtains the highest acceptability index and becomes the recommended alternative, although its acceptability index was smaller than that of alternative x^1 before the third alternative was introduced: a rank reversal has occurred.

Positive reports from applications (see, e.g., [37]) suggest that DMs may feel comfortable with the SMAA method. Yet the possibility of rank reversals casts some doubt on its validity as a decision support methodology. Another source of potential concern is that the early variants of SMAA do not encourage the DM to learn about her preferences by making explicit preference statements. Thus, although the weight set $W(x^j)$ that yields support for alternative x^j may be larger than the other weight sets for the alternatives $W(x^k), k \neq j$, this set may consist of weights that are not aligned with the DM's (unstated) preferences. Thus, further validity checks may be needed to ensure that the weights on which is the acceptability index is based are compatible with the DM's preferences (cf. [69]).

If rank reversals are deemed unacceptable, the above discussion suggests that unitary performance measures should be given precedence over other decision rules in the derivation of decision recommendations. The other decision rules may still be useful for other purposes: for example, the computation of the maximum loss of value for alternatives highlights just how much aggregate value the decision maker may forego by choosing an alternative when dominance results do not hold. This measure also gives an upper bound on how much more additional value could be, at best, attained by continuing preference elicitation efforts.

5.3 Case Studies and Decision Support Tools

The literature on preference programming has gradually matured from purely methodological contributions towards the deployment of these methods in high-impact applications. In an early case study, Anandaligam [4] describes how incomplete preference information can be harnessed in the comparison of strategies for mitigating the harmful consequences of acid raid. Hämäläinen et al. [22] consider the use of preference programming in assisting groups of decision makers in the comparison of energy policy options. Hämäläinen and Pöyhönen [25] report experiences from the development of policies for traffic management, highlighting the impacts of preference programming on the decision outcome and the decision support process. Hämäläinen et al. [24] describe a multi-stakeholder participatory process where incomplete preference information was employed to support the comparison of alternative countermeasures in nuclear emergency planning.

Cristiano et al. [12] provide support for the optimal design of a surgical product by accommodating incomplete preference information in quality function deployment. Gustafsson et al. [19] apply the PRIME method to the valuation of a high technology company and illustrate how preference programming can be used in scenario-based forecasting problems. Salo and Liesiö [64] describe a series of workshops where the RICH method [65] was employed to help Scandinavian research managers establish priorities for research and technology development activities in an international research program. Ojanen et al. [48] prioritize alternative risk management measures by developing a hierarchical representation of relevant criteria and by soliciting ordinal preference statements from two groups of decision makers (i.e., client perspective, utility perspective). Alanne et al. [3] assess building technologies by using the PAIRS method [60] to account for economic and technological uncertainties.

Because preference programming methods involve more computations than conventional approaches based on the elicitation of point estimates, the availability of adequate software tools is an essential precondition for their deployment. At present,

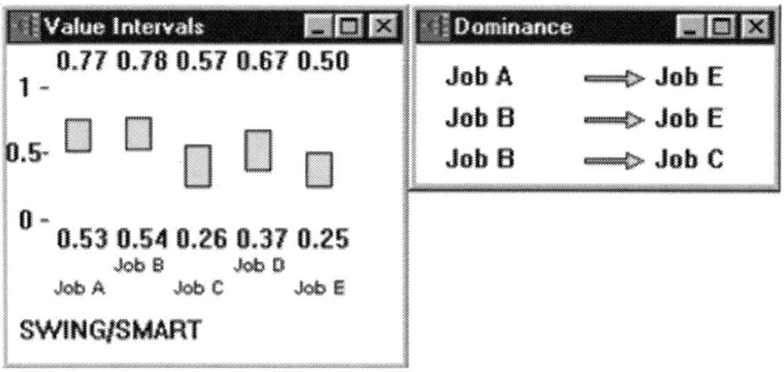

Fig. 5.3 A screenshot from WinPRE with overall value intervals and dominance structures

there are several software tools for preference programming. WINPRE (Workbench for INteractive PREference Programming) [1] provides computerized support for both PAIRS [60] and the use of interval valued-statements in the AHP [61] (see Figure 5.3 for examples of the user interface). The RINGS system [31] allows the DMs to analyze range-based information in multi-attribute group decision making. PRIME Decisions [19] features *elicitation tours* which assist the DM in the specification of interval-valued ratio statements. VIP Analysis and its extensions [13, 14] support groups of decision makers who seek to reach a consensus.

RICH Decisions [2] is a web-based decision support tool which admits and processes incomplete ordinal preference information in accordance with the RICH method [41, 65]. A web-front end is also offered by RPM Decisions [3] which provides support for the Robust Portfolio Modeling methodology [40, 39], designed for the project portfolio selection in the presence multiple attributes and possibly incomplete information about attribute weights and project scores. Interactive web-functionalities are also provided by the Even Swaps-software which applics preference programming to develop suggestions for what swaps the DM should consider next when using the Even Swaps-method [44, 45].

5.4 Experiences from Applications

Preference programming methods have already been applied across very different domains. Indeed, experiences from these studies warrant some remarks about the benefits of these methods and preconditions for their successful deployment:

[1] See http://www.sal.tkk.fi/English/Downloadables/winpre.html

[2] See http://www.rich.hut.fi/

[3] See http://http.rpm.tkk.fi/

- Preference programming methods make it possible to check the implications of incomplete information even at the earliest phases of the analysis. Thus, the preferred alternative(s) can be possibly identified more easily and quickly while the resulting recommendations are still robust and methodological sound. Another important benefit is that these methods enable *iterative decision support processes* where tentative results can be provided early on, which is useful because such results help engage the DMs into the decision support process. Moreover, the subsequent phases of information elicitation can be focused on those attributes and alternatives about which the additional information is likely to contribute most to the development of more conclusive results.
- The numerous ways in which incomplete information can be elicited from the DM makes it necessary to plan and implement *elicitation processes* through which this information is acquired. Without such planning, the elicitation process may appear unstructured, or it may fail to ensure that all alternatives are treated with the same degree of thoroughness. Sufficient prior planning is also motivated by effectiveness, because the number of numerical estimates elicited from the DM may be larger than in conventional approaches: for instance, when stating interval-valued score information, the DM needs to state two crisp numbers as opposed to a single point estimate.
- If the preference model remains incomplete after the initial preference elicitation efforts (in the sense that the sets of feasible parameter remain large; see [56] for measures of incompleteness), dominance structures are unlikely to identify a single dominating alternative, particularly if there are many alternatives whose scores are correlated. In such situations, it is pertinent to examine different decision rules in order to gain complementary insights in facilitated decision workshops, for instance (e.g., [25]).

Overall, preference programming holds considerable potential in decision problems where reliable information about the DMs' preferences or the alternatives' impacts cannot be readily obtained. Such settings include, for instance, the evaluation of risk mitigation strategies in highly uncertain domains where the application of the *precautionary principle* is warranted (see, e.g., [24, 59, 68]). Here, instead of recommending seemingly 'optimal' alternatives, the limitations of information in such settings need to be recognized, for instance by giving precedence to *robust* alternatives that perform satisfactorily across the full range of plausible parameter values.

The recognition that the costs of information elicitation can be significant makes it possible to characterize novel uses of preference programming. For instance, these methods can be employed for the purpose of *screening* a large number of alternatives to a smaller set of non-dominated alternatives. Such a screening process can start with an initial phase where some information about all alternatives is first analyzed before proceeding to a more detailed analysis of the remaining non-dominated alternatives. Specifically, if the number of initial alternatives is large and the costs of information elicitation are relatively high, a phased analysis if this kind is likely to be more cost-effective than a process which seeks to acquire complete information about all alternatives at the outset.

5.5 Guidelines for Applications

The fundamental philosophy in the preference programming methods is quite simple. First, make an effort to elicit as much information as is reasonably possible. Second, check if the elicited information makes it possible to identify a dominating alternative, or an alternative that can be selected with a reasonable degree of confidence (as measured, say, by the maximum possible loss of value). If this is the case, present this alternative as the recommended decision. Otherwise, seek possibilities for eliciting additional score and weight information by adopting elicitation strategies that are likely to reduce the set of non-dominated alternatives.

The phases in the above process highlight that there are close links between the steps of (i) eliciting preference information, (ii) computing dominance structures and decision rules and (iii) terminating the decision support process. For instance, recommendations based on decision rules will be contingent on how many or few statements have been elicited from the DM up until the point where these rules are applied. As an example, assume that the set of available information is not balanced (meaning that some parameters are almost exactly specified while there is hardly any information about others). Then, there is a possibility that one alternative may be (dis)favorably evaluated in comparison with others, only because the information elicitation process has not yet progressed to the point where statements about its parameters are elicited. In consequence, attention must be paid to questions of how elicitation questions are posed to the DM, and when and how intermediate results are presented.

- An attempt should be made to obtain equally 'complete' score information about all decision alternatives, in the sense that the DM is equally confident in that their preference specifications contain the 'true' scores. Such an interpretation can be encouraged by interpreting the lower and upper bounds in terms of symmetric confidence intervals to the preference statements, for example. One may also apply fuzzy mathematics in the aggregation of such confidence intervals (see, e.g., [57]).
- The same level of 'thoroughness' should be pursued also when assessing the relative importance of attributes. Otherwise, for instance, there is a possibility that an alternative will appear weak if it has its highest scores (relative to those of other alternatives) on attributes about which less weight information has been provided. In this case, the computations in applying the maximin decision rule would assign little weight to these attributes so that the alternative would have a small maximin value–even if its standing might improve when more information is elicited about these attributes.
- From the viewpoint of transparency, it may be advisable *not* to mix different types of preference elicitation questions in weight elicitation, because this may define a feasible region whose geometric structure is less symmetric than what would be obtained by using questions of the same type (e.g., interval-valued ratio statements). Another benefit of restricting the number of question types is that

this may results in an elicitation process that can be more readily understood by the DMs (see, e.g., [47]).

5.6 Outstanding Research Questions

While the methodological contributions and reported case studies offer important insights into the uses of preference programming, there are nevertheless several areas that call for further research:

- **Development of elicitation approaches:** Further work is needed on how incomplete preference information can be best elicited from the DMs. Some advances have been made by organizing the elicitation process into structured subsequences (see, e.g., [47]). The PRIME Decisions tool encourages the DM to complete several elicitation tours which consist of sequences of elicitation tasks [19]. In effect, effective and defensible elicitation strategies can be built from such these kinds of 'building blocks' with the help of which the decision support process can be aligned with the above guidelines. Related research should also address to what extent preference programming approaches may mitigate behavioral biases in the elicitation of attribute weights or possibly even create new ones (see, e.g., [23, 72]).
- **Impacts in different contexts and uses:** An explicit *ante* consideration of how much better decisions (say, as measured by the expected aggregate value or proportion of optimal choices) can be reached through preference programming. Advances in this area can be supported, among others, through simulation studies that analyze which elicitation approaches and decision rules perform best, subject to varying assumptions about the number of attributes and alternatives, distribution of attribute weights, correlations among alternatives, and costs of preference elicitation, among others. The results in [6, 59, 65] exemplify results from this kind of research, even though they focus mostly on ratio statements and ordinal preference information. Quite importantly, preference programming methods also enable various *ex post* sensitivity analyses with regard to all model parameters [47]).
- **Approaches for group decision support:** Many authors have argued that preference programming methods are particularly suitable for group decision making (see, e.g., [25, 22, 32]). There is, however, call for empirical evidence on how these methods can be best deployed in group settings. Interestingly enough, the group context also makes possible to introduce entirely new decision making principles. For instance, the group members may agree that each member shall acquire at least one third of the total value of his or her personally preferred alternative. After the introduction of such cross-cutting requirements, the aggregate group value can then be maximized, in the assurance that the resulting recommendation will comply with the principles that the group has set for itself.
- **Development of software tools and reflective case studies:** The computations and visualizations in preference programming methods typically require dedi-

cated software tools. Although several such tools exist (e.g., WinPRE, PRIME Decisions, RICH Decisions; [20]), further attention must be given to tool development. The integration of such tools with other IT systems may lead to new applications: for instance, one could envisage computerized search agents that would use preference programming methods to identify items that would be of most interest to potential buyers. Furthermore, because the benefits of preference programming techniques are ultimately realized in the context of applications, there is a need for reflective case studies. Among other things, such studies need to address the qualitative impacts that the application of preference programming methods may have on the decision support process (e.g., satisfaction with the process; commitment to the decision).

5.7 Conclusion

We have reviewed preference programming methods which accommodate incomplete preference information in hierarchical weighting models and synthesize such information into well-founded decision recommendations. By building on experiences from reported case studies, we have also developed guidelines for the deployment of these methods. These guidelines help ensure, among other things, that the consecutive phases of preference elicitation and preference synthesis are properly interlinked, and that all alternatives are treated equally during the development of decision recommendations.

More specifically, preference programming methods seem particularly suitable in decision problems where the elicitation of complete information is either impossible or involves prohibitively high costs, or where the DMs are simply more prepared to characterize their preferences through interval statements rather than through exact point estimates. These methods also offer possibilities for carrying global sensitivity analyses [46], and in group decision making they help incorporate the preferences of all group members who can thus be engaged in an interactive decision support process [22]. Moreover, the use of preference programming during the early phases of the analysis can be motivated by an attempt to reduce the set of relevant decision alternatives so that subsequent elicitation efforts can be focused on the remaining non-dominated alternatives. At best, such *screening* processes may offer much better overall cost-benefit characteristics than conventional approaches. We expect that these advantages, together with the improved availability of decision support tools, will contribute to the wider use of preference programming methods.

Acknowledgements This research has been supported by the Academy of Finland. We are grateful to Juuso Liesiö, Pekka Mild and Antti Punkka for their comments on this paper.

References

1. B.S. Ahn. Multiattribute decision aid with extended ISMAUT. *IEEE Transactions on Systems, Man and Cybernetics, Part A*, 36:507–520, 2006.
2. B.S. Ahn, K.S. Park, C.H. Han, and J.K. Kim. Multi-attribute decision aid under incomplete information and hierarchical structure. *European Journal of Operational Research*, 125:431–439, 2000.
3. K. Alanne, A. Salo, A. Saari, and S.I. Gustafsson. Multi-criteria evaluation of residential energy supply systems. *Energy and Buildings*, 39(12):1218–1226, 2007.
4. G. Anandaligam. A multiagent multiattribute approach for conflict resolution in acid rain impact mitigation. *IEEE Transactions on Systems, Man, and Cybernetics*, 19:1142–1153, 1989.
5. A. Arbel. Approximate articulation of preference and priority derivation. *European Journal of Operational Research*, 43:317–326, 1989.
6. F.H. Barron and B.E. Barrett. Decision quality using ranked attribute weights. *Management Science*, 42(11):1515–1523, 1996.
7. V. Belton and T. Gear. On a short-coming of Saaty's method of analytic hierarchies. *Omega*, 11:228–230, 1983.
8. V. Belton and T.J. Stewart. *Multiple Criteria Decision Analysis: An Integrated Approach*. Kluwer Academic Publishers, Boston, 2001.
9. E. Carrizosa, E. Conde, F.R. Fernández, and J. Puerto. Multi-criteria analysis with partial information about the weighting coefficients. *European Journal of Operational Research*, 81:291–301, 1995.
10. D.H. Choi, B.S. Ahn, and S.H. Kim. Multicriteria group decision making under incomplete preference judgments: Using fuzzy logic with a linguistic quantifier. *International Journal of Intelligent Systems*, 22(6):641–660, 2007.
11. E.U. Choo, B. Schoner, and W.C. Wedley. Interpretation of criteria weights in multicriteria decision making. *Computers & Industrial Engineering*, 37(3):527–541, 1999.
12. J.J. Cristiano and C.C. White III. Application of multiattribute decision analysis to quality function deployment to target setting. *IEEE Transactions on Systems, Man, and Cybernetics - Part C*, 31(3):366–382, 2001.
13. L.C. Dias and J.N. Climaco. Additive aggregation with variable interdependent parameters: The vip analysis software. *Journal of the Operational Research Society*, 51(9):1070–1082, 2000.
14. L.C. Dias and J.N. Climaco. Dealing with imprecise information in group multicriteria decisions: A methodology and a gdss architecture. *European Journal of Operational Research*, 160:291–307, 2005.
15. J.S. Dyer. Remarks on the analytic hierarchy process. *Management Science*, 36(3):249–258, 1990.
16. E. Edwards. How to use multiattribute utility measurement for social decisionmaking. *IEEE Transactions on Systems, Man, and Cybernetics*, 7:326–340, 1977.
17. W. Edwards and F.H. Barron. SMARTS and SMARTER: Improved simple methods of multiattribute utility measurement. *Organizational Behavior and Human Decision Processes*, 60:306–325, 1994.
18. Y.S. Eum, K.S. Park, and S.H. Kim. Establishing dominance and potential optimality in multicriteria analysis with imprecise weight and value. *Computers & Operations Research*, 28:397–409, 2001.
19. J. Gustafsson, A. Salo, and T. Gustafsson. PRIME decisions: An interactive tool for value tree analysis. In M. Köksalan and S. Zionts, editors, *Multiple Criteria Decision Making in the New Millennium*, pages 165–176. Springer-Verlag, Berlin, 2001.
20. R.P. Hämäläinen. Decisionarium – aiding decisions, negotiating and collecting opinions on the web. *Journal of Multi-Criteria Decision Analysis*, 12(2-3):101–110, 2003.
21. R.P. Hämäläinen. Reversing the perspective on the applications of decision analysis. *Decision Analysis*, 1(1):26–31, 2004.

22. R.P. Hämäläinen, Salo A., and K. Pöysti. Observations about consensus seeking in a multiple criteria environment. In *Proceedings of the Twenty-Fifth Hawaii International Conference on System Sciences*, volume IV, pages 190–198. IEEE Computer Society Press, January 1992.
23. R.P. Hämäläinen and S. Alaja. The threat of weighting biases in environmental decision analysis. *Ecological Economics*, 68:556–569, 2008.
24. R.P. Hämäläinen, M. Lindstedt, and K. Sinkko. Multi-attribute risk analysis in nuclear emergency management. *Risk Analysis*, 20(4):455–468, 2000.
25. R.P. Hämäläinen and M. Pöyhönen. On-line group decision support by preference programming in traffic planning. *Group Decision and Negotiation*, 5:485–500, 1996.
26. C.H. Han, J.K. Kim, and S.H. Choi. Prioritizing engineering characteristics in quality function deployment with incomplete information: A linear partial ordering approach. *International Journal of Production Economics*, 91(3):235–249, 2004.
27. G.B. Hazen. Partial information, dominance, and potential optimality in multiattribute utility theory. *Operations Research*, 34:296–310, 1986.
28. D.L. Keefer, C.W. Kirkwood, and J.L. Corner. Perspective on decision analysis applications, 1990–2001. *Decision Analysis*, 1(1):4–22, 2004.
29. R.L. Keeney. *Value Focused-Thinking: A Path to Creative Decisionmaking*. Harvard University Press, Cambridge, MA, 1992.
30. R.L. Keeney and H. Raiffa. *Decisions with Multiple Objectives: Preferences and Value Tradeoffs*. John Wiley & Sons, New York, 1976.
31. J.K. Kim and S.H. Choi. A utility range-based interactive group support system for multiattribute decision making. *Computers & Operations Research*, 28:485–503, 2001.
32. S.H. Kim and B.S. Ahn. Interactive group decision making procedure under incomplete information. *European Journal of Operational Research*, 116:498–507, 1999.
33. S.H. Kim and C.H. Han. An interactive procedure for multi-attribute group decision making with incomplete information. *Computers & Operations Research*, 26:755–772, 1999.
34. S.H. Kim and C.H. Han. Establishing dominance between alternatives with incomplete information in a hierarchically structured attribute tree. *European Journal of Operational Research*, 122:79–90, 2000.
35. R. Lahdelma, J. Hokkanen, and P. Salminen. SMAA - Stochastic multiobjective acceptability analysis. *European Journal of Operational Research*, 106:137–143, 1998.
36. R. Lahdelma, K. Miettinen, and P. Salminen. Ordinal criteria in stochastic multicriteria acceptability analysis (SMAA). *European Journal of Operational Research*, 147(1):117–127, 2003.
37. R. Lahdelma, P. Salminen, and J. Hokkanen. Locating a waste treatment facility by using stochastic multicriteria acceptability analysis with ordinal criteria. *European Journal of Operational Research*, 142:345–356, 2002.
38. K.S. Lee, K.S. Park, Y.S. Eum, and K. Park. Extended methods for identifying dominance and potential optimality in multi-criteria analysis with incomplete information. *European Journal of Operational Research*, 134:557–563, 2001.
39. J. Liesiö, P. Mild, and A. A. Salo. Robust portfolio modeling with incomplete cost information and project interdependencies. *European Journal of Operational Research*, 190(3):679–695, 2008.
40. J. Liesiö, P. Mild, and A. Salo. Preference programming for robust portfolio modeling and project selection. *European Journal of Operational Research*, 181(3):1488–1505, 2007.
41. J. Liesiö, A. Punkka, and A. Salo. RICH decisions v1.01. Helsinki University of Technology, Systems Analysis Laboratory, 2003.
42. A.M. Mármol, J. Puerto, and F.R. Fernández. The use of partial information on weights in multicriteria decision problems. *Journal of Multi-Criteria Decision Analysis*, 7:322–329, 1998.
43. A. Mateos, S. Rios-Insua, and A. Jimenez. Dominance, potential optimality and alternative ranking in imprecise multi-attribute decision making. *Journal of the Operational Research Society*, 58:326–336, 2007.
44. J. Mustajoki and R.P. Hämäläinen. A preference programming approach to make the even swaps method even easier. *Decision Analysis*, 2(2):110–123, 2005.

45. J. Mustajoki and R.P. Hämäläinen. Smart-swaps - Decision support for the smart choices process with the even swaps method. *Decision Support Systems*, 44(1):313–325, 2007.
46. J. Mustajoki, R.P. Hämäläinen, and M.R.K. Lindstedt. Using intervals for global sensitivity and worst case analyses in multiattribute value trees. *European Journal of Operational Research*, 174(1):278–292, 2006.
47. J. Mustajoki, R.P. Hämäläinen, and A. Salo. Decision support by interval SMART/SWING - Methods to incorporate uncertainty into multiattribute analysis. *Decision Sciences*, 36(2):317–339, 2005.
48. O. Ojanen, S. Makkonen, and A. Salo. A multi-criteria framework for the selection of risk analysis methods at energy utilities. *International Journal of Risk Assessment and Management*, 5(1):16–35, 2005.
49. K.S. Park. Mathematical programming models for characterizing dominance and potential optimality when multicriteria alternative values and weights are simultaneously incomplete. *IEEE Transactions on Systems, Man and Cybernetics Part A*, 34:601–614, 2004.
50. K.S. Park and S.H. Kim. Tools for interactive decision making with incompletely identified information. *European Journal of Operational Research*, 98:111–123, 1997.
51. K.S. Park, S.H. Kim, and W.C. Yoon. An extended model for establishing dominance in multiattribute decisionmaking. *Journal of the Operational Research Society*, 47(11):1415–1420, 1996.
52. J. Puerto, A.M. Mármol, L. Monroy, and F.R. Fernández. Decision criteria with partial information. *International Transactions in Operational Research*, 7:51–65, 2000.
53. D. Rios Insua and S. French. A framework for sensitivity analysis in discrete multi-objective decision making. *European Journal of Operational Research*, 54:176–190, 1991.
54. T.L. Saaty. *The Analytic Hierarchy Process*. McGraw-Hill, New York, 1980.
55. A.P. Sage and C.C. III White. ARIADNE: A knowledge-based interactive system for planning and decision support. *IEEE Transactions on Systems, Man, and Cybernetics*, 14(1):35–47, 1984.
56. A. Salo. Interactive decision aiding for group decision support. *European Journal of Operational Research*, 84:134–149, 1995.
57. A. Salo. On fuzzy ratio comparisons in hierarchical weighting models. *Fuzzy Sets and Systems*, 84:21–32, 1996.
58. A. Salo. Tighter estimates for the posteriors of imprecise prior and conditional probabilities. *IEEE Transactions in Systems, Man, and Cybernetics - Part A*, 26(6):820–825, 1996.
59. A. Salo. On the role of decision analytic modelling. In A. Stirling, editor, *On Science and Precaution in the Management of Technological Risk*, volume II, pages 123–141. Institute of Prospective Technological Studies, Joint Research Centre of the European Commission, Report EUR 19056/EN/2, 2001. Downloadable at ftp://ftp.jrc.es/pub/EURdoc/eur19056IIen.pdf.
60. A. Salo and R.P. Hämäläinen. Preference assessment by imprecise ratio statements (PAIRS). *Operations Research*, 40:1053–1061, 1992.
61. A. Salo and R.P. Hämäläinen. Preference programming through approximate ratio comparisons. *European Journal of Operational Research*, 82:458–475, 1995.
62. A. Salo and R.P. Hämäläinen. On the measurement of preferences in the analytic hierarchy process. *Journal of Multi-Criteria Decision Analysis*, 6(6):309–319, 1997.
63. A. Salo and R.P. Hämäläinen. Preference ratios in multiattribute evaluation (PRIME) - Elicitation and decision procedures under incomplete information. *IEEE Transactions on Systems, Man, and Cybernetics - Part A*, 31(6):533–545, 2001.
64. A. Salo and J. Liesiö. A case study in participatory priority-setting for a scandinavian research program. *International Journal of Information Technology & Decision Making*, 5(1):65–88, 2006.
65. A. Salo and A. Punkka. Rank inclusion in criteria hierarchies. *European Journal of Operational Research*, 163(2):338–356, 2005.
66. T.J. Stewart and F.B. Losa. Towards reconciling outranking and value measurement practice. *European Journal of Operational Research*, 145:645–659, 2003.

67. W.G. Stillwell, D.A. Seaver, and E. Edwards. A comparison of weight approximation techniques in multiattribute utility decision making. *Organizational Behavior and Human Performance*, 28:62–77, 1981.
68. A. Stirling. *On the Role of Decision Analytic Modelling*, volume I. Institute of Prospective Technological Studies, Joint Research Centre of the European Commission, Report EUR 19056/EN/2, 1999. Downloadable at ftp://ftp.jrc.es/pub/EURdoc/eur19056en.pdf.
69. L.V. Tavares. A review of major paradigms for the design of civil engineering systems. *European Journal of Operational Research*, 119:1–13, 1999.
70. M. Weber. A method of multiattribute decision making with incomplete information. *Euro-Management Science*, 39:431–445, 1985.
71. M. Weber. Decision making with incomplete information. *European Journal of Operational Research*, 23:44–57, 1987.
72. M. Weber and K. Borcherding. Behavioral influences on weight judgments in multiattribute decision making. *European Journal of Operational Research*, 67:1–12, 1993.
73. C.C. White III, A.P. Sage, and S. Dozono. A model of multiattribute decisionmaking and trade-off weight determination under uncertainty. *IEEE Transactions on Systems, Man, and Cybernetics*, 14(2):223–229, 1984.

Chapter 6
New Trends in Aggregation-Disaggregation Approaches

Yannis Siskos and Evangelos Grigoroudis

Abstract The aggregation-disaggregation approaches as an important field of multicriteria decision-aid systems aim to infer global preference models from preference structures, as directly expressed by one or more decision-makers. The main objective of this chapter is to present new research developments of aggregation-disaggregation models and discuss related research topics. These recent developments cover a wide variety of topics, like post-optimality analysis, robustness analysis, group and collective decision-making. They focus mainly on the UTA family of models and highlight their most important advantages: they are flexible in the modeling process of a decision problem, they may provide analytical results that are able to analyze the behavior of the decision-maker, and they can offer alternative ways to reduce the preferential inconsistencies between the decision-maker and the results of the disaggregation model. Finally, future research topics in the context of preference disaggregation approaches are outlined in this chapter.

6.1 Introduction

Preference disaggregation constitutes an important Multiple Criteria Decision Aid (MCDA) philosophy aiming to assess/infer global preference models from given preference structures and to address decision-aiding activities through operational models within the aforementioned framework. In other words, the preference disaggregation approach refers to the analysis (disaggregation) of the global preferences (judgment policy) of the Decision-Maker (DM) in order to identify the criteria ag-

Yannis Siskos
University of Piraeus, Department of Informatics, 80 Karaoli & Dimitriou Str., 18534 Piraeus, Greece e-mail: ysiskos@unipi.gr

Evangelos Grigoroudis
Technical University of Crete, Decision Support Systems Laboratory, University Campus, 73100 Chania, Greece e-mail: vangelis@ergasya.tuc.gr

C. Zopounidis and P.M. Pardalos (eds.), *Handbook of Multicriteria Analysis*, Applied Optimization 103, DOI 10.1007/978-3-540-92828-7_6, © Springer-Verlag Berlin Heidelberg 2010

gregation model that underlies the preference result. From the previous it is clear that contrary to the traditional aggregation paradigm, where the criteria aggregation model is known a priori and the global preference is unknown, the philosophy of preference disaggregation aims to infer the preference models from given global preferences.

Although several approaches have developed in the context of aggregation-disaggregation paradigm, UTA methods [14, 40] may be considered as the main initiative and the most representative example of preference disaggregation theory. UTA methods are regression-based approaches that have been developed as an alternative to multiattribute utility theory (MAUT).

The philosophy of aggregation-disaggregation is explicitly presented in Figure 6.1, where the emphasis on the analysis of the behavior and the cognitive style of the DM is clear. In the context of UTA methods, special iterative interactive procedures are used, where the components of the problem and the DM's global judgment policy are analyzed and then they are aggregated into a value system. The goal of this approach is to aid the DM to improve his/her knowledge about the decision situation and his/her way of preferring that entails a consistent decision to be achieved.

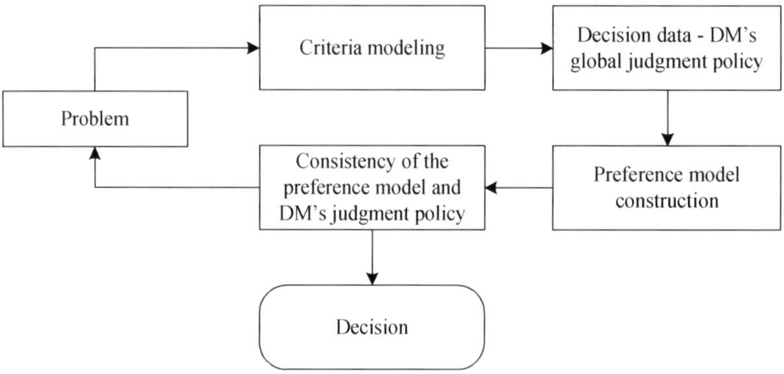

Fig. 6.1 The aggregation-disaggregation approach [39]

As known, in the general MCDA context the main objective is usually to analyze a set A of potential actions (or objects, alternatives, decisions) in terms of multiple criteria in order to model all the possible impacts, consequences or attributes related to this set A. However, in the aggregation-disaggregation approach, it is often necessary to use a set reference actions A_R in order to clarify the DM's global preference [15]. This reference set may be a set of past decision alternatives, a subset of decision actions, especially when A is large ($A_R \subset A$), or a set of fictitious actions, consisting of performances on the criteria, which can be easily judged by the DM to perform global comparisons [38].

Moreover, following the general modeling methodology of MCDA problems, a consistent family of criteria $\{g_1, g_2, \ldots, g_n\}$ should be assessed [26]. Each criterion is a non-decreasing real valued function defined on A, as follows:

$$g_i : A \rightarrow [g_{i*}, g_i^*] \subset \mathbb{R}/a \rightarrow g(a) \in \mathbb{R} \tag{6.1}$$

where $[g_{i*}, g_i^*]$ is the criterion evaluation scale, g_{i*} and g_i^* are the worst and the best level of the i-th criterion respectively, $g_i(a)$ is the evaluation or performance of action a on the i-th criterion and $\mathbf{g}(a)$ is the vector of performances of action a on the n criteria.

Therefore, the preference structure on a set of actions, which is necessary in a preference disaggregation approach, may have the following form, based on the aforementioned definitions:

$$\begin{cases} g_i(a) > g_i(b) \Leftrightarrow a \succ b \ (a \text{ is preferred to } b) \\ g_i(a) = g_i(b) \Leftrightarrow a \sim b \ (a \text{ is indifferent to } b) \end{cases} \tag{6.2}$$

This preference structure has a form of a weak-order, although alternative aggregation-disaggregation approaches may adopt different problem statements. In any case, the DM is asked to externalize and/or confirm his/her global preferences on the set A_R taking into account the performances of the reference actions on all criteria. So, in the UTA family of models, the problem is to adjust additive value or utility functions based on multiple criteria, in such a way that the resulting structure would be as consistent as possible with the initial structure.

Goal programming techniques have always played an important role in the development of preference disaggregation models. In fact, the history of the disaggregation principle in multidimensional/multicriteria analyses begins with the use of this special form of linear programming. Between mid 50s and mid 70's the most important research efforts refer to the development of linear or non-linear multidimensional regression analyses [4, 18, 20, 42, 45], while later works studied the case of ordinal criteria in order to assess/infer preference/aggregation models [14, 31, 40, 46]. Jacquet-Lagrèze and Siskos [15] and Siskos et al. [38] present an analytical review of the history of aggregation-disaggregation principle.

The main objective of this chapter is to present new research developments of aggregation-disaggregation approaches. Although these recent research efforts cover a wide variety of research topics, the chapter focuses on particular issues that have recently drawn significant attention in the literature, like post-optimality analysis, robustness analysis, group and collective decision-making.

The chapter is organized into 5 sections. Section 6.2 presents briefly the UTA model, as well as its significant proposed extensions. Post-optimality and robustness analysis in UTA-type models is discussed in section 6.3, while section 6.4 refers to the presentation of UTA-based group and collective decision models. Finally, section 6.5 summarizes some concluding remarks and outlines future research topics in the context of preference disaggregation approaches.

6.2 The UTA Family of Models

6.2.1 UTA and UTASTAR Methods

The UTA method initially proposed by Jacquet-Lagrèze and Siskos [14] originates an entire family of preference disaggregation models during the last thirty years. As already noted, the main objective of the method is to infer one or more additive value functions from a given ranking on a reference set A_R.

The original UTA method assumes an additive value function of the following form:

$$u(\mathbf{g}) = \sum_{i=1}^{n} u_i(g_i) \tag{6.3}$$

subject to normalization constraints:

$$\begin{cases} \sum_{i=1}^{n} u_i(g_i^*) = 1 \\ u_i(g_{i*}) = 0 \quad \forall i = 1,2,\ldots,n \end{cases} \tag{6.4}$$

where u_i $(i = 1,2,...,n)$ are the marginal value functions.

As discussed by Siskos et al. [38], this additive formula satisfies the axioms of comparability, reflexivity, transitivity of choices, continuity, and strict dominance, since $u_i(g_i) \geq 0$ holds and $du_i/dg_i > 0$ is assumed (see [19] for a detailed discussion about the properties of an additive utility model).

The method estimates the aforementioned value functions using linear goal programming techniques so that the ranking(s) obtained through these functions on A_R is (are) as consistent as possible with the one given by the DM. Thus, introducing a potential error $\sigma(a)$ relative to A_R, the value of each action may be written as:

$$u'[\mathbf{g}(a)] = \sum_{i=1}^{n} u_i[g_i(a)] + \sigma(a) \quad \forall a \in A_R \tag{6.5}$$

The implementation of the UTA algorithm requires the use of linear interpolation in order to approximate the marginal value functions u_i in a piecewise linear form. Moreover, taking into account the DM's ranking on $A_R = \{a_1, a_2, ..., a_m\}$, the reference actions are "rearranged" from the best (a_1) to the worst action (a_m).

As emphasized by Jacquet-Lagrèze and Siskos [14], DM's ranking has the form of weak order R, and thus, given the transitivity of R, it is possible to avoid unnecessary comparisons on A_R. Thus, assuming that

$$\Delta(a_k, a_{k+1}) = u'[(\mathbf{g}(a_k)] - u'[(\mathbf{g}(a_{k+1})] \tag{6.6}$$

the comparison of every pair of consecutive actions (a_k, a_{k+1}) gives the following conditions:

$$\begin{cases} \Delta(a_k, a_{k+1}) \geq \delta \ \text{ iff } a_k \succ a_{k+1} \\ \Delta(a_k, a_{k+1}) = 0 \ \text{ iff } a_k \sim a_{k+1} \end{cases} \tag{6.7}$$

where δ is a small positive number so as to discriminate significantly two successive equivalence classes of R.

Taking into account the previous assumptions and notations, the following Linear Program (LP) is used in order to estimate the marginal value functions:

$$\begin{cases} [\min]\ F = \sum\limits_{a \in A_R} \sigma(a) & \\ \text{s.t.} \quad \left.\begin{array}{l} \Delta(a_k, a_{k+1}) \geq \delta \text{ if } a_k \succ a_{k+1} \\ \Delta(a_k, a_{k+1}) = 0 \text{ if } a_k \sim a_{k+1} \end{array}\right\} & \forall k \\ u_i(g_i^{j+1}) - u_i(g_i^j) \geq s_i & \forall i \text{ and } j \\ \sum\limits_{i=1}^{n} u_i(g_i^*) = 1 & \\ u_i(g_{i*}) = 0,\ u_i(g_i^j) \geq 0,\ \sigma(a) \geq 0\ \forall a \in A_R,\ \forall i \text{ and } j & \end{cases} \tag{6.8}$$

where the last two constraints refer to the monotonicity and normalization constraints of u_i, respectively (with $s_i \geq 0$ indifference thresholds defined on each criterion).

On the other hand, the UTASTAR method proposed by Siskos and Yannacopoulos [40] may be considered as an improved version of the original UTA model, since it proposes two important modifications in the UTA algorithm:

- Double error function: the single error $\sigma(a)$ is replaced by a double positive error term (i.e. $\sigma^+(a)$ and $\sigma^-(a)$ being the overestimation and the underestimation error, respectively) in order to assure the minimization of the objective function of LP (6.8).
- Transformation of the variables: the original $u_i(g_i^j)$ variables of LP (6.8) are replaced by the new transformation variables w_{ij}, which represent the successive steps of the marginal value functions u_i, in order to reduce the size of LP (6.8) by removing the monotonicity constraints of u_i.

Siskos and Yannacopoulos [40] note that the UTASTAR algorithm perform better compared to the original UTA method based on a variety of experimental data and taking into account a number of comparison indicators (e.g. number of the necessary simplex iterations for arriving at the optimal solution, Kendall's τ between the initial weak order and the one produced by the estimated model, and the total sum of errors as the indicator of dispersion of the observations).

6.2.2 *Extensions of the UTA Method*

There are several variants and extensions of the UTA/UTASTAR method that try to model different forms of DM's preferences, apply different optimality criteria in the aforementioned LP formulation, or adopt the method in different decision problems.

As presented in the previous section, the LP formulation is a simple but powerful modeling approach that allows considering alternative types of global preference expressed by the DM. For example, Jacquet-Lagrèze and Siskos [14] propose to infer $u[\mathbf{g}(a)]$ from pairwise comparisons among the actions of the A_R. The intensity of the DM's preferences may also be considered in the formulation of LP (6.8) by adding a series of constraints of the following type [38]:

$$\Delta(a,b) - \Delta(b,c) \geq \varphi \tag{6.9}$$

where φ is a measure of preference intensity, which implies that the preference of alternative a over alternative b is stronger than the preference of b over c.

Similar modeling approaches have been proposed by Despotis and Zopounidis [6] and Oral and Ketanni [23], where a ratio scale is used to express intensity of preferences.

Regarding the different optimality criteria, it should be emphasized that this problem is discussed by Jacquet-Lagrèze and Siskos [14] in the development of the original UTA method. They propose, for example, the following alternatives:

1. Maximize the Kendall's τ between the ranking provided by the DM and the ranking given by the model (i.e. minimize the number of violated pairs between these rankings).
2. Weight the potential errors in F taking into account a different degree of confidence in each ranked action.

Alternative desired properties of u_i may also lead to different variations of the UTA/UTASTAR method. In this context, Despotis and Zopounidis [6] present extensions of the method in the case of non-monotonic marginal value functions or other additional properties of the assessed value functions (e.g. concavity).

An analytical presentation and discussion of other variants of the UTA and UTASTAR methods may be found in Jacquet-Lagrèze and Siskos [15] and Siskos et al. [38]. The most important of these extensions include:

- The stochastic UTA method developed in the framework of multicriteria decision-aid under uncertainty, where the adopted aggregation model has the form of a von Neumann-Morgenstern additive utility function [33, 34].
- The UTA-type sorting methods, which are developed in the context of problem statement β (sorting the actions into predefined and preference-ordered categories), like the UTADIS family of models [7, 47, 49] and the MHDIS method [48].
- The incorporation of the UTA method in the solution process of multiobjective programming problems (see for example the works of Stewart [43], Jacquet-Lagrèze et al. [12], Siskos and Despotis [35]).
- The MACBETH method (Measuring Attractiveness by a Categorical Based Evaluation Technique) proposed by Bana e Costa and Vansnick [1], which infers a single value function from pairwise comparisons externalized from the DM on a single criterion in terms of criterion attractiveness. The same procedure is repeated for each criterion and, finally for the whole set of criteria in order to infer the criteria weights. The overall evaluation model is an additive value model.

Finally, it should be emphasized that the main principles of the aggregation-disaggregation approach may be adopted in other MCDA fields (e.g. outranking relation methods, ordinal regression analysis), where the problem is to extract DM's preferences (value system, model parameters, etc) in a consistent way. In general, as mentioned by Siskos et al. [38] this philosophy is also employed in other non-classical MCDA approaches, like rough sets, machine learning, and neural networks, in order to infer some form of a decision model (a set of decision rules or a network) from given decision results involving assignment examples, ordinal or measurable judgments.

6.3 Post-optimality Analysis and Robustness

6.3.1 Post-optimality Analysis

The stability analysis is considered as an important part of the algorithm of the UTA methods, since all these approaches are based on a LP modeling and thus often the problem of multiple or near optimal solutions appears.

In the classical approach of the UTA/UTASTAR method the stability analysis is considered as a post-optimality analysis problem, based on a heuristic method for near optimal solutions search [37]. These solutions have some desired properties, while the heuristic technique is based on the following:

- In several cases, the optimal solutions are not the most interesting, given the uncertainty of the model parameters and the preferences of the decision-maker [44].
- The number of the optimal or near optimal solutions is often huge. Therefore an exhaustive search method (reverse simplex, Manas-Nedoma algorithms) requires a lot of computational effort.

In particular, if the optimum $F^* = 0$, the polyhedron of admissible solutions for u_i is not empty and many value functions lead to a perfect representation of the weak order R. Even when the optimal value $F^* > 0$, other solutions, less good for F, can improve other satisfactory criteria, like Kendall's τ. In any case, as emphasized by Jacquet-Lagrèze and Siskos [14], it is crucial to explore the post-optimal solutions space defined by the polyhedron:

$$\begin{cases} F \leq F^* + k(F^*) \\ \text{all the constraints of LP (6.8)} \end{cases} \tag{6.10}$$

where $k(F^*)$ is a positive threshold, which is a small proportion of F^*.

The previous polyhedron is partially explored in the original UTA method by solving the following LPs:

$$\begin{cases} [\min]\ u_i(g_i^*) \\ \text{in} \\ \text{polyhedron (6.10)} \end{cases} \text{and} \begin{cases} [\max]\ u_i(g_i^*) \\ \text{in} \\ \text{polyhedron (6.10)} \end{cases} \quad \forall i = 1,2,\ldots,n \qquad (6.11)$$

As noted by Siskos et al. [38], the solutions of the above LPs give the internal variation of the weight of all criteria g_i, and consequently give an idea of the importance of these criteria in the DM's preference system. The final solution of the problem is calculated as the average of the previous LPs; this average solution is less representative, if a large variation of the provided solutions appears.

However, the efficiency of the aforementioned optimization procedure is based on the number and the meaning of the criteria introduced in the model. As mentioned by [32], when the number of criteria is small, the previous $2n$ LPs may be solved, otherwise it is possible to solve only n LPs (maximization of $u_i(g_i^*)$). Nevertheless, if there is an a priori typology of the criteria, i.e. when the criteria can be grouped into different classes determining different policies of the DM, then it is possible to minimize or maximize the sum of the weighting factors of the criteria for each policy. In this way, the value systems obtained by the post-optimality analysis are able to show the strengths and weaknesses of these policies, in relation to the global policy and to the DM's behavior.

Other approaches in the post-optimality analysis process of the UTA methods may be also found in the literature. These approaches propose the use of alternative optimality criteria during the exploration of the polyhedron (6.10), like the minimization of the errors' dispersion, i.e. Tchebycheff criterion [5], or the optimal assessment of the δ and s parameters in the context of the UTAMP models [2, 3].

6.3.2 Robustness in UTA-type Models

Roy [27] recently considers the robustness as a tool of resistance of decision analysts against the phenomena of approximations and ignorance zones. In fact, robustness appears as a tool to analyze the gap between the "true" DM's model and the one resulting from a computational mechanism. It is important to note that the robustness analysis should be distinguished from the sensitivity analysis, which is marginal and depends each time on the changes of one or more parameters. Moreover, it should be emphasized that robustness refers mainly to the decision model, in the light of the assertion "robust models produce a fortiori robust results". However, robustness should also refer to the results and the decision support activities (e.g. conclusions, argumentation).

From the previous comments, it is clear that robustness should be measured and controlled in any decision-aid activity. However, this need poses a number of new problems referring to the measurement of the robustness of a decision model, the development of appropriate robustness indicators, and the potential improvement of robustness. Moreover, this measurement process should always take into account the different perspectives of robustness:

1. Analyst's point of view (is a decision model reliable?)
2. DM's point of view (is a decision model acceptable?)

As already mentioned, in the UTA family of models, robustness deals with LP, which is the main mechanism to infer decision models. In particular robustness refers to the post/near-optimality analysis, as presented in the previous section.

The general methodological framework for applying robustness analysis in the context of preference disaggregation approaches is presented in Figure 6.2 and consists of the following main steps:

1. The applied preference disaggregation method is used to infer a representative additive value model based on A_R. This step is discussed in the previous sections of this chapter.
2. The inconsistencies between the DM's preferences and the results of the disaggregation method are identified and removed using interactive techniques with the DM. An example methodological approach for this is given in Figure 6.3, while further details may be found in [36, 41].
3. A robustness measure is established.
4. If the robustness measure, established in step 3, is judged satisfactory by the analyst, the model is proposed to the DM for application on the set A and the process is terminated. Otherwise, the process goes to step 5.
5. Alternative rules of robustness analysis are examined and the process goes back to step 3.

Particularly for the assessment of the robustness measures (step 3), it should be noted that the robustness of the decision model depends on the post-optimality analysis results, and especially on the form and the extent of the polyhedron of multiple/near optimal value functions. In order to handle this polyhedron, the following heuristic is applied: during post-optimality analysis $2n$ LPs are formulated and solved, which maximize and minimize repeatedly $u_i(g_i^*)$. The observed variance in the post-optimality matrix indicates the degree of instability of the results. Thus, following the approach of Grigoroudis and Siskos [11] an Average Stability Index (ASI) may be assessed as the mean value of the normalized standard deviation of the estimated values $u_i(g_i^*)$. Alternatively, instead of exploring only the extreme values of $u_i(g_i^*)$, the post-optimality analysis may investigate every value of each criterion $u_i(g_i^j)$. In this case, during the post-optimality stage, $T = 2\sum_i(\alpha_i - 1)$ LPs are formulated and solved, which maximize and minimize repeatedly $u_i(g_i^j)$ and the ASI for the i-th criterion is assessed as follows:

$$ASI(i) = 1 - \frac{1}{\alpha_i - 1} \sum_{j=1}^{\alpha_i - 1} \frac{\sqrt{T \sum_{k=1}^{T} \left(u_i^{jk}\right)^2 - \left(\sum_{k=1}^{T} u_i^{jk}\right)^2}}{\frac{T}{\alpha_i - 1}\sqrt{\alpha_i - 2}} \tag{6.12}$$

where α_i is the number of points that are estimated in the interval $[g_{i*}, g_i^*]$ and u_i^{jk} is the estimated value of $u_i(g_i^j)$ in the k-th post-optimality analysis LP ($j = 1, 2, ..., \alpha_i$).

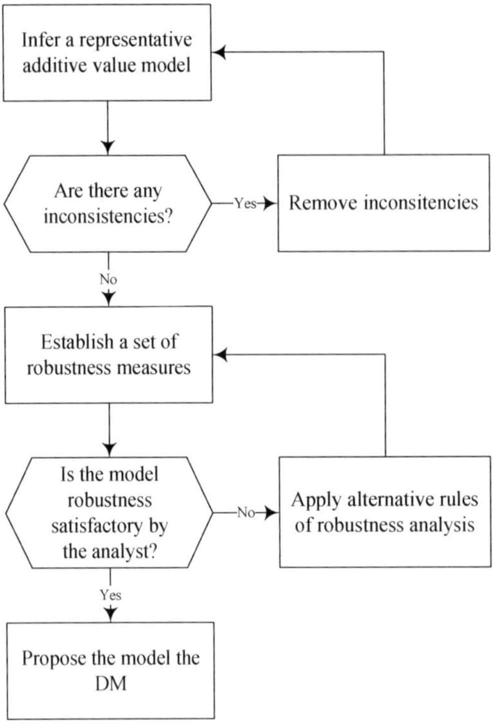

Fig. 6.2 Robustness analysis in preference disaggregation approaches

Using formula (6.12), the global robustness measure may be assessed as the average of the individual $ASI(i)$ values:

$$ASI = \frac{1}{n}\sum_{i=1}^{n} ASI(i) \tag{6.13}$$

It should be noted that all the previous ASI measures are normalized in the interval $[0,1]$, and thus high levels of robustness are achieved when ASI is close to 1.

On the other hand, the alternative rules that should be applied if the analyst is not satisfied with the value of the ASI measures (step 4), may include the following:

- Addition of new global preference judgments (e.g. pairwise comparisons, preference intensities as mentioned in section 6.2.2, or even new reference actions).
- Visualization of the observed value variations to support the DM in choosing his/her own model (see the example below).
- Enumeration and management of the hyperpolyhedron vertices (Manas-Nedoma algorithm, Tarry's method, etc.) in post-optimality analysis.

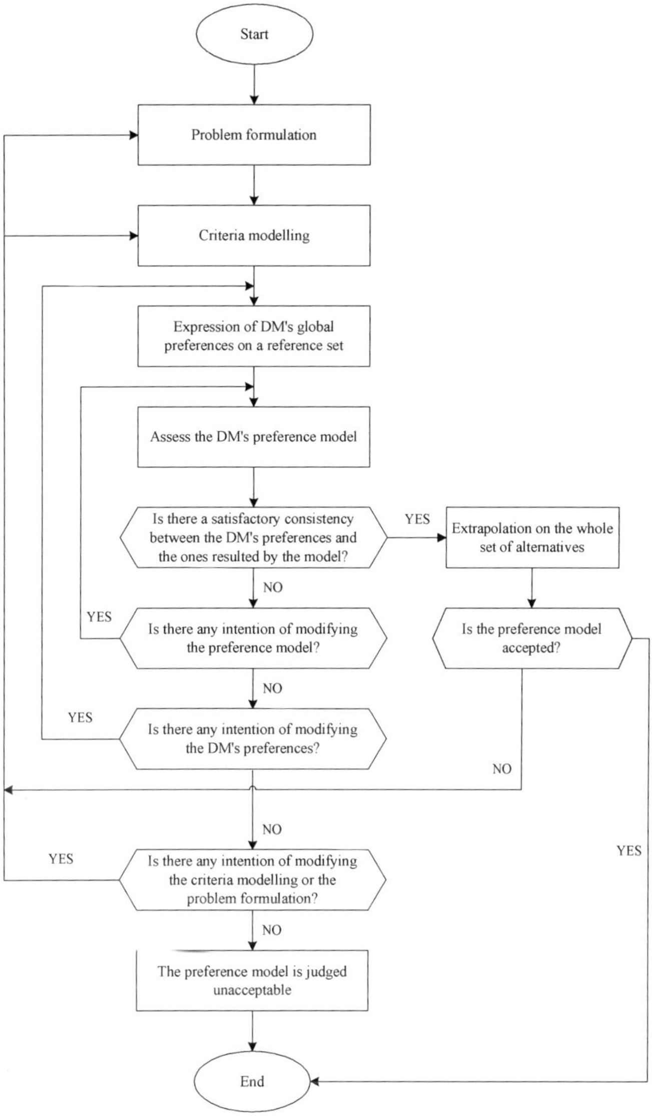

Fig. 6.3 Removal of inconsistencies in UTA methods [15]

- Building new preference relations on the set A during the extrapolation phase (see the next section).
- Computation of the barycentre as a representative model.

In order to illustrate the aforementioned methodological approach, let us consider a simple example where an ordinal evaluation scale is used in a single criterion case. Suppose that the evaluation scale has the following form: b=Bad, m=Medium, g=Good, and e=Excellent. The problem is to estimate the DM's value function u on the scale $\{$b, m, g, e$\}$. Of course, this naïve example is not realistic, but it is presented here in order to illustrate the proposed robustness approach.

Taking into account the normalization constraints of u, we have $u(\mathrm{b}) = 0$ and $u(\mathrm{e}) = 1$, while $u(\mathrm{m})$ and $u(\mathrm{g})$ are the model parameters that should be estimated. Suppose also, that the DM's preferences may be modeled according to the following relations:

1. The value of $u(\mathrm{m})$ should not be more than 20% and less than 10%:

$$0.1 \leq u(\mathrm{m}) \leq 0.2$$

2. The difference between "good" to "excellent" is at least 2 times more important than the difference between "medium" and "good":

$$\frac{u(\mathrm{e}) - u(\mathrm{g})}{u(\mathrm{g}) - u(\mathrm{m})} \geq 2 \Leftrightarrow 1 - u(\mathrm{g}) \geq 2u(\mathrm{g}) - 2u(\mathrm{m}) \Leftrightarrow 3u(\mathrm{g}) - 2u(\mathrm{m}) \leq 1$$

3. The indifference threshold on g is selected so as $s = 0.01$, thus:

$$u(g^{j+1}) - u(g^j) \geq s \Rightarrow \begin{cases} u(\mathrm{m}) - u(\mathrm{b}) \geq 0.01 \\ u(\mathrm{g}) - u(\mathrm{m}) \geq 0.01 \\ u(\mathrm{e}) - u(\mathrm{g}) \geq 0.01 \end{cases} \Rightarrow \begin{cases} u(\mathrm{m}) \geq 0.01 \text{ (redundant)} \\ u(\mathrm{g}) - u(\mathrm{m}) \geq 0.01 \\ 1 - u(\mathrm{g}) \geq 0.01 \text{ (redundant)} \end{cases}$$

$$\Rightarrow u(\mathrm{g}) - u(\mathrm{m}) \geq 0.01$$

Using the previous constraints and introducing the error variables, the preference disaggregation LP has the following form:

$$\begin{array}{llllll} \min z = z_1 + z_2 + z_3 \\ \text{s.t.} & -u(\mathrm{m}) + & u(\mathrm{g}) & & & \geq 0.01 \\ & u(\mathrm{m}) & & + z_1 & & \geq 0.1 \\ & u(\mathrm{m}) & & & - z_2 & \leq 0.2 \\ & -2u(\mathrm{m}) + & 3u(\mathrm{g}) & & & - z_3 \leq 1 \\ & u(\mathrm{m}), & u(\mathrm{g}), & z_1, & z_2, & z_3 \geq 0 \end{array}$$

The previous LP gives $z^* = 0$, while it should be emphasized the existence of optimal solutions, as indicated by the polygon ABCD in Figure 6.4. The solutions obtained during the post-optimality analysis, where 4 LPs are formulated and solved, are presented in Table 6.1. Also, the barycentral solution, along with the variation of the value function in post-optimality analysis is given in Figure 6.5. Finally, using formula (6.12), the proposed robustness measure for the examined criterion is calculated as $ASI = 0.8059$.

Although this value of ASI may be considered as acceptable by the analyst, suppose that the DM is able to give new global preference judgments. For example, if the value of "good" is assumed to be no less than 40%, the new constraint

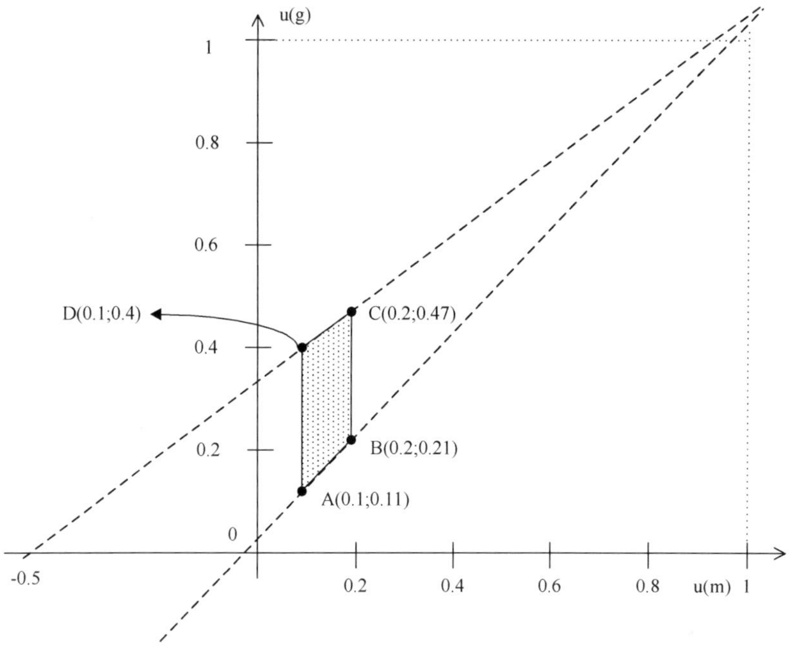

Fig. 6.4 Multiple optimal solutions for the numerical example

Table 6.1 Post-optimality analysis results

	u(m)	u(g)
[min] u(m)	0.10	0.40
[max] u(m)	0.20	0.21
[min] u(g)	0.10	0.11
[max] u(g)	0.20	0.47

$u(\text{m}) \geq 0.4$ should be added in the previous LP. The new optimal solution is also $z^* = 0$, but the optimal solution space, as shown in Figure 6.6 is now defined by the triangle CDE. In this case, the *ASI* increases from 80.59% to 91.97%. Figure 6.7 presents the revised barycentral solution for the simple numerical example, which is more robust than the initial one. In any case, this Figure is able to visualize the variability of the value function in order to support the DM in choosing his/her own model.

Consequently, robustness may be considered as a gap between the "true" DM's model and the one resulting from a computational mechanism. Robustness is also a decision support tool to decide about:

- the decision model and
- the answers to the decision problematic.

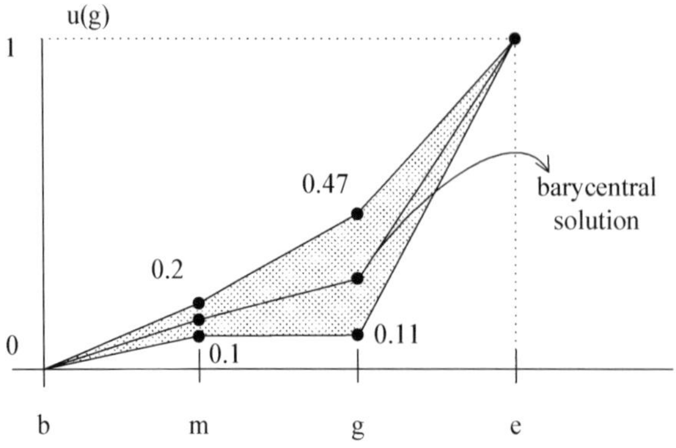

Fig. 6.5 Visualization of multiple decision models for the numerical example

Although research about robustness could continue taking into account the methodological issues highlighted in this chapter, the proposed stability index (*ASI*) may provide a helpful robustness measure from post-optimality analysis stage in any preference disaggregation method.

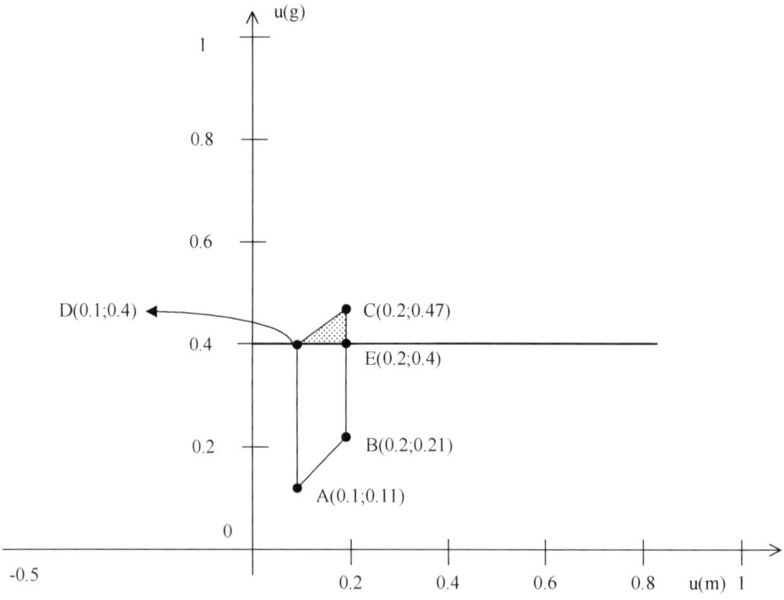

Fig. 6.6 Revised optimal solution space for the numerical example

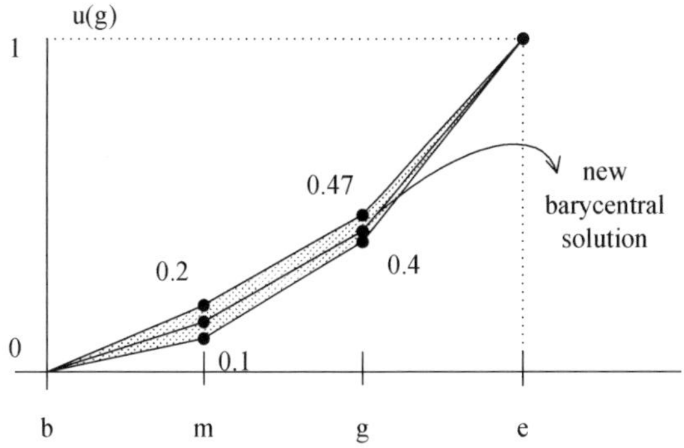

Fig. 6.7 Revised decision models' space for the numerical example

6.3.3 Building Preference Relations

An important alternative rule applicable in the proposed robustness analysis of aggregation-disaggregation approaches refers to the assessment of preference relations by exploiting the results of post-optimality analysis. Thus, such a rule is able to deal with the problem of multiple optimal solutions.

In this context, Siskos [32] proposes the assessment of fuzzy outranking relations based on the results of the UTA models. In this approach, the term "fuzzy" refers to the fuzzy aspect of the DM's preferences. In particular, these fuzzy relations analyze the preference relation R and they are based on a system of additive value functions estimated by a UTA model. It should be mentioned that a fuzzy outranking relation is a fuzzy subset of the set of all pairs of actions, i.e. a fuzzy subset of the set $A \times A$, characterized by a membership function $d(A \times A)$ called the degree of credibility of the outranking [25]. The main objective of the approach is to evaluate this degree of superiority of one action to another, according to the information given by the criteria and the a priori preferences of the DM.

As already noted, post-optimality analysis offers different possibilities to restitute the weak order R by additive value functions. Thereby, Siskos [32] introduces the notion of an additive value system, which is represented by a set of indexed value functions $\mathscr{U} = \{u^1, u^2, \ldots, u^i, \ldots\}$ and must satisfy the following rules:

- u_i must give a satisfactory degree of consistency between the weak order that it defines over A_R and the initial weak order R.
- The value functions of the system must be as characteristic as possible of the polyhedron (6.10).

It should be emphasized that using the heuristic approach of exploring the multiple/near optimal solutions space presented in the previous section (see LPs (6.11)), the convex set of value functions defined by the polyhedron (6.10) becomes a discrete and finite set.

So, the problem is to synthesize the results on the set A, taking into account this system of additive value functions $\mathcal{U}$ in order to obtain a general rule of decision. According to Siskos [32], this synthesis may be done in a simple way, considering the majority or concordance rule on the value functions of the system. In particular, he suggests calculating the percentage of value functions for which an action is better or equivalent than another.

The degree of credibility, which is actually the membership function of the fuzzy relation in $A \times A$, is defined by the following formula [32]:

$$d(a,b) = \frac{\left| u/u^i[\mathbf{g}(a)] - u^i[\mathbf{g}(b)] \geq 0 \right|}{|\mathcal{U}|} \tag{6.14}$$

where $\left| u/u^i[\mathbf{g}(a)] - u^i[\mathbf{g}(b)] \geq 0 \right|$ is the number of value functions for which $a \succ b$ or $a \sim b$ and $|\mathcal{U}|$ is the number of value functions of the system $\mathcal{U}$.

It is easy to see that $0 \leq d(a,b) \leq 1$, while it should be mentioned that the previous fuzzy relation enables to measure the outranking degree of one action by another, using only the ordinal structure determined over A by the additive value functions of the system $\mathcal{U}$.

Fuzzy preference relations have been extensively studied in the literature in order to analyze their properties and examine the conditions under which they can be applied in decision-making process. So, taking into account the aforementioned definitions and assumptions, the following fuzzy preference relations may be introduced [24]:

1. Fuzzy indifference relation

$$\mu^e(a,b) = \min\{d(a,b), d(b,a)\} \tag{6.15}$$

2. Fuzzy strict preference relation

$$\mu^s(a,b) = \begin{cases} d(a,b) - d(b,a) & \text{if } d(a,b) \geq d(b,a) \\ 0 & \text{otherwise} \end{cases} \tag{6.16}$$

Using the definition of μ^s, the fuzzy nondomination degree of an action is given by:

$$\begin{aligned} \mu^{ND}(a) &= \min_{b \in A}\{1 - \mu^s(b,a)\} = 1 - \max_{b \in A}\{\mu^s(b,a)\} \\ &= 1 - \max_{b \in A}\{d(b,a) - d(a,b)\} \end{aligned} \tag{6.17}$$

The value $\mu^{ND}(a)$ represents the degree to which the action a is dominated by no one of the other actions in A [24].

So, under problematic α, the best action a^* may be selected by maximizing $\mu^{ND}(a)$ on A. If $\mu^{ND}(a)^* = 1$, then a full robustness is achieved, otherwise the robustness of a^* is characterized by a value between 0 and 1. Similarly, under problematic γ, the ranking of actions from A can be made according to the values of the $\mu^{ND}(a)$ indicator.

The previous methodological approach is also adopted by the UTA ^{GMS} and GRIP models, which are also based on the whole set of additive value functions "compatible" with the given preference information.

In particular, the UTA^{GMS} method requires a set of pairwise comparisons on A_R as the main input preference information. Using LP techniques, the method extrapolates the results on A by assessing two relations:

- The necessary weak preference relation which holds for any two alternatives $a, b \in A$, if and only if all compatible value functions give to a a value greater than the value given to b.
- The possible weak preference relation which holds for this pair $a, b \in A$, if and only if at least one compatible value function give to a a value greater than the value given to b.

The previous preference relations follow the main principles of the aforementioned fuzzy outranking relations proposed by Siskos [32]. In fact they may be considered as two special cases of the this fuzzy preference relation, with $d(a,b) = 1$ for the necessary relation and $0 < d(a,b) \leq 1$ for the possible relation.

These preference relations are used the UTA^{GMS} method in order to produce two rankings on the set A, such that for any pair of solutions $a, b \in A$: (1) in the necessary ranking, a is ranked at least as good as b, if and only if, $u(a) \geq u(b)$ for all value functions compatible with the preference information and (2) in the possible ranking, a is ranked at least as good as b, if and only if, $u(a) \geq u(b)$ for at least one value function compatible with the preference information.

As noted by Greco et al. [9] the necessary ranking can be considered as robust with respect to the preference information: any pair of solutions is compared in the same way whatever the additive value function compatible with the preference information is. So, when no preference information is given, the necessary ranking boils down to the dominance relation, and the possible ranking is a complete relation. The addition of new pairwise comparisons on A_R is able to enrich the necessary ranking and impoverish the possible ranking, so that they converge with the incorporation of this preference information. Thus, the method is intended to be used interactively, with an increasing reference set A_R and a progressive statement of pairwise comparisons.

In the same context, Figueira et al. [8] propose the GRIP (Generalized Regression with Intensities of Preference) method as an extension of the UTA^{GMS} approach, which infers this set of compatible additive value functions, taking into account not only a preorder on a set of alternatives, but also the intensities of preference among alternatives. These comparisons may be expressed comprehensively (on all criteria) and/or partially (on each criterion).

The previous methods, although able to deal with the robustness problem, cannot always provide a "final solution" to the DM. For this reason, Greco et al. [10] propose a procedure to explore the set of compatible value functions and identify the "most representative" one. Their idea is to select among compatible value functions the one that better highlights the necessary ranking (maximize the difference of evaluations between actions for which there is a preference in the necessary ranking). Alternatively, they propose to minimize the difference of evaluations between actions for which there is not a preference in the necessary ranking.

6.4 Group and Collective Decision Approaches

Most of the real-world decision-making problems involve multiple actors having different viewpoints on the way the problem should be handled and the decision to be made [17]. In these situations it is common to encounter conflict between the opinions and desires of the group members. This conflict may arise because multiple DMs have different value and informational systems (objectives, criteria, preference relations, communication support, etc). This is also noted by Roy [26] as "distinct value systems", e.g. different ethical and or ideological beliefs, different specific objectives, or different roles within an organization. In this context, MCDA methods have been used in numerous previous studies in order to represent the multiple viewpoints of the problem, to aggregate the preferences of the multiple DMs, or to organize the decision process (see [22] for a detailed review of MCDA methods in group decision support systems).

The family of the UTA methods has been also used in several studies of conflict resolution in multi-actor decision situations [15]. These studies refer to the development and application of group decision or negotiation support systems [16, 28, 29, 30], or conflict resolution approaches for single actors [13]. Beside UTA methods, Matsatsinis and Samaras [22] review several other aggregation-disaggregation approaches incorporated in group decision support systems.

While group decision approaches aim to achieve consensus among the group of DMs or at least attempt to reduce the amount of conflict by compensation, collective decision methods focus on the aggregation of the DMs' preferences. Therefore, in the latter case, the collective results are able to determine preferential inconsistencies among the DMs, and to define potential interactions (trade-off process) that may achieve a higher group and/or individual consistency level.

The UTA method may be extended in the case of multiple DMs, taking into account different input information (criteria values) and preferences for a group of DMs. Two alternative approaches may be found in the literature:

1. Application of the UTA/UTASTAR methods in order to optimally infer marginal value functions of individual DMs; the approach enables each DM to analyze his/her behavior according to the general framework of preference disaggregation.
2. Application of the UTA/UTASTAR methods in order to assess a set of collective additive value functions; these value functions are as consistent as possible with the preferences of the whole set of DMs, and thus they are able to aggregate individual value systems.

In the context of the first approach, Matsatsinis et al. [21] propose a general methodology for collective decision-making combining different MCDA approaches. As shown in Figure 6.8, in the first step of the methodology, the UTASTAR algorithm is implemented in order to assess individual's preference systems. Then, the values of the alternatives are aggregated with some averaging operator (normalized relative utility values). However, such representations of group preferences were found no to guarantee neither a consensus nor a good compromise, since individual assessments may be considerably different. Therefore, Matsatsinis et al. [21] incorporate in their proposed methodology several criteria in order to measure the DMs' satisfaction over the aggregated rank-order of alternatives.

The UTA/UTASTAR method may also be applied in the problem of inferring collective preference systems. Consider for example the case of q DMs evaluating m alternatives (a_k with $k = 1,2,...,m$) according to a set of n criteria (g_i with $i = 1,2,...,n$) which are assessed in the interval $[g_i* = g_i^1, g_i^2, ..., g_i^* = g_i^{\alpha_i}]$. Furthermore, suppose that $g_i^r(a_k)$ is the evaluation of the r-th DM for the k-th alternative on the i-th criterion and $R^r(a_k)$ is the ranking of the the k-th alternative given by the r-th DM.

Using the previous notations, formula (6.5), which represents the global value of actions in terms of marginal values, may be rewritten as follows:

$$u'[\mathbf{g}^r(a)] = \sum_{i=1}^{n} u_i[g_i^r(a)] - \sigma_r^+(a) + \sigma_r^-(a) \forall a \in A_R \tag{6.18}$$

where a double error function is included, similar to the UTASTAR method.

Respectively, taking into account the multiple DMs, formula (6.6) becomes:

$$\Delta^r(a_k, a_{k+1}) = u'[(\mathbf{g}^r(a_k)] - u'[(\mathbf{g}^r(a_{k+1})] \tag{6.19}$$

and thus, the LP (6.8) may be written as follows:

$$\left\{ \begin{array}{lll} [\min]\ F_1 = \sum_{r=1}^{q} \sum_{k=1}^{m} \left[\sigma_r^+(a) + \sigma_r^-(a)\right] & & \\ \text{s.t.} \quad \left. \begin{array}{l} \Delta^r(a_k, a_{k+1}) \geq \delta \text{ if } a_k \succ a_{k+1} \\ \Delta^r(a_k, a_{k+1}) = 0 \text{ if } a_k \sim a_{k+1} \end{array} \right\} & \forall k, r & \\ \quad u_i(g_i^{j+1}) - u_i(g_i^j) \geq s_i & \forall i \text{ and } j & \\ \quad \sum_{i=1}^{n} u_i(g_i^*) = 1 & & \\ \quad u_i(g_{i*}) = 0,\ u_i(g_i^j) \geq 0,\ \sigma_r^+(a),\ \sigma_r^-(a) \geq 0\ \forall i, j, k, r & & \end{array} \right. \tag{6.20}$$

This LP minimizes the sum of (absolute) errors for all DMs, which in several cases may not provide a "compromise" solution. Therefore, different optimality criteria in the previous LP formulation may be considered, given that the main objective of such methodology is to minimize potential individual deviation from the inferred group preference system.

For example, the following LP minimizes the maximum sum of errors for every DM (i.e. variance of errors to DMs):

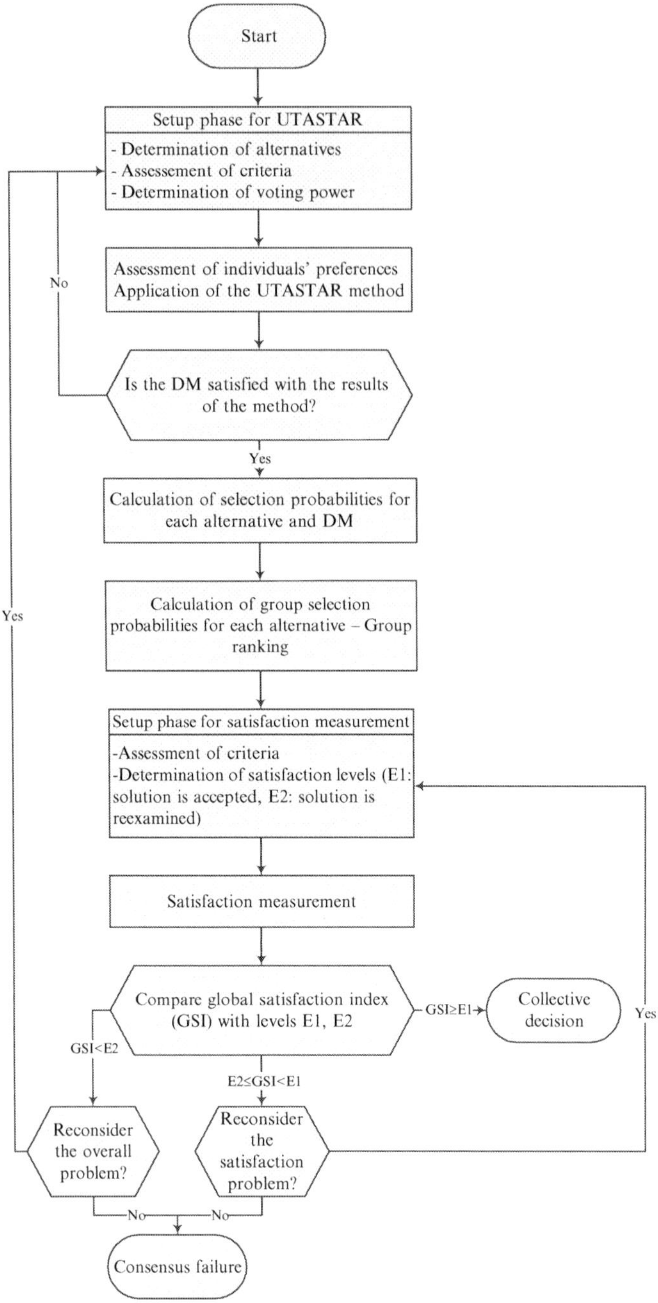

Fig. 6.8 Preference disaggregation approach for collective decision-making [21]

$$\begin{cases} [\min]\ F_2 = \sigma_{\max} \\ \text{s.t.} \quad \text{all the constraints of LP (6.20)} \\ \qquad \sum_{k=1}^{m} [\sigma_r^+(a_k) + \sigma_r^-(a_k)] \leq \sigma_{\max} \quad \forall r \end{cases} \tag{6.21}$$

Another approach is to consider the number of violated pairs of R^r (equivalent to Kendall's τ). In this case, similarly to Jacquet-Lagrèze and Siskos [14], a set of new binary variables should be introduced:

$$\gamma_{ab}^r = \begin{cases} 0 \text{ if } u[\mathbf{g}^r(a)] - u[\mathbf{g}^r(b)] \geq \delta & \text{(the judgment is respected)} \\ 1 \text{ otherwise} & \text{(the judgment is violated)} \end{cases} \tag{6.22}$$

Thus the following LP minimizes the sum of violated judgments for all DMs:

$$\begin{cases} [\min]\ F_3 = \sum_{r=1}^{q} \sum_{(a.b)\in R^r} \gamma_{ab}^r \\ \text{s.t.} \quad \sum_{i=1}^{n} \{u_i[g_i^r(a_k)] - u_i[g_i^r(a_{k+1})]\} + M\gamma_{ab}^r \geq \delta \quad \text{if } a_k^r \succ a_{k+1}^r \\ \qquad \left.\begin{array}{l} \sum_{i=1}^{n} \{u_i[g_i^r(a_k)] - u_i[g_i^r(a_{k+1})]\} + M\gamma_{ab}^r \geq 0 \\ \sum_{i=1}^{n} \{u_i[g_i^r(a_k)] - u_i[g_i^r(a_{k+1})]\} + M\gamma_{ba}^r \leq 0 \end{array}\right\} \text{if } a_k^r \sim a_{k+1}^r \quad \forall k, r \\ \qquad u_i(g_i^{j+1}) - u_i(g_i^j) \geq s_i \qquad \forall i \text{ and } j \\ \qquad \sum_{i=1}^{n} u_i(g_i^*) = 1 \\ \qquad u_i(g_{i*}) = 0,\ u_i(g_i^j) \geq 0,\ \gamma_{ab}^r \in \{0,1\} \qquad \forall i, j, r, (a,b) \in R^r \end{cases} \tag{6.23}$$

where M is a large number.

Finally, similarly to LP (6.21), the minimization of the maximum sum of violated judgments for every DM may considered by the following LP:

$$\begin{cases} [\min]\ F_4 = \gamma_{\max} \\ \text{s.t.} \quad \text{all the constraints of LP (6.23)} \\ \qquad \sum_{(a.b)\in R^r} \gamma_{ab}^r \leq \gamma_{\max} \quad \forall r \end{cases} \tag{6.24}$$

These alternative LP formulations may provide different collective results, and thus provide alternative bases for the compensation process among the DM's, which is usually applied in order to achieve consensus. In order to illustrate the previous modeling approaches, consider the simple example of Table 6.2, where 3 DMs evaluate a set of 7 alternatives (cars) using 6 criteria. As shown, the criteria evaluations are the same for all DMs, except for the "design" criterion, where the DMs express different preferences (this particular criterion is evaluated using a 5-point ordinal scale). Also, Table 6.3 shows the different rankings given by the set of DMs.

Table 6.2 DMs' evaluations

Alternatives	Horse power (CV)	Max Speed (km/h)	Acceleration (0-100km/h)	Consumption (lt/100km)	Design DM1	Design DM2	Design DM3	Price (€)
Daewoo Matiz	75	152	16	7	****	*	**	9000
Opel Agila	80	155	13	7	****	****	****	10500
Hyundai Atos	55	142	15	6.5	**	*****	*****	8400
Daihatsu Cuore	60	140	13	5	*	**	*	7500
Ford CA	70	155	15	6	*****	***	****	8600
Suzuki Wagon	50	145	19	6	*	****	**	9000
Fiat Seicento	55	150	14	6.5	***	***	***	8300

Table 6.3 DMs' rankings

Alternatives	Ranking DM1	Ranking DM2	Ranking DM3
Daewoo Matiz	6	4	6
Opel Agila	1	5	3
Hyundai Atos	3	2	1
Daihatsu Cuore	4	1	2
Ford CA	5	2	4
Suzuki Wagon	2	4	3
Fiat Seicento	6	3	5

The previous alternative LP models give different optimization results and rankings for the set of DMs as shown in Tables 6.4–6.5. These results correspond to different collective solutions, and thus they are able to determine preferential inconsistencies among the DMs, and define potential interactions (trade-off process) that may achieve a higher group and/or individual consistency level. For example, Figure 6.9 presents the alternative estimated values for the set of alternatives for every DM.

Trade-off analysis can be used in order to reduce preferential inconsistencies among the DMs. For example, the DMs' preferences that may be modified include the criteria evaluations $g_i^r(a_k)$ or the ranking of alternatives $R^r(a_k)$. The process can be easily implemented using the following approach:

1. Search error variables $(\sigma_r^+, \sigma_r^-, \gamma_{ab}^r)$ with non zero values.
2. Find $g_i^r(a_k)$ and/or $R^r(a_k)$ that should be modified.
3. Propose changes that reduce inconsistencies, while creating new ones.

In any case, it should be emphasized that the UTA method can be easily extended in the case of multiple actors, taking advantage of the flexibility of the LP modeling. Moreover, the different optimality criteria may lead to different results and thus they may offer alternative solutions to start a negotiation dialogue.

Table 6.4 Optimization results for the alternative LPs

Optimality criteria	SAE (F_1)	MAE (F_2)	SVJ (F_3)	MVJ (F_4)
Sum of errors	0.892	1.05	2.206	1.90
Maximum error per DM	0.392	0.35	1.277	1.15
Sum of violated judgments	14	12	6	6
Maximum violated judgments per DM	5	4	3	2

Table 6.5 Rankings from the alternative LPs

	Ranking (SAE)			Ranking (MAE)			Ranking (SVJ)			Ranking (MVJ)		
Alternatives	DM1	DM2	DM3	DM1	DM2	DM3	DM1	DM2	DM3	DM1	DM2	DM3
Daewoo Matiz	7	7	7	7	7	7	2	6	5	2	2	2
Opel Agila	2	3	3	4	4	4	6	7	7	7	7	7
Hyundai Atos	4	4	4	3	3	3	5	1	1	6	1	1
Daihatsu Cuore	1	1	1	2	2	2	4	2	4	4	5	5
Ford CA	5	5	5	5	5	5	1	3	2	1	4	3
Suzuki Wagon	3	2	2	1	1	1	7	5	6	3	3	4
Fiat Seicento	6	6	6	6	6	6	3	4	3	5	6	6

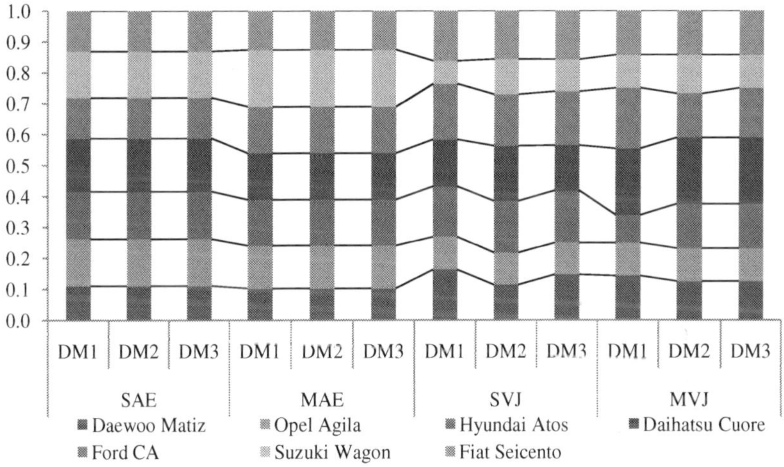

Fig. 6.9 Alternatives' values for every DM

6.5 Conclusions and Future Research

The aggregation-disaggregation philosophy is not only an important field of MCDA, but its principles can be found in other decision-making areas. The main aim of all these approaches is to infer global preference models from preference structures, as directly expressed by one or more DMs. In this context, the UTA methods not only adopt the preference disaggregation principles, but they may also be considered as the main initiative and the most representative example of preference disaggregation theory.

The new research developments of aggregation-disaggregation approaches presented in this chapter cover a variety of topics, like post-optimality analysis, robustness analysis, group and collective decision-making. They focus mainly on the UTA family of models and highlight their most important advantages (e.g. flexible modeling, analytical results, and alternative ways to reduce preferential inconsistencies).

Besides the numerous previous studies, additional research efforts are necessary in order to further exploit the potentials of the preference disaggregation philosophy within the context of MCDA. These efforts may include the development of more sophisticated aggregation models, the further exploitation of the provided results, or the adoption of aggregation-disaggregation philosophy in other decision-making approaches.

References

1. C.A. Bana e Costa and J.C. Vansnick. MACBETH: An interactive path towards the construction of cardinal value functions. *International Transactions in Operations Research*, 1(4):489–500, 1994.
2. M. Beuthe and G. Scannella. Applications comparées des méthodes d'analyse multicritère UTA. *RAIRO Recherche Opérationelle*, 30(3):293–315, 1996.
3. M. Beuthe and G. Scannella. Comparative analysis of UTA multicriteria methods. *European Journal of Operational Research*, 130(2):246–262, 2001.
4. A. Charnes, W. Cooper, and R.O. Ferguson. Optimal estimation of executive compensation by linear programming. *Management Science*, 1(2):138–151, 1955.
5. D.K. Despotis, D. Yannacopoulos, and C. Zopounidis. A review of the UTA multicriteria method and some improvements. *Foundations of Computing and Decision Sciences*, 15(2):63–76, 1990.
6. D.K. Despotis and C. Zopounidis. Building additive utilities in the presence of non-monotonic preference. In P.M. Pardalos, Y. Siskos, and C. Zopounidis, editors, *Advances in Multicriteria Analysis*, pages 101–114. Kluwer Academic Publisher, Dordrecht, 1993.
7. M. Doumpos and C. Zopounidis. *Multicriteria Decision Aid Classification Methods*. Kluwer Academic Publishers, Dordrecht, 2002.
8. J.R. Figueira, S. Greco, and R. Słowiński. Building a set of additive value functions representing a reference preorder and intensities of preference: GRIP method. *European Journal of Operation Research*, 195(2):460–486, 2009.
9. S. Greco, V. Mousseau, and R. Słowiński. Ordinal regression revisited: Multiple criteria ranking using a set of additive value functions. *European Journal of Operation Research*, 191(2):416–436, 2008.
10. S. Greco, R. Słowiński, J.R. Figueira, and V. Mousseau. Robust ordinal regression. In M. Ehrgott, S. Greco, and J. Figueira, editors, *Trends in Multiple Criteria Decision Analysis*. Springer, Berlin, 2010. forthcoming.
11. E. Grigoroudis and Y. Siskos. Preference disaggregation for measuring and analysing customer satisfaction: The MUSA method. *European Journal of Operational Research*, 143(1):148–170, 2002.
12. E. Jacquet-Lagrèze, R. Meziani, and R. Słowiński. MOLP with an interactive assessment of a piecewise linear utility function. *European Journal of Operational Research*, 31(3):350–357, 1987.

13. E. Jacquet-Lagrèze and M.F. Shakun. Decision support systems for semistructured buying decisions. *European Journal of Operational Research*, 16:48–56, 1984.
14. E. Jacquet-Lagrèze and Y. Siskos. Assessing a set of additive utility functions for multicriteria decision making: The UTA method. *European Journal of Operational Research*, 10(2):151–164, 1982.
15. E. Jacquet-Lagrèze and Y. Siskos. Preference disaggregation: 20 years of MCDA experience. *European Journal of Operational Research*, 130(2):233–245, 2001.
16. M. Jarke, M.T. Jelassi, and M.F. Shakun. MEDIATOR: Toward a negotiation support system. *European Journal of Operational Research*, 31(3):314–334, 1987.
17. T. Jelassi, G. Kersten, and S. Zionts. An introduction to group decision and negotiation support. In C.A. Bana e Costa, editor, *Readings in Multiple Criteria Decision Aid*, pages 537–568. Springer, Berlin, 1990.
18. O.J. Karst. Linear curve fitting using least deviations. *Journal of the American Statistical Association*, 53:118–132, 1958.
19. R.L. Keeney and H. Raiffa. *Decisions with Multiple Objectives: Preferences and Value Tradeoffs*. John Wiley and Sons, New York, 1976.
20. J.E. Kelley. An application of linear programming to curve fitting. *Journal of Industrial and Applied Mathematics*, 6(1):15–22, 1958.
21. N.F. Matsatsinis, E. Grigoroudis, and A. Samaras. Aggregation and disaggregation of preferences for collective decision-making. *Group Decision and Negotiation*, 14(3):217–232, 2005.
22. N.F. Matsatsinis and A. Samaras. MCDA and preference disaggregation in group decision support systems. *European Journal of Operational Research*, 130(2):414–429, 2001.
23. M. Oral and O. Kettani. Modelling the process of multiattribute choice. *Journal of the Operational Research Society*, 40(3):281–291, 1989.
24. S.A. Orlovsky. Decision-making with a fuzzy preference relation. *Fuzzy Sets and Systems*, 1(3):155–167, 1978.
25. B. Roy. Partial preference analysis and decision-aid: The fuzzy outranking relation concept. In D.E. Bell, R.L. Keeney, and H. Raiffa, editors, *Conflicting Objectives in Decisions*, pages 40–75. Wiley, New York, 1977.
26. B. Roy. *Multicriteria Methodology for Decision Aiding*. Kluwer Academic Publishers, Dordrecht, 1996.
27. B. Roy. Robustness in operational research and decision aiding: A multi-faceted issue. *European Journal of Operational Research*, 200(3):629–638, 2010.
28. M.F. Shakun. *Evolutionary Systems Design: Policymaking Under Complexity and Group Decision Support Systems*. Holden-Day, San Francisco, CA, 1988.
29. M.F. Shakun. Group decision and negotiation support in evolving, non-shared information contexts. *Theory and Decision*, 28:275–288, 1990.
30. M.F. Shakun. Airline buyout: Evolutionary systems design and problem restructuring in group decision and negotiation. *Management Science*, 37(10):1291–1303, 1991.
31. J. Siskos. Comment modéliser les préférences au moyen de fonctions d'utilité additives. *RAIRO Recherche Opérationelle*, 14:53–82, 1980.
32. J. Siskos. A way to deal with fuzzy preferences in multicriteria decision problems. *European Journal of Operational Research*, 10(3):314–324, 1982.
33. J. Siskos. Analyse de systèmes de décision multicritère en univers aléatoire. *Foundations of Control Engineering*, 10(3–4):193–212, 1983.
34. J. Siskos and N. Assimakopoulos. Multicriteria highway planning: A case study. *Mathematical and Computer Modelling*, 12(l(10-11)):1401–1410, 1989.
35. J. Siskos and D.K. Despotis. A DSS oriented method for multiobjective linear programming problems. *Decision Support Systems*, 5(1):47–55, 1989.
36. J. Siskos, A. Spyridakos, and D. Yannacopoulos. MINORA: A multicriteria decision aiding system for discrete alternatives. *Journal of Information Science and Technology*, 2(2):136–149, 1993.
37. Y. Siskos. Le traitement des solutions quasi-optimales en programmation linéaire continue: Une synthèse. *RAIRO Recherche Opérationelle*, 18:382–401, 1984.

38. Y. Siskos, E. Grigoroudis, and N.F. Matsatsinis. UTA methods. In J. Figueira, S. Greco, and M. Ehrgott, editors, *Multiple Criteria Analysis: State of the Art Surveys*, pages 297–344. 2005.
39. Y. Siskos and A. Spyridakos. Intelligent multicriteria decision support: Overview and perspectives. *European Journal of Operational Research*, 113(2):236–246, 1999.
40. Y. Siskos and D. Yannacopoulos. UTASTAR: An ordinal regression method for building additive value functions. *Investigação Operacional*, 5(1):39–53, 1985.
41. Y. Siskos, C. Zopounidis, and A. Pouliezos. An integrated dss for financing firms by an industrial development bank in greece. *Decision Support Systems*, 112(2):151–168, 1994.
42. V. Srinivasan and A.D. Shocker. Linear programming techniques for multidimensional analysis of preferences. *Psychometrika*, 38(3):337–396, 1973.
43. T.J. Stewart. Pruning of decision alternatives in multiple criteria decision making, based on the UTA method for estimating utilities. *European Journal of Operational Research*, 28(1):79–88, 1987.
44. C. Van de Panne. *Methods for Linear and Quadratic Programming*. North-Holland Publishing Company, Amsterdam, 1975.
45. H.M. Wagner. Linear programming techniques for regression analysis. *Journal of the American Statistical Association*, 54:206–212, 1959.
46. F.W. Young, J. De Leeuw, and Y. Takane. Regression with qualitative and quantitative variables: An alternating least squares method with optimal scaling features. *Phychometrica*, 41(4):505–529, 1976.
47. C. Zopounidis and M. Doumpos. A multicriteria decision aid methodology for the assessment of country risk. *European Research on Management and Business Economics*, 3(3):13–33, 1997.
48. C. Zopounidis and M. Doumpos. Building additive utilities for multigroup hierarchical discrimination: The MHDIS method. *Optimization Methods and Software*, 14(3):219–240, 2000.
49. C. Zopounidis and M. Doumpos. A preference disaggregation decision support system for financial classification problems. *European Journal of Operation Research*, 130(2):402–413, 2001.

Chapter 7
Disaggregation Analysis and Statistical Learning: An Integrated Framework for Multicriteria Decision Support

Michael Doumpos and Constantin Zopounidis

Abstract Disaggregation methods have become popular in multicriteria decision aiding (MCDA) for eliciting preferential information and constructing decision models from decision examples. From a statistical point of view, data mining and machine learning are also involved with similar problems, mainly with regard to identifying patterns and extracting knowledge from data. Recent research has also focused on the introduction of specific domain knowledge in machine learning algorithms. Thus, the connections between disaggregation methods in MCDA and traditional machine learning tools are becoming stronger. In this chapter the relationships between the two fields are explored. The differences and similarities between the two approaches are identified and a review is given regarding the integration of the two fields.

7.1 Introduction

Decision-making under multiple criteria or uncertainty is a subjective task that depends on the system of preferences of the decision-maker (DM). Multicriteria decision aid (MCDA) provides a broad set of methodologies suitable for such situations, where conflicting criteria, goals, objectives, and points of view, have to be taken into consideration. Among others, MCDA is involved with problem structuring, preference modeling, the construction and characterization of different forms of criteria aggregation models, as well as the design of interactive solution and decision aid/support procedures.

In many cases, the decision situation involves a finite set of actions or alternatives that need to be evaluated following a choice, ranking, sorting or description decision problematic [116]. Within this context, the evaluation process is based on a combination of all the criteria describing the performance of the alternatives. Such a

Department of Production Engineering and Management, Technical University of Crete, University Campus, 73100 Chania, Greece e-mail: mdoumpos; kostas@dpem.tuc.gr

C. Zopounidis and P.M. Pardalos (eds.), *Handbook of Multicriteria Analysis*, Applied Optimization 103, DOI 10.1007/978-3-540-92828-7_7, © Springer-Verlag Berlin Heidelberg 2010

combination, however, cannot be meaningful within a given decision context, unless it is able to represent (with some acceptable accuracy) the DM's judgment policy. This can be achieved in two quite different ways.

The first, is a "forward" approach based on interactive, structured communication sessions between the analyst and the DM, during which the analyst elicits specific information about the DM's preferences (e.g., weights, trade-offs, aspiration levels, etc.). The success of this approach is heavily based on the willingness of the DM to participate actively in the process, as well as the ability of the analyst to guide the interactive process in order to address the DM's cognitive limitations. This kind of approach is widely used in situations involving decisions of strategic character.

However, depending on the selected criteria aggregation model, a considerable amount of information may be needed by the DM. In "repetitive" decisions, where time limitations exist, the above direct approach may not be applicable. Disaggregation methods [72] are very helpful in this context. Disaggregation methods use regression-like techniques to infer a decision model from a set of decision examples on some reference alternatives, so that the model is as consistent as possible with the actual evaluation of the alternatives by the DM. This model inference approach provides a starting basis for the decision-aiding process. If the obtained model's parameters are in accordance with the actual preferential system of the DM, then the model can be directly applied to new decision instances. On the other hand, if the model is consistent with the sample decisions, but its parameters are inconsistent with the DM's preferential system (which may happen if, for example, the decision examples are inadequate), then the DM has a starting basis upon which he/she can provide recommendations to the analyst about the calibration of the model in the form of constraints about the parameters of the model. Thus, starting with a model that is consistent with a set of reference examples, an interactive model calibration process is invoked.

Similarly to disaggregation analysis, statistical learning and data mining are also involved with learning from examples [61, 62]. Many advances have been made within these fields for regression, classification and clustering problems. Recently there has been a growing interesting among machine learning researchers towards preference modeling and decision-making. Some interest has also been developed by MCDA researchers on exploiting the advances in machine learning.

Given the growing interest on the integration of the two fields, the objective of this chapter is to explore their connections, to highlight their similarities and differences and analyze the potential from their integration towards providing improved decision support.

The rest of the chapter is organized as follows: We begin with an introduction to disaggregation paradigm of MCDA in section 7.2, followed by an introduction to statistical learning and data mining (section 7.3). Then, section 7.4 discusses the differences and the similarities between the two fields, whereas section 7.5 provides a literature review on the interactions between them. Finally, section 7.6 concludes the chapter and discusses some future research directions.

7.2 The Disaggregation Approach in MCDA

7.2.1 General Framework

Disaggregation analysis (DA) provides a general methodological framework for the analysis of the actual decisions taken by a DM so that an appropriate model can be constructed representing the DM's system of preferences, as consistently as possible. The main input used in this process is a reference set of alternatives evaluated by the DM. The reference set may consist of past decisions, a subset of the alternatives under consideration, or a set of fictitious alternatives which can be easily judged by the DM [72]. Depending on the decision problematic, the evaluation of the reference alternatives may be expressed by defining an order structure (total, weak, partial, etc. [106]) or by classifying them into appropriate classes.

Formally, let $\mathscr{D}(X)$ denote the DM's evaluation of a set X consisting of m reference alternatives described over n criteria (the description of alternative i on criterion j will henceforth be denoted by x_{ij}). The DM's evaluation is assumed to be based (implicitly) on a decision model f_β defined by some parameters β, which represents the actual preferential system of the DM. Different classes of models can be considered. Typical examples include:

- Value functions defined such that $V(\mathbf{x}) > V(\mathbf{y})$ iff alternative $\mathbf{x}$ is preferred over alternative $\mathbf{y}$ and $V(\mathbf{x}) = V(\mathbf{y})$ in cases of indifference [77]. The parameters of a value function model involve the criteria tradeoffs and the form of the marginal value functions.
- Outranking relations defined such that $\mathbf{x} S \mathbf{y}$ iff alternative $\mathbf{x}$ is at least as good as alternative $\mathbf{y}$ [115]. Depending on the specific method used, the parameters of an outranking model, may involve the weights of the criteria, as well as preference, indifference and veto thresholds, etc.
- "If ... then ..." decision rules [53]. In this case the parameters of the model involve the conditions and the conclusions associated to each rule.

The objective of DA is to infer the "optimal" parameters $\widehat{\beta}^*$ that approximate, as accurately as possible, the actual preferential system of the DM as represented in the unknown set of parameters β, i.e.:

$$\widehat{\beta}^* = \arg\min_{\widehat{\beta} \in \mathscr{A}} \|\widehat{\beta} - \beta\| \tag{7.1}$$

where $\mathscr{A}$ is a set of feasible values for the parameters $\widehat{\beta}$. With the obtained parameters, the evaluations performed with the corresponding decision model $f_{\widehat{\beta}^*}$ will be consistent with the evaluations actually performed by the DM for any set of alternatives.

However, problem (7.1) cannot be solved explicitly because β is unknown. Instead, an empirical estimation approach is employed using the DM's evaluation of the reference alternatives to proxy β. Thus, the general form of the optimization problem is now expressed as follows:

$$\widehat{\beta}^* = \arg\min_{\widehat{\beta}\in\mathscr{A}} L[\mathscr{D}(X), \widehat{\mathscr{D}}(X, f_{\widehat{\beta}})] \tag{7.2}$$

where $\widehat{\mathscr{D}}(X, f_{\widehat{\beta}})$ denotes the recommendations of the model $f_{\widehat{\beta}}$ for the alternatives in X and $L(\cdot)$ is a function that measures the differences between $\mathscr{D}(X)$ and $\widehat{\mathscr{D}}(X, f_{\widehat{\beta}})$.

Through the solution of (7.2), it is implicitly assumed that the decision model's estimated parameters $\widehat{\beta}^*$ represent the actual preferential system of the DM within some acceptable error threshold $\varepsilon > 0$, i.e., $\|\widehat{\beta}^* - \beta\| < \varepsilon$. This, however, may not be true for a number of reasons related to the quality of the reference set (e.g., too small, noisy, etc.). Thus, problems (7.1) and (7.2) are not necessarily equivalent in a realistic setting.

7.2.2 Methods and Implementations

The general framework of DA is materialized in several MCDA methods that enable the development of decision models in different forms. This section focus on two popular paradigms, which involve functional and relational models. Symbolic models have also become quite popular recently. However, given their close connections with machine learning methods, the discussion of this modeling form is given later in section 7.5.1.2.

7.2.2.1 Functional Models

Value functions are the most widely used type of functional models in MCDA. A value function aggregates all the criteria into an overall performance measure V defined such that:

$$\begin{aligned} V(\mathbf{x}) > V(\mathbf{y}) \Leftrightarrow \mathbf{x} \succ \mathbf{y} \\ V(\mathbf{x}) = V(\mathbf{y}) \Leftrightarrow \mathbf{x} \sim \mathbf{y} \end{aligned} \tag{7.3}$$

where $\succ$ and $\sim$ denote the preference and indifference relations, respectively. A value function may expressed in different forms, depending on the criteria independence conditions that describe the DM's preferences [77]. Due to its simplicity, the most widely used form of value function is the additive one:

$$V(\mathbf{x}) = \sum_{j=1}^{n} w_j v_j(x_j) \tag{7.4}$$

where $w_1, \ldots, w_n$ are non-negative constants representing the criteria tradeoffs ($w_1 + \cdots + w_n = 1$) and $v_1(x_1), \ldots, v_n(x_n)$ are the marginal value functions of the criteria, usually scaled such that $v_j(x_{j*}) = 0$ and $v_j(x_j^*) = 1$, where x_{j*} and x_j^* are the least and the most preferred level of criterion j, respectively.

Such a model can be used to rank a set of alternatives or to classify them in predefined groups. In the ranking case, the relationships (7.3) provide a straightforward way to compare the alternatives. In the classification case, the simplest approach is to define groups $G_1, G_2, \ldots, G_q$ in the value scale with the following rule:

$$t_k \leq V(\mathbf{x}) < t_{k-1} \Leftrightarrow \mathbf{x} \in G_k \tag{7.5}$$

where $0 = t_q < t_{q-1} < \cdots < t_1 < t_0 = 1$ are thresholds that distinguish the groups.

The construction of a value function from a set of reference examples can be performed using mathematical programming techniques. For example, in an ordinal regression setting, the DM's defines a weak-order of the alternatives in the reference set, by ranking them from the best one (alternative $\mathbf{x}_1$) to the worst one (alternative $\mathbf{x}_m$). Then, the general form of the optimization problem can be expressed as in the case of the UTA method [71] as follows:

$$\begin{aligned}
\min\ & \sum_{i=1}^{m} \sigma_i \\
\text{s.t.}\ & \sum_{j=1}^{n} [v_j(x_{ij}) - v_j(x_{i+1,j})] + \sigma_i - \sigma_{i+1} \geq \delta,\ \forall \mathbf{x}_i \succ \mathbf{x}_{i+1} \\
& \sum_{j=1}^{n} [v_j(x_{ij}) - v_j(x_{i+1,j})] + \sigma_i - \sigma_{i+1} = 0,\ \forall \mathbf{x}_i \sim \mathbf{x}_{i+1} \\
& v_j(x_j) \text{ non-decreasing, with } v_j(x_{j*}) = 0 \text{ and } \sum_{j=1}^{n} v_j(x_j^*) = 1 \\
& \sigma_i \geq 0, \quad \forall i
\end{aligned} \tag{7.6}$$

The solution of this optimization problem provides an additive value function that reproduces the DM's ranking of the reference alternatives as accurately as possible. The differences between the model's recommendations and the DM's weak-order are measured by the error variables $\sigma_1, \ldots, \sigma_m$. In this case the value function is expressed in pure additive form as:

$$V(\mathbf{x}) = v_1(x_1) + \cdots + v_n(x_n) \tag{7.7}$$

where the marginal value functions are now scaled such that $v_j(x_{j*}) = 0$ and $v_j(x_j^*) = w_j$. By modeling the marginal values as piecewise linear functions, the above optimization problem can be re-expressed in linear programming form (for the details see [71]).

Several variants of the UTA method for ordinal regression problems have been presented. Siskos et al. [125] provide a detailed review of different formulations, whereas Beuthe and Scannella [13] present a comparative analysis. Some recent extensions are presented by Figueira et al. [41] and Greco et al. [56]. Formulations for classification problems have also been developed, such as the UTADIS method and its variants [34, 37, 71, 80], the MHDIS method [150], and other similar approaches [19, 28, 81].

The optimization processes most often used involve linear and integer programming formulations (LP, IP). LP models are usually used to minimize some predefined norm (L_1 or L_∞) of real-valued error variables representing the violations of (7.3) or (7.5). The optimization problem (7.6) is an example using the L_1 norm of the error variables $\sigma_1, \ldots, \sigma_m$ for the reference alternatives. IP formulations on the other hand, consider more direct measures of the number of disagreements between the recommendations of the estimated decision model and the actual evaluation of the reference alternatives by the DM. The Kendall's τ rank correlation coefficient is a typical example of such a measure.

It is also worth mentioning the considerable recent research on extending this modeling framework, which is based on simple form of value functions, towards more general preference modeling forms that allow the consideration of interaction between the criteria. The use of the Choquet integral as an aggregation function has proved quite useful and convenient towards this direction. Marichal and Roubens [93] first introduced a methodology implementing this approach in a disaggregation context. Some works on this topic can be found in the works of Angilella et al. [5] and Kojadinovic [78, 79], while a review of this topic has been presented by Grabisch et al. [51].

7.2.2.2 Relational Models

The evaluations performed on the basis of value functions are transitive and complete. In several cases, however, preferences do not satisfy these properties. Intransitivity is often observed and furthermore the alternatives can be incomparable. Relational models enable the modeling of such situations. The outranking relations theory of MCDA [115] describes such models, with close connections to social choice theory.

Typically, an outranking relation S between a pair of alternatives $\mathbf{x}$ and $\mathbf{y}$ is defined as:

$$\mathbf{x} S \mathbf{y} \Leftrightarrow \mathbf{x} \text{ is at least as good as } \mathbf{y} \tag{7.8}$$

Outranking techniques operate in two stages. The first stage involves the pairwise comparison of the alternatives. Then, an algorithmic procedure is used in the second stage to derive the evaluation results from the pairwise comparisons of the first stage.

There are several outranking methods that implement the above framework in different ways. The most widely used include the families of ELECTRE [115] and PROMETHEE methods [10, 14]. Martel and Matarazzo [94] provide a comprehensive review of other outranking approaches. Other non-outranking relational models based on distances have been presented by Chen et al. [23, 24].

In contrast to a value function approach, outranking models usually require too many parameters, which define the decision model in a complex non-linear way. This fact poses a significant computational burden in eliciting the preferential parameters from decision examples. With some simplifying assumptions this issue can be resolved. For instance, Mousseau et al. [98], Dias et al. [29], Ngo The and Mousseau [101], and Dias and Mousseau [30] developed several LP simplifications and heuristics to infer some of the parameters of pessimistic ELECTRE TRI models, while assuming the others fixed. Conventional optimization approaches (LP and quadratic programming) are generally applicable for simpler forms of outranking/relational models that implement a compensatory approach (see for instance [23, 24, 36].

A first attempt to develop an "holistic" approach for more complex outranking models was presented by Mousseau and Slowinski [99] for the ELECTRE TRI method. Similarly to the previous studies, they assumed the pessimistic assignment rule, and developed a non-linear, non-convex optimization formulation to infer all the parameters of a classification decision model from a set of assignment examples. Later, Doumpos and Zopounidis [35] presented an alternative approach combining heuristic rules with LP formulations. Recently, metaheuristics and evolutionary approaches have been used. Goletsis et al. [49] used a genetic algorithm for the development of an outranking model in a two-group problem involving ischemic beat classification. Fernandez et al. [40] used a multiobjective genetic optimization approach for constructing an outranking classification model, whereas Belacel et al. [11] used the reduced variable neighborhood search metaheuristic to infer the parameters of the PROAFTN method from a set of reference examples, and Doumpos et al. [33] used the differential evolution algorithm to develop classification models based on the ELECTRE TRI.

7.3 Statistical Learning and Data Mining

7.3.1 General Framework

Hand et al. [61] define data mining as *"the analysis of (often large) observational data sets to find unsuspected relationships and to summarize the data in novel ways that are both understandable and useful to the data owner"*.

Statistical learning plays an important role in the data mining process, by describing the theory that underlies the identification of such relationships and providing the necessary analysis techniques. According to Vapnik [135, 136] the process of learning from examples includes three main components:

1. A set X of data vectors $\mathbf{x}$ drawn independently from a probability distribution $P(\mathbf{x})$. This distribution is assumed to be unknown, thus implying that there is no control on how the data are observed [128].
2. An output y from a set Y, which is defined for every input $\mathbf{x}$ according to an unknown conditional distribution function $P(y \mid \mathbf{x})$. This implies that the relationship between the input data and the outputs is unknown.

3. A learning method (machine), which is able to assign a function $f_\beta : X \to Y$, where β are some parameters of the unknown function.

The best function f_β is the one that best approximates the actual outputs, i.e., the one that minimizes:

$$\int L[y, f_\beta(\mathbf{x})] dP(\mathbf{x}, y) \tag{7.9}$$

where $L[y, f_\beta(\mathbf{x})]$ is a function of the differences between the actual output y and the estimate $f_\beta(\mathbf{x})$,[1] and $P(\mathbf{x}, y) = P(\mathbf{x})P(y \mid \mathbf{x})$ is the joint probability distribution of $\mathbf{x}$ and y. However, this joint distribution is unknown and the only available information is contained in a training set of m objects $\{(\mathbf{x}_1, y_1), \ldots, (\mathbf{x}_m, y_m)\}$, which are assumed to be generated independently from this unknown distribution. Thus, the objective (7.9) is substituted by its empirical estimate:

$$\frac{1}{m} \sum_{i=1}^{m} L[y_i, f_\beta(\mathbf{x}_i)] \tag{7.10}$$

For a class of functions f_β of a given complexity, the minimization of (7.10) leads to the minimization of an upper bound for (7.9).

7.3.2 Methods

One of the main research directions in statistical learning involves the study of the theoretical properties of the optimization problem (7.10) in order to derive data independent bounds of the generalization performance of learning machines (for a complete coverage of the theoretical aspects of statistical learning, see [136]). The other main direction involves the development of efficient learning methods and algorithms. While a full review of all methods and algorithms is out of the scope of this chapter (detailed presentations are available in [61, 62]), a brief outline of some popular schemes is given below.

7.3.3 Neural Networks

Neural networks have been one of the most popular approaches in statistical learning and data mining. Neural networks are based on an "artificial" representation of the human brain, through a directed acyclic graph with nodes (neurons) organized into layers. In a typical feed-forward architecture, there is an layer of input nodes, a layer of output nodes, and a series of intermediate layers. The input nodes correspond to

---

[1] The specification of the loss function L depends on the problem under consideration. For instance, in a regression setting it may correspond to the mean squared error, whereas in a classification context it may represent the accuracy rate.

the information that is available for every input vector (the attributes/independent variables), whereas the output nodes provide the recommendations of the network. The nodes in the intermediate (hidden) layers are parallel processing units that define the input-output relationship. Every neuron at a given layer receives as input the weighted average of the outputs of the neurons at the preceding layer and maps it to an output signal through a predefined transformation function. Depending on the topology of the network and the selection of the neurons' transformation functions, a neural network can model real functions of arbitrary complexity. This flexibility has made neural networks a very popular modeling approach in addressing complex real-world problems in engineering and management.

Training a neural network involves the optimization of the connections' weights. In a supervised learning context, the optimization is based on a training set, in accordance with the general framework of statistical learning. Unconstrained non-linear optimization algorithms are commonly used in this context [63]. Evolutionary techniques have also been recently used [1].

7.3.4 Decision Trees and Rule-Based Models

Symbolic models expressed as decision trees and rule sets are quite popular among machine learning researchers and practitioners, mainly due to their interpretability. Typically, the nodes of a decision tree represent a series of (usually) binary splits defined on the independent variables, while the recommendations of the model are given at the terminal nodes of the tree. Decision tree models can also be expressed in a rule-based format of the form of "If ... then ..." decision rules. The first part of a given rule examines the necessary conditions required for the conclusion part to be valid. The conclusion provides the recommendation (output) of the rule. Except for the easy interpretation and use of such models by DMs and analysts, other advantages also include their ability to handle different types of data (quantitative or qualitative), the handling of missing data, as well as their applications in discovering interesting relations between variables in large databases (e.g., through the development of association rules [4]).

Some typical examples of algorithms used to build decision trees and rule-based models include, among others, ID3 [112] and its successor C4.5 [113], as well as CART [18], CHAID [76], and rough sets [108].

7.3.5 Support Vector Machines

Support vector machines (SVMs) have become increasingly popular among the statistical learning community. SVMs implement the structural risk minimization principle taking into consideration the empirical loss (7.10), while controlling the complexity of the model through a Tikhonov regularization approach. The empirical error can be minimized with a highly complex model, but in such cases the model is usually unstable to the selection of the training set, and consequently its generalizing ability is poor. SVMs introduce this tradeoff in the analysis providing a unified framework for both linear and non-linear models.

SVMs are usually realized in a binary classification setting, but they are also applicable in multi-group classification, regression, and clustering problems. In the simplest case involving a binary classification task, the two groups are defined by the canonical hyperplanes $a+\mathbf{wx}=\pm 1$ (where a is a constant term and $\mathbf{w}$ is the normal vector for the hyperplane), such that $a+\mathbf{wx}\geq 1$ for the positive examples and $a+\mathbf{wx}\leq -1$ for the negative ones. The distance (separating margin) between the two hyperplane is then $2/\|\mathbf{w}\|$ and it is related to an upper bound of the probability that an observation will be misclassified (the higher the margin the lower the misclassification probability [120]). Extending, this reasoning by introducing the classification errors, leads to the following convex quadratic programming formulation:

$$\begin{aligned} \min\ & \frac{1}{2}\mathbf{w}^{\top}\mathbf{w}+C\sum_{i=1}^{m}\sigma_i \\ \text{s.t.}\ & y_i(a+\mathbf{wx}_i)+\sigma_i\geq 1, \ \forall i \\ & \sigma_i\geq 0, \quad \forall i \\ & a,\mathbf{w}\in\mathbb{R} \end{aligned} \tag{7.11}$$

where $y_i=\pm 1$ denotes the class label for observation i, σ_i is the corresponding slack variable defined such that $\sigma_i>0$ iff $y_i(a+\mathbf{wx}_i)<1$, and $C>0$ is a user-defined constant that defines the trade-off between the two conflicting objectives (margin maximization and error minimization).

The generalization to the nonlinear case is achieved by mapping the problem data to a higher dimensional space H (feature space) through a transformation of the form $\mathbf{x}_i\mathbf{x}_j^{\top}=\phi(\mathbf{x}_i)\phi^{\top}(\mathbf{x}_j)$. The mapping function ϕ is implicitly defined through a symmetric positive definite kernel function $K(\mathbf{x}_i,\mathbf{x}_j)=\phi(\mathbf{x}_i)\phi^{\top}(\mathbf{x}_j)$. The representation of the data using the kernel function enables the development of a linear model in the feature space H.

For large training sets several computational procedures have been proposed to enable the fast training of SVM models. Most of these procedures are based on a decomposition scheme, where the optimization problem is split into a series of smaller subproblems. Other algorithms are based on reformulations of the optimization problem that enable the development of the model through the solution of a set of linear equations. Linear programming formulations have also been used.

A detailed presentation of SVMs, the theory of kernel methods, the existing optimization tools, and applications, can be found in the book of Schölkopf and Smola [120] as well as the review paper of Campbell [20].

7.3.6 *Ensembles*

Individual learning algorithms often exhibit instability to the training data. This may lead to high bias/variance and poor generalizing performance. A popular approach to address this issue is to combine multiple models, thus forming ensemble predictors. This is not a new issue, since the first works on the combination of regression forecasts can be traced back to the work of Bates and Granger [9]. Recently, theoretical evidences have been presented (no free lunch theorems [146, 147, 148]) showing that there is no method that is universally better than others in terms of its predictive performance. On the basis of such findings, it is natural to investigate the potential of a method/model combination framework. Of course, combined models are also subject to the no free lunch theorem. However, the development of combined models aims at the reduction of the bias and/or variance of the individual models, which is expected to be useful in improving the results in real-world situations. The combination is most useful when the predictions of the models which are combined have low correlations to each other.

Since the 1990s there has been a considerable growth in the research on model combination approaches and several algorithm-independent approaches have been proposed that exploit the instability of statistical learning models and the differences between methods. Some approaches combine multiple models of the same learning method, each developed using different perturbations of the training set [15, 16, 43], while other approaches enable the combination of models from multiple methods [45, 145]. A review of different ensemble approaches can be found in [31].

7.4 Similarities and Differences

Disaggregation analysis and statistical learning have evolved significantly over the past two decades, as two separate fields. Nevertheless, the similarities between the two fields are obvious, since both consider the problem of learning a decision/prediction model from data. Within this context, it is worth noting that the minimization of the empirical loss (7.10) in statistical learning methods is actually identical to the optimization problem (7.2), which is commonly used in disaggregation methods. This may lead to the conclusion that both fields actually address the same problem. But there are a series of noticeable differences.

1. *Model interpretability*: The interpretability of MCDA models is of outmost importance. Interpretable and easy to understand decision models enable the DM's active participation in the decision-aiding process, they provide insights on the characteristics of the alternatives, and the DMs often feel more confident with them in their daily practice. On the other hand, statistical learning theory has mostly focused on the development of models of high predicting ability. In most cases these models are too complex to interpret (e.g., neural networks, non-linear

SVMs, ensembles, etc.), and consequently their operation has often been described as a "black box".
2. *Data dimensionality*: Real world applications of statistical learning methods involve large (often massive) data sets and considerable research has been devoted to the development of computationally efficient algorithms that scale up well with the size of the training data. DA methods, on the other, usually assume that only a small reference set is available, since it is difficult for the DMs to express their global preferences on too many alternatives.
3. *Inconsistencies*: In DA, data inconsistencies are usually explicitly treated during the model development process [50, 88, 96, 97]. This is done through interactive procedures whose objective is to reveal the inconsistencies to the DM, to support their resolution, and to enhance the DM's understanding of the problem data. Contrary to this approach, data mining treats inconsistencies in the training data as "hard cases", i.e., observations which are simply difficult to learn and predict. Only outliers are treated as real inconsistencies.
4. *Model validation*: While model validation is used in both DA and statistical learning to check the quality of the model, the implementation of the validation process differs. In DA it is usually assumed that the analyst cooperates with the DM and the validation is an interactive process, during which the DM checks the validity of the parameters' estimation results. The generalizing ability of the model, is on the other hand, the core issue in the validation stage of all statistical learning models. This is tested using additional data sets, outside the training sample, or through resampling methods (e.g., cross-validation, bootstrap, etc.).

The differences between statistical learning and disaggregation methods in MCDA have also been discussed by Waegeman et al. [138]. In addition to the above points, the authors have also noted some other interesting issues:

1. *The role of the DM*: In MCDA, the DM participates actively in the decision modeling process and interacts with the analyst in order to achieve the best calibration of the decision model. This interactive process is dynamic in nature, in that the DM's preferences may change as he/she gains insight to the problem data and its characteristics. Statistical learning and data mining on the other hand, assume that only a statistical sample, whereas specific inputs from the DM are not.
2. *Regularization*: The traditional MCDA disaggregation methods usually do not take into account the trade-off between model complexity and model performance. On the other hand, regularization has become a crucial issue in statistical learning processes.
3. *Data type*: The data considered in an MCDA setting involve the description of the alternatives over a number of criteria. In addition to this type of data setting, statistical learning is also concerned with more complex structures, which are often encountered in areas such as text mining, image analysis, and signal processing.

Overall, these differences really seem quite fundamental; an indeed they are. But the consideration of these differences, should take into account the crucial aspect of the scope of the application of DA methods as opposed to the common uses of data mining.

DA methods are used in a MCDA context to facilitate the decision support process. In particular, the main objective of eliciting preferential information through decision examples is to facilitate the DM in gaining insight into: (1) the characteristics of the problem data (alternatives and criteria), (2) the implications of the judgments that (s)he implicitly makes, (3) the characteristics and limitations of the modeling process, (4) the interpretations of the results, and ultimately (4) the actions that need to be taken in order to obtain good decisions through a practical model.

On the other hand, modern statistical learning and data mining adopt an *algorithmic modeling culture* as described by Breiman [17], in which the focus is shifted from data models to the characteristics and predictive performance of learning algorithms. In this framework, the data generation process in a real problem is a "black box whose insides are complex, mysterious and, at least, partly unknowable" [17], thus leading to the important issue of developing efficient algorithms that provide accurate predictions of the observed outcome from some given input data.

7.5 Interactions

Despite the development of MCDA and statistical learning/data mining as separate fields, and the differences outlined in the previous section, there have been several attempts to integrate concepts and methods from the two fields. This section reviews this emerging research stream and its potential towards the development of improved decision support methodologies. The interactions of the two fields are examined in two opposite directions. The first involves the use of statistical learning and data mining techniques in a decision aiding context through disaggregation analysis and preference learning. The second direction involves the implementation of MCDA concepts in a statistical learning framework and the development of hybrid methodologies.

7.5.1 Using Statistical Learning Methods for Disaggregation Analysis and MCDA

7.5.1.1 Neural Networks

One of the main advantages of neural network (NN) models is their ability to model highly complex problems, with an unknown underlying structure. This characteristic has important implications for MCDA, mainly with respect to modeling general preference structures.

Within this context, NNs have been successfully used for learning generalized MCDA models from decision examples. Wang and Malakooti [141], and Malakooti and Zhou [91] used feedforward NNs to learn an arbitrary value function for ranking a set of alternatives, as well as to learn a relational multicriteria model based on pairwise comparisons (binary relations) among the alternatives. The main advantage of this NN-based approach, is that the resulting decision models are free of the various independence assumptions, which are often implied by commonly used value function models (see [77]). Thus, the model is independent of functional form and quite stable to parameter perturbations. The authors examined the conditions that characterize the monotonicity of the NN model, as well as its convexity/concavity properties. The monotonicity condition of the form (7.3) is a fundamental property for any rational decision model. On the other hand, the convexity/concavity properties are very useful for calibrating the model development (training) process in order to ensure that the final model complies with the DM's preference policy. Experimental simulation results showed that NN trained models performed very well in representing various forms of decision models, outperforming other popular model development techniques based on linear programming formulations. Wang et al. [142] applied a similar NN model to a job shop production system problem.

In a different framework compared to the aforementioned studies, Stam et al. [127] used NNs within the context of the analytic hierarchy process (AHP) [117]. AHP is based on a hierarchical structuring of the decision problem, with the overall goal on the top of the hierarchy and the alternatives at the bottom. With this hierarchical structure, the DM is asked to perform pairwise comparisons of the elements at each level of the hierarchy with respect to the elements of the preceding (higher) level. The principal eigenvalues and the corresponding normalized eigenvectors of the resulting reciprocal pairwise comparison matrices are then used to obtain preference ratings for the alternatives. This eigenvector-based approach has received much criticism (see for example [8]). Stam at el. investigated two different NN structures for accurately approximating the preferences ratings of the alternatives, within the context of imprecise preference judgments by the DM. They showed that a modified Hopfield network has very close connections to the mechanics of the AHP, but found that this network formulation cannot provide good results in estimating the mapping from a positive reciprocal pairwise comparison matrix to its preference rating vector. On the other hand, a feed-forward NN model was found to provide very good approximations of the preference ratings in the presence of impreciseness. This NN model was actually superior to the standard principal eigenvector method.

Similar NN-based methodologies have also be used to address dynamic MCDA problems (where the DM's preferences change over time) [90], to learn fuzzy preferences [139, 140, 143] and outranking relations [67], to provide support in group decision making problems [143], as well as in multicriteria clustering [89].

NNs have also been employed for preference representation and learning in multiobjective optimization (MOP). The main goal in a MOP problem is to identify the set of non-dominated solutions (Pareto optimal solutions) and then to select the most appropriate one that best fits the DM's preferences. Within this context, Sun et al. [130] proposed a feed-forward NN model, which is trained to represent the DM's preference structure. The training of the model is performed using a representative sample of non-dominated solutions, which are evaluated by the DM. The flexibility of NNs enables them to model complex preference structures, even highly nonlinear ones. Thus, the trained NN model is used to formulate the objective function of a nonlinear programming problem, which is solved in order to find a solution that maximizes the output of the trained NN. A similar NN optimization formulation has also been proposed by Chen and Lin [21], while Shimizu et al. [122] presented a web-based implementation integrating a NN model with AHP. Such approaches are generally similar to techniques proposed in the MOP literature based on traditional value function models (see for example [124]). Despite the good results obtained with this approach and its robustness to the NN's architecture, the solution of the nonlinear optimization problem having the NN's output as the objective is often cumbersome. To overcome this difficulty, Sun et al. [131] presented a hybrid methodology combining the feed-forward NN model with the interactive weighted Tchebycheff procedure (IWTP) [129]. In this case a trained NN model is used to evaluate a set of nondominated solutions and to select the ones that are most likely to be of interest to the DM, thus supporting the interactive search procedure. Results on various test problems characterized with different underlying value functions (linear and nonlinear) indicated that the use of the NN-based approach provided improved results compared to IWTP and a NN-based optimization model. Other NN architectures have also been used as optimizers in MOP problems [48, 87, 95, 144] and hybrid evaluation systems [6, 111, 114, 121].

7.5.1.2 Rule-based Models

Rule-based and decision tree models are very popular within the machine learning research community. The symbolic nature of such models makes them easy to understand, which usually is a very important characteristic in decision aiding problems. During the last decade significant research has been devoted on the use of such approaches as preference modeling tools in MCDA and disaggregation analysis.

Within this framework there has been proposed a complete and well-axiomatized methodology for constructing decision rule preference models from decision examples, based on the rough sets theory [108, 109]. Rough sets have been initially introduced as a methodology to describe dependencies between attributes, to evaluate the significance of attributes and to deal with inconsistent data in multicriteria decision problems. However, over the past decade significant research has been conducted on the use of the rough set approach as a methodology for preference modeling in multicriteria decision problems [52, 53]. The main novelty of this approach concerns the possibility of handling criteria, i.e. attributes with preference ordered domains, and preference ordered classes in the analysis of sorting examples and the induction of decision rules. The rough approximations of decision classes involve the dominance relation, instead of the indiscernibility relation considered in the basic rough sets approach. They are build of reference alternatives given in the decision examples. Decision rules derived from these approximations constitute a preference model. Each "if ... then ..." decision rule is composed of a condition part specifying a partial profile on a subset of criteria to which an alternative is compared using the dominance relation, and a decision part suggesting an assignment of the alternative to "at least" or "at most" a given class.

The decision rule preference model has also been considered in terms of conjoint measurement [55] and Bayesian decision theory [57]. A representation theorem [55] for multicriteria sorting states an equivalence of simple cancellation property, a general discriminant (sorting) function and a specific outranking relation, on the one hand, and the decision rule model on the other hand. It is also shown that the decision rule model resulting from the dominance-based rough set approach has an advantage over the usual functional and relational models because it permits handling inconsistent sorting examples. The inconsistency in sorting examples is not unusual due to instability of preference, incomplete determination of criteria and hesitation of the DM.

An important feature of this methodology is that its applicability is not restricted to multicriteria classification problems, but is also extended to ranking and choice decision problems [42, 53], as well as to MOP problems [56]. It also provides the ability to work with missing data and to handle cases that involved both criteria and attributes (whose domains are not preference ordered; see [54]).

A similar approach that implements symbolic models has also been presented by Dombi and Zsiros [32], while Hammer et al. [60] modified the LAD method (logical analysis of data), which is based on the theory of boolean functions, to address multicriteria classification problems through the introduction of Pareto-optimal patterns.

7.5.1.3 Kernel Methods and Margin-Based Approaches

Kernel methods become an important research direction in statistical learning and they are now widely used for classification and regression models, as well as for density estimation. Kernel methods map the problem data to a high dimensional space (feature space), thus enabling the development of complex nonlinear decision and prediction models, using linear estimation methods [65, 119]. This kind of representation is based on a positive definite kernel function, which corresponds to a dot product in the feature space. The main novelty of the introduction of the kernel function is that it makes the explicit computation of the feature space unnecessary.

SVMs are one of the most common implementations of the theory of kernel methods, with numerous application in pattern recognition problems. Recently, they have also been used within the context of preference learning for approximating arbitrary utility/value functions and preference aggregation.

Herbrich et al. [64] illustrated the use of kernel approaches, within the context of SVM formulations, for representing value/ranking functions of the generalized form $V(\mathbf{x}) = \mathbf{w}\phi(\mathbf{x})$, where ϕ is a possibly infinite-dimensional and in general unknown feature mapping. The authors derived bounds on the generalizing performance of the estimated ranking models, based on the margin separating objects in consecutive ranks. A similar approach was also explored by Joachims [75] in the RankSVM algorithm, which has been used to improve the retrieval quality of search engines.

Waegeman et al. [138] extend this approach to relational models. In this case, the preference model of the form $f(\mathbf{x}_i,\mathbf{x}_j) = \mathbf{w}\phi(\mathbf{x}_i,\mathbf{x}_j)$ is developed to represent the preference of alternative i compared to alternative j. This framework is general enough to accommodate special modeling forms. For instance, it includes value models as a special case, and similar techniques can also be used to kernelize Choquet integrals. As an example, Waegeman et al. illustrated the potential of this framework in the case of valued concordance relations, which are used in the ELECTRE methods.

Together with the use the kernel approach for the development of generalized decision models, an additional important feature of SVM formulations, is the implementation of the regularization concept to handle the complexity of the models. Evgeniou et al. [39] gave an interpretation of this regularization approach within the context of estimating a linear value function $V(\mathbf{x}) = \mathbf{wx}$ used for ranking purposes (ordinal regression). From a geometric point of view, assuming the simplest case where the reference data are representable by such a linear model with no errors, the value function that minimizes $\|\mathbf{w}\|$ corresponds to the most robust solution in the feasible space defined by constraints of the form:

$$\mathbf{w}(\mathbf{x}_i - \mathbf{x}_j) \geq 1, \quad \forall\, \mathbf{x}_i \succ \mathbf{x}_j$$

The term "robust" in this case refers to the solution that is the center of the largest sphere in the polyhedron defined by the constraints [39]. In the general case, when the reference data include some inconsistent comparisons, Evgeniou et al. intuitively explained the minimization of the violations of the above constraints together with the minimization of $\|\mathbf{w}\|$, as the search of a decision model that minimizes the errors compared to the DM's judgments, while satisfying the correct comparisons as much as possible. The authors also describe the generalization of this framework to nonlinear (polynomial) value function models.

Doumpos and Zopounidis [37] analyzed a similar methodology for the construction of additive value functions, using the L_1 norm of the parameters of the function. They showed that formulating an augmented linear objective function considering both the errors of the model and its complexity, leads to interesting insights on the quality of the reference set. Experimental results on both ranking and classification problems showed that such a modification improves both the generalizing performance of the obtained decision models and their robustness.

In a different context, Dembczynski et al. [28] combined concepts from the dominance-based rough set approach and SVMs towards the development of additive value function models in classification problems. Contrary to the methodology of Doumpos and Zopounidis [37], the authors used a regularization term based on the L_2 norm, and illustrated how the formulation can be expressed in kernel form.

The margin maximization principle, which is implemented in SVMs as a regularization mechanism, has also been explored in connection to some ensemble algorithms, such as boosting [43, 118]. Within this context, Freund et al. [44] developed the RankBoost algorithm, which provides a single linear ordering of the given set of objects by combining a set of given linear orderings on a set of ranking features. Instead of using the evaluation criteria as ranking features, Freund et al. developed an algorithm to combine multiple "weak" rankings defined over the criteria, in a weighted additive model (similar to an additive value function).

7.5.2 MCDA Concepts in Statistical Learning Models

7.5.2.1 MCDA Methodologies for Building Statistical Learning Models

As mentioned in section 7.3.1, the development of statistical learning models is based on the minimization of a loss function measured on the basis of a set of training samples. This general setting, however, can be implemented in many different ways. The variety of loss functions employed in different models indicates that model development is not based on a straightforward, universally accepted criterion. For instance, in regression problems measures such as the mean squared error or the mean absolute error may lead to completely different models. In classification problems, similar L_1, L_2 and L_∞ norms have been used, together with measures such as the classification error rate or the area under the receiver operating characteristic curve. The introduction of the regularization terms also adds complexity and degrees of freedom on the model development process.

Similar "multicriteria" issues also arise in specification of the predictor variables (feature selection and extraction), the construction of ensemble models, the pruning and selection of decision rules, as well as the extension of case-based reasoning models on the basis of generalized distance metrics.

The existing research on all the above topics is indeed quite rich. Some indicative works include:

- Learning through multiobjective optimization [38, 47, 59, 74, 83, 84, 100, 132].
- Feature selection and feature extraction [46, 68, 103, 149].
- Construction of ensemble models [58, 69, 70, 73].
- Pruning and use of decision rules [105, 126].
- Case-based reasoning [85, 86, 107].

7.5.2.2 Model Performance Evaluation

The approaches discussed in the previous subsection aim towards the consideration of multiple performance measures at the model construction (optimization) phase. Obviously, multiple performance measures can also be used when evaluating the suitability and performance of a given set of models. For instance, Osei-Bryson [104] proposed a multicriteria methodology to evaluate decision trees based on criteria related to their discriminatory power, simplicity, and stability, taking into account the DM's subjective judgements on the relative importance of these criteria. In a later study Osei-Bryson applied this modeling framework to the problem of pruning decision trees [105]. In a similar context, Choi et al. [25] and Chen [22] introduced multiple criteria for the evaluation of association rules using MCDA approaches. In a neural network setting, Das [26] used the TOPSIS multicriteria method to evaluate neural network models using multiple performance measures (e.g., mean squared error, the Akaike's information criterion, the Bayesian information criterion, cross-validation accuracy, etc.), whereas Ni et al. [102] used the PROMETHEE multicriteria method to evaluate different neural network architectures developed for the prediction of carbamate concentrations in ternary mixtures.

7.5.2.3 Monotonicity in Predictive Modeling

Monotonicity plays a crucial role in decision modeling and aiding. In simple terms, given two alternatives such that $\mathbf{x}_i \geq \mathbf{x}_j$, the monotonicity principle implies that alternative j cannot be preferred over alternative i. Assuming, for instance, a functional decision model (e.g., value/ranking function) $f(\mathbf{x})$, the monotonicity condition requires that $f(\cdot)$ is monotone with respect to the inputs, i.e., $\mathbf{x}_i \geq \mathbf{x}_j \Rightarrow f(\mathbf{x}_i) \geq f(\mathbf{x}_j)$.

Monotonicity is usually not taken into consideration in a data mining context. However, in several cases, the users of prediction models would like the models not only to predict well, but also to make sense within the context of a specific application domain. Furthermore, studies have shown that the introduction of specific domain knowledge into the statistical learning process, may actually improve the generalizing ability of the obtained models [3, 92, 133], by reducing overfitting, minimizing the effect of noisy data, and controlling the complexity of the model.

Within this context, monotonicity can be considered as a special form of domain knowledge. In simple linear decision models of the form $f(\mathbf{x}) = \mathbf{w}\mathbf{x}$, monotonicity can be easily introduced by imposing the condition $\mathbf{w} \geq \mathbf{0}$, which requires the scaling constants to be non-negative. For generalized non-linear decision models, however, the introduction of monotonic conditions is more involved.

Ben-David et al. [12] were among the first to explore this issue within the context of rule-based models. Some recent works on learning monotonic rule-based models and decision trees can be found in [27, 82, 110, 134]. Studies involving other learning models, such as neural networks and SVMs, include [2, 7, 66, 123, 137].

7.6 Conclusions

Data mining/statistical learning and the disaggregation approach of MCDA, both study similar problems in a different context. Data mining has focused on the development of generalized prediction models from a statistical point of view and statistical learning has focused on the theory of the learning process aiming at the development of scalable algorithms for accurate predictive modeling with large and complex data sets. On the other hand, the disaggregation approach of MCDA has mainly focused on the development of comprehensible decision models from small data sets, whose main objective is to support decision aiding through an interactive model calibration process.

Despite the conceptual and modeling differences in the two paradigms, there are clear connections. This chapter highlighted these connections together with the existing differences. The literature review also shows that the interactions between the two fields have already been explored, thus enabling the development of new improved techniques, which can be used either for predictive purposed in a pure data mining context or for aiding the DMs in complex decision problems.

The road ahead should mainly focus on exploring further ways to integrate the two fields. The use of statistical learning approaches for modeling new types of preference models, the addition of comprehensibility into data mining tools, the issues of validation, regularization, and robustness of preference models developed through DA, the scalability of disaggregation methods to large data sets, and the applications of new models into innovative fields, are only some indicative topics where future research can focus.

References

1. H.A. Abbass. Speeding up backpropagation using multiobjective evolutionary algorithms. *Neural Computation*, 15(11):2705–2726, 2003.
2. Y.S. Abu-Mostafa. Learning from hints in neural networks. *Journal of Complexity*, 6(2):192–198, 1990.
3. Y.S. Abu-Mostafa. Machines that learn from hints. *Scientific American*, 272(4):64–69, 1995.
4. R. Agrawal, T. Imieliński, and A. Swami. Mining association rules between sets of items in large databases. In P. Buneman and S. Jajodia, editors, *Proceedings of the 1993 ACM SIGMOD International Conference on Management of Data*, pages 207–216, New York, NY, 1993. ACM.
5. S. Angilella, S. Greco, and B. Matarazzo. Non-additive robust ordinal regression: A multiple criteria decision model based on the Choquet integral. *European Journal of Operational Research*, 201:277–288, 2010.
6. O.U. Araz. A simulation based multi-criteria scheduling approach of dual-resource constrained manufacturing systems with neural networks. *Lecture Notes in Computer Science*, 3809:1047–1052, 2005.
7. N.P. Archer and S. Wang. Application of the back propagation neural network algorithm with monotonicity constraints for two-group classification problems. *Decision Sciences*, 24(1):60–75, 1993.
8. C.A. Bana e Costa and J-C. Vansnick. A critical analysis of the eigenvalue method used to derive priorities in AHP. *European Journal of Operational Research*, 187(3):1422–1428, 2008.
9. J.M. Bates and Granger C.W.J. The combination of forecasts. *Operational Research Quarterly*, 20(4):451–468, 1969.
10. M. Behzadian, R.B. Kazemzadeh, A. Albadvi, and M. Aghdasi. PROMETHEE: A comprehensive literature review on methodologies and applications. *European Journal of Operational Research*, 2009. In press.
11. N. Belacel, H. Bhasker Raval, and A.P. Punnenc. Learning multicriteria fuzzy classification method PROAFTN from data. *Computers and Operations Research*, 34:1885–1898, 2007.
12. A. Ben-David, L. Sterling, and Y.-H. Pao. Learning and classification of monotonic ordinal concepts. *Computational Intelligence*, 5:45–49, 1989.

13. M. Beuthe and G. Scannella. Comparative analysis of UTA multicriteria methods. *European Journal of Operational Research*, 130:246–262, 2001.
14. J-P. Brans and B. Mareschal. PROMETHEE methods. In J. Figueira, S. Greco, and M. Ehrgott, editors, *Multiple Criteria Decision Analysis-State of the Art Surveys*, pages 163–195. Springer, Boston, 2005.
15. L. Breiman. Bagging predictors. *Machine Learning*, 24(2):123–140, 1996.
16. L. Breiman. Random forests. *Machine Learning*, 45:5–32, 2001.
17. L. Breiman. Statistical modeling: The two cultures. *Statistical Science*, 16:199–231, 2001.
18. L. Breiman, J.H. Friedman, R.A. Olsen, and C.J. Stone. *Classification and Regression Trees*. Wadsworth, Pacific Grove, CA, 1984.
19. V. Bugera, H. Konno, and S. Uryasev. Credit cards scoring with quadratic utility function. *Journal of Multi-Criteria Decision Analysis*, 11:197–211, 2002.
20. C. Campbell. Kernel methods: A survey of current techniques. *Neurocomputing*, 48:63–84, 2002.
21. J. Chen and S. Lin. An interactive neural network-based approach for solving multiple criteria decision-making problems. *Decision Support Systems*, 36:137–146, 2003.
22. M-C. Chen. Ranking discovered rules from data mining with multiple criteria by data envelopment analysis. *Expert Systems with Applications*, 33:1110–1116, 2007.
23. Y. Chen, K.W. Hipel, and M.D. Kilgour. Multiple-criteria sorting using case-based distance models with an application in water resources management. *IEEE Transactions on Systems, Man, and Cybernetics-Part A*, 37(5):680–691, 2007.
24. Y. Chen, M.D. Kilgour, and K.W. Hipel. A case-based distance method for screening in multiple-criteria decision aid. *Omega*, 36(3):373–383, 2008.
25. D.H. Choi, B.S. Ahn, and S.H. Kim. Prioritization of association rules in data mining: Multiple criteria decision approach. *Expert Systems with Applications*, 29(4):876878, 2005.
26. P. Das. In search of best alternatives: a topsis driven mcdm procedure for neural network modeling. *Neural Computing and Applications*, 2009. In press.
27. J. Dembczynski, W. Kotlowski, and R. Slowinski. Ensemble of decision rules for ordinal classification with monotonicity constraints. *Lecture Notes in Computer Science*, 5009:260–267, 2008.
28. K. Dembczynski, W. Kotlowski, and R. Slowinski. Additive preference model with piecewise linear components resulting from dominance-based rough set approximations. *Lecture Notes in Computer Science*, 4029:499–508, 2006.
29. L. Dias, V. Mousseau, J. Figueira, and J. Clímaco. An aggregation/disaggregation approach to obtain robust conclusions with ELECTRE TRI. *European Journal of Operational Research*, 138(2):332–348, 2002.
30. L.C. Dias and V. Mousseau. Inferring Electre's veto-related parameters from outranking examples. *European Journal of Operational Research*, 170(1):172–191, 2006.
31. T.G. Dietterich. Ensemble methods in machine learning. *Lecture Notes in Computer Science*, 1857:1–15, 2000.
32. J. Dombi and A. Zsiros. Learning multicriteria classification models from examples: Decision rules in continuous space. *European Journal of Operational Research*, 160:663–675, 2005.
33. M. Doumpos, Y. Marinakis, M. Marinaki, and C. Zopounidis. An evolutionary approach to construction of outranking models for multicriteria classification: The case of the ELECTRE TRI method. *European Journal of Operational Research*, 199(2):496–505, 2009.
34. M. Doumpos and C. Zopounidis. *Multicriteria Decision Aid Classification Methods*. Springer, New York, 2002.
35. M. Doumpos and C. Zopounidis. On the development of an outranking relation for ordinal classification problems: An experimental investigation of a new methodology. *Optimization Methods and Software*, 17(2):293–317, 2002.
36. M. Doumpos and C. Zopounidis. A multicriteria classification approach based on pairwise comparisons. *European Journal of Operational Research*, 158:378–389, 2004.

37. M. Doumpos and C. Zopounidis. Regularized estimation for preference disaggregation in multiple criteria decision making. *Compututational Optimization and Applications*, 38:61–80, 2007.
38. R.M. Everson and J.E. Fieldsend. Multi-class roc analysis from a multi-objective optimisation perspective. *Pattern Recognition Letters*, 27:918–927, 2006.
39. T. Evgeniou, C. Boussios, and G. Zacharia. Generalized robust conjoint estimation. *Marketing Science*, 24(3):415–429, 2005.
40. E. Fernandez, J. Navarroa, and S. Bernal. Multicriteria sorting using a valued indifference relation under a preference disaggregation paradigm. *European Journal of Operational Research*, 198(2):602–609, 2009.
41. J.R. Figueira, S. Greco, and R. Slowinski. Building a set of additive value functions representing a reference preorder and intensities of preference: GRIP method. *European Journal of Operational Research*, 195:460–486, 2009.
42. P. Fortemps, S. Greco, and R. Slowinski. Multicriteria decision support using rules that represent rough-graded preference relations. *European Journal of Operational Research*, 188:206–223, 2008.
43. Y. Freund. Boosting a weak learning algorithm by majority. *Information and Computation*, 121(2):256–285, 1995.
44. Y. Freund, R. Iyer, R.E. Schapire, and Y. Singer. An efficient boosting algorithm for combining preferences. *Journal of Machine Learning Research*, 4:933–969, 2003.
45. J. Gama and P. Brazdil. Cascade generalization. *Machine Learning*, 41:315–343, 2000.
46. J. García-Nieto, E. Alba, L. Jourdan, and E. Talbi. Sensitivity and specificity based multiobjective approach for feature selection: Application to cancer diagnosis. *Information Processing Letters*, 109(16):887–896, 2009.
47. N. García-Pedrajas and D. Ortiz-Boyer. A cooperative constructive method for neural networks for pattern recognition. *Pattern Recognition*, 40:8098, 2007.
48. M.R. Gholamian, S.M.T. Fatemi Ghomi, and M. Ghazanfari. A hybrid intelligent system for multiobjective decision making problems. *Computers and Industrial Engineering*, 51:26–43, 2006.
49. Y. Goletsis, C. Papaloukas, D.I. Fotiadis, A. Likas, and L.K. Michalis. Automated ischemic beat classification using genetic algorithms and multicriteria decision analysis. *IEEE Transactions on Biomedical Engineering*, 51(10):1717–1725, 2004.
50. J. Gonzalez-Pachon and C. Romero. A method for dealing with inconsistencies in pairwise comparisons. *European Journal of Operational Research*, 158(2):351–361, 2004.
51. M. Grabisch, I. Kojadinovic, and P. Meyer. A review of methods for capacity identification in Choquet integral based multi-attribute utility theory: Applications of the Kappalab R package. *European Journal of Operational Research*, 186:766–785, 2008.
52. S. Greco, B. Matarazzo, and R. Slowinski. Rough approximation of a preference relation by dominance relations. *European Journal of Operational Research*, 117:63–83, 1999.
53. S. Greco, B. Matarazzo, and R. Slowinski. Rough sets theory for multicriteria decision analysis. *European Journal of Operational Research*, 129:1–47, 2001.
54. S. Greco, B. Matarazzo, and R. Slowinski. Rough sets methodology for sorting problems in presence of multiple attributes and criteria. *European Journal of Operational Research*, 138:247–259, 2002.
55. S. Greco, B. Matarazzo, and R. Slowinski. Axiomatic characterization of a general utility function and its particular cases in terms of conjoint measurement and rough-set decision rules. *European Journal of Operational Research*, 158(2):271–292, 2004.
56. S. Greco, V. Mousseau, and R. Slowinski. Ordinal regression revisited: Multiple criteria ranking using a set of additive value functions. *European Journal of Operational Research*, 191(2):415–435, 2008.
57. S. Greco, R. Slowinski, and Y. Yao. Bayesian decision theory for dominance-based rough set approach. In J. Yao, P. Lingras, W-Z. Wu, M. Szczuka, N.J. Cercone, and S. Ślęzak, editors, *Rough Sets and Knowledge Technology*, pages 134–141. Springer, Berlin, 2007.

58. M. Guijarro and G. Pajares. On combining classifiers through a fuzzy multicriteria decision making approach: Applied to natural textured images. *Expert Systems with Applications*, 36:7262–7269, 2009.
59. A. Guillén, H. Pomares, J. González, I. Rojas, O. Valenzuela, and B. Prieto. Parallel multi-objective memetic rbfnns design and feature selection for function approximation problems. *Neurocomputing*, 72(16-18):3541–3555, 2009.
60. P.L. Hammer, A. Kogan, B. Simeone, and S. Szedmák. Pareto-optimal patterns in logical analysis of data. *Discrete Applied Mathematics*, 144:79–102, 2004.
61. D. Hand, H. Mannila, and P. Smyth. *Principles of Data Mining*. MIT Press, Cambridge, 2001.
62. T. Hastie, R. Tibshirani, and J. Friedman. *The Elements of Statistical Learning: Data Mining, Inference, and Prediction*. Springer, New York, 2001.
63. S. Haykin. *Neural Networks: A Comprehensive Foundation*. Prentice Hall, Singapore, 2nd edition, 1999.
64. R. Herbrich, T. Graepel, and K. Obermayer. Large margin rank boundaries for ordinal regression. In A.J. Smola, P.L. Bartlett, B. Schölkopf, and D. Schuurmans, editors, *Advances in Large Margin Classifiers*, pages 115–132. MIT Press, Cambridge, MA, 2000.
65. T. Hofmann, B. Schölkopf, and A.J. Smola. Kernel methods in machine learning. *The Annals of Statistics*, 36(3):1171–1220, 2008.
66. A. Howard and T. Jebara. Learning monotonic transformations for classification. In J.C. Platt, D. Koller, Y. Singer, and S. Roweis, editors, *Advances in Neural Information Processing Systems 20*, pages 681–688. MIT Press, Cambridge, MA, 2008.
67. Y-C. Hu. Bankruptcy prediction using ELECTRE-based single-layer perceptron. *Neurocomputing*, 72:3150–3157, 2009.
68. B. Huang, B. Buckley, and T.-M. Kechadi. Multi-objective feature selection by using NSGA-II for customer churn prediction in telecommunications. *Expert Systems with Applications*, 2007. In press.
69. E. Hüllermeier and K. Brinker. Learning valued preference structures for solving classification problems. *Fuzzy Sets and Systems*, 159(18):2337–2352, 2008.
70. E. Hüllermeier and S. Vanderlooy. Combining predictions in pairwise classification: An optimal adaptive voting strategy and its relation to weighted voting. *Pattern Recognition*, 43(1):128–142, 2010.
71. E. Jacquet-Lagrèze and Y. Siskos. Assessing a set of additive utility functions for multicriteria decision making: The UTA method. *European Journal of Operational Research*, 10:151–164, 1982.
72. E. Jacquet-Lagrèze and Y. Siskos. Preference disaggregation: 20 years of MCDA experience. *European Journal of Operational Research*, 130:233–245, 2001.
73. T. Jiao, J. Peng, and T. Terlaky. A confidence voting process for ranking problems based on support vector machines. *Annals of Operations Research*, 166:23–38, 2009.
74. J. Jin. *Multi-Objective Machine Learning*. Springer, Berlin Heidelberg, 2006.
75. T. Joachims. Optimizing search engines using clickthrough data. In *ACM SIGKDD Conference on Knowledge Discovery and Data Mining (KDD)*, pages 133–142, 2002.
76. G.V. Kass. An exploratory technique for investigating large quantities of categorical data. *Journal of Applied Statistics*, 29:119–127, 1980.
77. R.L. Keeney and H. Raiffa. *Decisions with Multiple Objectives: Preferences and Value Trade-offs*. Cambridge University Press, Cambridge, 1993.
78. I. Kojadinovic. Estimation of the weights of interacting criteria from the set of profiles by means of information-theoretic functionals. *European Journal of Operational Research*, 155:741–751, 2004.
79. I. Kojadinovic. Minimum variance capacity identification. *European Journal of Operational Research*, 177:498–514, 2007.
80. M. Köksalan and S.B. Özpeynirci. An interactive sorting method for additive utility functions. *Computers & Operations Research*, 36(9):2565–2572, 2009.
81. M. Köksalan and C. Ulu. An interactive approach for placing alternatives in preference classes. *European Journal of Operational Research*, 144:429–439, 2003.

82. W. Kotlowski and R. Slowinski. Rule learning with monotonicity constraints. In *ICML '09: Proceedings of the 26th Annual International Conference on Machine Learning*, pages 537–544, New York, NY, USA, 2009. ACM.
83. K. Kottathra and Y. Attikiouzel. A novel multicriteria optimization algorithm for the structure determination of multilayer feedforward neural networks. *Journal of Network and Computer Applications*, 19:135–147, 1996.
84. G. Kou, X. Liu, Y. Peng, Y. Shi, M. Wise, and W. Xu. Multiple criteria linear programming approach to data mining: Models, algorithm designs and software development. *Optimization Methods and Software*, 18(4):453–473, 2003.
85. H. Li and J. Sun. Hybridizing principles of the ELECTRE method with case-based reasoning for data mining: ELECTRE-CBR-I and ELECTRE-CBR-II. *European Journal of Operational Research*, 197(1):214–224, 2009.
86. H. Li and J. Sun. Business failure prediction using hybrid^2 case-based reasoning (H^2CBR). *Computers and Operations Research*, 37(1):137–151, 2010.
87. Y. Li, K. Ida, M. Gen, and R. Kobuchi. Neural network approach for multicriteria solid transportation problem. *Computers and Industrial Engineering*, 33(3-4):465–468, 1997.
88. J. Ma, Z.P. Fan, Y.P. Jiang, J.Y. Mao, and L. Ma. A method for repairing the inconsistency of fuzzy preference relations. *Fuzzy Sets and Systems*, 157(1):20–33, 2006.
89. B. Malakooti and V. Raman. Clustering and selection of multiple criteria alternatives using unsupervised and supervised neural networks. *Journal of Intelligent Manufacturing*, 11:435–451, 2000.
90. B. Malakooti and Y. Zhou. A recursive ann for solving adaptive multiple criteria problems. *Pure Mathematics and Applications Series C*, 2(2-4):165–176, 1991.
91. B. Malakooti and Y.Q. Zhou. Feedforward artificial neural networks for solving discrete multiple criteria decision making problems. *Management Science*, 40(11):1542–1561, 1994.
92. O.L. Mangasarian and E.W. Wild. Nonlinear knowledge-based classification. *IEEE Transactions on Neural Networks*, 19(10):1826–1832, 2008.
93. J.-L. Marichal and M. Roubens. Determination of weights of interacting criteria from a reference set. *European Journal of Operational Research*, 124:641–650, 2000.
94. J-M. Martel and B. Matarazzo. Other outranking approaches. In J. Figueira, S. Greco, and M. Ehrgott, editors, *Multiple Criteria Decision Analysis-State of the Art Surveys*, pages 197–262. Springer, Boston, 2005.
95. P.R. McMullen. A kohonen self-organizing map approach to addressing a multiobjective mixed-model JIT sequencing problem. *International Journal of Production Economics*, 72:59–71, 2001.
96. V. Mousseau, L.C. Dias, and J. Figueira. Dealing with inconsistent judgments in multiple criteria sorting models. *4OR*, 4(3):145–158, 2006.
97. V. Mousseau, L.C. Dias, J. Figueira, C. Gomes, and J.N. Clímaco. Resolving inconsistencies among constraints on the parameters of an MCDA model. *European Journal of Operational Research*, 147(1):72–93, 2003.
98. V. Mousseau, J. Figueira, and J.-Ph. Naux. Using assignment examples to infer weights for ELECTRE TRI method: Some experimental results. *European Journal of Operational Research*, 130:263–275, 2001.
99. V. Mousseau and R. Slowinski. Inferring an ELECTRE-TRI model from assignment examples. *Journal of Global Optimization*, 12(2):157–174, 1998.
100. H. Nakayama, Y.B. Yun, T. Asada, and M. Yoon. MOP/GP models for machine learning. *European Journal of Operational Research*, 166:756–768, 2005.
101. A. Ngo The and V. Mousseau. Using assignment examples to infer category limits for the ELECTRE TRI method. *Journal of Multi-Criteria Decision Analysis*, 11:2943, 2002.
102. Y. Ni, C. Huang, and S. Kokot. Application of multivariate calibration and artificial neural networks to simultaneous kinetic-spectrophotometric determination of carbamate pesticides. *Chemometrics and Intelligent Laboratory Systems*, 71:177193, 2004.
103. L.S. Oliveira, M. Morita, and R. Sabourin. Feature selection for ensembles applied to handwriting recognition. *International Journal on Document Analysis and Recognition*, 8(4):262–279, 2006.

104. K-M. Osei-Bryson. Evaluation of decision trees: A multi-criteria approach. *Computers and Operations Research*, 31:1933–1945, 2004.
105. K-M. Osei-Bryson. Post-pruning in decision tree induction using multiple performance measures. *Computers and Operations Research*, 34:3331–3345, 2007.
106. M. Öztürk, A. Tsoukiàs, and Ph. Vincke. Preference modelling. In J. Figueira, S. Greco, and M. Ehrgott, editors, *Multiple Criteria Decision Analysis-State of the Art Surveys*, pages 27–71. Springer, Boston, 2005.
107. C.-P. Park and I. Han. A case-based reasoning with the feature weights derived by analytic hierarchy process for bankruptcy prediction. *Expert Systems with Applications*, 23(3):255–264, 2002.
108. Z. Pawlak. Rough sets. *International Journal of Information and Computer Sciences*, 11:341–356, 1982.
109. Z. Pawlak and R. Slowinski. Rough set approach to multi-attribute decision analysis. *European Journal of Operational Research*, 72:443–459, 1994.
110. R. Potharst and A.J. Feelders. Classification trees for problems with monotonicity constraints. *ACM SIGKDD Explorations Newsletter*, 4(1):1–10, 2002.
111. L. Qu and Y. Chen. A hybrid MCDM method for route selection of multimodal transportation network. *Lecture Notes in Computer Science*, 5263:374–383, 2008.
112. J.R. Quinlan. Induction of decision trees. *Machine Learning*, 1:81–106, 1986.
113. J.R. Quinlan. *C4.5: Programs for Machine Learning*. Morgan Kaufmann Publishers, Los Altos, California, 1993.
114. K.S. Raju, D.N. Kumar, and L. Duckstein. Artificial neural networks and multicriterion analysis for sustainable irrigation planning. *Computers and Operations Research*, 33:1138–1153, 2006.
115. B. Roy. The outranking approach and the foundations of ELECTRE methods. *Theory and Decision*, 31:49–73, 1991.
116. B. Roy. *Multicriteria Methodology for Decision Aiding*. Springer, New York, 1996.
117. T.L. Saaty. *Fundamentals of the Analytic Hierarchy Process*. RWS Publications, Pittsburgh, PA, 2006.
118. R.E. Schapire, Y. Freund, P. Bartlett, and W.S. Lee. Boosting the margin: A new explanation for the effectiveness of voting methods. *The Annals of Statistics*, 26(5):1651–1686, 1998.
119. B. Schölkopf and A. Smola. *Learning with Kernels: Support Vector Machines, Regularization, Optimization and Beyond*. MIT Press, Cambridge, 2002.
120. B. Schölkopf and A.J. Smola. *Learning with Kernels: Support Vector Machines, Regularization, Optimization and Beyond*. MIT Press, Cambridge, Massachusetts, 2002.
121. J.-B. Sheu. A hybrid neuro-fuzzy analytical approach to mode choice of global logistics management. *European Journal of Operational Research*, 189(3):971–986, 2008.
122. Y. Shimizu, Y. Tanaka, and A. Kawada. Multi-objective optimization system, MOON^2 on the internet. *Computers and Chemical Engineering*, 28:821–828, 2004.
123. J. Sill. Monotonic networks. In M.I. Jordan, M.J. Kearns, and S.A. Solla, editors, *Advances in Neural Information Processing Systems 10*, pages 661–667. MIT Press, Cambridge, MA, 1997.
124. J. Siskos and D.K. Despotis. A DSS oriented method for multiobjective linear programming problems. *Decision Support Systems*, 5:47–55, 1989.
125. J. Siskos, E. Grigoroudis, and N.F. Matsatsinis. UTA methods. In J. Figueira, S. Greco, and M. Ehrgott, editors, *Multiple Criteria Decision Analysis-State of the Art Surveys*, pages 297–343. Springer, Boston, 2005.
126. R. Slowinski and J. Stefanowski. Rough classification with valued closeness relation. In E. Diday, Y. Lechevallier, M. Schader, P. Bertrand, and B. Burtschy, editors, *New Approaches in Classification and Data Analysis*, pages 482–489. Springer, Berlin, 1994.
127. A. Stam, M. Sun, and M. haines. Artificial neural network representations for hierarchical preference structures. *Computers and Operations Research*, 23(12):1191–1201, 1996.
128. I. Steinwart and A. Christmann. *Support Vector Machines*. Springer, New York, 2008.
129. R.E. Steuer and E.U. Choo. An interactive weighted Tchebycheff procedure for multiple objective programming. *Mathematical Programming*, 26(1):326–344, 1983.

130. M. Sun, A. Stam, and R.E. Steuer. Solving multiple objective programming problems using feed-forward artificial neural networks: The interactive FFANN procedure. *Management Science*, 42(6):835–849, 1996.
131. M. Sun, A. Stam, and R.E. Steuer. Interactive multiple objective programming using Tchebycheff programs and artificial neural networks. *Computers and Operations Research*, 27(7-8):601–620, 2000.
132. R.A. Teixeira, A.P. Braga, R.H.C. Takahashi, and R.R. Saldanha. Improving generalization of mlps with multi-objective optimization. *Neurocomputing*, 35(14):189194, 2000.
133. G.G. Towell and J.W. Shavlik. Knowledge-based artificial neural networks. *Artificial Intelligence*, 70(1-2):119–165, 1994.
134. R. van de Kamp, A. Feelders, and N. Barile. Isotonic classification trees. *Lecture Notes in Computer Science*, 5772:405–416, 2009.
135. V.N. Vapnik. An overview of statistical learning theory. *IEEE Transactions on Neural Networks*, 10(5):988–999, 1999.
136. V.N. Vapnik. *The Nature of Statistical Learning Theory*. Springer, New York, 2nd edition, 2000.
137. M. Velikova, H. Daniels, and A. Feelders. Mixtures of monotone networks for prediction. *International Journal of Computational Intelligence*, 3(3):205–214, 2006.
138. W. Waegeman, B. De Baets, and B. Boullart. Kernel-based learning methods for preference aggregation. *4OR*, 7:169–189, 2009.
139. J. Wang. A neural network approach to modeling fuzzy preference relations for multiple criteria decision making. *Computers and Operations Research*, 21(9):991–1000, 1994.
140. J. Wang. A neural network approach to multiple criteria decision making based on fuzzy preference information. *Information Sciences*, 78:293–302, 1994.
141. J. Wang and B. Malakooti. A feedforward neural network for multiple criteria decision making. *Computers and Operations Research*, 19(2):151–167, 1992.
142. J. Wang, J.-Q. Yang, and H. Lee. Multicriteria order acceptance decision support in over-demanded job shops: A neural network approach. *Mathematical and Computer Modelling*, 19(5):1–19, 1994.
143. S. Wang and N.P. Archer. A neural network technique in modeling multiple criteria multiple person decision making. *Computers and Operations Research*, 21(2):127–142, 1994.
144. Y. Wang. Multicriteria neural network approach to turbulent image reconstruction. *Optics Communications*, 143:279–286, 1997.
145. D.H. Wolpert. Stacked generalization. *Neural Networks*, 5:241–259, 1992.
146. D.H. Wolpert. The existence of a priori distinctions between learning algorithms. *Neural Computation*, 8(7):1391–1420, 1996.
147. D.H. Wolpert. The lack of a priori distinctions between learning algorithms. *Neural Computation*, 8(7):1341–1390, 1996.
148. D.H. Wolpert and W.G. Macready. No free lunch theorems for optimization. *IEEE Transactions on Evolutionary Computation*, 1(1):67–82, 1997.
149. Y. Zhang and P.I. Rockett. Domain-independent feature extraction for multi-classification using multi-objective genetic programming. *Pattern Analysis and Applications*, 2009. In press.
150. C. Zopounidis and M. Doumpos. Building additive utilities for multi-group hierarchical discrimination: The MHDIS method. *Optimization Methods and Software*, 14(3):219–240, 2000.

Part III
Multiobjective Optimization

Chapter 8
Multiobjective Optimization, Systems Design and De Novo Programming

Milan Zeleny

Abstract In this chapter we explore some topics beyond traditional MCDM. First we explain in the simplest possible terms what multiobjective optimization is, and define the subject matter of this chapter. We discuss the role of tradeoffs and draw a distinction between tradeoffs-based versus tradeoffs-free thinking. Next, we introduce the concept of optimization and optimal systems design. Then we build the foundation of *De novo programming*, dealing with designing optimal systems in linear cases. Finally, we provide some numerical examples and discuss additional applications where optimal design and multiobjective optimization can be used.

8.1 Introduction

In decision making, we have to draw a distinction between the *decision producer* (or provider) and the *decision consumer* (or customer). Any decision maker makes decisions either for himself (being producer or consumer) or for others (consumers). A decision maker can also consume decisions of producers or produce decisions for consumers.

The vantage points of producers and consumers are very different - to the point of becoming opposing or conflicting. Observe preferences in attaining basic criteria (like cost, quality and speed) of objects, products or alternatives:

Milan Zeleny
Graduate School of Business, Fordham University, New York, USA
e-mail: mzeleny@fordham.edu

C. Zopounidis and P.M. Pardalos (eds.), *Handbook of Multicriteria Analysis*, Applied Optimization 103, DOI 10.1007/978-3-540-92828-7_8, © Springer-Verlag Berlin Heidelberg 2010

Decision producer	Decision consumer
Cost	Cost
OR quality	**AND** quality
OR speed	**AND** speed
OR reliability	**AND** reliability
OR ...	**AND** ...

Clearly, the consumer follows AND strategy (wants it *all*) while the producer follows OR strategy (offers either one *or* the other). There is not a single consumer in the world who would prefer the left column OR to the right column AND.

This is more serious than it appears. There are some who would argue that companies should follow the OR strategy (e.g. [7]) rather than the AND strategy, assuming clearly producer's rather than consumer's standpoint. Ignoring the consumers (customers) while propagating what is good for producers only, could spell downfall of such decision makers, individuals, corporations or institutions.

The above case provides not only the argument for single versus multiple criteria but, more importantly, for tradeoffs-based versus tradeoffs-free choice of alternatives. The comparison refers not to the mere number of criteria but mainly to the nature of criteria performance and improvement. While the producer offers tradeoffs, like cheaper *or* better quality products, the consumer always wants products that are cheaper *and* of better quality. The tradeoffs aversion is self-evident.

It is therefore advisable that we adopt and follow *modern* consumer strategy of decision making, making feasible only what should be feasible and discarding that which should not. Only then we can supply both consumers and producers with *modern* decision making [9].

8.2 What is Multiobjective Optimization?

Let us consider a basket of apples. Which one is the heaviest? How do we find the heaviest apple? The *criterion* is clearly weight. How do we proceed?

First, we have to *measure*. We weigh each apple, either precisely or approximately by judgment, and assign each apple a number: its precise or estimated weight. Second, we have to *search*. Compare and rank the obtained weights, choose a search procedure or algorithm and carry it out.

Such measurement & search is necessary *and* sufficient for solving the problem. It cannot fail, barring errors, to find the heaviest apple. This is *not* a problem of decision making or optimization, but a problem of measurement and search. It should be clear that the problem of *computation* (measurement & search) is merely a technological problem: no value or economic judgments are necessary, no tradeoffs are involved, and no decisions have to be taken. If I want to marry the richest girl in a lineup of candidates (basket of apples), the only thing I have to do is to measure and search.

Let us consider the same basket of apples again. We now look for the heaviest *and* the sweetest apple. Our *criteria* are clearly weight and sweetness. What has changed? How do we proceed now?

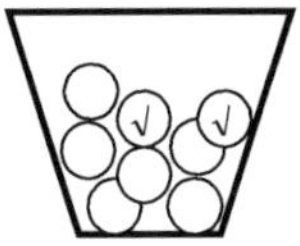

The heaviest apple clearly remains unchanged. We have to measure the sweetness either by the amount of sugar, by tasting, or some similar way. Then we search for the sweetest apple. In the end we obtain the sweetest and the heaviest apple. Both measurement and search are still necessary, but *not sufficient* to solve (or answer) the problem. Which apple do we choose? We have to go *beyond measurement & search* and perform additional function: we have to *decide*. Decision making is the function to be performed *after* the measurement and search have been completed. The difference between the two situations is the distinction between a *single* criterion versus *multiple* criteria. Problems with a single objective function, no matter how complex, important or aggregate that function is, are *not* problems of decision making or optimization, but of measurement and search only. No tradeoffs are involved [14].

All problems of decision making or optimization must have multiple criteria, i.e. be characterized by tradeoffs. (But not all problems with multiple criteria have to be decision-making problems.)

If I choose to marry the richest *and* the most beautiful girl from the lineup (assume I cannot marry both), then measurement and search will not be sufficient. I still have to decide: either the richest one, or the most beautiful one, or a third one - with the best combination of attributes. However, if there is one girl which is both the richest and the most beautiful (or one apple the heaviest and the sweetest) then the choice is clear: there are no tradeoffs, measurement & search is sufficient and I do not have to decide - even in the case of multiple criteria.

So, what is decision making? *Decision making is a function aimed at either resolving or dissolving the conflict of tradeoffs.* No tradeoffs $\rightarrow$ no decision making needed.

Clearly, replacing multiple objectives by a single composite or aggregate objective function, like for example the utility function, reduces initial multidimensionality to a single dimension, removes tradeoffs, and replaces the decision-making problem by a technical problem of measurement and search. We simply combine weight and sweetness into a sweet/weight superfunction, i.e. add sweet apples and heavy oranges (or money and beauty) - and search, and search ...

If we look into the previous basket again, with the same criteria of weight and sweetness, we see that there is one heaviest and one sweetest apple, tradeoffs exist between them, and the need for tradeoffs evaluation and decision making becomes necessary. We can limit our choice to such apples which retain tradeoffs among themselves, after eliminating all apples which are *dominated*, i.e. there is at least one apple in the basket which improves their weight and sweetness without any need for trading off. Resulting subset of apples is called Pareto-optimal, non-dominated, efficient, etc., depending on the nature of problem formulation. Our final decision will always be one of the nondominated apples.

The problem with such tradeoffs-based choice is that it will never be the best (optimal) but always a *compromise* for any decision maker. Each person evaluates tradeoffs differently, each person would choose a different nondominated apple, but no person prefers such tradeoffs-based choice.

Let us look at our basket in a different way: instead of concentrating on the criteria (weight and sweetness), let us focus on the apples themselves. Is this really a good basket of apples? Can we find better apples? Can we improve the basket? Or, design the best basket of apples for the money? How would that change our decision making?

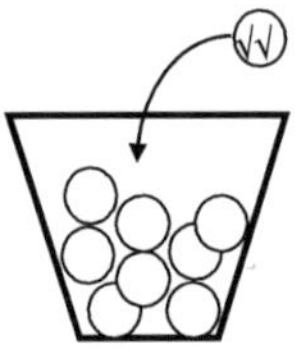

What if we find that an apple which would be both the heaviest and the sweetest at the same time could be added to the basket? This particular apple would then be preferred by all decision makers who want to maximize both criteria. The process of measurement & search would safely pick such apple. There are no tradeoffs, this "ideal" apple now dominates the entire basket and becomes an obvious choice for every rational decision maker.

As we said earlier: If there is one girl which is both the richest and the most beautiful (or one apple, both the heaviest and the sweetest) then the choice is clear, there are no tradeoffs, measurement & search is sufficient and I do not have to decide - even in the case of multiple criteria. Suddenly, there are no tradeoffs between weight and sweetness.

8.2.1 What is Optimization?

Traditionally, by optimal solution or optimization we implicitly understood maximizing (or minimizing) a single, pre-specified objective function (criterion) with respect to a given, fixed set of decision alternatives (feasibility constraints). Both the criterion and decision alternatives are given, the (optimal) solution only remains to be explicated (computed). In addition, *multiple criteria* should be explicitly considered in order to preserve the sense of practicality.

Such notion of optimization is constructed as a computational valuation of a given set of alternatives with respect to a given set of criteria and objectives [3]. This is *computation* or, more precisely, *computational explication* of the formulation-embedded solution. It essentially amounts to a search for the best apple from a given basket of apples. No optimization takes place; all is given *a priori* and remains to be explicated. Measurement and search is neither decision making nor optimization.

Computation also differs from systems design, analysis or management, i.e. from optimal design, optimization or optimality [16]. Computation is a mathematical tool aiming for digital characterization and explication of given, fixed structures and configurations. Systems design is all about shapes, topology and configuration, less so about computation or computational optimization.

The very notion of *a priori* feasibility is therefore dubious because the purpose of system design is to expand and redefine feasibility, not to accept it axiomatically. Only in pre-determined, pre-given or mandated (dictated) systems - i.e. systems non-optimal by definition - would mere computation be adequate. However, true optimality requires change, re-formulation and re-design in order to identify something better than the already "given".

Innovation is not about doing the same thing better, but about doing things differently and - most importantly - doing different things. In systems design it is not the efficiency but the effectiveness which is of interest. There are several rules that must be respected:

1. What is strictly determined or given *a priori* cannot be subject to subsequent optimization and thus, clearly, does not need to be optimized: *it is given.*
2. What is not yet firmly given must still be selected, chosen or identified and is therefore, by definition, subject to optimization.
3. Consequently, different optimality concepts can be derived from distinctions between what is given and what is yet to be determined in problem solving, systems design and optimization or decision-making.

Consequently, there are at least *eight distinct optimality concepts*, all mutually irreducible, all characterized by different applications, interpretations and mathematical/computational formalisms - as displayed in Table 8.1.

Table 8.1 Eight concepts of optimality

	Number of criteria	
Given	Single	Multiple
Criteria & alternatives	Traditional "optimality"	MCDM
Criteria only	Optimal design (De Novo programming)	Optimal design (De Novo programming)
Alternatives only	Optimal valuation (Limited equilibrium)	Optimal valuation (Limited equilibrium)
"Value complex" only	Cognitive equilibrium (Matching)	Cognitive equilibrium (Matching)

In Table 8.1, the eight key optimality concepts [17] are summarized according to a simplest classification: single versus multiple criteria (the horizontal) against the extent of the 'given': ranging from 'all-but' to 'none except' (the vertical). The traditional concept of optimality, characterized by many 'givens' and a single criterion, naturally appears to be the most remote from the optimal conditions or circumstances for problem solving as is represented by *cognitive equilibrium* (the diagonal) of multiple criteria.

The problems of the second row of Table 8.1 have already been addressed by the De Novo programming methodology, which we outline in section 8.3 as a short linear-case summary. Before that we have to elucidate appropriate solution concepts of Pareto optimality, efficiency frontier or nondominated solutions - all based on the notion of tradeoffs.

8.3 Tradeoffs-Based versus Tradeoffs-Free Thinking

The notion of tradeoffs starts from understanding that there can be *no tradeoffs* in cases of a single criterion: we cannot "trade off" more for less of the same thing [1].

Consequently, tradeoffs emerge only in cases of *multiple criteria*.

We should emphasize that tradeoffs *emerge*: they are not *a priori* natural or fixed properties of criteria, attributes or objectives. *Tradeoffs are imputed* by the set of *scarce means* (feasible set of alternatives) and its properties. It would be erroneous to treat tradeoffs as being the real properties of specific criteria, objectives or dimensions.

Whether or not there are tradeoffs depends not on alternative ends but only on scarce means. Although no single-criterion situation can have tradeoffs and therefore is not a subject of decision making, not all multiple-criteria cases will be characterized by tradeoffs: tradeoffs emerge on the basis of the means (feasible set of alternatives) configuration.

Tradeoffs are the properties of the means, not of criteria or objectives.

Popular statements about criteria, like "there are tradeoffs between cost and quality", are often erroneously accepted at their face value, as facts of reality.

What are criteria?

Criteria are simply measures or measure "tapes" for evaluating (measuring) objects of reality (things, alternatives, options, or strategies). There is a fundamental difference between *measures* and *measured objects*.

There can be no tradeoffs between measures (or measure tapes). Measures of cost and quality do not produce tradeoffs, the set of evaluated (measured) choices (alternatives, options) does. It is the configuration (size, shape and structure) of the feasible set (the measured "object") of alternatives, options and strategies that produces or brings forth any tradeoffs.

In Figure 8.1 we look at two conflicting objectives, f_1 and f_2, to be *both* maximized over the variable design space. The point of the picture is to show that the

---

[1] Aggregating multiple criteria (or attributes) into a single super-function (like utility function) forms a single aggregate criterion and therefore does not pertain to decision making, as no tradeoffs along the same function (regardless its importance or complexity) are possible.

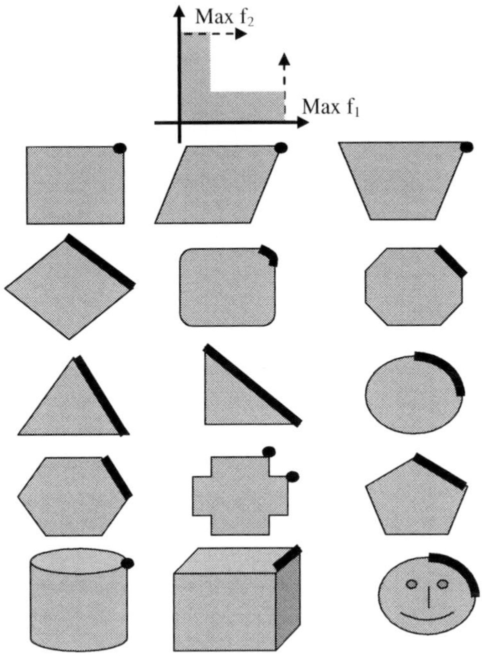

Fig. 8.1 Optimallity and Pareto-optimal solutions as functions of the feasible set

conflict, tradeoffs or any other forms of relationships between criteria and objectives are not inner attributes of those measures, criteria or objectives, but outer attributes of the *objects they allege to measure*, in this case feasible sets, sets of constraints, system designs, design topologies, etc.

It is then apparent, that the tradeoff boundary and its shapes, like the nondominated set, Pareto-optimal solutions, efficiency frontier, productivity frontier, etc., are the property of the set of options (of the objects of measurement), and not of the set of measures. This is significant because in order to truly maximize an objective function(s) one has to *optimize the feasible set* - all the rest is valuation and computation.

Because different configurations of means (different feasible sets) give rise to different solution configurations (different tradeoff or nondominated sets), the question of securing the best or optimal decision faces the following challenge:

Any decision can undoubtedly be improved through changing the configuration of means (reshaping feasible sets of alternatives) while it clearly cannot be improved through computing over an *a priori* given and fixed set of alternatives. Consequently, *modern* decision analysis should be more about reshaping the means in order to attain a tradeoffs-free design as closely as possible, rather than struggling with unnecessary tradeoffs brought forth by inadequate design of means.

8.3.1 Pareto-Efficiency

Most of solution concepts in MCDM come from the old idea of *Pareto-efficiency*. Any solution is deemed efficient if it is impossible to move to another solution which would improve at least one criterion and make no criterion worse.

How is it possible to define economic efficiency through tradeoffs? That is, at equilibrium, one side, person or criterion can gain only if the other loses? Such zero-sum thinking lies at the core of current financial crisis.

How can any two parties enter into a free-market transaction without both realizing a gain? Would anybody freely enter a transaction when one side must lose while the other gains? How could efficient allocation mean that scarce resources are squandered through inefficiency?

The key is in a careful wording of the Pareto principle: it holds true if and only if consumer tastes, resources, and technology are *given*. Of course, they never are. Even the production possibility frontier (see Figure 8.2) can be drawn only if the resources are *a priori* fixed and given. In the reality of free markets, individuals, firms and economies *continually* produce, purchase, destroy and sell resources, incessantly creating and re-creating the conditions where *both* sides of a transaction can benefit. Although resources are often limited, they are never given or fixed locally: their optimal composition (or portfolio) is sought through entrepreneurial action. The existence of tradeoffs is the sign of inefficiency, not efficiency.

At the Pareto-efficient point no free transaction is possible because the gain of one person (or one criterion) implies the loss to another.

8.4 Optimal Systems Design

A key question in economics is whether the individual, firm or economy could produce more of some goods while producing no less of other goods? The answer is always yes and so resources are always wasted in a given (non-optimal) production configuration. Such *productive efficiency* depends on how resources are selected, purchased, organized and coordinated. Are they assembled and operated in an optimal manner? Resources are only given (and thus treated as *sunk costs*) in centrally planned economies. In free markets they must be produced and purchased to create an *optimal portfolio of resources* for economic agents.

In Figure 8.2, observe that even as the productivity frontier shifts outwards (due to technological improvements), the firms are forced to face the customer-unfriendly tradeoffs on an ever larger scale. The gain of one must be the loss to another, improving one criterion must degrade another criterion. No rational decision maker would enter the transaction with assured loss.

In Figure 8.2, the tradeoffs between value and cost are assumed to exist *a priori*: only then can the frontier be drawn in this way. No differentiation of means and goals is present; companies cannot design their own resource configurations by

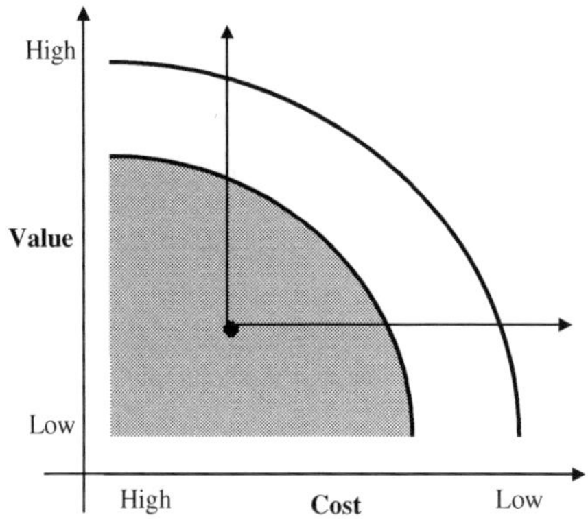

Fig. 8.2 Tradeoffs - based improvement

engaging in different activities and different ways of carrying them out. This is not how the real world works.

In Figure 8.3 we represent how economic agents (companies) redesign and reengineer their own processes and operations (reallocate their resources), so that the tradeoffs are eliminated and a tradeoffs-free environment reinstated so that it can be continually expanded and improved. The shaded area (the set of alternatives or activities) represents a distinct advantage and improvement over the shaded area of Figure 8.2. The situation in Figure 8.3 provides long-term strategic advantage, while the situation in Figure 8.2 requires continuous operational improvements and tradeoff choices, without ever fully satisfying the customer or decision maker.

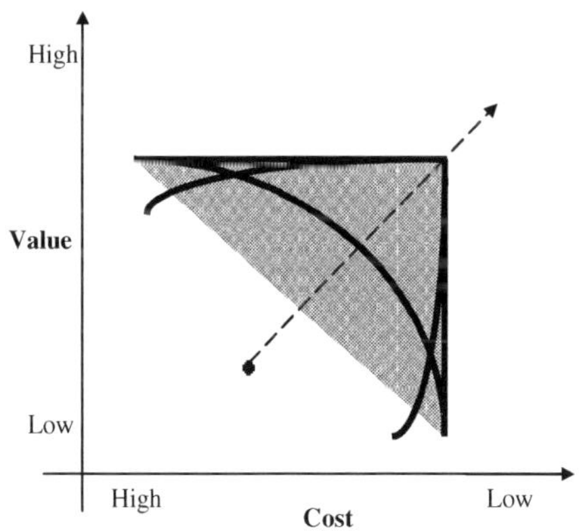

Fig. 8.3 Tradeoffs - free improvement

8.4.1 Profit Maximization

The difference between *optimizing the given* (tradeoffs-based) system and *designing the optimal* (tradeoffs-free) system is of crucial importance in economics. It goes to the very core of free-market assumptions: profit or utility maximization by firms and individuals.

A firm does not *just* maximize profits, but must *organize and coordinate* its resources properly (optimally) so that it can do so (maximize profits).

Although the literature is rich about definitions and calculations of profit functions, it remains silent about *how* profits should be maximized. Should firms just do their best or second best? Should they "maximize" profits even while some resources are wasted? Should the resources be organized in the profit-maximizing configuration or will *any* given resource configuration suffice?

Then there is the problem of *ceteris paribus*: can we calculate the marginal product of a production factor while holding constant the input of all other factors? Is the assumption of "holding all factors constant except one" rational? The fact is that the factors of production are *not independent* and we cannot change one while holding all others unchanged. All factors of production are interdependent and can only be changed in synchrony, together as a system - as a *portfolio* of resources. Factors of production form a matrix: all entries have to be adjusted as a whole, not *per partes*.

It should be clear that rational economic agents can maximize profits in *at least* two (see Table 8.1 for all eight options) fundamentally different and mutually exclusive modes:

1. Manage (operate) a *given* system - so that a profit function is maximized.
2. Design an *optimal system* - so that its management (operation) leads to maximum profits.

This distinction is independent of the actual formula for profit definition or calculation. This is a fundamental distinction between a system given *a priori*, and a system designed *a posteriori*, i.e. as a result of the process of optimization.

These two forms of profit maximization are not identical. In the first case, one requires doing one's best and squeezing the maximum profits from a *given* system. This is known as profit maximization. In the second case, one designs (re-engineers) resources of a profit-maximizing system so that doing one's best leads to maximum profits. This is also profit maximization.

How can profit maximization, being so fundamental, mean two entirely different things? The knowledge, skills and expertise required are entirely different in the two cases: to coordinate and manage the already given is fundamentally different from coordinating and managing designing of the optimal.

Because the second case is, *ceteris paribus*, always superior to the first one, we are facing two strategically different concepts of profit maximization. It *does* matter - in business, economics or management - which particular mode of profit maximization the individuals, corporations or economic cultures prefer: free markets are bound to reward those who consistently adhere to *the second* mode of profit maximization - the optimal design of profit-maximizing systems - while punishing those who just struggle to do their best with their "worst".

8.5 Foundations of De Novo Programming

The traditional resource allocation problem in economics is modeled via standard *linear programming* formulation of the single-objective product-mix problem as follows:

$$\max \quad \mathbf{cx} \quad \text{s.t.} \quad \mathbf{Ax} \leq \mathbf{b}, \quad \mathbf{x} \geq \mathbf{0} \tag{8.1}$$

That is, given the levels of m resources, $\mathbf{b} = (b_1, ..., b_m)$, determine the production levels, $\mathbf{x} = (x_1, ..., x_n)$, of n products, so as to maximize the value of the product mix $\mathbf{cx} = \sum_j c_j x_j$. Because all components of $\mathbf{b}$ are determined *a priori*, problem (8.1) deals with the *"optimization" of a given system.*

When the purpose is to *design an optimal system*, the following formulation is of interest:

$$\max \quad \mathbf{cx} \quad \text{s.t.} \quad \mathbf{Ax} - \boldsymbol{\beta} \leq \mathbf{0}, \quad \mathbf{p}\boldsymbol{\beta} \leq B, \quad \mathbf{x}, \boldsymbol{\beta} \geq \mathbf{0} \tag{8.2}$$

That is, given the unit prices of m resources, $\mathbf{p} = (p_1, ..., p_m)$, and the total available budget B, allocate the budget so that the resulting *portfolio of resources* $\boldsymbol{\beta} = (\beta_1, ..., \beta_m)$ maximizes the value of the product mix. We assume that $\mathbf{A} > \mathbf{0}$, $\mathbf{p} > \mathbf{0}$ in (8.2).

It can be shown [2] that the optimal design problem (8.2) is equivalent to a continuous "knapsack" problem (8.3) below:

$$\max \quad \mathbf{cx} \quad \text{s.t.} \quad \mathbf{Cx} \leq B, \quad \mathbf{x} \geq \mathbf{0} \tag{8.3}$$

where $\mathbf{C} = [C_1, ..., C_n] = \mathbf{pA}$. Since the "knapsack" solution is

$$\mathbf{x}^* = [0, ..., B/C_k, ..., 0]^\top \tag{8.4}$$

where

$$c_k/C_k = \max_j (c_j/C_j), \tag{8.5}$$

the optimal solution to (8.3) is given by (8.4) and

$$\boldsymbol{\beta}^* = \mathbf{Ax}^*. \tag{8.6}$$

Numerical Example. A simple numerical demonstration of the optimal-design procedure is seen in the following LP problem:

$$\begin{aligned} \max\ & 400x_1 + 300x_2 \\ \text{s.t.}\ & 4x_1 \leq b_1 \\ & 2x_1 + 6x_2 \leq b_2 \\ & 12x_1 + 4x_2 \leq b_3 \\ & 3x_2 \leq b_4 \\ & 4x_1 + 4x_2 \leq b_5 \end{aligned}$$

where $p_1 = 30$, $p_2 = 40$, $p_3 = 9.5$, $p_4 = 20$ and $p_5 = 10$ are market prices ($ per unit) of the resources b_1 through b_5 respectively. Let also $B = 2600$ and $x_1 \leq 6$ (e.g., maximum demand limitation). Two basic approaches can be used to determine b_1 through b_5 optimally:

(a) *Accounting approach*. In an optimally designed system, all of the constraint inequalities must become equalities. Then,

$$30b_1 + 40b_2 + 9.5b_3 + 20b_4 + 10b_5 = 2600.$$

Substituting the left-hand sides of constraints, we get:

$$30(4x_1) + 40(2x_1 + 6x_2) + 9.5(12x_1 + 4x_2) + 20(3x_2) + 10(4x_1 + 4x_2) = 2600,$$

which is reduced to $354x_1 + 378x_2 = 2600$ and thus $x_1 = 7.35 - 1.07x_2$. Maximize $400x_1 + 300x_2 = 400(7.35 - 1.07x_2) = 2940 - 426x_2$ by making $x_2 = 0$. So, $x_1 = 7.35$, $x_2 = 0$ and $b_1 = 29.4$, $b_2 = 14.7$, $b_3 = 88$, $b_4 = 0$, $b_5 = 29.4$. If x_1 cannot exceed 6, then $x_1 = 6$ and $x_2 = (7.35 - 6)/1.07 = 1.26$ and thus $b_1 = 24$, $b_2 = 19.56$, $b_3 = 77.04$, $b_4 = 3.78$, $b_5 = 29.04$.

(b) *De novo approach*. Solve the simple LP-knapsack problem:

$$\begin{aligned} \max\ & 400x_1 + 300x_2 \\ \text{s.t.}\ & 354x_1 + 378x_2 \leq 2600 \end{aligned}$$

by choosing the largest of the two ratios $400/354$ and $300/378$ and making the corresponding variable (here x_1) as large as possible: $x_1 = 2600/354 = 7.35$. Because $x_1 = 6$, then $x_2 = 2600 - 354(6) = 476/378 = 1.26$; the same result as above is obtained.

In the next section we summarize *De novo programming* [17] which designs the optimal portfolio of resources in dependency on market prices and an investment budget with respect to multiple objectives, and redesigns the shape of the feasible set so that the tradeoffs are fully eliminated. In economic multiobjective problems, the existence of tradeoffs is always a sure sign of suboptimality, poor system performance and consumer dissatisfaction.

8.5.1 Multiple Criteria De Novo Programming

Let us formulate a linear programming problem:

$$\max \quad \mathbf{z} = \mathbf{Cx} \quad \text{s.t.} \quad \mathbf{Ax} - \mathbf{b} \leq \mathbf{0}, \quad \mathbf{pb} \leq B, \quad \mathbf{x} \geq \mathbf{0}, \tag{8.7}$$

where $\mathbf{C} \in \mathbb{R}^{q\times n}$ and $\mathbf{A} \in \mathbb{R}^{m\times n}$ are matrices of dimensions $q \times n$ and $m \times n$, respectively, $\mathbf{b} \in \mathbb{R}^m$ is the m-dimensional unknown vector of resources, $\mathbf{x} \in \mathbb{R}^n$ is the n-dimensional vector of decision variables, $\mathbf{p} \in \mathbb{R}^m$ is the vector of the unit prices of m resources, and B is the given total available budget.

Solving problem (8.7) means finding the optimal allocation of B so that the corresponding resource portfolio $\mathbf{b}$ maximizes simultaneously the values $\mathbf{z} = \mathbf{Cx}$ of the product mix $\mathbf{x}$. Obviously, we can transform problem (8.7) into:

$$\max \quad \mathbf{z} = \mathbf{Cx} \quad \text{s.t.} \quad \mathbf{vx} \leq B, \quad \mathbf{x} \geq \mathbf{0}, \tag{8.8}$$

where $\mathbf{z} = (z_1, ..., z_q) \in \mathbb{R}^q$ and $\mathbf{v} = (v_1, ..., v_n) = \mathbf{pA} \in \mathbb{R}^n$.

Let $z_{k*} = \max z_k$, $k = 1, ..., q$, be the optimal value for the kth objective of problem (8.8) subject to $\mathbf{vx} \leq B$, $\mathbf{x} \geq \mathbf{0}$. Let $\mathbf{z}^* = (z_{1*}, ..., z_{q*})$ be the q-objective value for the ideal system with respect to B. Then, a *metaoptimum problem* can be constructed as follows:

$$\min \quad \mathbf{vx} \quad \text{s.t.} \quad \mathbf{Cx} \geq \mathbf{z}^*, \quad \mathbf{x} \geq \mathbf{0}. \tag{8.9}$$

Solving Problem (8.9) yields $\mathbf{x}^*$, $B^*(=\mathbf{vx}^*)$ and $\mathbf{b}^*(=\mathbf{Ax}^*)$. The value B^* identifies the minimum budget to achieve $\mathbf{z}^*$ through $\mathbf{x}^*$ and $\mathbf{b}^*$.

Since $B^* \geq B$, the *optimum-path ratio* for achieving the ideal performance $\mathbf{z}^*$ for a given budget level B is defined as:

$$r^* = B/B^* \tag{8.10}$$

We establish the optimal system design as $(\mathbf{x}, \mathbf{b}, \mathbf{z})$, where $\mathbf{x} = r^*\mathbf{x}^*$, $\mathbf{b} = r^*\mathbf{b}^*$ and $\mathbf{z} = r^*\mathbf{z}^*$. The optimum-path ratio r^* provides an effective and fast tool for the efficient optimal redesign of large-scale linear systems.

There are two additional types of budgets (other than B and B^*). One is B_j^k, the budget level for producing the optimal x_j^k with respect to the kth objective, referring back to the single-objective De Novo programming problem.

The other, B^{**}, refers to the case $q \leq n$ (the number of objectives is less than the number of variables). If $\mathbf{x}^{**}$ is the degenerate optimal solution, then $B^{**} = \mathbf{vx}^{**}$. It can be shown that $B^{**} \geq B^* \geq B \geq B_j^k$, for $k = 1, ..., q$.

Shi [8] introduced six types of optimum-path ratios:

$$r_1 = B^*/B^{**}; \qquad r_2 = B/B^{**}; \qquad r_3 = \sum \lambda_k B_j^k/B^{**};$$
$$r_4 = r^* = B/B^*; \; r_5 = \sum \lambda_k B_j^k/B^*; \; r_6 = \sum \lambda_k B_j^k/B,$$

leading to six different policy considerations and optimal system designs.

Numerical example. The following numerical example is adapted from Zeleny [15]:

$$
\begin{array}{llrcrcrl}
\max z_1 = & 50x_1 & + & 100x_2 & + & 17.5x_3 & \\
z_2 = & 92x_1 & + & 75x_2 & + & 50x_3 & \\
z_3 = & 25x_1 & + & 100x_2 & + & 75x_3 & \\
\text{s.t.} & 12x_1 & + & 17x_2 & & & \leq b_1 \\
& 3x_1 & + & 9x_2 & + & 8x_3 & \leq b_2 \\
& 10x_1 & + & 13x_2 & + & 15x_3 & \leq b_3 \\
& 6x_1 & + & & & 16x_3 & \leq b_4 \\
& & & 12x_2 & + & 7x_3 & \leq b_5 \\
& 9.5x_1 & + & 9.5x_2 & + & 4x_3 & \leq b_6
\end{array}
\tag{8.11}
$$

We assume, for simplicity, that the objective functions z_1, z_2, and z_3 are equally important. We are to identify the optimal resource levels of b_1 through b_6 when the vector $\mathbf{p}$ of current unit prices of resources is $p_1 = 0.75$, $p_2 = 0.60$, $p_3 = 0.35$, $p_4 = 0.50$, $p_5 = 1.15$ and $p_6 = 0.65$. The initial budget $B = \$4658.75$.

We calculate the maxima $\mathbf{z}^* = (10916.813; 18257.933; 12174.433)$ with respect to the given B (\$4658.75). The feasibility of $\mathbf{z}^*$ can only be assured by the metaoptimum solution $\mathbf{x}^* = (131.341, 29.683, 78.976)$ at the cost of $B^* = \$6616.5631$.

Because the optimal-path ratio $r^* = 4658.75/6616.5631 = 70.41$, the resulting $\mathbf{x} = (92.48, 20.90, 55.61)$ and $\mathbf{z} = (7686.87; 12855.89; 8572.40)$. It follows that the optimal portfolio $\mathbf{b}$, with respect to $B = \$4658.75$, can be calculated by substituting $\mathbf{x}$ into the constraints (8.11). We obtain the optimal portfolio of resources as:

$$
\begin{aligned}
b_1 &= 1465.06 \\
b_2 &= 910.42 \\
b_3 &= 2030.65 \\
b_4 &= 1444.64 \\
b_5 &= 640.07 \\
b_6 &= 1299.55
\end{aligned}
\tag{8.12}
$$

If we spend precisely $B = \$4658.8825$ (approx. \$4658.75) we can purchase the optimum portfolio of resources (8.12) at current market prices, allowing us to produce $\mathbf{x}$ and realize $\mathbf{z}$ in criteria performance. No better solution can be realized for given amount of money.

8.6 Examples of Applications

The *tradeoffs-free* decision-making, optimization and de novo design have a significant number of methodological applications in MCDM-related problems. We shall review several more interesting cases.

All such applications have the tradeoffs-free alternative (ideal point, zenith, prominent alternative, reference point, etc.) in common. The purpose is to make such generally desirable alternative (where *all* rational utility functions are maximized) feasible through system redesign and optimal reallocation of resources.

8.6.1 Compromise Programming

The original CP-method [12, 13] is designed to minimize the distances of the criteria evaluation vectors of feasible alternatives from the ideal point. If one cannot have the best solution, one should aim for at least the second best.

CP first calculates a combined n-dimensional distance for each feasible alternative as for example:

$$F_{CP} = \left[\sum_{i=1}^{n} a_i^p\right]^{1/p} \qquad 1 \leq p \leq \infty$$

where a_i represents the normalized differences between component i of the *ideal point* and the evaluation of an alternative with respect to criterion i $(1,2,...,n)$. Normalized difference means that the resulting value is in the range [0, 1]. The alternative with the smallest combined distance is the best proxy for the ideal one. We can modify the CP method by maximizing the distance from the worst point (nadir).

The structure of the CP remains the same, but instead of minimizing the distance from the *zenith* we maximize the distance from the *nadir*.[2] The pull towards the ideal is of course preferable to the push from the nadir because the zenith approach pulls towards tradeoffs-free alternative (worthy of making feasible) while the nadir approach pushes towards tradeoffs-based alternatives (feasible already).

The value of p in the distance formula expresses how strongly each difference value is emphasized. The case of $p = 1$ corresponds to simple average; the case of $p = 2$ implies a simple squared distance weighting, while if $p = \infty$, only the largest deviation is considered, i.e. moving from minimizing the sum of individual regrets to minimizing the maximum regret.

In all such cases, inferior, tradeoffs-based solutions are computed and recommended in the end even though designing towards the feasibility of tradeoffs-free options is always preferred by the customer.

8.6.2 Risk Management

One of the typical paradigms where tradeoffs are also *a priori* assumed is the risk management model of *portfolio selection* due to Markowitz [5, 6]. Artificial tradeoffs between investment returns and investment risk have plagued rational financial analysis ever since its inception. Efforts to measure risk by a single-dimensional,

[2] We may in fact perform both procedures because the resulting compromise set is then corrrespondingly reduced.

technological form (like e.g. standard deviation or variance) have demonstrated the lack of understanding of risk to the extent of bringing forth a worldwide crisis in financial investment praxis. How can one even hope to measure risk with a single number? Markowitz's standard deviation cannot measure risk since it *penalizes* gains as well as losses.

Let us quote E. Lerner [4] on this issue:

> "Some financial writers, unfortunately, have come to look upon the standard deviation of the distribution of returns as a measure not only of the variability (which it is) but of the risk inherent in a project (which it is not). In everyday usage, risk means the probability of a loss or the probability that a return will be lower than some target level."

It is clear that financial "decision making" cannot avoid periodic financial collapses without clearly defining and measuring risk. Risk certainly is related to the *probabilities of gain and loss* with respect to specified target levels. The theory of *stochastic dominance* leads to proper estimates of such probabilities.

In Figure 8.4, we show how the tradeoffs between *risk*, $R = \sigma^2$, and *expected return*, $E(X)$, form the (E,R)-nondominated set or efficiency frontier. Tradeoffs between variance and expected return are implanted on a *single* stochastic dimension of *returns*, characterized by relevant probability distributions and their moments (like expected value, variance, and so forth). No matter how many moments one uses to describe stochastic returns, it is still *a single dimension* of returns, albeit uncertain or stochastic.

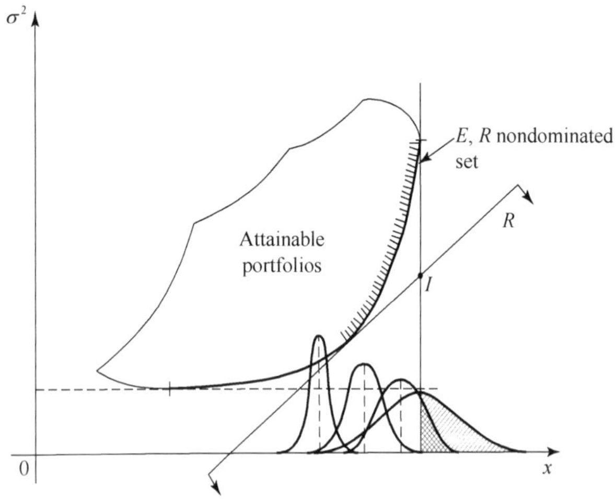

Fig. 8.4 Portfolio "riskiness" is related to the *relative* positioning of the distributions, not their variances

Even if we take a hypothetical measure of risk R which decreases with smaller variance σ^2 and *also* with larger E, we get so called E, R *efficiency frontier* of nondominated or Pareto-optimal prospects.

In Figure 8.4 observe that the ideal portfolio ***I*** remains infeasible (see [1]) even though a portfolio redesign should either secure its feasibility or minimize the distance, as is the case with compromise programming.

The measure of risk should be multidimensional, based on probabilities of specific achievements rather than on the context-free distributions and their moments. Colson and Zeleny [1] explore two three-dimensional measures of risk (prospect ranking vectors): one for the assumption of known distribution,

$$PVR_{im} = \left(1 - R_{im}^{(1)}, \overline{x}_i, 1 - R_i^{(2)}\right)$$

and the other for unknown probability distributions.

$$PVR_{im} \overset{P}{\Longleftrightarrow} \left(\frac{1}{k_{im}^{\prime 2} + 1}, \overline{x}_i, \frac{1}{k_i^2 + 1}\right)$$

Such risk measures would bring portfolio analysis directly within the MCDM domain, especially its optimal-design *de novo* environment.

8.6.3 Conflict Dissolution

Conflict occurs when two or more distinct alternatives, selected as the means for achieving stated objectives, are mutually exclusive.

Next we show that conflict can be characterized as being induced by the mutual exclusivity of distinct alternatives selected by decision agents. We shall argue for a concise and general definition of conflict: *Conflict is the absence of a prominent alternative* [18].

Such prominent alternative is a *tradeoffs-free solution* (or as close as to tradeoffs-free as possible).

Let us graphically represent a generic conflict of two decision agents with a single objective *a*, as in Figure 8.4:

We can similarly formulate the problem for one decision maker and two criteria (*a* and *b*), and so on. Observe that *M* and *F* (like male and female) maximize criterion *a* (a place to go for vacations) at points 2 and 1 respectively. Even though they have a common objective and there are no cognitive differences, no mistakes or insufficient communication, *there is a conflict*.

The prominent alternative A is either non-existent or it has not been considered by either M or F. It is the absence of A which causes the conflict to emerge. Should A become feasible, the conflict would be *fully dissolved*. The heavily traced boundary of X represents a region of compromise, a *bargaining set*. Observe that no *compromise solution*, including both extremes 1 and 2, removes or resolves the underlying conflict. Conflict resolution via compromise is only a temporary disguise of the absence of A. At any compromise located on the heavy boundary of X there is at least

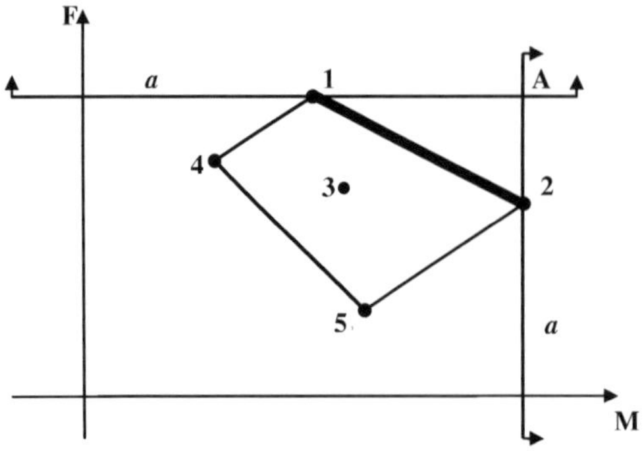

Fig. 8.5 Conflict dissolution via the insertion of A

one decision agent (or at least one objective) which remains unsatisfied in relation to what is actually achievable.

The only way to dissolve conflict is to consider, find or create A. The only way to reduce the intensity of conflict is to generate alternatives that are "closer" to A.

If we cannot achieve A, we should at least attempt to move as close to it as possible, as in compromise programming. The unattainable ideal should not serve as an excuse for trying to achieve what is simply given. Ignoring the ideal and settling down to what is "good enough", like in Simon's *satisficing* [10, 11], does not remove the conflict and it is incompatible with good management. It is not good enough.

8.6.4 Added Value

One of the most common examples of agent tradeoffs is the win-lose tension between the producer and the consumer. Because the *price paid* remains the main economic category, its increase brings higher added value (and profits) to the producer, but at the cost of lower added value to the consumer. Of course, the lowering of the price paid has the opposite effect. There are constant tensions between producer and consumer in trading off a total added value between themselves, instead of increasing its apportioned levels to both.

In any business transaction, value has to be *added to both* participants (or sides): the producer *and* the consumer. Adding value is what makes the transaction satisfactory and sustainable [18].

There are two kinds of value to be created: *value for the producer* and *value for the consumer*. Both parties must benefit: the producer - in order to make it; the consumer - in order to buy it. In the global age it is precisely this producer-consumer *value competition* that is emerging as the hardest and the busiest battleground.

In Figure 8.6 we outline the process of creating new value. First, the consumer pays for the service or product: the *price paid*. The producer subtracts the *cost incurred*, including all direct and indirect materials and services purchased. The dif-

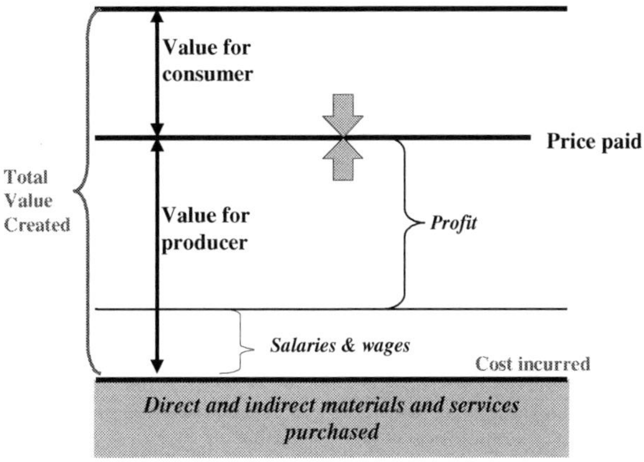

Fig. 8.6 Adding value to the consumer and the producer

ference is the *added value* for the producer. This added value can also be interpreted as the *value of knowledge* (see [18, 19]) engaged in producing the service or product. In order to pay salaries and wages, the production process and its coordination must generate such added value. Added value is the only source of corporate wages, salaries and profits.

If the added value does not *cover* the salaries and wages, then they must be correspondingly lowered. If no added value has been created, then the value of knowledge is zero and no payment can be attributed to it. The producer must add enough value in order to *cover* at least its workers and managers, their salaries and wages. If even more value has been created, then *profits* can be realized, up to the price received.

The consumer, of course, must be willing and ready to pay more for the service/product than he actually paid. The *maximum price* the consumer would be willing to pay must exceed the price the producer has asked for. The difference is the added *value for the consumer.*

If there is no value added for the consumer - the maximum price is lower than the price to be paid - then the consumer would not buy the service or product (like GM automobile). In a competitive market, the consumer pays only for the added value he receives.

The best free-market strategy is to create value for both sides, not at the expense (or tradeoff) of the other side. Such strategy mean minimizing the *cost incurred* and maximizing the *maximum price* - while keeping *the price paid* relatively stable - all at the same time.

8.7 Conclusions

As a result of the financial and economic crisis there will be significant paradigmatic changes in the way businesses are being run, value created and decisions made. The tradeoffs-based solution concepts of win-lose type will be replaced by the tradeoffs-free concepts of the win-win type. We cannot afford to gain only what somebody else has lost.

Instead of choosing the "best" among bad tradeoffs-based alternatives, we have to learn how to design the best systems in order to deliver the tradeoffs-free alternatives that consumers want and economies need.

Correspondingly, the field of MCDM and its multiobjective optimization has to shift from merely optimizing the given to designing the optimal.

References

1. G. Colson and M. Zeleny. *Uncertain Prospects Ranking and Portfolio Analysis under the Conditions of Partial Information*. Verlag Anton Hain, Meisenheim, 1979.
2. M. Hessel and M. Zeleny. Optimal system design: Towards new interpretation of shadow prices in linear programming. *Computers and Operations Research*, 14(4):265–271, 1987.
3. T.C. Koopmans. *Activity Analysis of Production and Allocation*. Wiley, New York, 1951.
4. E. Lerner. *Managerial Finance*. Harcourt Brace Jovanovich, New York, 1971.
5. H. Markowitz. Portfolio selection. *Journal of Finance*, 7(1):77–91, 1952.
6. M. Markowitz. *Portfolio Selection: Efficient Diversification of Investments*. Wiley, New York, 1959.
7. M.E. Porter. What is strategy? *Harvard Business Review*, pages 61–78, 1996. November-December.
8. Y. Shi. Studies on optimum-path ratios in De Novo programming problems. *Computers and Mathematics with Applications*, 29:43–50, 1995.
9. Y. Shi, D.L. Olson, and A. Stam. *Advances in Multiple Criteria Decision Making and Human Systems Management: Knowledge and Wisdom*. IOS Press, Amsterdam, 2007.
10. H.A. Simon. A behavioral model of rational choice. *Quarterly Journal of Economics*, 69:99–118, 1955.
11. H.A. Simon. *Administrative Behavior*. The Free Press, New York, 3rd edition, 1976.
12. M. Zeleny. Compromise programming. In J.L. Cochrane and M. Zeleny, editors, *Multiple Criteria Decision Making*, pages 263–301. University of South Carolina Press, Columbia, S.C., 1973.
13. M. Zeleny. The theory of the displaced ideal. In M. Zeleny, editor, *Multiple Criteria Decision Making*, pages 153–206. Springer-Verlag, Berlin, 1976.
14. M. Zeleny. *Multiple Criteria Decision Making*. McGraw-Hill, New York, 1982.
15. M. Zeleny. Optimal system design with multiple criteria: De Novo programming approach. *Engineering Costs and Production Economics*, 10:89–94, 1986.
16. M. Zeleny. *Information Technology in Business*. Thomson, London, 2001.
17. M. Zeleny. The evolution of optimality: De Novo programming. In C.A. Coello Coello, A. Hernández Aguirre, and E. Zitzler, editors, *Evolutionary Multi-Criterion Optimization*, pages 1–13. Springer-Verlag, Berlin-Heidelberg, 2005.
18. M. Zeleny. *Human Systems Management: Integrating Knowledge, Management and Systems*. World Scientific, Hackensack, NJ, 2008.
19. M. Zeleny. The KM-MCDM interface in decision design: Tradeoffs-free conflict dissolution. *International Journal of Applied Decision Sciences*, 1(1):3–23, 2008.

Chapter 9
Interactive Multiple Objective Programming Methods

Pekka Korhonen and Jyrki Wallenius

Abstract We provide an introduction to interactive methods in multiple objective programming[1]. Our discussion focuses on the principles to implement such methods. Our purpose is not to review existing procedures, but to provide some examples to illustrate the underlying main ideas. Furthermore, we discuss two available software systems developed to implement interactive methods.

9.1 Introduction

In most decision and planning situations a *Decision Maker* (DM) will have to consider multiple criteria implicitly or explicitly. The term *Multiple Criteria Decision Making* (MCDM)[2] refers to decision and planning problems involving multiple (generally conflicting) criteria. For an MCDM-problem it is typical that no unique solution for the problem exists. The solution is determined by the preferences of the DM. *Solving a Multiple Criteria Problem* means that a DM will choose one *"reasonable"* alternative which pleases him/her most. The word "reasonable" is defined more precisely by using the terms *efficiency* or *nondominance*.

Solving an MCDM-problem requires a dialogue with the DM. The main idea is simple: the system generates reasonable solutions (alternatives), and the DM will make choices. The DM's choices are used by the algorithm to generate more solutions until the DM will reach the solution that pleases him/her most. Helping DMs deal with multiple criteria decision and planning problems has been the subject of intense studies since the beginning of the 1970's (see, e.g., [8, 14, 20, 50, 65]),

Helsinki School of Economics, Department of Business Technology, Runeberginkatu 22–24, 00100 Helsinki, Finland

[1] An earlier version of this paper was published by P. Korhonen under the name "Interactive Methods" in *Multiple Criteria Decision Analysis: State of the Art Surveys*, by José Figueira, Salvatore Greco, and Matthias Ehrgott, Springer, 2005.

[2] Zionts [64].

C. Zopounidis and P.M. Pardalos (eds.), *Handbook of Multicriteria Analysis*, Applied Optimization 103, DOI 10.1007/978-3-540-92828-7_9,

but many theoretical concepts were defined much earlier (see, e.g., [31, 48]). In the 1970's, the research focused on the theory of multiple objective mathematical programming and the development of procedures and algorithms for solving such problems. Many ideas originated from mathematical programming. The algorithms were programmed for mainframe computers and were used mainly for illustrative purposes.

During the 1980's, a clear shift towards multiple criteria decision support occurred. Accordingly, more research focused on the user interface, on the behavioural aspects of decision-making, and on supporting the entire decision-making process from problem structuring to solution implementation. Several *Multiple Criteria Decision Making Support Systems* (MCDSS) were developed where a graphical presentation was an essential part of the system (see, e.g., [2, 6, 7, 29, 33, 35, 43, 44]). Some articles were also published where the behavioural realism of the MCDSSs was critically evaluated (see, e.g., [36, 39, 42, 61]).

In practice, MCDM-problems are not often so well-structured that they can be considered purely as choice problems. Before a decision problem is ready to be *"solved"*, the following questions require a lot of preliminary work: How to structure the problem? How to find the relevant criteria? How to handle uncertainty? These questions are by no means outside the interest area of MCDM-researchers.

In this article we take a narrower perspective and focus on an essential support problem in Multiple Criteria Decision Making: How to assist a DM to find the "best" solution from among a set of available "reasonable" alternatives, when the alternatives are evaluated by using several criteria? The criteria are assumed to be given and the alternatives are assumed to be defined explicitly or implicitly by means of a mathematical model.

This article consists of seven sections. In Section 9.2, we give a brief introduction to some basic definitions, and in Section 9.3, we consider the main principles to implement interactive methods. How to generate nondominated alternatives is considered in Section 9.4. The properties of Multiple Criteria Decision Support Systems (MCDSS) are discussed in Section 9.5. Section 9.6 considers stopping conditions, and in Section 9.7, we present two examples of interactive systems: VIG and VIMDA. Concluding remarks are given in Section 9.8.

9.2 Basic Definitions and Some Theory

The Multiple Objective Programming (MOP) problem in the criterion space can be characterized as a vector maximization problem as follows:

$$\begin{aligned} \max \quad & \mathbf{q} = (q_1, q_2, \ldots, q_k) \\ \text{s.t.} \quad & \mathbf{q} \in Q, \end{aligned} \tag{9.1}$$

where set $Q \subset \mathbb{R}^k$ is the *feasible region* in a k-dimensional criterion space $\mathbb{R}^k$. Set Q is of special interest. Most considerations in multiple objective programming are made in the criterion space.

Set Q may be convex/nonconvex, bounded/unbounded, precisely known or unknown, consist of a finite or infinite number of alternatives, etc. When Q consists of a finite number of elements which are explicitly known in the beginning of the solution process, we have the class of problems which may be called *(Multiple Criteria) Evaluation Problems*. Sometimes such problems are referred to as *Discrete Multiple Criteria Problems* or *Selection Problems* (for a survey, see e.g. [47]).

When the number of alternatives in Q is nondenumerable, the alternatives are usually defined using a mathematical model, and the problem is called continuous. In this case the alternatives are only implicitly known. This kind of problem is referred to as a *Multiple Criteria Design Problem* or a *Continuous Multiple Criteria Problem* (for a survey, see e.g. [46, 57]).

In this case, the set Q is specified by means of decision variables as is usually done in single optimization problems:

$$\begin{aligned} \max \quad & \mathbf{q} = \mathbf{f}(\mathbf{x}) = (f_1(\mathbf{x}), \ldots, f_k(\mathbf{x})) \\ \text{s.t.} \quad & \mathbf{x} \in X \end{aligned} \tag{9.2}$$

where $X \subset \mathbb{R}^n$ is a *feasible set* and $\mathbf{f}: \mathbb{R}^n \to \mathbb{R}^k$. The space $\mathbb{R}^n$ is called *decision variable space*. The functions f_i, $i = 1, 2, \ldots, k$ are *objective functions*. The feasible region Q can now be written as $Q = \{\mathbf{q} \mid \mathbf{q} = \mathbf{f}(\mathbf{x}), \mathbf{x} \in X\}$.

Conceptually the multiple objective mathematical programming problem may be regarded as a *value (or utility) function* maximization program:

$$\begin{aligned} \max \quad & v(\mathbf{q}) \\ \text{s.t.} \quad & \mathbf{q} \in Q, \end{aligned} \tag{9.3}$$

where v is strictly increasing in each argument and real-valued, and defined at least in the feasible region Q. It maps the feasible region into a one-dimensional *value* space. Function v specifies the DM's preference structure over the feasible region. However, the key assumption in multiple objective programming is that v is **unknown**. Even if the value function v is not explicitly known, the solution methods are heavily dependent on the additional assumptions we make about its form and stability. Those assumptions vary from very strict assumptions to "no assumptions". We may also assume that v exists, but is not stable during the solution process.

Regardless of the assumptions concerning v, possible solutions to our problem are alternatives that can be solutions to (9.1) for some value function $v: Q \to \mathbb{R}$. Those solutions are called *efficient* or *nondominated* depending on the space where the alternatives are considered. The term nondominated is used in criterion space and efficient in decision variable space. (Some researchers use the terms interchangeably.) Any choice from among the set of nondominated (efficient) solutions is possible, given that we have no additional information about the DM's preference structure.

Nondominated and weakly nondominated solutions are defined as follows:

Definition 9.1. In (9.1), $\mathbf{q}^* \in Q$ is *nondominated* iff there does not exist another $\mathbf{q} \in Q$ such that $\mathbf{q} \geq \mathbf{q}^*$ and $\mathbf{q} \neq \mathbf{q}^*$ (i.e. $q_i \geq q_i^*$ for all $i = 1,2,...,k$ and $q_i \neq q_i^*$ for some i).

Definition 9.2. In (9.1), $\mathbf{q}^* \in Q$ is *weakly nondominated* iff there does not exist another $\mathbf{q} \in Q$ such that $\mathbf{q} > \mathbf{q}^*$ (i.e. $q_i \geq q_i^*, i = 1,2,...,k$).

Correspondingly, efficient and weakly efficient solutions are defined as follows:

Definition 9.3. In (9.2), $\mathbf{x}^* \in X$ is *efficient* iff there does not exist another $\mathbf{x} \in X$ such that $\mathbf{q} = \mathbf{f}(\mathbf{x}) \geq \mathbf{q}^* = \mathbf{f}(\mathbf{x}^*)$ and $\mathbf{q} = \mathbf{f}(\mathbf{x}) \neq \mathbf{q}^* = \mathbf{f}(\mathbf{x}^*)$.

Definition 9.4. In (9.2), $\mathbf{x}^* \in X$ is *weakly efficient* iff there does not exist another $\mathbf{x} \in X$ such that $\mathbf{q} = \mathbf{f}(\mathbf{x}) > \mathbf{q}^* = \mathbf{f}(\mathbf{x}^*)$.

The set of all nondominated (efficient) solutions is called the nondominated (efficient) frontier. The final ("best") solution $\mathbf{q} \in Q$ of the problem (9.1) is called the *Most Preferred Solution* (MPS). It is the solution preferred by the DM to all other solutions. At the conceptual level, we may think that it is the solution that maximizes an (unknown) value function in problem (9.3). How to find the maximum of an unknown function is the key problem in multiple objective programming.

In the following, we use the term MCDM to refer to a multiple criteria problem generally and the term MOP to emphasize the mathematical formulation of the problem. In case the mathematical model is linear, we use the term multiple objective linear programming (MOLP).

9.3 Principles for Implementing Interactive Methods

In MCDM we assume that the DM seeks to find a solution which is preferred to all other solutions. We do not need to consider dominated solutions, because for each dominated solution there exists at least one (nondominated) solution which is better at least on one criterion and worse on no criterion. Although we are able to reduce the number of alternatives in this way, it is not realistic to assume that the DM is able to compare all nondominated solutions simultaneously, and pick the best. That is why the above characterization of the MPS is not very operational.

Another way is to approach the problem through the value function. We can distinguish the following principles, which most MCDM approaches use implicitly or explicitly in solving MCDM problems. The following considerations are based on Korhonen et al. [37] where you can find additional details.

1. Assume the existence of a value function v, and assess it explicitly.
2. Assume the existence of a stable value function v, but do not attempt to assess it explicitly. Make assumptions of its general functional form.

3. Assume the existence of a value function v, but do not assume it stable. Let it change with time, and assume that the DM's final choice is based on its specific form.
4. Do not assume the existence of a value function v.

Those principles lead to the following general approaches to solving MCDM problems.

Approach 1: Prior Articulation of Preferences

The value function v is explicitly constructed by means of preference information received from a DM. The multiple attribute utility theory (MAUT) provides a classical example [28] of this approach. Using Keeney-Raiffa type of interaction, the following steps can be identified:

a) assumptions are made about the underlying value function, in particular, its functional form;
b) the parameters of the value function are assessed using elaborate interview techniques;
c) the internal consistency of the DM's responses is checked;
d) the value function is used to determine a value score for each alternative. Those scores are used to determine the MPS or to rank alternatives.

In other words, once an explicit value function has been assessed, the function determines a "value" or a "score" for each alternative making it possible to rank (a finite number of) the decision alternatives or to find the alternative having an optimal value. Currently, the popular AHP (the Analytic Hierarchy Process) developed by Saaty [51] is also based on these principles. In the AHP, the value function is assumed to be linear. MACBETH [4] is another example of an approach, in which a cardinal value function is constructed in an interactive manner.

Interaction between the system and the DM is needed in step b. If inconsistency is revealed in step c, in the MAUT, the DM is asked to revise his/her preference information; in the AHP (s)he is informed about inconsistency, but not required to correct for it. One of the basic assumptions in the AHP is that a DM cannot always be fully consistent.

Actually, this approach is widely used in practical decision making. A weighted average (sum) may be the most common way to aggregate several criteria. The value function is thus implicitly assumed to be linear. However, the users are seldom aware of many implicit assumptions they have made concerning the meaning of weights, the dependence of criteria, the functional form of the value function, etc.

A classical example of the use of weights to aggregate preferences is goal programming. Since Charnes and Cooper [10, 11] developed goal programming, it has been a very popular method to deal with multiple criteria. In Archimedean goal programming, the weighted-sums of the deviational variables are used to find the solutions that best satisfy the goals. The deviational variables measure overachievement and underachievement from the target or threshold levels. See also Ignizio [22, 23].

For other methods based on the use of weights, see e.g. Antunes and Climaco [1] and Borges and Antunes [9].

Approach 2: Interactive Articulation of Preferences

i) Based on an Implicit Value Function

The value function is neither assumed to be known nor elicited explicitly. DM's responses to specific questions are used to guide the solution process towards an "optimal" or "most preferred" solution. The DM is assumed to behave according to some specific underlying value function. Classical examples are Geoffrion et al. [20], Zionts and Wallenius [65], and Steuer [56].

Geoffrion et al. presented their approach in a general framework, where objective functions were assumed to be differentiable and concave, and the feasible set X is compact and convex. The value function was assumed concave increasing. The idea was adopted from the well-known Frank-Wolfe algorithm. In the GDF-method [20], the direction of improvement was chosen by estimating the gradient of a concave value function on the basis of marginal rates of substitution, which a DM evaluated. Zionts and Wallenius [65] developed a method for MOLP-problems. In their method, the value function was also assumed to be linear. (In 1983, the authors extended the method to deal with concave value functions [66].) The weights of the value function were determined on the basis of the preference information provided by the DM. The DM compared two alternatives at a time, and expressed his/her preference over them. Based on the idea that a value function has a higher value at a more preferred alternative, the method generated a sequence of inequalities, which finally specified the most preferred (extreme point) solution - in theory. In practice, the responses of the DM often lead to conflicting information. As a solution to this problem, Zionts and Wallenius [65] proposed ignoring the oldest responses. Quite plausibly, the DM learns during the search process, and will be able to give more precise information about his/her "true" preferences over time.

The Interval Criterion Weights / Vector-Maximum Approach [56] is based on the idea to restrict the region where the optimal solution of the value function may lie. The DM's preferences are used to reduce the possible weights of the objective functions. Steuer uses the term "criterion cone reduction method" to describe the class of methods his method belongs to.

The following steps typically characterize this approach:

a) assumptions are made of the functional form of the underlying value function;
b) an initial solution is provided for evaluation;
c) the DM is asked to give preference information which is used to update knowledge about the underlying value function;

d) an improved solution or improved solutions are generated;
e) iteration is stopped when an "optimal" solution has been identified.

ii) Based on No Stable Value Function

Approaches belonging to this category are typically based on the idea of generating nondominated solutions for the DM's evaluation without making any specific assumptions concerning the value function. The DM is free to search the efficient frontier and stop at any time (s)he likes. For instance, change of mind and learning are allowed. A quite common approach is to let the DM freely express aspiration levels for the objectives, and to let the system generate feasible solutions to him/her. The previous responses are used to generate hopefully more preferred solutions. This projection is usually accomplished via minimizing a so called achievement scalarizing function [58, 62].

Typically the following steps are included:

- present the DM with an efficient solution and provide him/her with as much information as possible about the nondominated region, in particular in the "neighborhood" of the current solution;
- ask the DM to provide preference information in the form of aspiration levels, weights, etc.;
- use the responses to generate a single nondominated solution or a set of nondominated solutions for the DM's evaluation;
- iterate until the DM stops, or some specified termination criteria for the search have been satisfied.

A typical example of the method, where no assumptions about the value function are made until the DM stops, is the method by Korhonen and Laakso [35]. The DM is free to search the efficient frontier, but at the moment (s)he likes to stop, the DM is helped to evaluate whether (s)he has found the most preferred solution or not. At this moment, specific assumptions about the functional form of the value function are introduced.

Another example of this type of method is the Light Beam Search (LBS) approach by Jaszkiewicz and Slowinski [24]. LBS enables the DM to analyze multiple objective decision problems by presenting samples of non-dominated points. The DM can control the search by either modifying the aspiration and reservation points, or by shifting the current point to a selected better point from its neighborhood. Michalowski and Szapiro [45] have also developed an interactive method which is based on the use of the aspiration and reference points.

Approach 3: Posterior Articulation of Preferences

This approach tries to find a good approximation to a nondominated frontier. The choice problem is secondary. The main idea is to provide information to the DM about possible solutions. The presentation and visualization of the frontier are key aspects. A classical example is the ADBASE system by Steuer [57]. ADBASE finds all nondominated extreme point solutions for an MOLP-problem. The original idea was to approximate a nondominated frontier by means of nondominated extreme point solutions. Unfortunately, the number of nondominated extreme points can become large even in problems of reasonable size. Nowadays, the approach has become popular in problems where the functional forms of the objective functions are too complex for traditional optimization methods. Genetic algorithms are widely used for those problems, see, e.g. Coello et al. [12]. The approach seems to work well in case of two objectives.

9.4 Generating Nondominated Solutions

Despite many variations among different methods of generating nondominated solutions, the ultimate principle is the same in all methods: a single objective optimization problem is solved to generate a new solution or set of solutions. The objective function of this single objective problem may be called a *scalarizing function* according to Wierzbicki [62]. It has typically the original objectives and a set of parameters as its arguments. The form of the scalarizing function as well as what parameters are used depends on the assumptions made concerning the DM's preference structure and behavior.

Two classes of parameters are widely used in multiple objective optimization: 1) *weighting coefficients for objective functions* and 2) *reference/ aspiration/ reservation levels for objective function values*. Based on those parameters, there exist several ways to specify a scalarizing function. An important requirement is that this function completely characterizes the set of nondominated solutions: "*for each parameter value, all solution vectors are nondominated, and for each nondominated criterion vector, there is at least one parameter value, which produces that specific criterion vector as a solution*" (for theoretical considerations, see [63]).

9.4.1 A Linear Scalarizing Function

A classic method to generate nondominated solutions is to use the weighted-sums of objective functions, i.e. to use the following linear scalarizing function:

$$\max \quad \{\boldsymbol{\lambda}^\top \mathbf{q} \mid \mathbf{q} \in Q\} \tag{9.4}$$

If $\boldsymbol{\lambda} > \mathbf{0}$, then the solution $\mathbf{q}$ of (9.4) is nondominated, but if we allow that $\boldsymbol{\lambda} \geq \mathbf{0}$, then the solution is weakly nondominated [19, 57] (pp. 215 and 221). Using the parameter set $\Lambda = \{\boldsymbol{\lambda} \mid \boldsymbol{\lambda} > \mathbf{0}\}$ in the weighted-sums linear program we can completely characterize the efficient set provided the constraint set is convex. However,

Λ is an open set, which causes difficulties in a mathematical optimization problem. If we use $\mathrm{cl}(\Lambda) = \{\boldsymbol{\lambda} \mid \boldsymbol{\lambda} \geq \mathbf{0}\}$ instead, the nondominance of $\mathbf{q}$ cannot be guaranteed. It is surely weakly-efficient, but not necessarily efficient. When the weighted-sums are used to specify a scalarizing function in MOLP problems, the optimal solution corresponding to nonextreme points of Q is never unique. The set of optimal solutions always consists of at least one extreme point, or the solution is unbounded. In early methods, a common feature was to operate with weight vectors $\boldsymbol{\lambda} \in \mathbb{R}^k$, limiting considerations to nondominated extreme points (see, e.g., [65]).

9.4.2 A Chebyshev-type Scalarizing Function

Currently, most solution methods are based on the use of a so-called Chebyshev-type scalarizing function originally proposed by Wierzbicki [62]. We refer to this function by the term *achievement (scalarizing) function*.The achievement (scalarizing) function projects any given (feasible or infeasible) point $\mathbf{g} \in \mathbb{R}^k$ onto the set of nondominated solutions. Point $\mathbf{g}$ is called a *reference point*, and its components represent the desired values of the objective functions. These values are called *aspiration levels*.

The simplest form of the achievement function is:

$$s(\mathbf{g},\mathbf{q},\mathbf{w}) = \max\left(\frac{g_k - q_k}{w_k}\right), \quad k \in K = \{1,2,...,k\} \tag{9.5}$$

where $\mathbf{w} > \mathbf{0} \in \mathbb{R}^k$ is a given vector of weights, $\mathbf{g} \in \mathbb{R}^k$, and $\mathbf{q} \in Q = \{\mathbf{f}(\mathbf{x}) \mid \mathbf{x} \in X\}$. By minimizing $s(\mathbf{g},\mathbf{q},\mathbf{w})$ subject to $\mathbf{q} \in Q$, we find a weakly nondominated solution vector $\mathbf{q}^*$ (see, e.g., [62, 63]). However, if the solution is unique for the problem, then $\mathbf{q}^*$ is nondominated. If $\mathbf{g} \in \mathbb{R}^k$ is feasible, then $\mathbf{g}^* \in \mathbb{R}^k$, $\mathbf{q}^* \geq \mathbf{g}$. To guarantee that only nondominated (instead of weakly nondominated) solutions will be generated, more complicated forms for the achievement function have to be used, for example:

$$s(\mathbf{g},\mathbf{q},\mathbf{w},\rho) = \max_{k \in K}\left(\frac{g_k - q_k}{w_k}\right) + \rho \sum_{i=1}^{k}(g_i - q_i) \tag{9.6}$$

where $\rho > 0$. In practice, we cannot operate with "any positive value" for ρ. We have to use a pre-specified value for it. Another way is to use a lexicographic formulation (see, e.g., [57], p. 292 - 296).

The application of the scalarizing function (9.6) is easy, because given $\mathbf{g} \in \mathbb{R}^k$, the minimum of $s(\mathbf{g},\mathbf{q},\mathbf{w},\rho)$ is found by solving the following problem:

$$\begin{aligned} \min \quad & \varepsilon + \rho \sum_{i=1}^{k} (g_i - q_i) \\ \text{s.t.} \quad & \mathbf{q} \in Q \\ & \varepsilon \geq (g_i - q_i)/w_i, \quad i = 1, 2, ..., k \end{aligned} \tag{9.7}$$

The problem (9.7) can be further written as:

$$\begin{aligned} \min \quad & \varepsilon + \rho \sum_{i=1}^{k} (g_i - q_i) \\ \text{s.t.} \quad & \mathbf{q} \in Q \\ & \mathbf{q} + \varepsilon \mathbf{w} - \mathbf{z} = \mathbf{g} \\ & \mathbf{z} \geq \mathbf{0} \end{aligned} \tag{9.8}$$

To illustrate the use of the achievement scalarizing function, consider a two-criteria problem with a feasible region having four extreme points $\{(0,0), (0,3), (2,3), (8,0)\}$, as shown in Fig. 9.1. In case X is a polyhedron, (9.8) is an LP-model.

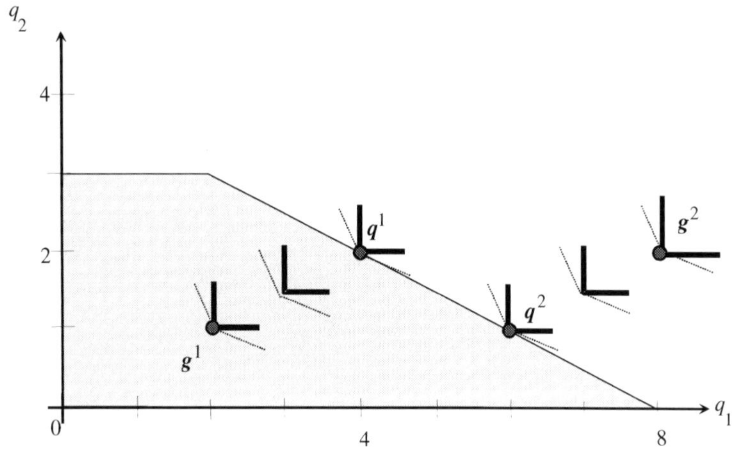

Fig. 9.1 Illustrating the projection of a feasible and an infeasible aspiration level point onto the nondominated surface

In Fig. 9.1, the thick solid lines describe the indifference curves when $\rho = 0$ in the achievement scalarizing function. The thin dotted lines stand for the case $\rho > 0$. Note that the line from $(2,3)$ to $(8,0)$ is nondominated and the line from $(0,3)$ to $(2,3)$ is weakly-nondominated, but dominated. Let us assume that the DM first specifies a feasible aspiration level point $\mathbf{g}^1 = (2,1)$. Using a weight vector $\mathbf{w} = [2,1]^\top$, the minimum value of the achievement scalarizing function (-1) is reached at point $\mathbf{q}^1 = (4,2)$ (cf. Fig. 9.1). Correspondingly, if an aspiration level point is infeasible, say $\mathbf{g}^2 = (8,2)$, then the minimum of the achievement scalarizing function (+1) is reached at point $\mathbf{q}^2 = (6,1)$. When a feasible point dominates an aspiration level point, then the value of the achievement scalarizing function is always negative; otherwise it is nonnegative. It is zero, if an aspiration level point is weakly nondominated.

As Fig. 9.1 illustrates, by varying aspiration levels different nondominated solutions are generated. Any nondominated point is a possible (unique) solution. Instead, a linear scalarizing function has not such property. Depending on the (strictly positive) weights used, the unique solutions are either point $(2,3)$ or point $(8,0)$. In case the ratio of the components of the weight vector is $\lambda_2/\lambda_1 = 2$, all nondominated solutions have the same value.

9.5 Solving Multiple Objective Problems

In MCDM we always assume the intervention of a DM at some stage in the solution process. The preference information can be gathered in advance, during the search process, or afterwards. In that sense, all MCDM-methods are interactive. However, generally the term *"interactive"* is used to refer to support systems, where the dialogue step and the computation step alternate until the final solution is reached. We will use this narrower interpretation in discussing interactive systems. The following steps typically appear in any interactive MCDM system:

- **Step 1**: Initial Solution(s)
- **Step 2**: Evaluation
- **Step 3**: Solution(s) Generation
- **Step 4**: Termination?

One or several solutions are generated in Step 1 and displayed to the DM. The DM considers the solution(s) and provides preference information in Step 2. Based on that information, new solutions are generated in Step 3. They are evaluated in Step 2. Steps 2 and 3 are repeated until the DM is willing to consider termination. The DM may simply stop the search or use termination conditions to help him/her to make a final decision.

The term *"support"* is also used in MCDM in a broad sense to refer to research dealing with the relationship between the problem and the decision-maker. For instance, the following questions may require support: How to structure the problem? How to find essential criteria? How to handle uncertainty? However, in this paper we consider the essential support problem in Multiple Criteria Decision Making: How to assist a DM to find the "best" solution from among a set of available alternatives?

9.5.1 Properties of a Good Multiple Criteria Decision Support System

There is no single criterion for evaluating multiple criteria decision support systems. Several relevant criteria can be introduced:

1. the system recognizes and generates nondominated solutions;
2. the system helps the DM feel convinced that the final solution is the most preferred one, or at least close enough;
3. the system helps the DM to have a "holistic" view over the nondominated frontier;
4. the system does not require too much time from the DM to find the final solution;
5. the communication between the DM and the system is not overly complicated;
6. the system provides reliable information about available alternatives;
7. the system makes it possible to evaluate the optimality conditions.

Provided that the problem is correctly specified, the final solution of a rational DM is always nondominated. Therefore it is important that the system is able to recognize and generate nondominated solutions. The system can operate with dominated solutions during the search process for behavioral reasons, but it has to lead the DM finally to a nondominated solution. The Geoffrion-Dyer-Feinberg method [20] is a typical example. Another example is the method by Arbel and Korhonen [3], in which the search is started from the nadir criterion values (worst criterion values over the nondominated set). To start from the worst solution enables the DM to proceed with a win-win strategy until the nondominated frontier is achieved. (S)he may change the search direction, but does not need to worsen any criterion value to gain on some other criterion value.

No system can provide a DM with a capability to compare all alternatives simultaneously. However, a good system can provide a holistic view over the alternatives and makes the DM convinced that his/her final choice is best or at least close enough. The user interface plays an important role in this aspect.

A good system does not waste the DM's time. Furthermore, the communication language should be easy. Irrelevant questions are boring to the DM. It is good to increase the intelligence of the system, but it is important to remember that the DM wants to keep the control of the system in his/her own hands. There are several ways to implement a "dialogue" between a system and the DM. For instance, Shin and Ravindran [54] list the following eight typical interaction styles (see also [59]):

- Binary pairwise comparisons - the DM must compare two-dimensional vectors at each iteration (see, e.g., [27]);
- Pairwise comparisons - the DM must compare a pair of k-dimensional vectors and specify a preference (see, e.g., [65]);
- Vector comparisons - the DM must compare a set of k-dimensional vectors and specify the best, the worst or the order of preference (see, e.g., [56]);
- Precise local tradeoff ratios - the DM must specify precise values of local tradeoff ratios at a given point (see, e.g., [20]).

- Interval local tradeoff ratios - the DM must specify an interval for each local tradeoff ratio (see, e.g., [52]).
- Comparative tradeoff ratios - the DM must specify his/her preference for a given tradeoff ratio (see, e.g., [26]).
- Index specification and value tradeoff - the DM must list the indices of objectives to be improved or sacrificed, and specify the amount [8];
- Aspiration levels (reference points) - the DM must specify or adjust the values of the objectives which indicate his/her optimistic wish concerning the outcomes of the objectives (see, e.g., [62]).

There are several ways to implement the above principles. The comparison information can be given to the DM in numeric or visual form. "One picture speaks more than one thousand words" is very true. Graphics can often be used to illustrate the effects of the DM's choices much more effectively than numbers.

A very important aspect is that the system provides reliable information about the available alternatives. For instance, if the system always produces the same nondominated solution for the DM's evaluation, (s)he is misled to believe that the solution is the only possible choice.

The optimality testing can also be implemented in interactive systems, but in that case, the DM has to be willing to accept some assumptions concerning the functional form of the value function. Such optimality conditions are considered in more detail in the next section.

9.5.2 Role of Interface

Mathematical programming models have been developed by mathematicians who are mainly interested in mathematical, not necessarily practical or behavioral aspects. Consequently, many systems make irrelevant assumptions about the behavior of a DM. A typical example is how "importance" is considered in models. When the DM says that one criterion is more important than another, a standard interpretation is that the more important criterion has to have a larger weight than the less important criterion. However, in most cases this is not a correct interpretation. The DM may be interested to reach a specific (reservation) level for a criterion value. That's why (s)he may experience that the criterion is important. Obviously, in some problems importance means the amount of attention the DM is going to pay to the criterion value at a certain moment. For instance, in a house buying problem, the DM may say that price is not important to him or her. The reason may simply be that the price of all the houses under consideration does not vary very much. That is why price does not require his/her primary attention.

Because we do not know very well how the brain of a DM works, it is important that we try to avoid unrealistic assumptions in our systems. From a behavioral point of view, a critical point is, when the system uses and interprets preference information received from the DM. The question the system asks the DM may be very clear and simple such as "Do you prefer this to that?", but the problem arises, when the system interprets this information. The purpose of the system is to produce alternatives which are probably closer to the most preferred solution than the previous ones. If we do not make any assumptions concerning the value function, there are no guidelines to generate a better solution. If the linear value function is a realistic assumption, then the problem is easy, provided that the DM is able to give consistent preference information. Even more general assumptions help the system to generate more preferable solutions. But whether those assumptions meet the requirements of behavioral realism is a difficult question with no definite answer.

A way to solve the above problems is to use a free search type of approach like VIG [32], in which the DM may move on the nondominated frontier, until (s)he is satisfied. In the approach, the search process is always in his/her own hands. We can avoid misinterpretation, but the approach is not without problems. Premature stopping may happen, because people experience sacrifices in criterion values more strongly than gains. This kind of behavior can be explained by prospect theory [25].

Currently, a very common way is to use graphics to illustrate nondominated solutions. In case of two criteria, the approach is very powerful, because all solutions (nondominated frontier) can be presented in one picture. To some extent, the idea also works in three dimensions, but the picture is not necessarily so illustrative than in two dimensions. In more than three dimensions, it is not possible generally to visualize the whole nondominated set. However, to illustrate single solution vectors graphically is helpful. A brief look at using various graphical techniques in the context of multiple criteria decision making is given in Korhonen and Wallenius [40].

9.6 Final Solution

The speed of (infinite) convergence and a stopping rule are important in mathematical programming algorithms. However, it is not the infinite convergence that matters in interactive procedures, because all interactive algorithms converge in a finite number of steps - actually in a few steps. Thus, the convergence in interactive methods has to be considered from the behavioral point of view. Hence we speak about *behavioral convergence*. In mathematical convergence, the main interest is in the general convergence. In the behavioral convergence, for instance, the initial rate of convergence is more interesting as pointed out already by Geoffrion et al. [20]. However, the initial rate of convergence is not the only aspect we have to take into account in the behavioral convergence. Human beings make errors in evaluations, not being able to give precise information, etc. In addition, the function we optimize is not usually known, and it may even change during the search process. In addition to studying the convergence process of an interactive method, it may be more useful to study its termination.

To test the "optimality" of the final solution in interactive methods, we may adopt ideas from mathematical programming with some exceptions. In mathematical programming, generally used optimality testing is based on the Karush-Kuhn-Tucker conditions (see, e.g., [5]). These conditions are based on the use of the gradients of the objective functions and of the functions that define the constraints. The gradients of the active constraints define a cone. Loosely speaking, when the gradient vector of the objective function is in the before mentioned cone, the optimality conditions are satisfied.

However, these kind of optimality conditions cannot be used in interactive methods. It is not realistic to assume that the DM could be able to compare the gradient vectors of implicitly defined value functions. Instead, the DM is able to compare feasible directions. A typical idea in interactive methods is based on that idea: if no direction of improvement exists, the solution is considered the most preferred one. If no assumptions are made about the value function, the idea is purely heuristic and based on the philosophy: "Because no better solution can be found, let's stop!" Provided that there is no feasible direction of improvement, the final solution is at least locally most preferred. A critical point is whether the DM has really been able to "see" all feasible directions. If we make assumptions about the functional form of the value function, we may reduce the number of the directions we have to consider to be sure that no direction of improvement exists. Zionts and Wallenius [65] assumed that the set Q is a polyhedron and the value function is linear. Based on this assumption it is sufficient to study only all adjacent efficient tradeoffs to prove the optimality of the current extreme point solution. Korhonen and Laakso [35] introduced general optimality conditions for a pseudoconcave value function. In case the set Q is a polyhedron, generally only k (= the number of objectives) directions are needed to check the optimality of the current solution.

9.7 Examples of Software Systems: VIG and VIMDA

9.7.1 VIG

Today many interactive systems use an aspiration level projection principle in generating nondominated solutions for the DM's evaluation. The projection is performed using Chebyshev-type achievement scalarizing functions as explained above. In section 9.4.2, we described how these functions can be controlled by varying the aspiration levels $\mathbf{g} \in \mathbb{R}^k$. Those functions may also be controlled by varying the weight vector $\mathbf{w} \in \mathbb{R}^k$ (keeping aspiration levels fixed). Instead of using aspiration levels, some algorithms ask the DM to specify the reservation levels for the criteria (see, e.g., [45]).

An achievement scalarizing function projects one aspiration (reservation) level point at a time onto the nondominated frontier. By parametrizing the function, it is possible to project an entire direction onto the nondominated frontier as originally proposed by Korhonen and Laakso [35]. The vector to be projected is called a *Reference Direction Vector* and the method *Reference Direction Method*, correspondingly. When a direction is projected onto the nondominated frontier, a curve traversing across the nondominated frontier is obtained. Then an interactive line search is performed along this curve. The idea enables the DM to make a continuous search on the nondominated frontier. The corresponding mathematical model is a simple modification from the original model (9.8) developed for projecting a single point:

$$\begin{aligned} \min \quad & \varepsilon + \rho \sum_{i=1}^{k} (g_i - q_i) \\ \text{s.t.} \quad & \mathbf{q} \in Q \\ & \mathbf{q} + \varepsilon \mathbf{w} - \mathbf{z} = \mathbf{g} + t\mathbf{r} \\ & \mathbf{z} \geq \mathbf{0}, \end{aligned} \tag{9.9}$$

where $t : 0 \to \infty$ and $\mathbf{r} \in \mathbb{R}^k$ is a reference direction. In the original approach, a reference direction was specified as a vector starting from the current solution and passing through the aspiration levels. The DM was asked to give aspiration levels for the criteria. In this way, the DM can specify where (s)he would like to go, and the system allows him/her to move in the desired direction. At any time, (s)he is free to change the search direction by specifying a new aspiration level vector.

The aspiration level vector describes the DM's preferences. A misspecification in its values is not drastic, because the DM can immediately respecify new aspiration levels for the criteria, if (s)he does not like the search direction. The idea of the reference direction approach can be applied in any multiple objective programming problem - in principle, but in practice only in the multiple objective linear programming context. For multiple objective linear programming problems, to generate a nondominated curve is straightforward by using parametric programming.

The original reference direction method has been further developed in many directions. First, Korhonen and Wallenius [38] improved upon the original procedure by making the specification of a reference direction dynamic. The dynamic version was called *Pareto Race*. In Pareto Race, the DM can freely move in any direction on the nondominated frontier (s)he likes, and no restrictive assumptions concerning the DM's behavior are made. Furthermore, the objectives and constraints are presented in a uniform manner. Thus, their role can also be changed during the search process. The software package which implements Pareto Race is called VIG.

In Pareto Race, a reference direction $\mathbf{r}$ is determined by the system on the basis of preference information received from the DM. By pressing number keys corresponding to the ordinal numbers of the objectives, the DM expresses which objectives (s)he would like to improve and how strongly. In this way (s)he implicitly specifies a reference direction. Figure 9.2 shows the Pareto Race interface for the search, embedded in the VIG software [32].

In Pareto Race, the user sees the objective function values on a display in numeric form and as bar graphs, as he/she travels along the nondominated frontier. The keyboard controls include an accelerator, gears, brakes, and a steering mechanism. The search on the nondominated frontier is analogous to driving a car. The DM can, e.g., increase/decrease his/her speed, make a turn and brake at any moment he/she likes.

To implement these features, Pareto Race uses certain mechanisms, which are controlled by the following keys:

(SPACE) BAR: An "Accelerator"
Proceed in the current direction at constant speed.

F1: "Gears (Backward)"
Increase speed in the backward direction.

F2: "Gears (Forward)"
Increase speed in the forward direction.

F3: "Fix"
Use the current value of objective i as the worst acceptable value.

F4: "Relax"
Relax the "bound" determined with key F3.

F5: "Brakes"
Reduce speed.

F10: "Exit"

num: "Turn"
Change the direction of motion by increasing the component of the reference direction corresponding to the goal's ordinal number $i \in [1, k]$ presented by DM.

An example of the Pareto Race screen is given in Fig. 9.2. The screen shot is from a numerical example described in the next section.

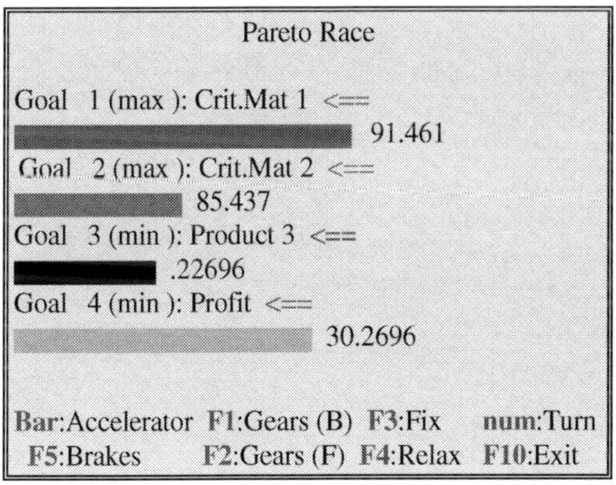

Fig. 9.2 An example of the Pareto race screen

Pareto Race does not specify restrictive behavioral assumptions for a DM. (S)he is free to conduct a search on the nondominated surface, until he/she believes that the solution found is his/her most preferred one.

Pareto Race was originally developed for solving moderate size problems consisting of less than 10 objectives, and a few hundred rows and variables. When the size of the problem becomes large, computing time makes the interactive mode inconvenient. To solve large-scale problems Korhonen et al. [41] proposed a method based on Pareto Race. An interactive local free search is first performed to find the most preferred direction. Based on the direction, a nondominated curve can be generated in a "batch mode" if desired. For an interesting extension of Pareto Race to solving nonlinear programming problems, see Eskelinen et al. [15].

9.7.2 VIMDA

When the MCDM-problem is a Multiple Criteria Evaluation Problem, i.e. Q consists of a finite number of elements which are explicitly known in the beginning of the solution process, the Pareto Race type of approach is not directly applicable, because the nondominated frontier is not "smooth". However, the reference direction method is a valid approach for generating nondominated alternatives for the DM's evaluation. Instead of an efficient curve, the path consists of a set of nondominated points which are displayed to the DM. These points can be shown to the DM, for instance, by using the visual representation shown in Figure 9.3.

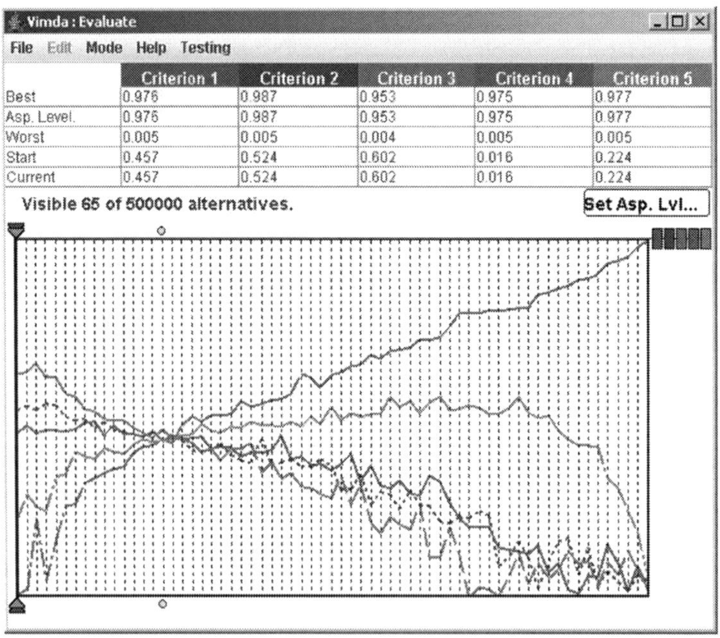

Fig. 9.3 Visual interface of VIMDA

The current alternative is shown on the left hand margin. Each line stands for one criterion. The vertical position of the line gives information about the value of the corresponding criterion. The ideal value of each criterion is the highest position on the screen and the worst value is the lowest position, respectively. The alternatives are evenly distributed horizontally. The vertical lines point to the positions of alternatives. The thick vertical line refers to the alternative for which the numerical values are shown on the top of the screen. The DM can control the position of the thick line by moving the cursor "left" and "right". The consecutive alternatives have been connected with lines for an enhanced visual representation.

The DM is asked to choose the most preferred alternative from the screen by moving the cursor to point to such an alternative. The chosen (most preferred) alternative becomes the current solution for the next iteration. Next, the user is asked to reconsider his/her aspiration levels, etc.

The process stops when the DM is satisfied with the solution. Actually, the mathematics underlying the approach is quite complex, because the projection of a direction to the set of finite points, requires solving a non-convex parametrization problem (see, e.g., [34]). However, the mathematics is hidden from the DM.

Figures 9.4 & 9.5 provide a visual illustration of the VIMDA algorithm in three dimensions. VIMDA algorithm starts from the south-eastern most point in Figure 9.4. The reference direction is the direction which starts from that point and passes through a given reference point, which is the north-western most point. The algorithm chooses the points minimizing an achievement scalarizing function [62] and eliminates points (light grey points in Figure 9.4) which are not necessary to consider further. The final projection is displayed in Figure 9.5. The current VIMDA-decision support system is able to solve problems having at most ten criteria and millions of points[3].

9.8 Concluding Remarks

An interactive approach is a very natural way to solve multiple criteria decision problems, because a DM is always an essential part of the solution process. The system generates "reasonable" (nondominated) solutions and the DM provides preference information to the system. The DM's responses are used to generate new potential most preferred alternatives.

In interactive methods, there are several critical points which require careful consideration:

- How preference information is gathered;
- How information is used to generate new alternatives for the DM's evaluation; and
- How the system generates alternatives.

[3] Programming credits are due to Mr. Jyri Ruutu.

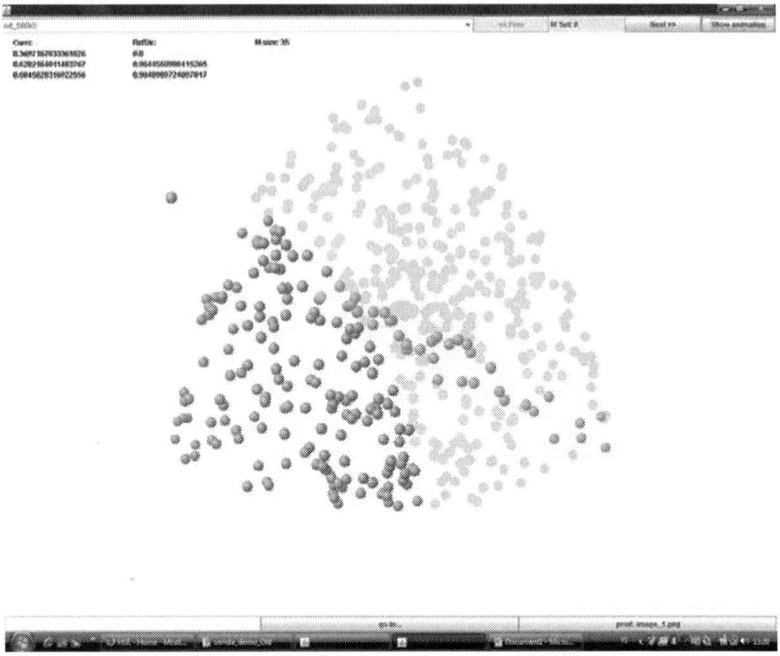

Fig. 9.4 Illustration of the VIMDA-algorithm in 3 dimensions

If the information gathering process is too complicated, the DM is unable to give reliable information and the system all the time "interprets" his/her responses erroneously. It is important that the way the system deals with the DM's responses is behaviorally realistic. Unrealistic interpretation of preference information generates alternatives which are not consistent with the DM's preferences. On the other hand, the system cannot too much restrict the choices. It is also important that the system enables the DM to search the whole nondominated frontier, not only a part of it.

Many interactive methods have been developed for MOP problems. Choosing the "best" interactive method is a multiple criteria decision problem in itself. Gardiner and Steuer [18] have proposed a unified interactive multiple-objective programming framework by putting together the best ideas developed in the MCDM-community. Obviously, in the future we have to focus on issues which improve our understanding of human behavior. The lack of behavioral realism in the MCDM systems may be one reason why the systems (with some exceptions such as the AHP; [51]) have not found widespread applications in practice, even though there is a great potential need for them. For additional future research questions in MCDM, the reader is asked to consult Wallenius et al. [60].

We conclude our chapter by citing some highly relevant research in Evolutionary Multi-Objective Optimization (EMO). Evolutionary optimization has emerged as a new field with strong ties to Multiple Criteria Decision Making [13]. Multiple cri-

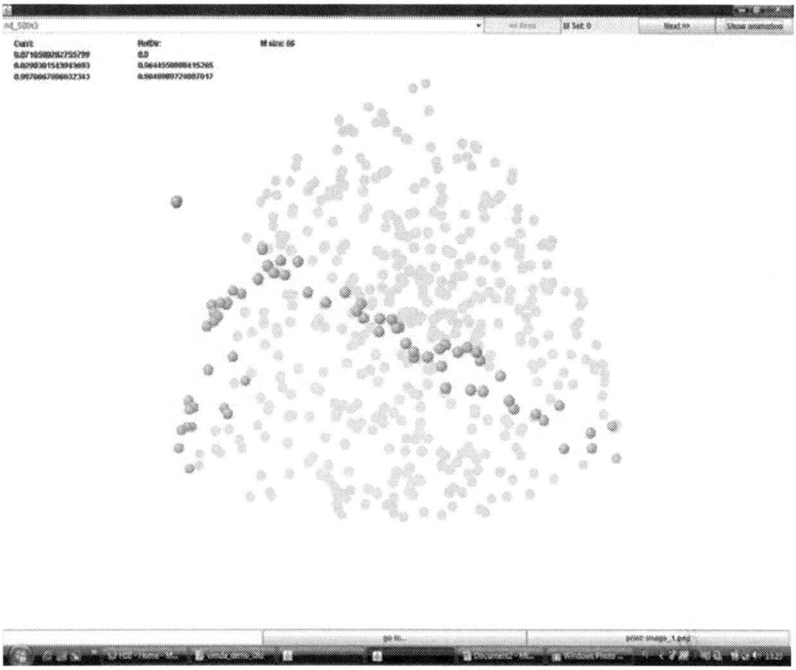

Fig. 9.5 A reference direction projection in VIMDA

teria optimization techniques were unable to solve many highly nonlinear problems that were common in engineering. Starting with an initial population, an evolutionary algorithm updates the population by mimicking processes from nature (survival of the fittest) and genetic operators to improve the average population from generation to generation in a stochastic manner. The goal is to converge to a good final population of solutions that represents the nondominated set. The first evolutionary multi-objective optimization algorithm is due to Schaffer [53]. However, it was not until about ten years later that three working evolutionary algorithms were suggested almost at the same time: MOGA by Fonseca and Fleming [16], NSGA by Srinivas and Deb [55], and NPGA by Horn et al. [21]. The main thrust in all these algorithms was to generate an approximation of the Pareto optimal frontier. Recently, much attention has been devoted to developing hybrid MCDM/EMO procedures, with the purpose of finding the most preferred solution or region of interest. One of the first was the paper by Phelps and Köksalan [49]. They developed an interactive genetic algorithm, and demonstrated its use on multiobjective knapsack and minimum spanning tree problems. Furthermore, Köksalan and Phelps [30] developed a multi-objective evolutionary algorithm to concentrate on a desired part of the efficient frontier using partial information on the preferences of the DM. Our recent research focuses on further building the bridge between EMO and MCDM. Our research uses a similar interactive genetic algorithm framework to that of Phelps and Köksalan [49]. However, our technique is more general since it is designed to support any quasi-concave value function of the objectives. See Fowler et al. [17].

References

1. C.H. Antunes and J.N. Climaco. Sensitivity analysis in MCDM using the weight space. *Operations Research Letters*, 12:187–196, 1992.
2. C.H. Antunes and J.N. Climaco. Decision aid for discrete alternative multiple criteria problems - A visual interactive approach. *Information and Decision Technologies*, 19:185–193, 1994.
3. A. Arbel and P. Korhonen. Using objective values to start multiple objective linear programming algorithms. *European Journal of Operational Research*, 128:587–596, 2001.
4. C.A. Bana e Costa and J.C. Vansnick. MACBETH - An interactive path towards the construction of cardinal value functions. *International Transactions in Operational Research*, 1:387–500, 1994.
5. M.S. Bazaraa, J.J. Jarvis, and H.D. Sherali. *Linear Programming and Network Flows*. Wiley, New York, 1990.
6. V. Belton and M. Elder. Visual interactive modeling. *European Journal of Operational Research*, 54:273–273, 1991.
7. V. Belton and S.P. Vickers. V.I.S.A. - VIM for MCDM. In *VIIth International Conference on MCDM*, August 1988.
8. R. Benayoun, J. de Montgolfier, J. Tergny, and O. Larichev. Linear programming with multiple objective functions: Step method (Stem). *Mathematical Programming*, 1:366–375, 1971.
9. A.R. Borges and C.H. Antunes. A weight space-based approach to fuzzy multiple-objective linear programming. *Decision Support Systems*, 34:427–443, 2003.
10. A. Charnes and W.W. Cooper. *Management Models and Industrial Applications of Linear Programming*, volume Vol. I, Appendix B. Wiley, New York, 1961.
11. A. Charnes, W.W. Cooper, and R.O. Ferguson. Optimal estimation of executive compensation by linear programming. *Management Science*, 1:138–151, 1955.
12. C. Coello, D. Van Veldhuizen, and G. Lamont. *Evolutionary Algorithms for Solving Multi-Objective Problems*. Kluwer, New York, 2002.
13. K. Deb. *Multi-Objective Optimization Using Evolutionary Algorithms*. Wiley, New York, 2001.
14. J. Dyer. A time-sharing computer program for the solution of the multiple criteria problem. *Management Science*, 19:1379–1383, 1973.
15. P. Eskelinen, K. Miettinen, K. Klamroth, and J. Hakanen. Pareto navigator for interactive nonlinear multiobjective optimization. *OR Spektrum*, 32:211–227, 2009.
16. C.M. Fonseca and P.J. Fleming. Genetic algorithms for multiobjective optimization: Formulation, discussion, and generalization. In *Proceedings of the Fifth International Conference on Genetic Algorithms*, pages 416–423, 1993.
17. J. Fowler, E. Gel, M. Köksalan, P. Korhonen, J. Marquis, and J. Wallenius. Interactive evolutionary multi-objective optimization for quasi-concave preference functions. 2009.
18. L.R. Gardiner and R.E. Steuer. Unified interactive multiple-objective programming - An open-architecture for accommodating new procedures. *Journal of the Operational Research Society*, 45:1456–1466, 1994.
19. S. Gass and T.L. Saaty. Parametric objective function (part II): Generalization. *Operations Research*, 3:395–401, 1955.
20. A. Geoffrion, J. Dyer, and A. Feinberg. An interactive approach for multi-criterion optimization, with an application to the operation of an academic department. *Management Science*, 19:357–368, 1972.

21. J. Horn, N. Nafploitis, and D. Goldberg. A niched Pareto genetic algorithm for multi-objective optimization. In *Proceedings of the First IEEE Conference on Evolutionary Computation*, pages 82–87, 1995.
22. J.P. Ignizio. *Goal Programming and Its Extensions*. D.C. Heath, Lexington, 1976.
23. J.P. Ignizio. Special issue on generalized goal programming. *Computers and Operations Research*, 10:277–389, 1983.
24. A. Jaszkiewicz and R. Slowinski. The light beam search approach - An overview of methodology and applications. *European Journal of Operational Research*, 113:300–314, 1999.
25. D. Kahneman and A. Tversky. Prospect theory: An analysis of decision under risk. *Econometrica*, 47:262–291, 1979.
26. I. Kaliszewski and W. Michalowski. Efficient solutions and bounds on tradeoffs. *Journal of Optimization Theory and Applications*, 94:381–394, 1997.
27. E.K. Karasakal and M. Koksalan. A simulated annealing approach to bicriteria scheduling problems on a single machine. *Journal of Heuristics*, 6:311–327, 2000.
28. R. Keeney and H. Raiffa. *Decisions with Multiple Objectives*. Wiley, New York, 1976.
29. R. Klimberg. GRADS - A new graphical display system for visualizing multiple criteria solutions. *Computers and Operations Research*, 19:707–711, 1992.
30. M. Köksalan and S. Phelps. An evolutionary metaheuristic for approximating preference-nondominated solutions. *INFORMS Journal on Computing*, 19:291–301, 2007.
31. T. Koopmans. Analysis of production as an efficient combination of activities. In T. Koopmans, editor, *Activity Analysis of Production and Allocation*, pages 33–97. John Wiley, New York, 1951.
32. P. Korhonen. VIG - A visual interactive support system for multiple criteria decision making. *Belgian Journal of Operations Research, Statistics and Computer Science*, 27:3–15, 1987.
33. P. Korhonen. Using harmonious houses for visual pairwise comparison of multiple criteria alternatives. *Decision Support Systems*, 7:47–54, 1991.
34. P. Korhonen and J. Karaivanova. An algorithm for projecting a reference direction onto the nondominated set of given points. *IEEE Transactions on Systems, Man, and Cybernetics - Part A: Systems and Humans*, 29:429–435, 1999.
35. P. Korhonen and J. Laakso. A visual interactive method for solving the multiple criteria problem. *European Journal of Operational Research*, 24:277–287, 1986.
36. P. Korhonen, H. Moskowitz, and J. Wallenius. Choice behavior in interactive multiple criteria decision-making. *Annals of Operations Research*, 23:161–179, 1990.
37. P. Korhonen, H. Moskowitz, and J. Wallenius. Multiple criteria decision support - A review. *European Journal of Operational Research*, 63:361–375, 1992.
38. P. Korhonen and J. Wallenius. A Pareto race. *Naval Research Logistics*, 35:615–623, 1988.
39. P. Korhonen and J. Wallenius. Observations regarding choice behavior in interactive multiple criteria decision-making environments: An experimental investigation. In A. Lewandowski and I. Stanchev, editors, *Methodology and Software for Interactive Decision Support*, pages 163–170. Springer, Berlin, 1989.
40. P. Korhonen and J. Wallenius. Visualization in the multiple objective decision-making framework. In J. Branke, K. Deb, K. Miettinen, and R. Slowinski, editors, *Multiobjective Optimization: Interactive and Evolutionary Approaches*, pages 195–214. Springer, Berlin, 2008.
41. P. Korhonen, J. Wallenius, and S. Zionts. A computer graphics-based decision support system for multiple objective linear programming. *European Journal of Operational Research*, 60:280–286, 1992.
42. O.I. Larichev and H.M. Moshkovich. Limits to decision-making ability in direct multiattribute alternative evaluation. *Organizational Behavior and Human Decision Processes*, 42:217–233, 1988.
43. A. Lewandowski and M. Granat. Dynamic BIPLOT as the interaction interface for aspiration based decision support systems. In P. Korhonen, A. Lewandowski, and J. Wallenius, editors, *Multiple Criteria Decision Support*, pages 229–241. Springer, Heidelberg, 1989.
44. B. Mareschal and J.P. Brans. Geometrical representations for MCDA. *European Journal of Operational Research*, 34:69–77, 1988.

45. W. Michalowski and T. Szapiro. A bi-reference procedure for interactive multiple criteria programming. *Operations Research*, 40:247–258, 1992.
46. K. Miettinen. *Nonlinear Multiobjective Optimization*. Kluwer, Berlin, 1999.
47. D. Olson. *Decision Aids for Selection Problems*. Springer, New York, 1996.
48. V. Pareto. *Manuel d'Economie Politique*. Giard, 1909.
49. S. Phelps and M. Köksalan. An interactive evolutionary metaheuristic for multiobjective combinatorial optimization. *Management Science*, 49:1726–1738, 2003.
50. B. Roy. Problems and methods with multiple objective functions. *Mathematical Programming*, 1:239–266, 1971.
51. T. Saaty. *The Analytic Hierarchy Process*. McGraw-Hill, New York, 1980.
52. A.A. Salo and R.P. Hamalainen. Preference assessment by imprecise ratio statements. *Operations Research*, 40:1053–1061, 1992.
53. J.D. Schaffer. *Some experiments in machine learning using vector evaluated genetic algorithms*. PhD thesis, Vanderbilt University, Nashville, Tennessee, 1984.
54. W.S. Shin and A. Ravindran. Interactive multiple objective optimization: Survey I - continuous case. *Computers and Operations Research*, 18:97–114, 1991.
55. N. Srinivas and Deb. K. Multi-objective function optimization using non-dominated sorting genetic algorithms. *Evolutionary Computation*, 2:221–248, 1994.
56. R.E. Steuer. An interactive multiple objective linear programming procedure. *TIMS Studies in the Management Sciences*, 6:225–239, 1977.
57. R.E. Steuer. *Multiple Criteria Optimization: Theory, Computation, and Application*. Wiley, New York, 1986.
58. R.E. Steuer and E.-U. Choo. An interactive weighted tchebycheff procedure for multiple objective programming. *Mathematical Programming*, 26:326–344, 1983.
59. H. Wallenius. Implementing interactive multiple criteria decision methods in public policy. Studies in Computer Science, Economics and Statistics 18, University of Jyväskylä, 1991.
60. J. Wallenius, J. Dyer, P. Fishburn, R. Steuer, S. Zionts, and K. Deb. MCDM/MAUT: Recent accomplishments and what lies ahead. *Management Science*, 54:1336–1349, 2008.
61. M. Weber and K. Borcherding. Behavioral-influences on weight judgments in multiattribute decision-making. *European Journal of Operational Research*, 67:1–12, 1993.
62. A. Wierzbicki. The use of reference objectives in multiobjective optimization. In G. Fandel and T. Gal, editors, *Multiple Objective Decision Making, Theory and Application*, pages 468–486. Springer, New York, 1980.
63. A. Wierzbicki. On the completeness and constructiveness of parametric characterizations to vector optimization problems. *OR Spektrum*, 8:73–87, 1986.
64. S. Zionts. MCDM, If not a Roman numeral, then what? *Interfaces*, 9:94–101, 1979.
65. S. Zionts and J. Wallenius. An interactive programming method for solving the multiple criteria problem. *Management Science*, 22:652–663, 1976.
66. S. Zionts and J. Wallenius. An interactive multiple objective linear programming method for a class of underlying nonlinear utility functions. *Management Science*, 29:519–529, 1983.

Chapter 10
On Multi-Objective Evolutionary Algorithms

Dalila B.M.M. Fontes and António Gaspar-Cunha

Abstract In this chapter Multi-Objective Evolutionary Algorithms (MOEAs) are introduced and some details discussed. A presentation of some of the concepts in which this type of algorithms are based on is given. Then, a summary of the main algorithms behind these approaches and their applications is provided, together with a brief discussion including their advantages and disadvantages, degree of applicability, and some known applications. Finally, future trends in this area and some possible paths for future research are pointed out.

10.1 Introduction

Most real-world optimization problems are multi-objective since they require the simultaneous satisfaction of several objectives. The most usual approach to deal with the multi-objective nature of these problems consists on congregating the various individual objectives into a unique function in order to form a single-objective optimization problem. In this case, it is necessary to define a priori a compromise between the objectives considered. If the relative importance of the criteria is changed a new optimization run needs to be carried out.

Dalila B.M.M. Fontes
LIAAD - INESC Porto L.A. and Faculdade de Economia, Universidade do Porto
Rua Dr. Roberto Frias, 4200-464 Porto, Portugal.
e-mail: fontes@fep.up.pt

António Gaspar-Cunha
IPC/I3N - Institute of Polymers and Composites
Department of Polymer Engineering, University of Minho
Campus de Azurém, 4800-058 Guimarães, Portugal.
e-mail: agc@dep.uminho.pt

C. Zopounidis and P.M. Pardalos (eds.), *Handbook of Multicriteria Analysis*, Applied Optimization 103, DOI 10.1007/978-3-540-92828-7_10,

Other possible approach takes advantages of the fact that Evolutionary Algorithms (EAs) work with a population of points processed in each iteration, yielding a set of non-dominated vectors, designated as Pareto optimal solutions. In this case, all the objectives are optimized simultaneously.

EAs mimic the process of natural evolution where an analogy between the mechanisms of natural selection and a learning (or optimization) process is made through the application of certain heuristic techniques [38]. These techniques can be classified into four main categories, Genetic Algorithms [42], Evolution Strategies [64], Evolutionary Programming [30] and Genetic Programming [52]. However, this classification is due to historical developments rather than to major functioning differences, since the basis of these techniques is essentially the same.

After using evolutionary techniques for single-objective optimization during more than two decades, the incorporation of more than one objective has finally become a popular area of research. As a consequence, many new evolutionary-based approaches and variations of existing techniques have recently been published in the literature [14]. The large number of applications [4, 8] and the continuously growing interest in this field are due to several advantages of EAs:

1. In-depth mathematical understanding of the problems to which they are applied to is not required;
2. Some of the solutions obtained by the EAs were previously out of range of the solutions obtained by other methods;
3. EAs can be applied to problems that cannot be solved by analytical mathematical techniques or that involve so many variables that other methods would take too long to solve them.
4. EAs can be applied to a high range of problems since they are robust;
5. EAs are relatively cheap and simple to implement;
6. It is easy to combine EAs with other techniques, such as local search and other heuristics (hybridization);
7. EAs are extremely adaptable due to the fact that the evolutionary mechanism is separate from the problem representation. Therefore, they can be transferred from problem to problem, that is, they are modular and portable;
8. EAs allow for the use of arbitrary constraints, simultaneous multiple objectives and the mixing of continuous and discrete parameters;
9. In addition, EAS are intrinsically parallel, i.e., they can be easily adapted to parallel computation.

Regarding multi-objective optimization problems they also have the advantage of working with a population of solutions rather than with a single solution. The ability to simultaneously search different regions of a solution space not only makes it possible to find a diverse set of solutions but also to address problems with non-convex, discontinuous, and multi-modal solutions spaces. These features enable the creation of Pareto fronts representing the trade-off between the criteria.

Multi-Objective Optimization (MOO) is undoubtedly a very important research topic both for scientists and practitioners, not only because of the multi-objective nature of most real-world problems but also because there are still many open questions in this area. The conflict of objectives entailed in MOO places the issue of compromise in a central position. Edgeworth [26] and Pareto [61] captured this notion mathematically in the criterion widely known as Pareto optimality [14]. Solutions belonging to the Pareto optimal set of a particular MOO problem perform better in one or more objectives and worst in the other objectives. In other words, solutions in the Pareto optimal set display different trade-offs.

Since, usually, no single solution optimizes simultaneously all objectives, decision making based on subjective human preference is an inherent aspect in solving MOO problems. Only a single solution out of the Pareto optimal set is required. Preference is the basis of tie-breaking between solutions in the Pareto optimal set. In the areas of Multi-criteria Decision Making (MCDM) and multi-objective Decision Aid (MCDA) a variety of frameworks capturing the decision maker(s) preferences have been proposed. Multi-attribute utility theory (MAUT) [49], Analytic Hierarchy Process [66] and outranking synthesis [78] are some of the most popular preference specification schemes. The multiplicity of preference articulation schemes highlights the complexity of human preference. Many approaches to this type of problems have been suggested, going all the way from naively combining objectives into one, to the use of game theory to coordinate the relative importance of each objective.

This chapter emphasizes the importance of looking at previous work in operations research, not only to get a good understanding of some of the existing techniques, but also to motivate the development of new EA-based approaches. Finally, some real applications are also described to provide the reader with a more complete idea of the range of applicability and the underlying motivation of this technique.

10.2 Multi-Objective Optimization

10.2.1 Definitions and Concepts

As soon as there are many (possibly conflicting) objectives to be optimized simultaneously, there is no longer a single optimal solution but rather a whole set of possible solutions of equivalent quality.

The general Multi-Objective Optimization Problem (MOOP) may be stated as finding the value for a set of n decision variables which must satisfy some constraints (J inequalities and K equalities) such that the M objective functions are optimized and can be modeled as follows:

$$
\begin{array}{llll}
(\mathscr{P}) & \text{Optimize} & f_m(x_i) & \text{for all } m = 1,2,\ldots,M \\
& \text{subject to} & & \\
& & g_j(x_i) \geq 0 & \text{for all } j = 1,2,\ldots,J \\
& & h_k(x_i) = 0 & \text{for all } k = 1,2,\ldots,K
\end{array}
$$

where $x_i = \{x_1, x_2, \ldots, x_n\}$ is the vector of decision variables.

In general the objectives are non-commensurable and in conflict with each other. Therefore, optimizing means finding a solution having values, for all objective functions, which satisfy the decision maker.

Generally speaking, in MOOPs two different solutions are related to each other in two possible ways: either one dominates the other or none of them is dominated. The set of Pareto solutions consists of good solutions, where none can be said to be better than the others, that is, the set of nondominated solutions. This concept is illustrated in Figure 10.1, where solutions 1, 2, 3 and 4 are non-dominated and constitute the Pareto front.

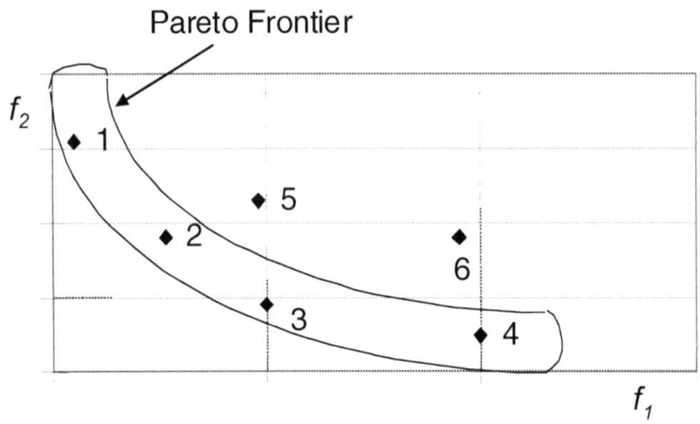

Fig. 10.1 Concept of non-dominance.

The optimal solution of a MOOP is not a single solution but a set of solutions composed by all the potential solutions such that the components of the objectives vector cannot be simultaneously improved. These solutions are known as Pareto optimal solutions, i.e., the set of non-dominated solutions. A solution is optimal when it is non-dominated by all other feasible solutions. In practice, it is generally impossible to know the actual optimal set and the corresponding Pareto optimal front, but, instead the optimization algorithms find an approximation to this set. The above mentioned concepts can be formally defined as follows:

Pareto Dominance:

Given the vector of objective functions $f_m = (f_1, \ldots, f_M)$ is said that candidate x^1 dominates x^2 (for minimizing), written as $x^1 \preceq x^2$, if

$$f_m(x^1) \leq f_m(x^2), \ \forall m \in \{1, \ldots, M\} \quad \text{and}$$
$$\exists m \in \{1, \ldots, M\} : f_m(x^1) < f_m(x^2). \tag{10.1}$$

Pareto Optimality:
For a MOP, a given solution x^* is Pareto optimal if and only if there is no vector $x \in \mathscr{F}$ ($\mathscr{F}$ is the set of feasible candidate solutions), so that

$$f_m(x) \leq f_m(x^*), \ \forall m \in \{1, \ldots, M\} \quad \text{and}$$
$$f_m(x) < f_m(x^*) \ \text{ for at least one objective function.} \tag{10.2}$$

Pareto Optimal Set:
For a MOP, the Pareto Optimal Set ($\mathscr{P}^*$) is defined as

$$\mathscr{P}^* := \{x \in \Omega | \neg \exists x' \in F(x\prime) \preceq f(x)\}. \tag{10.3}$$

Pareto Front:
For a MOP and Pareto Optimal Set ($\mathscr{P}^*$), the Pareto Front ($\mathscr{PF}^*$) is defined as

$$\mathscr{PF}^* := \{f_m(x) = (f_1(x), f_2(x), \cdots, f_M(x) | x \in \mathscr{P}\}. \tag{10.4}$$

In extending the ideas of single-objective EAs to multi-objective cases, three major problems must be addressed (Figure 10.2).

1. How to accomplish **fitness assignment** and selection in order to guide the search towards the Pareto optimal set;
2. How to maintain a **diverse population** in order to prevent premature convergence and achieve a well distributed, wide spread trade-off front;
3. How to prevent, during the successive generations, that some good solutions are lost.

It should be noticed that the objective function itself no longer qualifies as fitness function since it is a vector of values and not a scalar value. Different approaches to relate the fitness function to the objective functions are discussed in the following section, further details can be found, for example, in [14, 11].

To maintain good diversity of the population it is necessary to have a density estimation operator, such as, for example, niching [16]. It consists basically in counting the number of neighborhoods around each solution in order to deteriorate the fitness of the different individuals, i.e., the fitness decreases for the individuals with more neighbors.

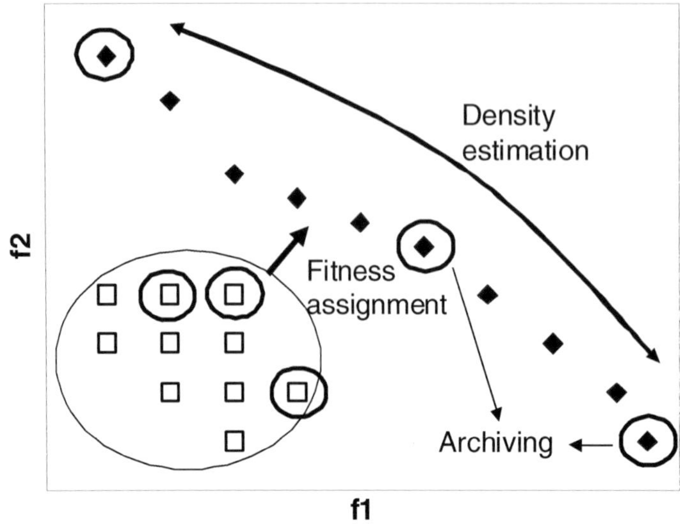

Fig. 10.2 Basic functions of a MOEA.

The third problem is usually solved by keeping the best solutions found so far in an archive in order to ensure that good individuals do not get lost, by mutation or recombination.

10.2.2 Addressing Multi-Objectives

Since there is no accepted definition of optimum as in single-objective optimization problems, it is difficult to compare reported results, as normally, the decision about the best answer corresponds to the so-called (human) decision maker. The decision maker preference for a particular solution is vague, base on perceptive information, and highly dependent on the application context. The vagueness and context-dependence of the decision makers preference structure have lead to the development of various mathematical models and techniques.

The approaches can be divided into two main categories. One that solves a single-objective problem, achieved by combining the objectives into a single-objective function, and another that searches for the Pareto optimal solutions set. In the former case, determination of a single-objective is possible with methods such as utility theory, weighted sum, etc., but the problem lies in the proper selection of the weights or utility functions to characterize the decision maker's preferences. Following the classification of Veldhuizen [77], Multi-objective Evolutionary Algorithms (MOEAs) may be a priori, interactive, or a posteriori algorithms based on the treatment of preference. A priori MOEAs involve preference specification prior to the optimization stage, and are traditionally implemented by aggregating objectives into a single fitness function with parameters reflecting the preference of the decision maker(s). Interactive MOEAs allow decision maker(s) to alter parameters during the search, effectively influencing the direction of the search. A posteriori MOEAs find the set of Pareto optimal solutions and relegate decision making based on human preference to a later stage.

In **a priori** algorithms, the decision maker states the preferences, which are then incorporated into the objective function through aggregation, prior to the optimization. This new formulation is then incorporated in the fitness function computation and cannot be changed throughout the optimization process. Aggregation of the objectives can be made in lexicographic order, that is the objectives are optimized in their order of the importance, or by linear/nonlinear combination of the objectives. In the latter case a single-objective function, reflecting the decision maker preferences, is obtained. This is the simpler approach to MOO, therefore a good choice if the preferences can be captured by the mathematical model and no practical computational difficulties arise. However, this is rarely the case, since often the non-commensurability of objectives makes it very hard, if not impossible to model them in a priori preference specification. In addition, this type of approach requires deep knowledge of the problem in hands, which usually is not possible. Furthermore, practical computational difficulties may also arise due to the non-convex nature of the Pareto front introduced through the combination of the objectives. The main drawback of this type of approach is that scaling amongst objectives is needed and small perturbations in the weights can lead to quite different solutions. In addition, the optimization method devised would return a single solution rather than a set of solutions that can be examined for trade-offs. See Figure 10.3 for an illustration of this approach. For a recent review on preference incorporation in multi-objective

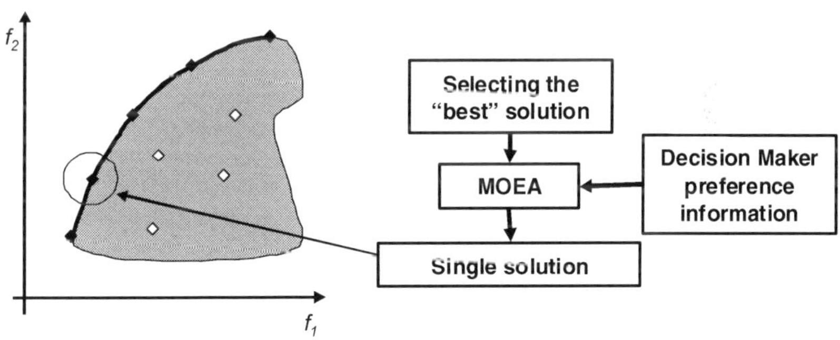

Fig. 10.3 Decision making in MOOP addressing preferences a priori.

evolutionary algorithms the reader is referred to the work by Abass and Sarker [63].

Interactive algorithms tend to be the most adequate approaches to MOO problems since they involve the decision maker in the optimization process. For this type of approach the decision maker only needs to specify a few parameters a priori. As the search progresses and more information on the problem becomes available, the decision maker can make better judgments and therefore parameter specifications. However, interactive approaches require intensive collaboration with the decision maker and quite often become a challenge due to the existence of more than one decision maker. Fonseca and Fleming [31] proposed the incorporation of an expert system to "replace" the decision maker. Nevertheless, the construction of an expert system still requires extensive problem knowledge and, as it is widely recognized, its success is highly dependent on the application context.

A posteriori algorithms tend to be the most popular since its application allows for independent optimization and decision making processes. The optimization and decision maker issues are separated by leaving the latter ones to a post-optimization stage. An illustration of such approach is provided in Figure 10.4. In these ap-

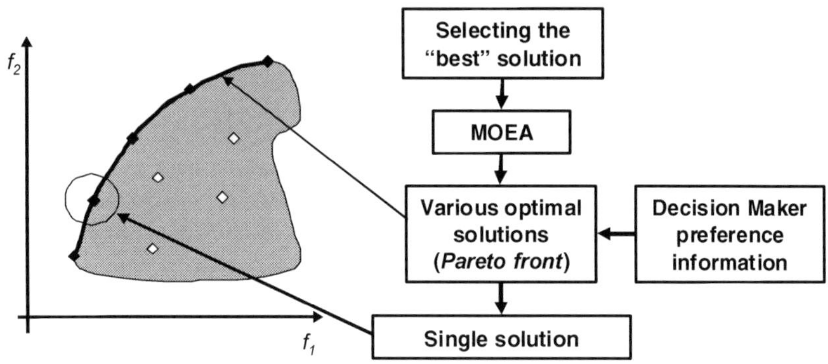

Fig. 10.4 Decision making in MOOP addressing preferences a posteriori.

proaches, the aim of the optimization process is, therefore, to find a set of well-distributed solutions along the Pareto front. Some challenging computational issues are avoided by only looking for such a set of solutions and leaving the choice of the preferred solution to the decision maker. However, ensuring that these solutions represent a wide range of trade-offs may be computationally expensive. Population-wide Pareto ranking, archiving strategy, and diversity preservation measures are features commonly found in the MOEAs which are computationally expensive. Efforts have been introduced to cut down on the computational burden, see ,e.g., [46]. In this case, the algorithm seeks to find the entire Pareto optimal solution set or, at least a representative subset of it. Pareto optimal solution sets are often preferred to single solutions because they can be practical when considering real-life problems since the final solution of the decision maker is always a trade-off.

Over the years several different alternative techniques have been proposed, some of which will be discussed below. However, identifying the entire Pareto optimal set, for many multi-objective problems, is practically impossible due to its size and, depending on the techniques, also due to its shape and properties. In addition, for many problems, especially for combinatorial optimization problems, proof of solution optimality is computationally infeasible. Thus, the approach commonly used is to search for a set of solutions that represent the Pareto optimal set. Therefore, such an approach should find a set of Pareto solutions such that [81]:

1. it is as close as possible to the true Pareto front,
2. the solutions are uniformly distributed and diverse, to cover a wide range of trade-offs,
3. it includes solutions at the extreme ends of the objective function space.

However, these characteristics are conflicting since (for a given computational time limit) the first one requires deep search on a particular region of the search space, the second leads to a distributed search effort, while the third requires the search to be directed to the extremes.

10.3 Multi-Objective Evolutionary Algorithms

After the seminal work of Schaffer [67, 68], a substantial number of different Multi-Objective Evolutionary Algorithms (MOEAs) have been proposed. Good reviews about this subject have been prepared by [14, 11]. Usually, these algorithms can be divided into three classes.

The first class is based on non-Pareto approaches, including techniques such as aggregating functions [14] and VEGA (Vector Evaluated Genetic Algorithm) [67]. In these cases, the decision maker's preferences are stated before the search (a priori), and the solution obtained is a single point. These techniques do not incorporate directly the concept of Pareto optimum, are unable to find some portions of the Pareto front, and are only capable of handling a small number of objectives. However, they are easy to implement.

The second class emerged after Goldbergs suggestion [38] that selection should be made using a non-dominated ranking scheme and that diversity should be maintained with the use of a sharing function, being based on the concept of Pareto Optimality. Some examples of such approach are referred to next. The algorithm proposed in [31] (MOGA) uses a ranking scheme where the rank of each individual corresponds to the number of individuals in the current population by which it is dominated. Fitness sharing is used in order to maintain diversity, together with a mating restriction scheme that avoids crossover between very distant individuals in the search space. Latter, Srinivas and Deb [74] implemented a Pareto based ranking scheme in the Non-dominated Sorting Genetic Algorithm (NSGA). They sort the population in various fronts. The non-dominated individuals belonging to the first front are more fit, hence they are removed from the population and the process is repeated until the entire population is classified. Then, Horn et al. [44] proposed a different algorithm (NPGA) that uses a tournament selection scheme based on the concept of Pareto dominance.

Lately, a third class of MOEAs based on the use of an elite-preserving operator, that suppresses the deterioration of the population fitness along the successive generations, has been proposed. These algorithms perform sequentially the three basic tasks of fitness assignment, density estimation and archiving. Deb and co-authors ([15, 17]) suggested an elitist non-dominated sorting GA (known as NSGA-II). The method uses simultaneously an elite preservation strategy and an explicit diversity preserving mechanism. First, an offspring population is created using the parent population, both of size N. These populations are combined together to form a single population of size $2N$. Then, a classification of the population using a non-dominated sorting is performed. Finally, the new population is filled with the individuals of the best fronts, until its size becomes equal to N. If the population becomes larger than N, a niching strategy is used to select the individuals of the last front. The algorithm proposed by Zitzler and Thiele [83], called Strength Pareto EA (SPEA), introduces elitism by maintaining an external population. Initially, a random population of size N and an empty external population of size N_e are created. In each generation the solutions belonging to the best front are copied to the external population. Then, the dominated solutions found in this modified population are deleted. When the number of solutions in the external population exceeds N_e, a clustering algorithm is used to eliminate the more crowded solutions. This algorithm was modified recently, in order to incorporate a fine-grained fitness assignment strategy, a density estimation technique and an enhanced archive truncation method - the SPEA2 algorithm [82]. Corne et al. [12] proposed PESA (Pareto Envelope-based Selection Algorithm), which uses a small internal population and a larger external population. Initially, an internal population and an empty external population are created. Then, the non-dominated points of the internal population are incorporated in the external population. When a stop criterion is reached, the result will be the non-dominated individuals of the external population. Otherwise, the individuals of the internal population are deleted and new ones are created by crossover and mutation, using as parents the individuals of the external population. Finally, Knowles and Corne [51] introduced an algorithm based on the use of an (1+1) evolution strategy and of an external archive of all the non-dominated solutions. Diversity is maintained though the use of an adaptive grid technique, which is based on a new crowding procedure where the objective space is divided recursively. According to the authors this technique has lower computational cost and the setting of the niche-size parameter is carried out in an adaptive mode [51].

10.3.1 Reduced Pareto Set Genetic Algorithm (RPSGA)

As stated before, in MOEAs only the selection phase of the traditional EA must be changed in order to be possible to deal with the multiple objectives (Figure 10.5).

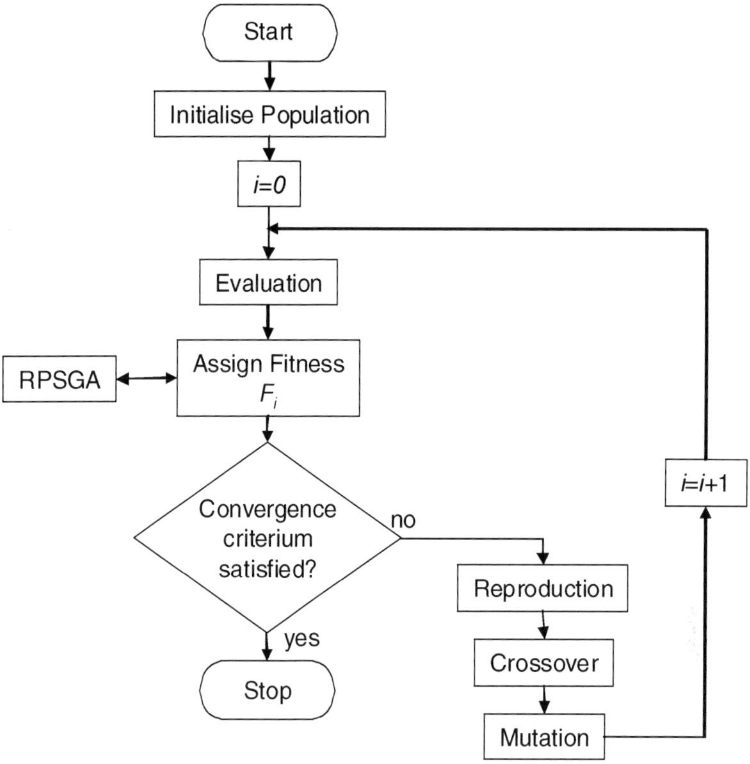

Fig. 10.5 Flowchart of a MOEA.

In this work a MOEA developed previously by one of the authors will be adopted to optimize the processes described in the next section. This algorithm is named Reduced Pareto Set Genetic Algorithm (RPSGA), involving the use of a clustering technique to reduce the number of solutions on the Pareto front [34, 36]. The main steps of this algorithm are illustrated below (Algorithm 1).

Algorithm 1

1. Generate a random initial population (internal).
2. Create an empty external population.
3. **While** not Stop-Condition **Do**
   a) Evaluate internal population.
   b) Calculate the Fitness of the individuals using clustering.
   c) Copy the best individuals to the external population.

d) **If** the external population becomes full **then**
Apply the clustering to this population.
Copy the best individuals to the internal population.
End If

e) Select the individuals for reproduction.

f) Apply crossover.

g) Apply mutation.

End While

Initially, an internal population of size N is randomly generated (step 1) and an empty external population created (step 2). At each generation, i.e., while a stop condition is not found (step 3) the following operations are performed in turn: i) The internal population is evaluated using the modeling routine (step 3a); ii) A clustering technique is applied to reduce the number of solutions on the efficient front and to calculate the fitness of the individuals of the internal population (step 3b) [36]; iii) A fixed number of best individuals are copied to an external population (step 3c); iv) If the external population is not totally full, the genetic operators of selection (step 3e), crossover (step 3f) and mutation (step 3g) are applied to the internal population to generate a new better population; v) When the external population becomes full (step d) the clustering technique is applied to sort the individuals of the external population, and a pre-defined number of the best individuals are incorporated into the internal population by replacing lowest fitness individuals. Detailed information about this algorithm can be found elsewhere [34, 36]. The influence of some important algorithm parameters, such as the size of the internal and of the external populations, the number of individuals copied to the external population and from the external population (to the internal population) in each generation, and the limits of the indifference of the clustering technique have been studied, see [36] for further details.

10.3.2 Recent Developments

Often, solving real optimization problems is very complex since the system performance and characteristics are influenced by more than one field of knowledge and, in addition, requires the use of powerful computational tools. Good examples of such type of problems are the optimization and design of, amongst others, aeroplanes, automobiles and building structures. The traditional way of tackling this type of problems consists of using approximation and decomposition techniques to split a problem into simpler blocks, which are individually solved. A global solution, to the original problem, is then obtained by integrating the solutions to the simpler blocks. This type of approach does not satisfy the actual needs as far as the increasing cost of the design life cycle is concerned. Simultaneously, the high efficiency of the numerical methods available for analyzing specific engineering problems (e.g., computational fluid dynamics and structural mechanics) and the existence of high performance computers enabled the possibility of numerically solving such problems. However, these advantages must go together with the development of more efficient and advanced approaches. Nevertheless, some difficulties persists. First, the result of a MOEA is a set of solutions, but in real problems only a single solution can be used. As a consequence, it will be necessary to provide additional information regarding the relative importance of every objective on the system. This is usually, accomplished by introducing, in the optimization system, the preferences of a Decision Maker [14]. Although some recent work enabled the development of an efficient decision making methodology based on the preferences of the decision maker [28, 29], additional developments are still needed.

In addition, robustness of the solutions should be seek [35, 27] since in real applications small changes on the design variables or on the environmental parameters may happen or may be imposed. Such changes should not affect, or at least affect only slightly, the quality of the proposed solution. Changes in a problem under consideration may arise due to several reasons:

1. parameter values may change due to, for example, data noise (originated by sources such as sensors) or environmental changes,
2. change in the design variables magnitude or on themselves (some may become parameters and new ones may appear),
3. uncertainty on or approximation of some values (parameters, assessment function, etc.),
4. dynamic nature of the problem or evaluation criteria.

From the above, it is clear that robustness is an important aspect to consider during optimization, nevertheless it is rarely included in traditional algorithms. Recently, problems of the second category have been addressed by the authors [35, 27].

One of the major difficulties in applying MOEAs to real problems is the large number of evaluations of the objective functions needed to obtain an acceptable solution - typically of the order of several thousands. Often these are time-consuming evaluations obtained by solving numerical codes with expensive methods like finite-differences or finite-elements. Therefore, reducing the number of evaluations needed, to reach an acceptable solution, is thus of major importance. Finding good approximate methods is even harder for multi-objective problems due to the number of objectives and to the possible interactions between them. Two different efficient methodologies were recently proposed, an Inverse Artificial Neural Network (IANN) approach [37] and a hybrid algorithm based on the use of a filter method as local search procedure [57].

Finally, another important issue that needs to be addressed is dimension of multi-objective problems since as the number of objectives grows, the number of incomparable solutions also grows. Therefore, the problem becomes much more difficult to solve, regarding the point of view of the EAs, since a large number of solutions move from one generation to the next, reducing the selection pressure. Moreover, with more than two objectives, the visualization of the compromises between different solutions becomes extremely complex. In order to tackle real problems with several objectives, it is necessary to investigate ways of reducing the number of objectives. Several approaches based on statistical techniques are reported in the literature [13].

Clearly, the research in multidisciplinary design optimization methodologies and its application to multi-objective engineering problems requires expertise in optimization (e.g., optimization methods, decision support, solution robustness, reduction of the computational requirements, and reduction of the number of objectives) and engineering tools (e.g., computational fluid dynamics, aerodynamics, structural mechanics, aesthetics, etc.). Further research in MOEAs will, certainly, encompass methodologies for dealing and linking these different tools.

10.4 Applications

The very large number of papers published in the last few years, either on international journals or conferences, dealing with applications of MOEAs can be considered a measure of its importance both for the practitioner and scientific communities. A good example is the book by Coello and Lamont [8], where applications on the areas of engineering, industry, economics and management, science and others have been presented.

10.4.1 Engineering

The use of MOEAs in engineering has been quite extensive. Therefore, engineering applications include many different problems, such as: design of welded beams, bulk carriers, airfoil, industrial magnetic devices, optimization of ground water monitoring networks, combinatorial logic circuits, autonomous vehicles navigation, control systems, polymer extrusion problems, truss optimization, city and regional planning, covering tour problem, routing and supersonic wings. Several of the above mentioned applications are described in [8].

Recent works still report new problems or improved methodologies for problems previously addressed. Examples of such works are, for example, the works by Gaspar-Cunha and Covas [36], Gong et al. [39] and Herrero et al. [41], to name just a few.

In [36] an automatic optimization methodology of the polymer extrusion process, using a Multiobjective Optimization Genetic Algorithms approach is proposed. The Reduce Pareto Set Genetic Algorithm with Elitism (RPSGAe), see [34], was applied to the optimization of the operating conditions and to the screw design of a polymer extrusion process, to automatically optimize, in terms of prescribed attributes, the values of important parameters, such as operating conditions and screw geometry. The results obtained for specific case studies have physical meaning and correspond to a successful optimization of the process.

In [39] it is proposed a differential evolution algorithm that adopts the orthogonal design method with quantization technique to generate the initial archive and evolutionary population. In order to keep the nondominated solutions found the authors use a secondary population, which is updated by a new relaxed form of Pareto dominance, at each generation. The authors found their method to be capable of yielding a good coverage and convergence to the true Pareto-optimal fronts for four engineering design problems, namely: Two-bar truss design, Welded beam design, Speed reducer design, and Disc brake design.

Herrero et al. [41] address the problem of designing nonlinear tracking filters under up to several hundreds performance specifications. The suitability of different evolutionary computation techniques for solving such a problem has been analyzed. The authors have found that the application of different MOEA techniques can lead to very different performances. Based on the results obtained and after trying several combination strategies to build the fitness function, the authors propose to build a fitness function based on an operator that selects worst cases of multiple specifications in different situations. Results are obtained for the design of an air traffic control (ATC) tracking filter that should accomplish a specific normative with 264 specifications. The results show good performance, both in terms of effectiveness and computational load.

Let us now illustrate the use of a MOEA to solve a feature extraction problem. The aim here is not to present all the details, but rather show how the RPSGAe algorithm can be used to solve such a problem. The problem to be addressed is a classification problem that can be solved using different methods such as logistics, support vector machines and artificial neural networks. The objective is to find the minimum number of features needed to obtain the maximum accuracy of the companies evaluation. Examples of the features considered are the number of employees, the fixed assets, the current ratio, the liquidity ratio, and the stock turnover days. For that purpose we consider a database with 30 features or attributes characterizing 1200 different companies, for a given year. Regarding the company evaluation we measure whether the company has survived or gone into bankruptcy. In the present study, and for illustration purposes, we have used the logistics and support vector machines methods with a gradient descent and holdout validation, having learning rate and training fraction of 0.01 and 0.6, respectively. The runs performed used the following RPSGA parameters (see [36] for more details): the main and the elitist populations had 100 and 200 individuals, respectively; a roulette wheel selection strategy was adopted; the crossover and mutation probabilities were, respectively, set to 80% and 5%, the number of ranks and the limits of indifference for the clustering technique were chosen to be 30 and 0.01, respectively. The results obtained are reported in Figure 10.6. As it can be seen, 100 generations of evolution lead to a considerable gain in accuracy while decreasing significantly the number of features needed. On the final population only 4 non-dominated solutions exist having, respectively 2, 3, 5 and 6 features, which are identified in Figure 10.7.

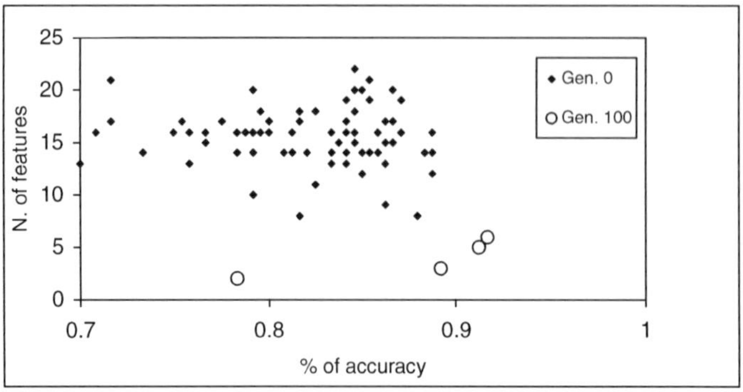

Fig. 10.6 MOEA results: initial population and non-dominated solutions after 100 generations.

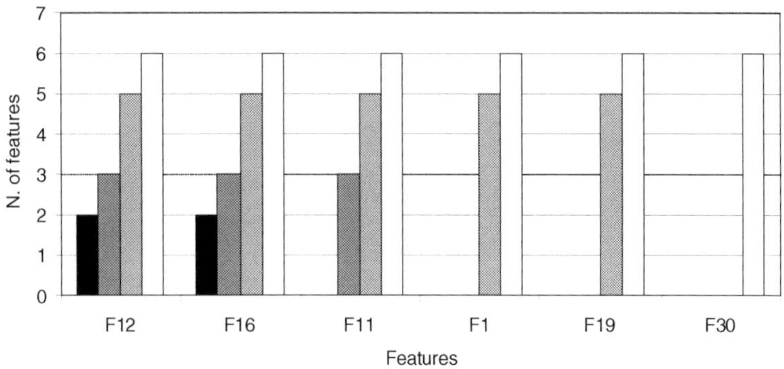

Fig. 10.7 Features used in the 4 non-dominated solutions after 100 generations.

10.4.2 Industrial

Industrial applications of EAs have been the subject of research for many years. Both the single-objective and multi-objective versions of several problem types have been addressed in the literature. In here, we look at recent work on multi-objective versions of cellular manufacturing systems, balancing assembly lines, and scheduling problems.

Cellular manufacturing attempts to bring the benefits of mass production to job-shop production systems. Recent surveys, see e.g., [79] indicate that the practical implementation of cellular manufacturing systems involves many conflicting objectives, however most of the literature is on the single-objective version of the problem. Even when the existence of multiple objectives is acknowledge the proposed solution methodologies, typically, aggregate objectives into a single one, see for example the reviews by Mansouri [56] and Dimopoulos [19]. A recent exception, although others exist, is that of Dimopoulos [20]. In this work the author proposes multi-objective genetic programming single-linkage cluster analysis (GP-SLCA) for the solution of the multi-objective cell-formation problem. The Multi-objective GP-SLCA combines an existing algorithm for the solution of single-objective cell formation problems [21] with NSGA-II, an elitist evolutionary multi-objective optimization technique, in order to automatically generate a set of non-dominated solutions for a given multi-objective cell-formation problem. Although the results presented in the article indicate that multi-objective GP-SLCA is a promising approach to the solution of the multi-objective cell-formation problem, the experimental basis that exists for this problem is small, and consequently there are no extensive comparative results.

The assembly line balancing problem (ALBP) is a decision problem arising when an assembly line has to be (re)-configured, and consists of determining the optimal partitioning of the assembly work among the workstations in accordance with some objectives [71]. These objectives usually take one of two forms: i) either minimising the number of workstations given the cycle time of the line, or ii) minimising the cycle time given the number of workstations. The multi-objective ALBP has attracted a considerable research attention in the last decade. There is a lack in the literature regarding the use of EAs for solving the multi-criteria ALBP of type ii above. Furthermore, as point out by Scholl and Becker [72], the computational testing of most EAs has been performed ignoring existing ALBP test bed. In a recent work, Nearchou [60] presents a differential evolution based approach , inspired on that of Murata [58], for solving the bi-criteria ALBP. The main objective was to minimize the cycle time of the line and secondary objectives to minimize balance delay time and workload smoothness index. The proposed method formulates the cost function of each individual solution as a weighted-sum of multiple objectives functions with self-adapted weights. It maintains and updates a separate population with diverse Pareto-optimal solutions, injects the actual evolving population with some Pareto-optimal solutions to preserve non-dominated solutions found over generations. The encoding scheme used maps real-valued vectors to integer strings corresponding to feasible solutions. The computational results reported are for benchmark problems taken from the open literature and are compared to that of two other previously proposed methods, namely, a weighted sum Pareto GA [58], and a Pareto-niched GA [50].

Production scheduling problems have been researched for many years, however the literature on multi-objective scheduling is notably sparser than on single-objective scheduling. The interest in multi-objective production scheduling, especially in the multi-objective deterministic problem has been sparked by some recent surveys. Particularly after the survey by Nagar et al. [59], which provides a good review of the research on this type of problems up to the early 1990s. They discuss the single machine bi-criteria problem, the single machine multi-criteria problem, the multiple machine bi-criteria problem, and the multiple machine multi-criteria problem. Later, Tkindt et al. [76] present a discussion on the one-machine job shop, the parallel-machine job shop, and the flow shop. For these problems more than 100 published papers have been listed. In a recent work, Hoogeveen [43] looks closer to the earliness-tardiness scheduling and the scheduling with the controllable processing time. More recently Lei [54] looks into multi-objective scheduling after 1995 and an extensively list of papers is provided. The author classifies the scheduling problems based on the nature of the problem, shop configuration, and the description method of uncertainty. Then, the main characteristics of the previous research are summarized, and finally the new trends in scheduling with multiple objectives are pointed out. The author also reviews some less researched problems.

10.4.3 Economics and Management

Economics and management are very promising research areas to apply Evolutionary Algorithms (EAs) for two kinds of reasons. On the one hand, problems within these areas are quite difficult, see for example Schlottmann and Seese [70] for proof of NP-completeness of some financial problems. On the other hand, although the use of EAs in these areas is not an emerging research area [7], the use of multi-objective EAs is still scarce and not many different problems have yet been addressed.

Regarding the use of MOEAs several problems have been addressed, such as time series forecasting, stock ranking, economic exploitation levels, risk management, forest management, space allocation. Recently, very good surveys have been written on applications of MOEAs to problems in economics and management, see [69, 24, 73, 75]. Here we only mention a few new approaches to the portfolio optimization problem, since it has been the most popular, just to show how active is the research on finding solutions to problems in economics and management using MOEAs.

The portfolio optimization problem is a well-known difficult problem occurring in the financial world. In this problem a collection of assets is chosen to be held by an institution or a private individual. The choice is done such that the expected return (mean profit) is maximized, while at the same time the risk is to be minimized. Since the optimal solution depends on the users risk aversion, various trade-offs are usually seek. In [6] the authors propose an approach that integrates an active set algorithm optimized for portfolio selection into a multi-objective evolutionary algorithm (MOEA). The MOEA provides some convex subsets of the set of all feasible portfolios and then a critical line algorithm is solved for each subset. Finally, the partial solutions are merged to obtain the solution to the original non-convex problem. The authors were able to show that their envelope-based MOEA significantly outperforms existing MOEAs, when solving some benchmark problems.

In another recent work Li proposes a multi-objective genetic programming system [55]. This system improves on the previous one, in two different ways, by taking advantage of the MOEAs. One the one hand, it improves on efficiency since a set of Pareto front solutions is obtained in one single execution. On the other hand, from the users perspective, it is simpler as it eliminates a number of user-supplied parameters previously required.

Other problems such as forest management have also been addressed by evolutionary algorithms, see for example [25] and the references therein.

10.4.4 Other Applications

In this section we include several application types that we designate by others. Such applications include for example spectroscopic data analysis, medical image processing, computer-aided diagnosis, treatment planning, machine learning, selection of attributes in data mining, regression and prediction, series forecasting, and biometric applications.

Since here we refer to several applications areas where vast research has been occurring we refer the reader to some recent surveys. In [62] the authors provide an overview of the application of evolutionary computation in the medical domains. First, six types of evolutionary algorithms are outlined (genetic algorithms, genetic programming, evolution strategies, evolutionary programming, classifier systems, and hybrid systems). Then their application to obtain solutions to medical problems, including diagnosis, prognosis, imaging, signal processing, planning, and scheduling, is discussed. Finally, the authors provide an extensive bibliography, classified both according to the medical task addressed and according to the evolutionary technique used. Another review is provided by Handl et al. [40]. In this work, the application of multi-objective optimization in the fields of bioinformatics and computational biology is reviewed. The literature reviewed in the survey, over 140 papers, is arranged by biological problem domain, namely classification problems (unsupervised, supervised, and semisupervised), inverse modeling problems (i.e. problems were it is intended to infer the original system from the observed data), sequence and structure alignment (which involves assessment of similarity and the identification of related sequences or structures), structure prediction and design, and system optimization and experimental design (investigate the degree of optimality of naturally occurring biochemical systems or to design optimal biochemical processes). Another important issue studied in this survey is the identification of five distinct problem contexts, giving rise to multiple objective problems.

A very interesting work, regarding the use of MOEAs, is that of Hruschka et al. [45]. This work provides an up-to-date overview of the use of evolutionary algorithms for clustering problems in different domains, such as image processing, computer security, and bioinformatics. It also provides a taxonomy that highlights some very important aspects in the context of evolutionary data clustering, namely, fixed or variable number of clusters, cluster-oriented or nonoriented operators, context-sensitive or context-insensitive operators, guided or unguided operators, binary, integer, or real encodings, centroid-based, medoid-based, label-based, tree-based, or graph-based representations, among others. The paper ends by addressing some important issues and open questions that can be subject of future research.

Two very successful scientific disciplines using evolutionary algorithms are data mining and machine learning. Data mining has emerged as a major research domain in the recent decades to extract implicit and useful knowledge. Initially, this knowledge extraction was computed and evaluated manually using statistical techniques. Subsequently, semi-automated data mining techniques emerged because of the advancement in technology. Such advancement was also in the form of storage which increased the demands for analysis. In such case, semi-automated techniques have become inefficient. Therefore, automated data mining techniques were introduced to synthesis knowledge efficiently. A critical literature review, highlighting the strengths and limitations of the automated techniques is given by Asghar and Iqbal [2]. Machine learning is concerned with the design and development of algorithms that allow computers to learn based on data. Therefore a research major focus is to automatically learn to recognize complex patterns and make intelligent decisions based on data. Machine learning is inherently multi-objective, since many of the applications where it is used are multi-objective and involve solving hard problems. However, until recently either only one of the objectives was adopted as the cost function or the multiple objectives were aggregated to a scalar cost function. Recently, this has been changing, mainly due to the great success of MOEAs [53]. MOEAs are now being used within machine learning techniques in order to find a number of non-dominated rule sets with respect to the several objectives, see for example the reviews in [48, 32]. For a more detailed account of the existing research on multi-objective learning, the reader is referred to [47].

10.5 Conclusions

This paper has presented a brief review of algorithms in the rapid growing area of Multi-objective Evolutionary Algorithms (MOEAs), as well as, some of their applications. Regarding the algorithms, their self robustness seems to be one of the main issues, since the conditions under which the solutions have been obtained are unlikely to be exactly the ones to be found during the implementation and usage of the method. This may happen due to several reasons, namely: data noise, changes in the design variables, environmental changes, quality measures or other changes with time, etc. Therefore, it is clear that it is an important aspect to be considered during optimization. However, it is rarely included in traditional algorithms.

Another important issue, that is still of major concern, is algorithmic efficiency. Recent research is looking at parallel implementation as a possible solution. Therefore, more in depth and detailed studies of the different aspects involved in parallelization need to be performed.

A recent trend, that most likely will continue to grow further is the use of nature inspired techniques, such as particle swarm optimization [10, 9, 65], differential evolution [1, 3, 80], ant colony systems [22, 23, 33], electromagnetism [5, 18], amongst others.

Reformulating some real problems, which are currently addressed as if they only have a single objective is likely to be one of the probable future trends in MOEAs. Thus, the research efforts should not only be put into the development of new algorithms but also on the adaptation of existing algorithms to new applications.

The main idea we would like to leave the reader with is that Evolutionary Algorithms are a viable alternative to solve difficult real-world problems in a reasonable amount of time. Sometimes, they might even be the only alternative providing good results. Given their heuristic nature there are no guarantees on the solution quality.

However, there is an overwhelming evidence showing their effectiveness to address complex real-world problems when compared to other heuristics, regardless of being deterministic or stochastic.

References

1. H.A. Abbass and R. Sarker. The Pareto differential evolution algorithm. *International Journal on Artificial Intelligence Tools*, 11(4):531–552, 2002.
2. S. Asghar and K. Iqbal. Automated data mining techniques:a critical literature review. In *Proceedings of the International Conference on Information Management and Engineering*, pages 75–76, 2009.
3. B.V. Babu and M.M.L. Jehan. Differential evolution for multi-objective optimization. In *The 2003 Congress on Evolutionary Computation*, volume 4, 2003.
4. D. Beasley. Possible applications of evolutionary computation. In T. Bäck, D.B. Fogel, and Z. Michalewicz, editors, *Handbook of Evolutionary Computation*, pages 4–19. IOP Publishing Ltd. and Oxford University Press, 1997.
5. S.I. Birbil and S.C. Fang. An electromagnetism-like mechanism for global optimization. *Journal of Global Optimization*, 25:263–282, 2003.
6. J. Branke, B. Scheckenbach, M. Stein, K. Deb, and H. Schmeck. Portfolio optimization with an envelope-based multi-objective evolutionary algorithm. *European Journal of Operational Research*, 199(3):684–693, 2009.
7. S.H. Chen. *Evolutionary Computation in Economics and Finance*. Physica-Verlag, 2002.
8. C.A.C. Coello and G.B. Lamont. *Applications of Multi-Objective Evolutionary Algorithms*. World Scientific, 2004.
9. C.A.C. Coello and M.S. Lechuga. MOPSO: a proposal for multiple objective particle swarm optimization. In *Proceedings of the 2002 Congress on Evolutionary Computation, 2002*, volume 2, pages 1051–1056, 2002.
10. C.A.C. Coello, G.T. Pulido, and M.S. Lechuga. Handling multiple objectives with particle swarm optimization. *IEEE Transactions on Evolutionary Computation*, 8(3):256–279, 2004.
11. C.A.C. Coello, D.A. Van Veldhuizen, and G.B. Lamont. *Evolutionary Algorithms for Solving Multi-Objective Problems*. Kluwer, Dordrecht, 2002.
12. D.W. Corne, J.D. Knowles, and M.J. Oates. The Pareto-Envelope based Selection Algorithm for Multiobjective Optimisation. In *Proceedings of the Parallel Problem Solving from Nature VI (PPSNVI)*, 2000.
13. L. Costa and P. Oliveira. Dimension reduction in multiobjective optimization. *PAMM*, 7(1):2060047–2060048, 2007.
14. K. Deb. *Multi-objective Optimization Using Evolutionary Algorithms*. John Wiley & Sons, Chichester, 2001.
15. K. Deb, S. Agrawal, A. Pratap, and T. Meyarivan. A fast elitist non-dominated sorting genetic algorithm for multi-objective optimisation: NSGAII. In *Proceedings of the Parallel Problem Solving from Nature VI (PPSNVI)*, pages 849–858, 2000.
16. K. Deb and D.E. Goldberg. An investigation of niche and species formation in genetic function optimization. In *Proc. Third Int. Conf. on Genetic Algorithms*, 1989.
17. K. Deb, A. Pratap, S. Agrawal, and T. Meyarivan. A fast and elitist multi-objective genetic algorithm: NSGAII. *IEEE Transactions on Evolutionary Computation*, 6:182–197, 2002.
18. D. Debels and M. Vanhoucke. The electromagnetism meta-heuristic applied to the resource-constrained project scheduling problem. *Lecture Notes in Computer Science*, 3871:259–270, 2006.
19. C. Dimopoulos. A review of evolutionary multiobjective optimization applications in the area of production research. In *Congress on Evolutionary Computation (CEC 04)*, pages 1487–1494, 2004.

20. C. Dimopoulos. Explicit consideration of multiple objectives in cellular manufacturing. *Engineering Optimization*, 39(5):551–565, 2007.
21. C. Dimopoulos and N. Mort. Evolving knowledge for the solution of clustering problems in cellular manufacturing. *International Journal of Production Research*, 42(19):4119–4133, 2004.
22. K.F. Doerner, W.J. Gutjahr, R.F. Hartl, C. Strauss, and C. Stummer. Pareto ant colony optimization: A metaheuristic approach to multiobjective portfolio selection. *Annals of Operations Research*, 131(1):79–99, 2004.
23. K.F. Doerner, W.J. Gutjahr, R.F. Hartl, C. Strauss, and C. Stummer. Pareto ant colony optimization with ILP preprocessing in multiobjective project portfolio selection. *European Journal of Operational Research*, 171(3):830–841, 2006.
24. E.I. Ducheyne, B. De Baets, and R.R. De Wulf. *Applications of Multi-Objective Evolutionary Algorithms*, volume 1 of *Advances in Natural Computation*, chapter Even Flow Scheduling Problems in Forest Management, pages 701–726. World Scientific Pub Co Inc, 2004.
25. E.I. Ducheyne, R.R. De Wulf, and B. De Baets. A spatial approach to forest-management optimization: linking GIS and multiple objective genetic algorithms. *International Journal of Geographical Information Science*, 20(8):917–928, 2006.
26. F.Y. Edgeworth. *Mathematical Physics*. P. Keagan, London, England, 1881.
27. J.C. Ferreira, C.M. Fonseca, J.A. Covas, and A. Gaspar-Cunha. *Advances in Evolutionary Algorithms*, chapter Evolutionary Multi-Objective Robust Optimization, pages 261–278. I-Tech Education and Publishing, 2008.
28. J.C. Ferreira, C.M. Fonseca, and A. Gaspar-Cunha. Methodology to select solutions from the Pareto-optimal set: A comparative study. In *GECCO 2007, Genetic and Evolutionary Computation Conference*, 2007.
29. J.C. Ferreira, C.M. Fonseca, and A Gaspar-Cunha. Selection of solutions in a multi-objective environment: Polymer extrusion a case study. In *Evolutionary Methods For Design, Optimization and Control*, 2008.
30. L.J. Fogel, A.J. Owens, and M.J. Walsh. *Artificial Intelligence Through Simulated Evolution*. Wiley, New York, 1996.
31. C.M. Fonseca and P.J. Fleming. Genetic algorithms for multiobjective optimization: Formulation, discussion and generalization. In S. Forrest, editor, *Multiple Criteria Problem Solving*, pages 416–423. Morgan Kauffman Publishers, San Mateo, California, 1993.
32. L. Galway, D. Charles, and M. Black. Machine learning in digital games: A survey. *Artificial Intelligence Review*, 29(2):123–161, 2009.
33. C. Garcia-Martinez, O. Cordon, and F. Herrera. A taxonomy and an empirical analysis of multiple objective ant colony optimization algorithms for the bi-criteria TSP. *European Journal of Operational Research*, 180(1):116–148, 2007.
34. A. Gaspar-Cunha. *Modeling and Optimisation of Single Screw Extrusion*. PhD thesis, University of Minho, Guimares, Portugal, 2000.
35. A. Gaspar-Cunha and J.A. Covas. Robustness in multi-objective optimization using evolutionary algorithms. *Computational Optimization and Applications*, 39(1):75–96, 2008.
36. A. Gaspar-Cunha and Covas J.A. *Metaheuristics for Multiobjective Optimisation*, chapter RPSGAe - A Multiobjective Genetic Algorithm with Elitism: Application to Polymer Extrusion, pages 221–249. Lecture Notes in Economics and Mathematical Systems. Springer, 2004.
37. A. Gaspar-Cunha and A.S. Vieira. A hybrid multi-objective evolutionary algorithm using an inverse neural network. In *Hybrid Metaheuristics (HM 2004) Workshop at ECAI 2004*, 2004.
38. D.E. Goldberg. *Genetic Algorithms in Search, Optimization and Machine Learning*. Addison-Wesley Longman Publishing Co., Inc., Boston, MA, USA, 1989.
39. W.Y. Gong, Z.H. Cai, and L. Zhu. An efficient multiobjective differential evolution algorithm for engineering design. *Structural And Multidisciplinary Optimization*, 38(2):137–157, 2009.
40. J. Handl, D.B. Kell, and J. Knowles. Multiobjective optimization in bioinformatics and computational biology. *IEEE/ACM Transactions On Computational Biology And Bioinformatics*, 4(2):279–292, 2007.

41. J.G. Herrero, A. Berlanga, and J.M.M. Lopez. Effective evolutionary algorithms for many-specifications attainment: Application to air traffic control tracking filters. *IEEE Transactions on Evolutionary Computation*, 13(1):151–168, 2009.
42. J.H. Holland. *Adaptation in Natural and Artificial Systems*. University of Michigan Press, 1975.
43. H. Hoogeveen. Multicriteria scheduling. *European Journal of Operational Research*, 167:592–623, 2005.
44. J. Horn, N. Nafpliotis, and D.E. Goldberg. A niched Pareto genetic algorithm for multiobjective optimization. In *Proceedings of the First IEEE Conference on Evolutionary Computation*, 1994.
45. E.R. Hruschka, R.J.G.B. Campello, A.A. Freitas, and A.C.P.L.F. de Carvalho. A survey of evolutionary algorithms for clustering. *IEEE Transactions On Systems, Man, and Cybernetics Part C*, 39(2):133–155, 2009.
46. M.T. Jensen. Reducing the run-time complexity of multi-objective EAs: The NSGA-II and other algorithms. *IEEE Transactions on Evolutionary Computation*, 7(5):503–515, 2003.
47. Y. Jin, editor. *Multi-Objective Machine Learning*. Springer-Verlag, New York, 2006.
48. Y. Jin and B. Sendhoff. Pareto-based multiobjective machine learning: An overview and case studies. *IEEE Transactions On Systems, Man, and Cybernetics Part C*, 38(3):397–415, 2008.
49. R.L. Keeney and H. Raiffa. *Decision with Multiple Objectives. Preferences and Value Trade-offs*. Cambridge University Press, Cambridge, UK, 1993.
50. Y.K. Kim, Y.-J. Kim, and Y. Kim. Genetic algorithms for assembly line balancing with various objectives. *Computers and Industrial Engineering*, 30(3):397–409, 1996.
51. J.D. Knowles and D.W. Corne. Approximating the non-dominated front using the Pareto archived evolutionary strategy. *Evolutionary Computation Journal*, 8(2):149–172, 2000.
52. J. Koza. *Genetic Programming: On the Programming of Computers by Means of Natural Selection*. MIT Press, 1992.
53. R. Kumar. *Applications of Multi-Objective Evolutionary Algorithms*, chapter On machine learning with multiobjective genetic optimization, pages 393–425. World Scientific, 2004.
54. D. Lei. Multi-objective production scheduling: A survey. *International Journal on Advanced Manufacturing and Technology*, 43:926–938, 2009.
55. J. Li. Enhancing financial decision making using multi-objective financial genetic programming. In *In IEEE Congress on Evolutionary Computation*, 2006.
56. S.A. Mansouri, S.M.M. Husseini, and Newman. S.T. A review of the modern approaches to multi-criteria cell design. *International Journal of Production Research*, 38(5):1201–1218, 2000.
57. F. Mendes, V. Sousa, M.F.P. Costa, and A. Gaspar-Cunha. Multi-objective memetic algorithm using pattern search filter methods. In *EU/Meeting 2009*, 2009.
58. T. Murata, H. Ishibuchi, and H. Tanaka. Multi-objective genetic algorithms and its application to flow-shop scheduling. *Computers and Industrial Engineering*, 30(4):957–968, 1996.
59. A. Nagar, J. Haddock, and S. Heragu. Multiple and bicriteria scheduling: A literature survey. *European Journal of Operational Research*, 81:88–104, 1995.
60. A.C. Nearchou. Multi-objective balancing of assembly lines by population heuristics. *International Journal of Production Research*, 46(8):2275–2279, 2008.
61. V. Pareto. *Cours dEconomie Politique*. vol. I and II, F. Rouge, Lausanne, 1896.
62. C.A. Peña Reyes and M. Sipper. Evolutionary computation in medicine: An overview. *Artificial Intelligence in Medicine*, 19:1–23, 2000.
63. L. Rachmawati and D. Srinivasan. Preference incorporation in multi-objective evolutionary algorithms: A survey. In *IEEE Congress on Evolutionary Computation, 2006*, pages 962–968, 2006.
64. I. Rechenberg. *Evolutionstrategie: Optimierung Technisher Systeme Nach Prinzipien der Biologischen Evolution*. Frommann-Hoolzboog Verlag, 1973.
65. M. Reyes-Sierra and C.A.C. Coello. Multi-objective particle swarm optimizers: A survey of the state-of-the-art. *International Journal of Computational Intelligence Research*, 2(3):287–308, 2006.

66. T.L. Saaty. *Multicriteria Decision Making: The Analytic Hierarchy Process - Planning, Priority Setting, Resource Allocation.* RWS Publications, Pittsburgh, PA, 1990.
67. J.D. Schaffer. *Multiple Objective Optimization with Vector Evaluated Genetic Algorithms.* PhD thesis, Vanderbilt University, Nashville, Tennessee, 1984.
68. J.D. Schaffer. Multiple objective optimization with vector evaluated genetic algorithms. In *Proceedings of the 1st International Conference on Genetic Algorithms and Their Applications*, pages 93–100. L. Erlbaum Associates Inc. Hillsdale, NJ, USA, 1985.
69. F. Schlottmann and D. Seese. *Applications of Multi-Objective Evolutionary Algorithms*, volume 1 of *Advances in Natural Computation*, chapter Financial Applications of Multi-Objective Evolutionary Algorithms: Recent Developments and Future Research Directions, pages 627–652. World Scientific Pub Co Inc, 2004.
70. F. Schlottmann and D. Seese. *Handbook of Computational and Numerical Methods in Finance*, chapter Modern Heuristics for Finance Problems:A Survey of Selected Methods and Applications, pages 331–360. Birkhäuser, 2004.
71. A. Scholl, editor. *Balancing and Sequencing of Assembly Lines.* Physica-Verlag, 1999.
72. A. Scholl and C. Becker. State of the art exact and heuristic solution procedures for simple assembly line balancing. *European Journal of Operational Research*, 168:666–693, 2006.
73. L.J.D. Silva and E.K. Burke. *Applications of Multi-Objective Evolutionary Algorithms*, volume 1 of *Advances in Natural Computation*, chapter Using Diversity to Guide the Search in Multi-objective Optimization, pages 727–747. World Scientific Pub Co Inc, 2004.
74. N. Srinivas and K. Deb. Multiobjective optimization using nondominated sorting in genetic algorithms. *Evolutionary Computation*, 2:221–248, 1995.
75. M.G.C. Tapia and C.A.C. Coello. Applications of multi-objective evolutionary algorithms in economics and finance: A survey. In *IEEE Congress on Evolutionary Computation, CEC*, pages 532–539, 2007.
76. V. TKindt, J. Billaut, and C. Proust. Multicriteria scheduling problems: A survey. *RAIRO Operations Research*, 35:143–163, 2001.
77. D.A. Van Veldhuizen and G.B. Lamont. Multiobjective evolutionary algorithms: Analyzing the state-of-the-art. *Evolutionary computation*, 8(2):125–148, 2000.
78. P. Vincke. Analysis of MCDA in Europe. *European Journal of Operational Research*, 25:160–168, 1995.
79. U. Wemmerlov and D.J. Johnson. Empirical findings in manufacturing cell design. *International Journal of Production Research*, 38:481–507, 2000.
80. F. Xue, A.C. Sanderson, and R.J. Graves. Pareto-based multi-objective differential evolution. In *Evolutionary Computation, 2003. CEC'03. The 2003 Congress on*, volume 2, 2003.
81. E. Zitzler, K. Deb, and L. Thiele. Comparison of multiobjective evolutionary algorithms: Empirical results. *Evolutionary Computation*, 8(2):173–195, 2000.
82. E. Zitzler, M. Laumanns, and L. Thiele. SPEA2: Improving the strength Pareto evolutionary algorithm. *TIK report no. 103, Swiss Federal Institute of Technology, Zurich, Switzerland*, 2001.
83. E. Zitzler and L. Thiele. Multiobjective evolutionary algorithms: A comparative study and the strength Pareto approach. *IEEE Transactions on Evolutionary Computation*, 3(4):257–271, 1999.

Chapter 11
Goal Programming: From Constrained Regression to Bounded Rationality Theories

Jacinto González-Pachón and Carlos Romero

Abstract The purpose of the paper is to provide a critical overview of the decision-making approach known as Goal Programming (GP). The paper starts by tracing the origins of GP back to work by Charnes and Cooper at the end of the 1950s in fields like non-parametric regression, and the analysis of contradictions in non-solvable linear programming problems. After chronicling its evolution from its original form into a powerful decision-making method, the GP approach is linked with the Simonian bounded rationality theories based upon the "satisficing" concept. In this way, several GP models are presented as fruitful vehicles for implementing this kind of "satisficing" philosophy. The last part of the paper presents some critical issues and extensions of the GP approach. The chapter ends by discussing potential extensions, as well as GP's role for solving complex real-world problems in the near future.

11.1 A Historical Sketch

The original idea of Goal Programming (GP) appears in a paper by Charnes, Cooper and Ferguson published in *Management Science* in 1955 [19]. The paper was aimed at developing an executive compensation formula for a division of a major company (*General Electric*). The need to introduce *a priori* requirements and sign conditions for the coefficients of some variables into the model made it impossible to solve by using classic regression analysis techniques. Given the insufficiency of classic statistical techniques, they formulated a "constrained regression" model, minimizing the sum of the absolute deviations. Since absolute deviation is a non-linear form that cannot be straightforwardly optimised, they linearised the model by introducing

---

Jacinto González-Pachón
Universidad Politecnica de Madrid, Spain e-mail: jgpachon@fi.upm.es

Carlos Romero
Universidad Politecnica de Madrid, Spain e-mail: carlos.romero@upm.es

C. Zopounidis and P.M. Pardalos (eds.), *Handbook of Multicriteria Analysis*, Applied Optimization 103, DOI 10.1007/978-3-540-92828-7_11,

negative and positive deviation variables for the first time in the literature. The seminal value of this paper is enormous for at least two reasons. First, it is the embryo of the future GP methodology, although the term GP is not explicitly used. Second, it is the outset of non-parametric regression methods.

Charnes and Cooper used the term GP explicitly for the first time six years later in Appendix B of their classic book *Management Models and Industrial Applications of Linear Programming* , under the heading of "Basic Existence Theorems and Goal Programming". Paradoxically, the two fathers of GP do not address a proper decision-making problem with multiple goals but analyse the contradictions in non-solvable linear programming problems. In other words, they use GP as an approach for determining a compromise solution to an infeasible linear programming problem.

In the early 1960s, Ignizio [32] had to solve a complex problem in the field of engineering design with the deployment of the antenna system for the Saturn/Apollo moon landing mission. This problem comprehended multiple goals, non-linear functions, as well as integer variables. With the help of a GP formulation, he was able to obtain implementable solutions.

Charnes et al. [20] demonstrated the potential of GP in financial and accounting problems. Ijiri [39] developed mathematical techniques, like the inverse generalized matrix to deal with preemptive priorities. Charnes et al. [18] formulated GP models to plan a company's advertising campaign. Finally, in 1969 Jaaskelainen [40] proposed a GP model for scheduling production, employment and inventories.

Two books published in the seventies and solely devoted to GP [33, 46] had a seminal influence on the development of this approach. In the 1970s other key papers appeared introducing refinements and extensions of the GP approach like: interactive GP, fuzzy GP, interval GP, multidimensional dual, algorithmic improvements, computer codes, etc. All these theoretical efforts fed a real explosion of applied papers, successfully addressing key decision-making problems through GP.

Some surveys and expository presentations of GP are, in chronological order, Ignizio [34], Zanakis and Gupta [67], Romero [53], Schniederjans [58], Tamiz et al. [63], Lee and Olson [47], Aouni and Kettani [3], Jones and Tamiz [43], Ignizio and Romero [38], and Caballero et al. [7].

11.2 Goal Programming and Bounded Rationality Theories

Modern GP is philosophically underpinned by the Simonian concept of satisficing that leads to a bounded rationality theory deeply rooted in psychology. This marks a clear departure from the classical theories based upon a perfect rationality paradigm. It should be noted that the term satisficing does not appear in English dictionaries. "Satisficing" is a Northumbrian term, chosen by Simon, to indicate the DM's desire to get "satisfying" and "sufficient" solutions to many real-world problems [60]. In short, "satisficing" is a merger of the words "satisfying" and "sufficing".

Let us develop the satisficing argument. Simon [59] conjectured that in today's complex organizations the environment is defined by incomplete information, limited resources, conflicts of interests, etc. In such an environment the decision-maker (DM) does not try to maximize anything, much less a well-defined objective function, as is assumed by classic theories of perfect rationality. Quite the contrary, within this kind of realistic environment, the DM tries to get as close as possible to a set of goals as determined by a set of satisficing targets, where satisficing means established figures that are perhaps not the "best" but are satisfactory and sufficient for the decision-making problem under consideration.

Although the Simonian satisficing solution seems to be the most fruitful philosophical groundwork for GP, it is not the only one. Section 11.4 of this chapter illustrates how GP can also be interpreted in terms of classic utility theory.

According to the satisficing philosophy, GP can be defined as an analytical approach devised to address decision-making problems where targets have been assigned to all the attributes and the DM is interested in minimizing, in one way or another, the non-achievement of the respective goals. As a consequence of the satisficing philosophy, the "goodness" of any solution to a decision-making problem is represented by an achievement function rather than a utility function or similar construct. This type of function measures the degree of non-achievement of the defined goals [38]. González-Pachón and Romero [27] give a formal derivation of the link between satisficing logic and GP, attacking the problem axiomatically.

11.3 Some Basic Goal Programming Models

Let us consider a decision-making problem involving goals. The structure of the generic ith goal reads as follows:

$$(g_i) \quad f_i(\mathbf{x}) + n_i - p_i = t_i \tag{11.1}$$

where:

$f_i(\mathbf{x})$ = mathematical expression for the ith attribute (i.e. a function of the vector $\mathbf{x}$ of decision variables).
t_i = target value for the ith attribute; i.e. the achievement level that the DM considers as satisficing for the ith attribute.
n_i = negative deviation variable; i.e. quantification of the under-achievement of the ith goal.
p_i = positive deviation variable; i.e. quantification of the over-achievement of the ith goal.

Once the goals have been formulated, the next step is to detect the unwanted deviation variables. These variables are unwanted in the sense that they are the ones a DM wants to minimize. To illustrate this idea, let us consider the following cases:

1. The goal derives from a "more is better" attribute (i.e., *satisfice* $f_i(\mathbf{x}) \geq t_i$). In this case, the DM does not want under-achievements with respect to target t_i. Consequently, the unwanted deviation variable would be the negative one (n_i) and would have to be minimized.
2. The goal derives from a "less is better" attribute (i.e., *satisfice* $f_i(\mathbf{x}) \leq t_i$). In this case, the DM does not want over-achievements with respect to target t_i. Consequently, the unwanted deviation variable would be the positive one (p_i) and would have to be minimized.
3. The goal derives from an attribute that needs to be achieved exactly (i.e., *satisfice* $f_i(\mathbf{x}) = t_i$). In this case, the DM wants neither over-achievements nor under-achievements with respect to target t_i. Hence, both the negative variable n_i and the positive one p_i are equally unwanted, making it necessary to minimize both deviation variables.

Let us assume that the unwanted deviation variables for a given problem are

$$p_1, n_2, \ldots, n_i, p_i, \ldots, p_q$$

The formulation of a GP model implies the minimization of a function of the former unwanted deviation variables:

$$\text{Min} \quad g(p_1, n_2, \ldots, n_i, p_i, \ldots, p_q) \tag{11.2}$$

The above function has a typical "less is better behaviour" and receives the name of achievement function. The arguments (i.e. the unwanted deviation variables) of (11.2) must be normalized. This type of normalization is required for two different types of reasons. First, the goals are generally measured in different units. Therefore, it makes no sense to apply a mathematical operator like the sum (e.g., to add together kilos of potatoes and pints of beer). Second, the value of the targets might be very different. Hence the minimization of (11.2) can lead to solutions biased towards goal with higher values being held for their targets. Finally, it is also necessary to introduce into (11.2) parameters reflecting the relative importance the DM attaches to the achievement of the different goals. Therefore, the achievement function (11.2) should read as follows:

$$\text{Min} \quad g\left(\frac{W_1 p_1}{K_1}, \frac{W_2 n_2}{K_2}, \ldots, \frac{(W_i n_i, W_i p_i)}{K_i}, \ldots, \frac{W_q p_q}{K_q}\right) \tag{11.3}$$

where W_i and K_i are the preferential weights and the normalizing factor attached to the generic ith goal, respectively. A suitable normalization factor is the target value of each goal; that is, $K_i = t_i$. Thus, all deviations are measured on a percentage scale. However, this normalization system is not applicable when any of the goals has a target value of zero. In this case, it is possible to resort to other normalization systems. See Tamiz et al. [64] (pages 572–573) and Kettani et al. [44] for a discussion of the different normalization techniques within a GP context.

Different methods can be used to minimize the achievement function, each one leading to a different GP variant. Let us introduce first the variant known as weighted

GP (WGP). The achievement function of a WGP model comprises the unwanted deviation variables, each weighted according to their importance [33]. Thus, we have:

$$\begin{aligned}
&\textbf{Achievement function:}\\
&\text{Min} \quad \sum_{i=1}^{q} (\alpha_i n_i + \beta_i p_i)\\
&\textbf{Goals and constraints:}\\
&f_i(\mathbf{x}) + n_i = t_i, \quad i \in \{1,\ldots,q\}\\
&\mathbf{x} \in \mathbf{F},\ \mathbf{n} \geq \mathbf{0},\ \mathbf{p} \geq \mathbf{0}
\end{aligned} \tag{11.4}$$

where $\alpha_i = W_i/K_i$ if n_i is unwanted, otherwise $\alpha_i = 0$, $\beta_i = W_i/K_i$ if p_i is unwanted, otherwise $\beta_i = 0$.

Let us now introduce the variant known as lexicographic GP (LGP). The achievement function of a LGP model is made up of an ordered vector whose dimension is equal to the Q number of preemptive priority levels defined in the model. Each component of this vector comprises the unwanted deviation variables of the goals placed at the corresponding priority level [37, 46]. Thus, we have:

$$\begin{aligned}
&\textbf{Achievement function:}\\
&\text{Lex min } a = \left[\sum_{i\in h_1} (\alpha_i n_i + \beta_i p_i), \ldots, \sum_{i\in h_r} (\alpha_i n_i + \beta_i p_i), \ldots, \sum_{i\in h_Q} (\alpha_i n_i + \beta_i p_i) \right]\\
&\textbf{Goals and constraints:}\\
&f_i(\mathbf{x}) + n_i - p_i = t_i, \quad i \in \{1,\ldots,q\},\ i \in h_r,\ r \in \{1,\ldots,Q\}\\
&\mathbf{x} \in \mathbf{F},\ \mathbf{n} \geq \mathbf{0},\ \mathbf{p} \geq \mathbf{0}
\end{aligned} \tag{11.5}$$

where h_r means the index set of goals placed at the rth priority level.

Finally, let us introduce the third classic variant called MINMAX (Chebyshev) GP (MGP). The achievement function of a MGP implies the minimization of the maximum deviation from any single goal [23]. Thus, we have:

$$\begin{aligned}
&\textbf{Achievement function:}\\
&\text{Min } D\\
&\textbf{Goals and constraints:}\\
&(\alpha_i n_i + \beta_i p_i) - D \leq 0\\
&f_i(\mathbf{x}) + n_i - p_i = t_i, \quad i \in \{1,\ldots,q\}\\
&\mathbf{x} \in \mathbf{F},\ \mathbf{n} \geq \mathbf{0},\ \mathbf{p} \geq \mathbf{0}
\end{aligned} \tag{11.6}$$

where the variable D represents the maximum weighted and normalized deviation.

11.4 A Utility Interpretation of a Goal Programming Model

GP, like any other decision-making approach, can be based on different philosophies or rationality theories. The primary philosophy underpinning GP is the Simonian concept of "satisficing" that leads to a bounded rationality theory, as the preceding sections show. However, this is not the only possible interpretation of GP. In this section GP will be analysed from the point of view of utility theory.

Let us start with LGP, where the non-compatibility between lexicographic orderings and utility functions is well known. In order to properly assess the effect of this property it is necessary to comprehend that the reason for this non-compatibility is exclusively due to the non-continuity of preferences underlying lexicographic orderings [54] (pp. 43–46).

Rather than disqualifying LGP because it implicitly assumes a non-continuous system of preferences, it would be worthwhile discussing whether or not the characteristics of the problem situation justify a system of continuous preferences. Hence, the possible problem associated with the use of the lexicographic variant lies not in its incompatibility with utility functions, but in the careless use of this approach. In contexts where the DM's preferences are clearly continuous, a compensatory GP model with utility support should be used.

It is relatively straightforward to demonstrate that a WGP model implies the maximization of a separable additive utility function in the attributes considered. This solution provides the maximum aggregated achievement between the different goals (e.g. [55]).

A MGP model implies the optimisation of a MINMAX utility function where the maximum deviation D is minimized. This type of function provides a solution that attaches the maximum importance to the most displaced goal with respect to its target. This achieves the most balanced solution for the achievement across the different goals (see [55, 64]).

11.5 Some Extensions of the Traditional Achievement Functions

Looking at GP models from the utility perspective discussed in the preceding section, we can say that from a preferential point of view, the WGP and the MGP solutions represent two opposite poles. Because the preferences underlying the weighted option are assumed to be separable, this variant can produce extremely biased results against one of the goals under consideration. On the other hand, because one of the goals is dominant, the MGP (Chebyshev) model sometimes provides results with poor aggregate performance across different goals. In short, the WGP solution implies the maximum aggregate achievement, while the MGP (Chebyshev) option provides the most balanced solution for achievement across the different goals. The extremity of both solutions can, in some cases, lead to solutions that are unacceptable for the DM. A possible solution for modelling this type of problem is a combination of the WGP and MGP models. This strikes a balance between the maximum aggregated achievement of the solution provided by the WGP model with the maximum balancedness of the solution provided by the MGP model. The result is the following Extended GP (EGP) model [11, 64]:

$$\begin{aligned}
&\textbf{Achievement function:}\\
&\text{Min } (1-\lambda)D+\lambda\sum_{i=1}^{q}(\alpha_i n_i+\beta_i p_i)\\
&\textbf{Goals and constraints:}\\
&(\alpha_i n_i+\beta_i p_i)-D\leq 0\\
&f_i(\mathbf{x})+n_i-p_i=t_i,\quad i\in\{1,\ldots,q\}\\
&\mathbf{x}\in\mathbf{F},\ \mathbf{n}\geq\mathbf{0},\ \mathbf{p}\geq\mathbf{0},\ \lambda\in[0,1]
\end{aligned}\tag{11.7}$$

where the different variables were previously defined and λ is a control parameter. For $\lambda = 0$, we have the MGP achievement function, and for $\lambda = 1$ the WGP achievement function. For other values of parameter λ belonging to the open interval $(0,1)$, the weighted combination of these two GP options can provide intermediate solutions, if they exist.

Two basic assumptions underlie all the achievement functions presented in the preceding sections:

1. The DM associates a precise target to each attribute.
2. Any unwanted deviation with respect to its target is penalized according to a constant marginal penalty; in other words, any marginal change is of equal importance no matter how distant it is from the target.

It is rather obvious that these assumptions are very strong and although they can suitably represent the preferences of certain DMs, they do not apply generally. Indeed, many DMs are not able to or are not interested in associating specific targets to certain attributes. Additionally, they may consider that the importance of marginal changes in the achievement of the goal depends upon its distance to the target.

Different achievement functions have been proposed in order to weaken the above assumptions. Chronologically, the first idea to address this problem was to conjecture that the DM feels satisfied when the achievement of a goal lies within the limits of a certain target interval $[a_i, b_i]$. This type of penalty function provides what is known as "Interval GP" (IGP) [16] or "goal range programming" [30]. The corresponding model implies a WGP formulation with a U-shaped penalty function with (1+1) sides. The analytical structure of the corresponding IGP is:

Achievement function:

$$\text{Min} \sum_{i=1}^{q} (\alpha_i \eta_i + \beta_i p_i)$$

Goals and constraints: (11.8)

$$f_i(\mathbf{x}) + n_i - p_i = a_i$$

$$f_i(\mathbf{x}) + \eta_i - \rho_i = b_i, \quad i \in \{1, \ldots, q\}$$

$$\mathbf{x} \in \mathbf{F},\ \mathbf{n} \geq \mathbf{0},\ \mathbf{p} \geq \mathbf{0},\ \boldsymbol{\eta} \geq \mathbf{0},\ \boldsymbol{\rho} \geq \mathbf{0}$$

Jones and Tamiz [42] suggested an efficient way of incorporating this type of penalty function into the achievement function of a GP model. For this type of modelling, however, all the achievement functions built with penalty systems underlie an assumption of separability of the DM's preferences It was argued above that this type of preference structure can produce biased results towards the achievement of some of the goals. For this reason, the DM might not interested in maximizing the aggregated achievement, but in getting the most balanced solution. To do this, some extensions and refinements of Jones and Tamiz's procedure have been proposed, like MINMAX (Chebyshev), Interval GP model [66] and Extended Interval GP [56].

Finally, note that refinements have recently been proposed in this direction. Thus, by using in part a computing procedure proposed by Li [48] for solving WGP problems, Chang [12, 13] introduced models that represent significant reductions in the number of auxiliary variables, as well as in the number of auxiliary constraints required to build the respective interval GP models.

11.6 Some Critical Issues and Extensions

This section addresses some key issues in GP for avoiding poor modelling practices, as well as for properly understanding the actual role of GP within a general MCDM context

11.6.1 Paretian Efficiency and GP

Within a GP context, efficiency is a condition that must hold any solution. In fact, if a GP solution is inefficient, then the achievement of at least one of the goals can be improved without impoverishing the achievement of the others. However, a standard GP formulation can produce inefficient solutions for all its variants. In the 1980s, this led to serious arguments against this approach. However, these criticisms were simply making a mountain out of a molehill. In fact, it has been demonstrated how GP models can, through minor refinements of the approach, assure the generation of efficient solutions.

Thus, Hannan [31] proposed a test to check whether or not a GP solution is efficient. The method can also establish the whole set of GP efficient solutions. Masud and Hwang [50] demonstrated that, in order to assure efficiency, it is enough to add an additional priority level to the GP model's achievement function, maximizing the sum of the wanted deviation variables. Tamiz and Jones [62] proposed a very general procedure for distinguishing the efficient from the non-efficient goals. This procedure can also restore the efficiency of the goals detected as non-efficient. Caballero et al. [8] developed procedures for generating efficient GP solutions for non-linear and convex models. Finally, Tamiz et al. [65] extended the issue of efficiency to integer and binary GP models.

In conclusion, the GP model's potential for generating an inefficient solution is not a real problem nowadays, since modern GP approaches can quite easily find a way around this prospective problem. In some areas, like engineering design, the efficient solutions can be very unstable. This high instability of the efficient solutions can make it sensible "to disregard" the issue of efficiency in some cases and to concentrate on the issue of GP solution stability (see [36]).

11.6.2 The Selection of Preferential Weights in GP

The preceding section introduced weights W_i reflecting the DM's preferences with respect to the generic ith goal. An important question is how to derive this type of weights in real applications. The following appear to be interesting procedures:

1. Establishing links between the Analytic Hierarchy Process (AHP) [57] and GP, as was suggested by Gass [24]. In this way, the weights derived from "pairwise" comparison matrices can be incorporated into a GP model.
2. Eliciting the preferential weights W_i through an interactive MCDM method. Lara and Romero [43] incorporate into a GP model preferential weights W_i, previously elicited using the Zionts-Wallenius interactive MCDM method [69].
3. Implementing a sensitivity analysis with the values of the preferential weights W_i in order to test the robustness of the GP solution to possible changes of value.

11.6.3 Redundancy in LGP

Let us now address another critical issue in GP: naive prioritization and redundancy in lexicographic models. It holds in all the algorithms solving LGP problems that if the mathematical programming problem corresponding to the ith component of the achievement function has no alternative optimal solutions, then the goals placed at priorities lower than the ith would be redundant. In other words, these goals do not play any real role in the optimisation process but become mere ornaments for the lexicographic model!

When the LGP model has a lot of priority levels, then lower priority goals are very likely to be redundant and will therefore be of no real use in the optimisation process. Such prioritisation is naive and should be avoided.

There being too many priority levels is not the only reason for the redundancy of goals in LGP. In fact, if the target values associated with the goals are very high (e.g. near their ideal values), then the likelihood of alternative optimal solutions is very small. Another possibility of redundancy is that there are many goals for which both deviational variables are unwanted. The exact achievement of a goal makes it much harder for there to be alternative optimal solutions and, consequently, the probability of redundant goals is high.

Goal redundancy is not just a theoretical possibility; it has important practical implications. Noteworthy in this sense is a research by Amador and Romero [1], testing more than twenty LGP applications reported in the literature for redundant goals. In all but one of the analysed cases at least one of the priority levels was redundant. In about 50% of the analysed cases, the number of redundant priorities was greater than or equal to two. Finally, in terms of aggregated results, more than a quarter of the goals considered were redundant.

11.6.4 Links Between GP and Other MCDM Approaches

It is common practice within the Multiple Criteria Decision Making (MCDM) field to present its different approaches separately, giving the impression that each approach is completely independent. However, this is not the case. In fact, there are significant connections between many MCDM methods. In this sense, the MULTIPLEX approach proposed by Ignizio [35] is a good example of a GP structure encompassing several single and multi-objective optimisation methods. Following on in this unifying direction, Romero [55] proposed a theoretical structure with the name of Extended Lexicographic Goal Programming (ELGP). If this structure is considered the *primary model*, then it is easy to demonstrate that a great many multi-criteria methods are just *secondary models* of ELGP. Thus, most of the multi-criteria methods can be straightforwardly deduced just by applying different parameter specifications to the above model.

The use of GP as a unifying framework looks interesting for at least the following reasons. The ELGP model stresses similarities between MCDM methods that can help reduce gaps between the advocates of different approaches. This unifying approach can become a useful teaching tool for introducing MCDM, thus avoiding the common presentation based upon a disconnected "pigeonhole" system of MCDM methods.

11.7 Other Topics

This section briefly presents some GP topics that are of clear theoretical and applied interest but have not been covered in the chapter for reasons of space and/or because some of these topics are still under development.

11.7.1 Interactive GP

The link between the interactive MCDM philosophy and GP can make GP more flexible, as well as increase the DM's level of involvement in the decision-making process. In this way, it could be easier to find a vector of target values leading to acceptable solutions for the DM. Some interesting GP proposals are: Spronk [61], Masud and Hwang [50] and Caballero et al. [9]. The introduction of the concept of meta-GP and its development and linkage within an interactive framework [52] is of interest in this respect. From a meta-GP perspective, then, the DM can establish targets on several achievement functions and use an interactive procedure to update these values. This alleviates the problem of selecting a suitable achievement function [10].

11.7.2 GP and Artificial Intelligence

The use of methods from the field of artificial intelligence (AI) for solving GP models with complex structures (non-linear goals, non-convexities, etc.) is an area of growing interest. Within the AI field, approaches like genetic algorithms, TABU search and neuronal networks look especially applicable. Jones et al. [41] is an extensive survey of applications of AI methods to the MCDM field and, particularly, to GP.

11.7.3 GP and the Aggregation of Individual Preferences

GP has proved to be a useful analytical tool for inducing models to aggregate individual preferences into a collective one. The basic idea underlying this type of approach is to define a consensus by minimizing a distance function. This function measures the distances between the information provided by the individual DMs and the unknown consensus. Different GP models have been formulated and resolved as a result of the distance function minimization process. Some results have been obtained for the complete ordinal case [25], the partial ordinal case [26], the cardinal case based upon utility functions [28] and the cardinal case based upon "pairwise" comparison matrices [29].

11.7.4 Stochastic GP

When any of the GP model parameters (target values, preferential weights, etc.) are not precisely known, the model moves from a certainty context to a stochastic/uncertainty context. Some attempts in the stochastic direction are Liu [49], Ballestero [5], and Aouni et al. [2]. Liu [49] proposes a procedure for solving stochastic GP problems with the help of genetic algorithms; Ballestero [5], presents a stochastic GP formulation within a mean-variance format; and Aouni et al. [2], model the DM's preferences within a stochastic GP. In the uncertainty direction, Rehman and Romero [51] proposed a procedure merging games against nature and GP, and Chang [14, 15] introduced the concept of Multi-Choice GP for working with target vectors instead of single figures.

11.7.5 Fuzzy GP

The basic idea underlying fuzzy GP is to represent some of the model parameters not as precise crisp numbers but as imprecise fuzzy numbers. The parameters that are usually fuzzyfied are the target values and the coefficients of the different goals. Several mathematical structures have been used to characterize the fuzzy parameters. The most widely used are triangular and trapezoidal fuzzy numbers. In this way, fuzzy GP aims to add the imprecision usually inherent in the information available into the models. Zimmerman [68] pioneered the work on fuzzy GP. Nowadays, all the different crisp GP variants have been successfully adapted to a fuzzy context (see [4, 6]).

11.7.6 GP and Data Envelopment Analysis

Data Envelopment Analysis (DEA) [21] is a linear programming-based, non-parametric approach widely used to analyse the efficiency of a set of organizational units, like branches of a bank or farms in an agricultural district. As Cooper [22] indicates, GP addresses management problems, while DEA targets problems related to the control and evaluation of activities. Even though GP and DEA have very different purposes, there are clear mathematical links between both approaches as Cooper [22] (pages 6–7) clearly demonstrate. These links are especially strong when the weighted GP variant is compared with the additive version of DEA.

11.8 Conclusions and Areas for Future Research

The enormous complexity of modern organizations make it very difficult to model, solve and analyse their real decision-making problems with methods of optimisation underpinned by classic theories of perfect rationality. In this context, the GP approach, underpinned by a bounded rationality theory, has represented an effective approach for solving decision-making problems in complex organizations. It is not bold to conjecture that the complexity of organizations will not drop in the near future and, consequently, GP will likely retain its prominent role for realistically addressing decision-making problems.

The following areas of research in GP look promising:

1. The combination of GP with AI approaches is an effective means to develop good solutions to very complex problems. In short, in many applied fields, the GP modelling effort leads to complex highly non-linear problems with high dimensions. This type of problems is not solvable by resorting to precise optimisation techniques. However, experience shows how metaheuristic procedures can output "good enough" solutions.
2. Connecting GP with other MCDM approaches. Good examples in this direction are the relationship between GP and AHP. Thus, preferential weights derived from the AHP approach can be fruitfully incorporated into a GP model. In the same way, GP can be a useful tool for deriving the weights from "pairwise" comparison matrices, as well as for dealing with inconsistencies within "pairwise" comparison scenarios.
3. Developing realistic and pragmatic interactive methods. This area can be a major aid in improving GP's inherent flexibility, allowing the DM to become involved in the problem-solving process.
4. Using GP to induce models in the field of social choice. Some initial results in this direction clearly show that this approach has enormous potential for determining social consensus.
5. The reliance on a single GP variant is not generally justified. It would be interesting, therefore, to research new achievement function forms by hybridising different variants. The recent idea of meta-goal programming offers an attractive prospective for addressing this type of problems.

Acknowledgements The authors wish to thank Rafael Caballero, Luis Díaz-Balteiro, María Victoria Rodríguez-Uría and Francisco Ruiz for helpful comments and suggestions. This research was funded by the Spanish "Ministerio de Educación y Ciencia" under project SEJ2005-04329/ECON. Thanks go to Rachel Elliot for reviewing the English.

Appendix: The Algebra Underlying Charnes and Cooper's Ideas

The idea behind this appendix is to state the algebra underlying the two basic pillars upon which GP has been built over the last fifty years.

Let us start with the first pillar in the paper by Charnes et al. [19]. The authors deal with a side constraint regression problem, with the following mathematical structure:

$$\begin{aligned} \text{Min } & \sum_{i=1}^{n} d_i^2 \\ \text{s.t. } & f_i(\mathbf{x}) + d_i = t_i, \quad i = 1,2,\ldots,n \\ & f(\mathbf{x}) \in \mathbf{F} \end{aligned} \tag{11.9}$$

where $f_i(\mathbf{x})$ is the function to be statistically fitted to the n observations $(t_1, t_2, ..., t_n)$ and $\mathbf{F}$ represents the set of side conditions defined over the parameters characterizing the function $f(\mathbf{x})$ to be fitted. Model (11.9) is a quadratic programming problem that was unsolvable in the mid-1950s. For that reason, instead of minimizing the sum of the square deviations, Charnes et al. [19] proposed minimizing the sum of the absolute deviations, as follows:

$$\begin{aligned} \text{Min } & \sum_{i=1}^{n} |t_i - f_i(\mathbf{x})| \\ \text{s.t. } & f(\mathbf{x}) \in \mathbf{F} \end{aligned} \tag{11.10}$$

However, as (11.10) implies the minimization of an absolute deviation, that is a non-linear form that it was impossible to compute at that time, Charnes et al. proposed linearizing the objective function of (11.10) by introducing the following change of variables:

$$n_i = \frac{1}{2}\left[|t_i - f_i(\mathbf{x})| + (t_i - f_i(\mathbf{x}))\right] \tag{11.11}$$

$$p_i = \frac{1}{2}\left[|t_i - f_i(\mathbf{x})| - (t_i - f_i(\mathbf{x}))\right] \tag{11.12}$$

By adding (11.11) and (11.12), and by subtracting (11.12) from (11.11), we have:

$$n_i + p_i = |t_i - f_i(\mathbf{x})| \tag{11.13}$$

$$n_i - p_i = t_i - f_i(\mathbf{x}) \tag{11.14}$$

Therefore, according to (11.13) and (11.14), the non-linear model (11.10) turns into the following LP model:

$$\begin{aligned} \text{Min } & \sum_{i=1}^{n} (n_i + p_i) \\ \text{s.t. } & f_i(\mathbf{x}) + n_i - p_i = t_i, \quad i = 1,2,\ldots,n \\ & f(\mathbf{x}) \in \mathbf{F} \end{aligned} \tag{11.15}$$

Let us move now to Appendix B of the classic book by Charnes and Cooper [17]. In section 5 of Appendix B under the heading **Goal Programming** they address the analysis of contradictions in non-solvable problems within a linear programming context. They illustrate the basic idea with the help of the following illustrative machine-loading problem:

$$\begin{aligned} 3x_1 + 2x_2 &\le 12 \\ 5x_1 \qquad &\le 10 \\ x_1 + \ x_2 &\ge 8 \\ -x_1 + \ x_2 &\ge 4 \\ x_1, x_2 &\ge 0 \end{aligned} \tag{11.16}$$

It is easy to check that there is no feasible solution to problem (11.16). Charnes and Cooper suggest considering the first two equations of (11.16) as proper constraints, and the last two equations as goals to be attained as closely as possible. Thus, model (11.16) turns into the following model:

$$\begin{aligned} \text{Min } & |8-(x_1+x_2)| + |4-(-x_1+x_2)| \\ \text{s.t. } & 3x_1 + 2x_2 \le 12 \\ & 5x_1 \qquad \le 10 \\ & x_1, x_2 \ge 0 \end{aligned} \tag{11.17}$$

Again by introducing the deviation variables (11.11) and (11.12), and by implementing the above arithmetic operations, the non-linear model (11.17) turns into the following linear structure:

$$\begin{aligned} \text{Min } & n_1 + n_2 \\ \text{s.t. } & 3x_1 + 2x_2 \le 12 \\ & 5x_1 \le 10 \\ & x_1 + x_2 + n_1 - p_1 = 8 \\ & -x_1 + x_2 + n_2 - p_2 = 4 \\ & x_1, x_2 \ge 0 \end{aligned} \tag{11.18}$$

Model (11.18) is a linear WGP model, for which the unwanted deviation variables appearing in the achievement function have not been normalized. This model was proposed by Charnes and Cooper almost fifty years ago!

References

1. F. Amador and C. Romero. Redundancy in lexicographic goal programming: An empirical approach. *European Journal of Operational Research*, 41:347–354, 1989.
2. B. Aouni, F. Ben Abdelaziz, and J.-M. Martel. Decision-maker's preference modeling in the stochastic goal programming. *European Journal of Operational Research*, 162:610–618, 2005.
3. B. Aouni and O. Kettani. Goal programming model: A glorious history and a promising future. *European Journal of Operational Research*, 133:225–231, 2001.
4. M. Arenas Parra, A. Bilbao Terol, and M.V. Rodríguez Uría. A fuzzy goal programming approach to portfolio selection. *European Journal of Operational Research*, 133:287–297, 2001.
5. E. Ballestero. Stochastic goal programming: A mean-variance approach. *European Journal of Operational Research*, 131:476–481, 2001.
6. A. Bilbao Terol. A new approach for multiobjective decision making based on fuzzy distance minimization. *Mathematical and Computer Modelling*, 47:808–826, 2008.

7. R. Caballero, T. Gómez, and F. Ruiz. Goal programming: Realistic targets for the near future. In *V International Conference on Multiobjective and Goal Programming (MOPGP06)*, Toulouse, France, June 2006.
8. R. Caballero, L. Rey, and F. Ruiz. Determination of satisfying and efficient solutions in convex multi-objective programming. *Optimization*, 37:125–137, 1996.
9. R. Caballero, L. Rey, and F. Ruiz. Lexicographic improvement of the target values in convex goal programming. *European Journal of Operational Research*, 107:644–655, 1998.
10. R. Caballero, F. Ruiz, M. Victoria Rodríguez Uría, and C. Romero. Interactive meta-goal programming. *European Journal of Operational Research*, 175:135–154, 2006.
11. E. Carrizosa and D. Romero-Morales. Combinig minsum and minmax: A goal programming approach. *Operations Research*, 49:169–174, 2001.
12. C.-T. Chang. Mixed binary interval goal programming. *Journal of the Operational Research Society*, 57:469–473, 2006.
13. C.-T. Chang. Efficient structures of achievement functions for goal programming models. *Asia-Pacific Journal of Operational Research*, 24:755–764, 2007.
14. C.-T. Chang. Multi-choice goal programming. *Omega*, 35:389–396, 2007.
15. C.-T. Chang. Revised multi-choice goal programming. *Applied Mathematical Modelling*, 32:2587–2596, 2008.
16. A. Charnes and B. Collomb. Optimal economic stabilization policy: Linear goal-interval programming models. *Socio-Economic Planning Sciences*, 6:431–435, 1972.
17. A. Charnes and W.W. Cooper. *Management Models and Industrial Applications of Linear Programming*. John Wiley and Sons, New York, 1961.
18. A. Charnes, W.W. Cooper, J.K. Devoe, D.B. Learner, and W. Reinecke. A goal programming model for media planning. *Management Science*, 14:423–430, 1968.
19. A. Charnes, W.W. Cooper, and R. Ferguson. Optimal estimation of executive compensation by linear programming. *Management Science*, 1:138–151, 1955.
20. A. Charnes, W.W. Cooper, and Y. Ijiri. Breakeven budgeting and programming to goals. *Journal of Accounting Research*, 1:16–43, 1963.
21. A. Charnes, W.W. Cooper, and E. Rhodes. Measuring the efficiency of decision making units. *European Journal of Operational Research*, 2:429–444, 1978.
22. W.W. Cooper. Origins, uses of, and relations between goal programming and data envelopment analysis. *Journal of Multi-Criteria Decision Analysis*, 13:3–11, 2005.
23. R.B. Flavell. A new goal programming formulation. *Omega*, 4:731–732, 1976.
24. S.I. Gass. A process for determining priority weights for large-scale linear goal programmes. *Journal of the Operational Research Society*, 37:779–785, 1986.
25. J. González-Pachón and C. Romero. Distance-based consensus methods: A goal programming approach. *Omega*, 27:341–347, 1999.
26. J. González-Pachón and C. Romero. Aggregation of partial ordinal rankings: an interval goal programming approach. *Computers and Operations Research*, 28:827–834, 2001.
27. J. González-Pachón and C. Romero. Satisficing logic and goal programming: Towards an axiomatic link. *INFOR*, 42:157–161, 2004.
28. J. González-Pachón and C. Romero. An analytical framework for aggregating multiattribute utility functions. *Journal of the Operational Research Society*, 57:1241–1247, 2006.
29. J. González-Pachón and C. Romero. Inferring consensus weights from pairwise comparison matrices without suitable properties. *Annals of Operations Research*, 154:123–132, 2007.
30. J. Gross and J. Talavage. A multiple-objective planning methodology for information service managers. *Information Processing and Management*, 15:155–167, 1979.
31. E.L. Hannan. Nondominance in goal programming. *INFOR*, 18:300–309, 1980.
32. J.P. Ignizio. S-II trajectory study and optimum antenna placement. Report SID-63, North American Aviation Corporation, Downwy, CA, 1963.
33. J.P. Ignizio. *Goal Programming and Extensions*. Lexington Books, Massachusetts, 1976.
34. J.P. Ignizio. A review of goal programming: A tool for multiobjective analysis. *Journal of the Operational Research Society*, 27:1109–1119, 1978.
35. J.P. Ignizio. Multiobjective mathematical programming via the MULTIPLEX model and algorithm. *European Journal of Operational Research*, 22:338–346, 1985.

36. J.P. Ignizio. Illusions of optimality. *Engineering Optimization*, 31:749–761, 1999.
37. J.P. Ignizio and J.H. Perlis. Sequential linear goal programming. *Computers and Operations Research*, 6:141–145, 1979.
38. J.P. Ignizio and C. Romero. Goal programming. In H. Bidgoli, editor, *Encyclopedia of Information Systems*, volume 2, pages 489–500. Academic Press, San Diego, CA, 2003.
39. Y. Ijiri. *Management Goals and Accounting for Control*. North-Holland, Amsterdam, 1965.
40. V. Jääskeläinen. A goal programming model of aggregate production. *Swedish Journal of Economics*, 6:14–19, 1969.
41. D.F. Jones, S.K. Mirrazavi, and M. Tamiz. Multi-objective meta-heuristics: An overview of the current-state-of-the art. *European Journal of Operational Research*, 137:1–9, 2002.
42. D.F. Jones and M. Tamiz. Expanding the flexibility of goal programming via preference modelling techniques. *Omega*, 23:41–48, 1995.
43. D.F. Jones and M. Tamiz. Goal programming in the period 1990–2000. In M. Ehrgott and X. Gandibleux, editors, *Multicriteria Optimization: State of the Art Annotated Bibliographic Survey*, volume 2, pages 129–170. Kluwer Academic Publishers, Boston, 2002.
44. O. Kettani, B. Aouni, and J.-M. Martel. The double role of the weight factor in the goal programming model. *European Journal of Operational Research*, 31:1833–1845, 2004.
45. P. Lara and C. Romero. Interactive multigoal programming model for determining livestock rations: An application to dairy cows in Andalusia, Spain. *Journal of the Operational Research Society*, 43:945–953, 1992.
46. S.M. Lee. *Goal Programming for Decision Analysis*. Auerbach, Philadelphia, 1972.
47. S.M. Lee and D. Olson. Goal programming. In T. Gal, T.J. Stewart, and T. Hanne, editors, *Multicriteria Decision Making: Advances in MCDM Models, Algorithms, Theory, and Applications*, volume 2, pages 1–33. Kluwer Academic Publishers, Boston, 1999.
48. H.L. Li. An efficient method for solving linear goal programming problems. *Journal of Optimization Theory and Applications*, 90:465–469, 1996.
49. B.D. Liu. Dependent chance goal programming and its genetic algorithm-based approach. *Mathematical and Computer Modelling*, 24:43–52, 1996.
50. A.S. Masud and C.L. Hwang. Interactive sequential goal programming. *Journal of the Operational Research Society*, 32:391–400, 1981.
51. T. Rehman and C. Romero. Formulating generalised "goal games" against nature: An illustration from decision-making under uncertainty in agriculture. *Applied Mathematics and Computation*, 175:486–496, 2006.
52. M.V. Rodríguez Uría, R. Caballero, F. Ruiz, and C. Romero. Meta-goal programming. *European Journal of Operational Research*, 136:422–429, 2002.
53. C. Romero. A survey of generalized goal programming (1970–1982). *European Journal of Operational Research*, 25:183–191, 1986.
54. C. Romero. *Handbook of Critical Issues in Goal Programming*. Pergamon Press, Oxford, 1991.
55. C. Romero. Extended lexicographic goal programming: A unifying approach. *Omega*, 21:63–71, 2001.
56. C. Romero. A general structure of achievement function for a goal programming model. *European Journal of Operational Research*, 153:675–686, 2004.
57. T.L. Saaty. *The Analytic Hierarchy Process*. McGraw-Hill, New York, 1980.
58. M.J. Schniederjans. *Goal Programming Methodology and Applications*. Kluwer Academic Publishers, Boston, 1995.
59. H.A. Simon. A behavioural model of rational choice. *Quarterly Journal of Economics*, 69:99–118, 1955.
60. H.A. Simon. Rational choice and the structure of the environment. *Psychological Review*, 63:129–138, 1956.
61. J. Spronk. *Interactive Multiple Goal Programming*. Martinus Nijhoff, The Hague, 1981.
62. M. Tamiz and D.F. Jones. Goal programming and Pareto efficiency. *Journal of Information and Optimization Sciences*, 17:291–307, 1996.
63. M. Tamiz, D.F. Jones, and E. El-Darzi. A review of goal programming and its applications. *Annals of Operations Research*, 58:39–53, 1995.

64. M. Tamiz, D.F. Jones, and C. Romero. Goal programming for decision making: An overview of the current state-of-the-art. *European Journal of Operational Research*, 111:569–581, 1998.
65. M. Tamiz, S.K. Mirrazavi, and D.F. Jones. Extensions of Pareto efficiency analysis to integer goal programming. *Omega*, 27:179–188, 1999.
66. B. Vitoriano and C. Romero. Extended interval goal programming. *Journal of the Operational Research Society*, 50:1280–1283, 1999.
67. S.H. Zanakis and S.K. Gupta. A categorized bibliographic survey of goal programming. *Omega*, 13:211–222, 1985.
68. H.J. Zimmerman. Fuzzy programming and linear programming with several objective functions. *Fuzzy Sets and Systems*, 1:45–55, 1978.
69. S. Zionts and J. Wallenius. An interactive multiple objective linear programming method for a class of underlying nonlinear utility functions. *Management Science*, 29:519–529, 1976.

Chapter 12
Interactive Decomposition-Coordination Methods for Complex Decision Problems

Alexander Engau

12.1 Introduction

Living in a vibrant and constantly changing world, in the last few decades we have witnessed tremendous advances in many important areas of human activity including medicine and drugs, public policy and service, engineering and economics, and new computing technologies. While rapid economic growth and substantial technological progress have resulted in a host of new opportunities for many private and public enterprises and organizations, in consequence, today we are also facing enormous competition in international trading, growing degrees of environmental pollution caused by continuing industrialization and urbanization, and unprecedented water and energy demands accelerating the steady shortage of natural resources and food. At the same time, the effects of globalization and the instantaneous exchange of data through modern telecommunication and information systems have radically changed the way we perceive and are able to respond to these and other increasingly complex challenges varying from homeland security and public health over climate, energy and transportation to natural resources and the environment. To solve these challenges, it is beyond doubt that the simultaneous consideration of a large number of interrelated aspects and criteria has become essential to make deliberate decisions and take responsible actions in both our professional as well as our private lives.

In order to structure, analyze, and ultimately resolve the complexity underlying these decision-making situations, an important and useful concept with analogies also in numerous other fields is the concept of decomposition. For example, in the biological and physical sciences decomposition refers to the processes by which organic tissue reduces to simpler forms of matter, and by which a complex object such as light (e.g., spectral decomposition) breaks into its individual components,

---

Alexander Engau
Department of Mathematical and Statistical Sciences, University of Colorado Denver, USA
Department of Management Sciences, Faculty of Engineering, University of Waterloo, Canada
e-mail: aengau@alumni.clemson.edu

C. Zopounidis and P.M. Pardalos (eds.), *Handbook of Multicriteria Analysis*, Applied Optimization 103, DOI 10.1007/978-3-540-92828-7_12, © Springer-Verlag Berlin Heidelberg 2010

with the purpose to promote new development and growth, and to help reveal structure and clarity, respectively. Similarly, in mathematics decompositions are used to represent certain objects in terms of two or more substructures with similar or related properties (e.g., prime, graph, or group decompositions) and facilitate the study and understanding of their structural or behavioral characteristics. A corresponding notion known commonly to emperors and computer scientists is the principle of "divide and conquer" to subdue kingdoms and to organize large and complex computer codes or programming problems into smaller parts or modules that are easier to understand, control, maintain, and solve. Finally, a "trias politica" in democratic governance divides its power into separate areas of responsibility to prevent conflict of interest and to achieve a harmonious functioning of the overall political system.

Considering these traditional reasons for decomposition, divide and conquer, and separation of powers in the natural and social sciences, it does not surprise that similar benefits can be obtained when making decisions in the engineering or managerial sciences by partitioning a large and complex system or decision problem into interrelated but independent subsystems or subproblems that reduce dimensionality, increase flexibility and model-reality verisimilitude, and facilitate overall modeling, computational, and decisional requirements. Intelligently distributing both work and decision authority among either concurrent or collaborative design teams from different engineering disciplines, or different departmental or business units in hierarchical organizations, however, it usually remains necessary to coordinate these different groups and guarantee that the individual solutions are consistent with each other and comply with overall resource requirements or objectives by manipulating interactions, resolving conflicts, and adjusting targets (goal coordination) or the associated model structure (model coordination). Hence, the purpose of this chapter is, first, to show how multicriteria decision problems can be modeled mathematically as multiobjective programs, second, to present a variety of optimization methods from the engineering and managerial literature to effectively solve the resulting problems using decomposition and coordination, and third, to discuss the interactive participation of one or a group of decision-makers in the overall solution process.

The remaining chapter is organized as follows. In Section 2, we first define the underlying mathematical model in Section 2.1 and briefly provide the most important concepts and results from multiple criteria decision-making and optimization that are used throughout our following discussion. An extensive although far from exhaustive overview of general decomposition methods that are of relevance to the topics discussed is included in the subsequent Section 2.2, and in spite of best intentions the author apologizes for any potential misquotations or major oversights in compiling this list of references. Combining two of the most prominent decomposition paradigms in design engineering and organizational management in Section 3, Sections 3.1 and 3.2 are focused on coordination methods in multidisciplinary (design) optimization and hierarchical multiobjective analysis (modeling, programming), respectively. To the author's knowledge, this is the first time that these different methods are formulated in a unified framework, i.e., for the same underlying model and using consistent notation to facilitate their detailed comparison and analysis. Some more recent approaches that have been developed by the author and his colleagues over the last few years are then described in Section 4 and highlight an interactive decision-making method based on tradeoffs and solution visualizations in Section 4.1, and the still emerging area of multiscenario multiobjective optimization in Section 4.2. Together with some final remarks, the chapter concludes in Section 5 with a short summary and outline of further work and possible research directions.

12.2 Decomposition in Decision-Making and Optimization

The decision-making problems that we study in this chapter are concerned with choosing a preferred decision from among a set of feasible ones, that are selected, evaluated, and compared with respect to a single or multiple objectives or decision criteria. In the presence of several and typically conflicting or incommensurable objectives, however, a unique best decision usually does not exist so that we need to weigh or trade off the different criteria to find an optimal compromise that satisfies fully articulated, gradually revealed, or initially unknown preferences by the decision-maker (DM). Respectively, we distinguish a priori, interactive, and a posteriori methods [106, 136] based on the timing of tradeoff decisions by DM within the decision analysis [66, 98], as compared to the repeated generation of one or more candidate solutions using mathematical programming and multicriteria optimization [50]. In this section, we address how both analysis and optimization can be facilitated using suitable decompositions of the underlying decision-making model.

12.2.1 Mathematical Model

Let $\mathbb{R}^n$, $\mathbb{R}^m$, and $\mathbb{R}^l$ be Euclidean vector spaces, and $\mathbf{f}: \mathbb{R}^n \to \mathbb{R}^m$ and $\mathbf{g}: \mathbb{R}^n \to \mathbb{R}^l$ be two vector-valued functions. The multiobjective programming (MOP) [1] model is

$$\text{MOP:}\ \underset{\mathbf{x}}{\text{Minimize}}\ \ \mathbf{f}(\mathbf{x}) = (f_1(\mathbf{x}), f_2(\mathbf{x}), \dots, f_m(\mathbf{x})) \tag{12.1a}$$

$$\text{subject to}\ \ \mathbf{g}(\mathbf{x}) = (g_1(\mathbf{x}), g_2(\mathbf{x}), \dots, g_l(\mathbf{x})) \leq \mathbf{0} \tag{12.1b}$$

where $\mathbf{x} = (x_1, x_2, \dots, x_n)$, $\mathbf{f} = (f_1, f_2, \dots, f_m)$, and $\mathbf{g} = (g_1, g_2, \dots, g_l)$ denote the decisions, objectives, and constraints, respectively. A decision $\mathbf{x}$ is said to be feasible for MOP if $\mathbf{g}(\mathbf{x}) \leq \mathbf{0}$, and (weakly) efficient if there is no other feasible decision $\mathbf{x}'$ such that $\mathbf{f}(\mathbf{x}') \leq \mathbf{f}(\mathbf{x})$ and $f_i(x') < f_i(\mathbf{x})$ for at least one (all) $i = 1, \dots, m$. An efficient decision $\mathbf{x}$ is said to be properly efficient if there exists $M > 0$ such that, whenever $f_i(\mathbf{x}') < f_i(\mathbf{x})$ for some i, there is j such that $f_j(\mathbf{x}) < f_j(\mathbf{x}')$ and $(f_i(\mathbf{x}) - f_i(\mathbf{x}'))/(f_j(\mathbf{x}') - f_j(\mathbf{x})) \leq M$. The sets of all feasible and (weakly, properly) efficient decisions are denoted by $X = \{\mathbf{x} \in \mathbb{R}^n : \mathbf{g}(\mathbf{x}) \leq \mathbf{0}\}$, $E_w(X, \mathbf{f})$, $E_p(X, \mathbf{f})$, and $E(X, \mathbf{f})$,

[1] We use **bold** symbols to distinguish vector variables from their individual components or scalars.

respectively, and satisfy that $E_p(X,\mathbf{f}) \subseteq E(X,\mathbf{f}) \subseteq E_w(X,\mathbf{f}) \subseteq X$. If $\mathbf{x}$ is (weakly, properly) efficient, then $\mathbf{f}(\mathbf{x})$ is also said to be (weakly, properly) Pareto optimal.

12.2.1.1 Generating Methods and Optimality Conditions

A great variety of solution methods exists for generating efficient decisions such as objective aggregation by a weighted sum, objective prioritization by constraining techniques and hierarchical orderings, or some form of norm or distance minimization to some reference point [49, 127, 94, 135]. Three representative methods, that we will continue to use also in later sections, are the weighted-sum method (WSM)

$$\text{WSM}(\mathbf{w})\text{:}\ \min_{\mathbf{x}\in X} \sum_{i=1}^{m} w_i f_i(\mathbf{x}) \tag{12.2}$$

where $\mathbf{w} \gneqq \mathbf{0}$ is a nonnegative weight vector, the epsilon-constraint method (ECM)

$$\text{ECM}_k(\boldsymbol{\varepsilon})\text{:}\ \min_{\mathbf{x}\in X} f_k(\mathbf{x}) \text{ subject to } f_i(\mathbf{x}) \leq \varepsilon_i,\ i=1,\ldots,m,\ i \neq k \tag{12.3}$$

where all but one objective f_k are converted to inequality constraints with upper bounds $\boldsymbol{\varepsilon}$, and the (weighted) Chebyshev-norm or reference-point method (RPM)

$$\text{RPM}(\mathbf{w},\mathbf{r})\text{:}\ \min_{\mathbf{x}\in X} \max_{1\leq i\leq m} w_i(f_i(\mathbf{x}) - r_i) \tag{12.4a}$$

$$\Longleftrightarrow \quad \min_{\mathbf{x}\in X} \alpha \text{ subject to } \alpha \geq w_i(f_i(\mathbf{x}) - r_i),\ i=1,\ldots,m \tag{12.4b}$$

with weights $\mathbf{w} \gneqq \mathbf{0}$ and some reference point $\mathbf{r}$ that satisfies $\mathbf{r} \leq \mathbf{f}(\mathbf{x})$ for all $\mathbf{x} \in X$.

Proposition 12.1 (Sufficient Conditions For Efficiency). *Let* $\mathbf{w}$, $\boldsymbol{\varepsilon}$, *and* $\mathbf{r}$ *be as described above, and* $\mathbf{x}$ *be an optimal solution to either problem WSM, ECM, or RPM.*

(a)If $\mathbf{x}$ *solves WSM(*$\mathbf{w}$*) with* $\mathbf{w} > 0$*, then* $\mathbf{x}$ *is properly efficient for MOP.*
(b)If $\mathbf{x}$ *solves ECM*$_k$*(*$\boldsymbol{\varepsilon}$*) for all* $k = 1,\ldots,m$*, then* $\mathbf{x}$ *is efficient for MOP.*
(c)If the solution $\mathbf{x}$ *is unique, then* $\mathbf{x}$ *is efficient for MOP.*

In any case, $\mathbf{x}$ *is weakly efficient for MOP.*

Proposition 12.2 (Necessary Conditions). *If* $\mathbf{x}$ *is an efficient decision for MOP, then* $\mathbf{x}$ *is optimal for RPM(*$\mathbf{w},\mathbf{r}$*) for some* $\mathbf{w} \gneqq 0$*, and for ECM*$_k$*(*$\boldsymbol{\varepsilon}$*) with* $\boldsymbol{\varepsilon} = \mathbf{f}(\mathbf{x})$ *for all* $k = 1,\ldots,m$*. If* $\mathbf{f}$ *and* $\mathbf{g}$ *are convex, then* $\mathbf{x}$ *is optimal for WSM(*$\mathbf{w}$*) for some* $\mathbf{w} \gneqq 0$.

12.2.1.2 Tradeoffs and Tradeoff Rates

Given two feasible decisions $\mathbf{x}$ and $\mathbf{x}'$ with $f_j(\mathbf{x}) - f_j(\mathbf{x}') \neq 0$, each difference ratio

$$t_{ij}(\mathbf{f},\mathbf{x},\mathbf{x}') = -\left(f_i(\mathbf{x}) - f_i(\mathbf{x}')\right) / \left(f_j(\mathbf{x}) - f_j(\mathbf{x}')\right) \tag{12.5}$$

is called a tradeoff between f_i and f_j. If $\mathbf{f}$ is continuously differentiable at $\mathbf{x}$, then

$$\tau_{ij}(\mathbf{f},\mathbf{x}) = \lim_{\mathbf{x}'\to\mathbf{x}} t_{ij}(\mathbf{f},\mathbf{x},\mathbf{x}') = -\partial f_i(\mathbf{x})/\partial f_j \tag{12.6}$$

is called the tradeoff rate between f_i and f_j at $\mathbf{x}$. Under suitable regularity and differentiability assumptions, these rates can be computed at every efficient decision that can be characterized according to Propositions 12.1 and 12.2 using a sensitivity analysis.

Proposition 12.3 (Tradeoff Rates [120, 156]). *Let $\mathbf{x} \in X$ be efficient for MOP. If $\mathbf{x}$ is generated by WSM, ECM$_k$, or RPM, respectively, then its tradeoff rates are given by*

$$\tau_{ij}(\mathbf{f},\mathbf{x}) = \begin{cases} -\partial f_i(\mathbf{x})/\partial f_j & = w_j/w_i \\ -(\partial f_k(\mathbf{x})/\partial f_j)/(\partial f_k(\mathbf{x})/\partial f_i) & = \lambda_{kj}/\lambda_{ki} \\ -\partial f_i(\mathbf{x})/\partial f_j & = \lambda_j w_j/\lambda_i w_i \end{cases} \tag{12.7}$$

where w_i are the weights for WSM and RPM, λ_{ki} are the Lagrangean multipliers associated with the constraints $f_i(\mathbf{x}) \leq \varepsilon_i$ in ECM$_k(\boldsymbol{\varepsilon})$, and λ_i are the multipliers associated with the constraints $\alpha \geq w_i(f_i(\mathbf{x}) - r_i)$ in RPM (12.4b), *for all $i = 1,\ldots,m$.*

12.2.2 Decomposition Methods

Given MOP (12.1), we primarily distinguish between model-based decompositions of the decision space $\mathbb{R}^n$ (decision decompositions) and the objective space $\mathbb{R}^m$ (objective decompositions). Whereas the former approaches are typically applied to problems with a large number of decision or optimization variables to accelerate or otherwise facilitate the generation of efficient decisions, the latter are most useful especially for large numbers of decision criteria that otherwise would require extensive tradeoff analyses which are often impractical to manage or cumbersome to conduct. Two related although less common classes of methods, that also emphasize either the computational or the preferential aspect of optimization or decision-making, respectively, propose to also decompose the underlying parameter space when computing efficient decision using one of the generating methods in Subsection 12.2.1.1, or the preference space of the DM when evaluating alternatives and making decisions on their associated tradeoffs [110]. Finally, and predominantly in the literature on structural and decentralized design, model decompositions of engineering systems (by mathematical representation) can also be contrasted to object (by physical components or parts), aspect (by knowledge domains or disciplines), sequential (by flow of information), or functional decompositions [134]. Here we briefly provide selected key references to some more specific decomposition schemes in both single and multiobjective programming, the former being of relevance when solving the single objective generating problems, and the later methods with a more prominent focus on interactions with the DM in the context of multicriteria decision-making.

12.2.2.1 Decomposition in Mathematical Programming and Optimization

Beginning with the seminal Dantzig-Wolfe [41] and Benders decomposition [12] for linear programs [40], sequential decomposition schemes like branch-and-bound [111] or cutting-plane methods [99] for integer and combinatorial optimization [209], and few related approaches for nonlinear problems [145, 159], a vast amount of extensions and new techniques has been proposed for the decomposition of mathematical programs with a special problem structure and for specific applications. The interested reader may consult the recent monograph [32] for an overview over several of these methods and their wide applicability in economics and engineering.

Decision Decompositions of MOP

Several of the traditional single objective decompositions schemes have also been modified for MOP with block-diagonal structure [211], cone constraints [39], and fuzzy numbers using interactive methods or dynamic programming [2, 90, 154, 157]. Besides some additional works that treat decompositions for general mathematical programs, but also address the specific case of decomposable MOP [17, 20], the majority of decision decompositions are proposed for specific application in hierarchical or multilevel optimization, that we review in more detail in Section 12.3.2.

Parameter Decompositions of MOP

Solution methods that are based on parameter decompositions were first proposed for linear MOP using weighted sums [176, 214, 215], and for convex and nonlinear problems using generalized Chebyshev norms [42, 97]. More recent approaches also use decompositions of the weight set to partition the outcome or objective space [13], or to find representations of the efficient set for linear MOP [14, 161, 162] and for convex bicriteria problems using a warmstarted interior-point algorithm [55, 67].

12.2.2.2 Decomposition in Multiple Criteria Decision-Making

Whereas the former approaches decompose MOP mainly to reduce the computational burden of generating a set of efficient decisions, the major challenge for the DM is the subsequent selection of a unique preferred solution. Often independent of the actual number of decision or optimization variables, thus, the following methods are primarily concerned with reducing the number of objectives or decision criteria.

Objective Decompositions of MOP

Several methods have been proposed to facilitate the choice of a preferred compromise decision and enable the DM to evaluate tradeoffs and specify preferences in an iterative, or interactive manner by dividing the overall decision problem into smaller and more manageable subproblems [21, 54, 57, 60, 76, 82, 114, 211] before coordinating the decomposed subproblems using an interactive decision-making procedure [10, 54, 60, 85, 106, 114, 136, 155, 190, 191, 198]. Two prominent examples of these approaches are the tradeoff-based decomposition of multiattribute utility functions [98] and the analytic hierarchy process [152], both of which have been successfully applied to a wide spectrum of practical decision-making situations [201].

Decompositions for Selected Applications

In addition to the above, a seemingly endless number of other ad-hoc decomposition approaches has been developed for specific applications in management, including resource allocation [116], transportation and routing in layout and location problems [100, 105], or plant location [65], as well as in engineering, for example, machine tool spindle systems [138], proprotor aerodynamics [187], wing design [197, 210], or the control of prosthetic arms [160]. Other innovative application areas include harvesting decision in a multispecies ecosystem [199] and, more recently, the deregulation of electricity markets by integration of distributed energy management systems [33].

12.3 Coordination of Decision Decompositions

In the following central section of this chapter, we review several approaches that decompose an initial large-scale problem into a collection of smaller-sized subproblems with autonomous decision authorities, which can be used for many engineering or managerial decision processes that involve design teams from different disciplines or different departmental units in hierarchical organizations. Namely, in Section 3.1 we first address decomposition-coordination methods for single-objective multidisciplinary (design) optimization, and then broaden our focus to hierarchical analysis (modeling, programming) methods with multiple objectives in Section 3.2.

12.3.1 Multidisciplinary Optimization

The field of multidisciplinary optimization (MDO) first evolved during the nineteen seventies to enable the simultaneous participation of multiple design teams from different disciplines for the development of large-scale engineering designs and systems. Originating in the aircraft and aerospace industry at NASA [173, 174], by now MDO has also expanded to many other industries that, as a consequence of globalization and worldwide outsourcing, typically depend on the ability to successfully manage and coordinate increasing numbers of decentralized design teams and distributed working groups [24, 25, 30, 69, 47, 204, 205]. Therefore, it is not surprising that a wide variety of new decomposition schemes has been developed in the specific context of MDO including traditional collaborative optimization [29, 77, 108, 146, 189], concurrent subspace optimization [15, 87] and analytical target cascading [101, 113], as well as various other methods [16, 28, 130, 143, 174, 175] some of which we will discuss in some more detail below. In particular, and although the vast majority of engineering problems is multiobjective in nature [178], these traditional MDO methods are developed for a single criterion only, which typically combines several objectives in the form of WSM (12.2). Only few extensions are proposed to also explicitly handle vector objective functions [95, 96, 102, 179, 180, 178, 192], including some methods that can be used to find an optimal decomposition based on those sets of criteria that can be computed in parallel or need to be evaluated in sequence [212], or using some other type of decomposition analysis and optimization [57, 129, 130, 146, 200]. For notational simplicity and consistency, however, at first we shall only focus on the following traditional, single-criterion formulation

$$\text{MDO:} \quad \underset{\mathbf{x},\mathbf{x}_1,\ldots,\mathbf{x}_k}{\text{Minimize}} \quad f(\mathbf{x},\mathbf{x}_1,\ldots,\mathbf{x}_k,\mathbf{y}_1,\ldots,\mathbf{y}_k) \tag{12.8a}$$

$$\text{subject to} \quad \mathbf{g}(\mathbf{x},\mathbf{x}_1,\ldots,\mathbf{x}_k,\mathbf{y}_1,\ldots,\mathbf{y}_k) \leq \mathbf{0} \tag{12.8b}$$

$$\mathbf{g}_i(\mathbf{x},\mathbf{x}_i,\mathbf{y}_i) \leq \mathbf{0}, \; i = 1,\ldots,k \tag{12.8c}$$

$$\text{where} \quad \mathbf{h}_i(\mathbf{x},\mathbf{x}_i,\mathbf{y}) - \mathbf{y}_i = \mathbf{0}, \; i = 1,\ldots,k. \tag{12.8d}$$

In this formulation, $\mathbf{x}$ and $\mathbf{x}_i$ are global and local decision (design, control) vectors, respectively, and $\mathbf{y}_i$ is a (coupling) state vector of discipline or subsystem i that depends on $\mathbf{x}$, $\mathbf{x}_i$, and the overall state vector $\mathbf{y} = (\mathbf{y}_1,\ldots,\mathbf{y}_k)$ through the coupling function $\mathbf{h}_i$. For each fixed allocation of decision variables $\mathbf{x}$ and $\mathbf{x}_i$, we assume that there exists a unique solution $\mathbf{y}$ to the nonlinear system of equations (12.8d) determining the system states. The vector functions $\mathbf{g}$ and $\mathbf{g}_i$ denote the overall system and subsystem constraints, respectively, whereas f denotes a single overall objective.

Given MDO, we say that a vector $(\mathbf{x},\mathbf{x}_i,\mathbf{y})$ is individual-discipline feasible (IDF) for discipline i if $\mathbf{y}_i = \mathbf{h}_i(\mathbf{x},\mathbf{x}_i,\mathbf{y})$, and $(\mathbf{x},\mathbf{x}_1,\ldots,\mathbf{x}_k,\mathbf{y})$ is said to be multiple-discipline feasible (MDF) if each $(\mathbf{x},\mathbf{x}_i,\mathbf{y})$ is individual-discipline feasible. In particular, then the state vector $\mathbf{y}$ is consistent and satisfies (12.8d) for all $i = 1,\ldots,k$. In general, however, the coupled system analysis $\mathbf{h}$ in (12.8d) is very complex and expensive to solve, and often relies on simulation and finite difference or element codes that need to be tied as external function evaluator to the system optimization. Hence, a major challenge in solving MDO is to properly coordinate the different disciplines to avoid solving this system more often than necessary, leading to different approaches that initially relax the MDF requirement and achieve a feasible solution in iterative single-level schemes, or by penalizing infeasibilities in bi- or multilevel methods.

12.3.1.1 Single-Level Discipline-Feasibility Methods

The following methods solve MDO in an iterative single-level scheme and differ in the way they handle the coupled system analysis as to maintain intermediate iterates that are multiple, individual, or no-discipline feasible. The resulting decomposition is thus based on the decision variables and state vectors of the individual disciplines.

Multiple-Discipline-Feasible (MDF)

Also known as monolithic or fully-integrated optimization (FIO), or iterative-loop or all-in-one method [35], this method solves MDO "as is" using the formulation (12.8). Consequently, for every allocation of decision variables $(\mathbf{x}, \mathbf{x}_i)$ in each iteration of the underlying optimization, a full system analysis is required to compute the associated state vector $\mathbf{y}$, generally consuming large amounts of computational resources. Nevertheless, this is the only method for which each intermediate iterate is MDF (thus the name), therefore allowing the early termination of the optimization while still providing a final solution that is overall feasible and hence realizable in practice.

Individual-Discipline-Feasible (IDF)

This method, which is also known as distributed analysis and optimization (DAO) [6, 7, 8, 9], avoids the computation of the state vector $\mathbf{y}$ using a full system analysis and uses a set of auxiliary surrogate variables $\bar{\mathbf{y}}$ that allow to decouple the MDO equations (12.8d). Then starting from initial states $\bar{\mathbf{y}}_i$ for all i, in each iteration we solve

$$\text{IDF: } \underset{\mathbf{x}, \mathbf{x}_1, \ldots, \mathbf{x}_k}{\text{Minimize}} \; f(\mathbf{x}, \mathbf{x}_1, \ldots, \mathbf{x}_k, \mathbf{y}_1, \ldots, \mathbf{y}_k) \tag{12.9a}$$

$$\text{subject to } \mathbf{g}(\mathbf{x}, \mathbf{x}_1, \ldots, \mathbf{x}_k, \mathbf{y}_1, \ldots, \mathbf{y}_k) \leq \mathbf{0} \tag{12.9b}$$

$$\mathbf{g}_i(\mathbf{x}, \mathbf{x}_i, \mathbf{y}_i) \leq \mathbf{0}, \; i = 1, \ldots, k \tag{12.9c}$$

$$\text{where } \mathbf{h}_i(\mathbf{x}, \mathbf{x}_i, \bar{\mathbf{y}}) - \mathbf{y}_i = \mathbf{0}, \; i = 1, ..., k \tag{12.9d}$$

in which each original state variable $\mathbf{y}_i$ is easily computable using a single function evaluation of $\mathbf{h}_i(\mathbf{x}, \mathbf{x}_i, \bar{\mathbf{y}})$. In particular, like in the MDF method these states are still related through the system constraints $\mathbf{g}$ but not anymore through the coupling functions $\mathbf{h}_i$. Hence, after each iteration of the optimization, we can update the surrogate variables $\bar{\mathbf{y}}_i = \mathbf{h}_i(\mathbf{x}, \mathbf{x}_i, \mathbf{y})$ using a decoupled state analysis and obtain new solutions for all disciplines that are IDF (thus the name), but usually not MDF before converging to an optimal (fixed-point) solution for which $\mathbf{y}_i = \mathbf{h}_i(\mathbf{x}, \mathbf{x}_i, \mathbf{y})$ for all $i = 1, \ldots, k$.

No-Discipline-Feasible (NDF)

Different from the two previous methods that solve either a full (in the case of MDF) or decoupled system analysis (in the case of IDF) after every iteration of the optimization, this method converts the MDO equations (12.8d) into explicit equality constraints and treats the state vectors $\mathbf{y}$ as additional decision or optimization variables. Thus also known as simultaneous analysis and design (SAND) with no further distinction between the two, this method solves the all-at-once (AAO) formulation

$$\text{AAO:} \quad \underset{\mathbf{x},\mathbf{x}_1,\ldots,\mathbf{x}_k,\mathbf{y}_1,\ldots,\mathbf{y}_k}{\text{Minimize}} \quad f(\mathbf{x}, \mathbf{x}_1, \ldots, \mathbf{x}_k, \mathbf{y}_1, \ldots, \mathbf{y}_k) \tag{12.10a}$$

$$\text{subject to} \quad \mathbf{g}(\mathbf{x}, \mathbf{x}_1, \ldots, \mathbf{x}_k, \mathbf{y}_1, \ldots, \mathbf{y}_k) \leq \mathbf{0} \tag{12.10b}$$

$$\mathbf{g}_i(\mathbf{x}, \mathbf{x}_i, \mathbf{y}_i) \leq \mathbf{0}, \; i = 1, \ldots, k, \tag{12.10c}$$

$$\mathbf{h}_i(\mathbf{x}, \mathbf{x}_i, \mathbf{y}) - \mathbf{y}_i = \mathbf{0}, \; i = 1, \ldots, k \tag{12.10d}$$

in which discipline feasibility is established as inherent part of the optimization. In particular, it allows to employ any suitable optimization technique including the possibility to further relax the equality constraints (12.10d) and permit nonzero residuals yielding intermediate iterates that are neither MDF nor IDF, in general. Like in the other methods, however, the final optimal solution to AAO will again be MDF.

12.3.1.2 Multi-Level Penalty-Based Methods

The following methods replace the single-level MDO formulation by a hierarchical scheme that eliminates all direct interactions among the local decision vectors $\mathbf{x}_i$ of the different disciplines and maintains intermediate iterates that are IDF with respect to their own disciplinary constraints $\mathbf{g}_i$ at the subproblem level, while hiding such details from a coordinating master problem that merely sets common targets and penalizes deviations at the system level and thus yields a consistent overall solution.

Collaborative Optimization (CO)

This method is originally developed in the doctoral thesis by Braun [18, 19] and since then has also become known as inexact penalty decomposition (IPD) [45, 46]. Using a bilevel formulation that promotes disciplinary autonomy while simultaneously achieving interdisciplinary compatibility, a top-level problem controls the system-level decision variables $\mathbf{x}$ and the interdisciplinary coupling variables $\mathbf{y}_i$, which then serve as targets for each subsystem that only controls its disciplinary decision variable $\mathbf{x}_i$ and a local surrogate variable $\hat{\mathbf{x}}_i$ replacing the global variable $\mathbf{x}$.

$$\text{System:} \quad \underset{\mathbf{x},\mathbf{x}_1,\dots,\mathbf{x}_k,\mathbf{y}_1,\dots,\mathbf{y}_k}{\text{Minimize}} \quad f(\mathbf{x},\mathbf{y}_1,\dots,\mathbf{y}_k) \tag{12.11a}$$

$$\text{subject to} \quad \|\mathbf{x}-\hat{\mathbf{x}}_i\| + \|\mathbf{y}_i-\hat{\mathbf{y}}_i\| = 0,\ i=1,\dots,k. \tag{12.11b}$$

$$\text{Subsystem:} \quad \underset{\hat{\mathbf{x}}_i,\mathbf{x}_i}{\text{Minimize}} \quad \|\mathbf{x}-\hat{\mathbf{x}}_i\| + \|\mathbf{y}_i-\hat{\mathbf{y}}_i\| \tag{12.11c}$$

$$\text{subject to} \quad \mathbf{g}_i(\hat{\mathbf{x}}_i,\mathbf{x}_i,\hat{\mathbf{y}}_i) \leq \mathbf{0},\ i=1,\dots,k, \tag{12.11d}$$

$$\text{where} \quad \mathbf{h}_i(\hat{\mathbf{x}}_i,\mathbf{x}_i,\mathbf{y}) - \hat{\mathbf{y}}_i = \mathbf{0},\ i=1,\dots,k. \tag{12.11e}$$

On the subsystem level, each discipline chooses its new local decisions $\mathbf{x}_i$ and proposes new global variables $\hat{\mathbf{x}}_i$ that minimize the Euclidean distance from its disciplinary feasible region to the system-level target values for global decisions $\mathbf{x}$ and coupling variables $\mathbf{y}_i$. Intuitively, this means that the individual disciplines collaborate (thus the method's name) in determining an optimal set of targets for both global decision variables and their associated system states. The corresponding local states $\hat{\mathbf{y}}_i = \mathbf{h}_i(\hat{\mathbf{x}}_i,\mathbf{x}_i,\mathbf{y})$ are obtained through a decoupled system analysis using the coupling variables $\mathbf{y}$ that are fixed at the system level and kept constant at the subsystem level. As a consequence, however, and like in the IDF method (12.9), intermediate iterates do not necessarily satisfy the system level consistency constraints (12.11b) and, thus, are not realizable in practice, in general. In addition, first-order (KKT) conditions for optimality usually do not hold for consistency constraints (12.11b) also giving rise to possible technical difficulties during the optimization process [141]. Nevertheless, this method has been successfully applied to a series of different problems also including some cases of multidisciplinary systems with multiple objectives [129, 189].

Analytical Target Cascading (ATC)

First developed in the doctoral thesis by Kim [101] and later applied to optimal system, automotive, and product design [34, 91, 92], this multilevel method inherently allows for multiple objectives and attempts to find a decision $\mathbf{x}$ whose outcome $\mathbf{f}(\mathbf{x})$ achieves or approximates some target $\mathbf{T}$ using a goal programming formulation

$$\text{ATC:} \quad \underset{\mathbf{x}}{\text{Minimize}} \quad \|\mathbf{R}-\mathbf{T}\| \tag{12.12a}$$

$$\text{subject to} \quad \mathbf{g}(\mathbf{x}) \leq \mathbf{0} \tag{12.12b}$$

$$\text{where} \quad \mathbf{R} = \mathbf{f}(\mathbf{x}). \tag{12.12c}$$

Note here that quite different from the original MDO formulation and closest in spirit to the AAO formulation (12.10) of the NDF method, ATC does not necessarily assume a special problem structure but initially combines all decision and state variables into one decision vector $\mathbf{x}$, and all system, subsystem, and coupling constraints into one vector inequality $\mathbf{g}$. Consequently, this method can also be applied to more general optimization problems and decision situations, and still make effective use of any underlying structure if it exists. Again considering MDO as our main problem of interest, for ATC we also decompose the overall problem into k

subsystems with local decision variables $\mathbf{x}_i$, associated state vectors $\mathbf{y}_i$, and system responses $\mathbf{R}_i$

$$
\begin{aligned}
\text{System: } & \underset{\mathbf{x},\mathbf{y}_i,\mathbf{R}_i,\boldsymbol{\delta},\boldsymbol{\varepsilon}}{\text{Minimize}} && \|\mathbf{R}-\mathbf{T}\| + \boldsymbol{\delta} + \boldsymbol{\varepsilon} && (12.13a)\\
& \text{subject to} && \sum_{i=1}^{k} \|\mathbf{v}_i \circ (\mathbf{y}_i - \hat{\mathbf{y}}_i)\| \leq \boldsymbol{\delta} && (12.13b)\\
& && \sum_{i=1}^{k} \|\mathbf{w}_i \circ (\mathbf{R}_i - \hat{\mathbf{R}}_i)\| \leq \boldsymbol{\varepsilon} && (12.13c)\\
& && \mathbf{g}(\mathbf{x},\mathbf{R}) \leq \mathbf{0} && (12.13d)\\
& \text{where} && \mathbf{R} = \mathbf{f}(\mathbf{x},\mathbf{R}_1,\ldots,\mathbf{R}_k) && (12.13e)\\
\text{Subsystem: } & \underset{\mathbf{x}_i,\hat{\mathbf{y}}_i,\hat{\mathbf{R}}_i}{\text{Minimize}} && \|\mathbf{v}_i \circ (\mathbf{y}_i - \hat{\mathbf{y}}_i)\| + \|\mathbf{w}_i \circ (\mathbf{R}_i - \hat{\mathbf{R}}_i)\| && (12.13f)\\
& \text{subject to} && \mathbf{g}_i(\mathbf{x}_i,\hat{\mathbf{y}}_i,\hat{\mathbf{R}}_i) \leq \mathbf{0} && (12.13g)\\
& \text{where} && \hat{\mathbf{R}}_i = \mathbf{r}_i(\mathbf{x}_i,\hat{\mathbf{y}}_i) && (12.13h)
\end{aligned}
$$

where the notation $\circ$ is used to denote the elementwise product between the penalty weights $\mathbf{v}_i$ and $\mathbf{w}_i$, and the deviations between the local coupling variables $\hat{\mathbf{y}}_i$, the subsystem responses $\hat{\mathbf{R}}_i$, and their top-level targets $\mathbf{y}_i$ and $\mathbf{R}_i$, respectively, that also form the objective on the subsystem level. The same deviations are then converted to relaxed consistency constraints with associated penalty terms $\boldsymbol{\delta}$ and $\boldsymbol{\varepsilon}$ added to the objective function of the system-level problem, that iteratively updates all global decision and state variables and coordinates target responses until converging to a consistent and overall optimal solution. Finally, using that inequalities (12.13b) and (12.13c) necessarily hold with equality at optimality, an alternative approach to ATC replaces the above bilevel method with the single-level all-in-one (AIO) formulation

$$
\begin{aligned}
\text{AIO: } & \underset{\mathbf{x},\mathbf{x}_i,\mathbf{y}_i,\hat{\mathbf{y}}_i,\mathbf{R}_i,\hat{\mathbf{R}}_i}{\text{Minimize}} && \|\mathbf{R}-\mathbf{T}\| + \sum_{i=1}^{k} \left(\|\mathbf{v}_i \circ (\mathbf{y}_i - \hat{\mathbf{y}}_i)\| + \|\mathbf{w}_i \circ (\mathbf{R}_i - \hat{\mathbf{R}}_i)\|\right) && (12.14a)\\
& \text{subject to} && \mathbf{g}(\mathbf{x},\mathbf{R}) \leq \mathbf{0} && (12.14b)\\
& && \mathbf{g}_i(\mathbf{x}_i,\hat{\mathbf{y}}_i,\hat{\mathbf{R}}_i) \leq \mathbf{0},\ i=1,\ldots,k && (12.14c)\\
& \text{where} && \mathbf{R} = \mathbf{f}(\mathbf{x},\mathbf{R}_1,\ldots,\mathbf{R}_k) && (12.14d)\\
& && \mathbf{R}_i = \mathbf{r}_i(\mathbf{x}_i,\hat{\mathbf{y}}_i),\ i=1,\ldots,k. && (12.14e)
\end{aligned}
$$

12.3.1.3 Global-Sensitivity-Based Methods

Among the many other existing approaches to solve MDO, several methods are based on the so-called global sensitivity equations (GSE) [172] that during the last few years have received significant attention by both researchers and practitioners. Since these methods depend on a variety of additional analytical concepts and mathematical tools whose description would be out of scope for the expository nature of our discussion, however, we refrain from presenting further details in this chapter but merely summarize the major characteristics of the two most common of these methods: concurrent subspace optimization, and bilevel integrated system synthesis.

Concurrent Subspace Optimization (CSSO)

This method uses a hierarchical decomposition-coordination scheme that in each iteration optimizes decomposed subsystems concurrently, and subsequently coordinates individual solutions and problem parameters to resolve subsystem conflicts and achieve overall compatibility and convergence. Different from other multi-level schemes, however, the CSSO coordination handles all decision variables simultaneously and repeatedly modifies their allocation to subsystems using sensitivity information: in each iteration, CSSO first calculates partial derivatives and solves the associated GSE to approximate their couplings using a first-order Taylor expansion, resulting in a decoupled system and allowing to construct a single cumulative-constraint function for each individual discipline. This function is used to calculate effectiveness coefficients for every discipline and every design variable, based on which each design variable is allocated to that discipline in which it has greatest impact. Based on a matrix of scaled cumulative-constraint sensitivities and additional responsibility, tradeoff, and switch coefficients, the subsystems are then solved concurrently and analyzed using another sensitivity analysis, after which the overall coordination problem resolves any remaining conflicts at the system level. Although dependent on several mathematical assumptions and first-order approximations that may cause convergence difficulties when applied to highly nonlinear problems [103], in each iteration of the optimization process CSSO provides a solution that is MDF, in theory. After its successful application to single objective problems [15], this method has also been extended to more general MOP [87, 88].

Bi-Level Integrated System Synthesis (BLISS)

This method was originally developed for the design of aerospace vehicles and later modified for more general applications in structural optimization, that comprise different interacting modules that are to be optimized independently and then coordinated by a final system-level optimization [175, 174]. With BLISS, the overall system is decomposed into a set of local subsystems dealing with the large number of local decision and design variables, and a global system-level optimization dealing with a relatively smaller number of global variables also in comparison to other MDO methods. Like the MDF method, for example, BLISS also performs a complete system analysis at the beginning of each iteration to only consider decisions that are MDF, and uses the same overall objective function for both system and subsystem problems similar to the synthesis between targets and responses in the bilevel formulation of ATC. Like CSSO based on GSE, the solution of the system-level problem is obtained using either the optimum sensitivity derivatives of the state variables with respect to the underlying system-level decision variables and Lagrangean multipliers of those constraints that are active at the solution of the disciplinary optimization, or a response surface constructed from the solutions of the global system analysis and the individual subsystems. Main benefits in practice are autonomy of individual substructures in especially large-scale design applications.

12.3.2 Hierarchical Multiobjective Analysis

Similar to the multidisciplinary optimization (MDO) methods in Section 12.3.1, that focus primarily on the simultaneous participation of different engineering design teams to accelerate or otherwise facilitate the solution, coordination, and integration of a set of decomposed subsystems into a consistent overall solution for a global and complex decision problem, there also exist various methods within the rich literature on traditional multilevel or hierarchical analysis (modeling, programming, optimization) methods to better understand, represent, solve, or improve decision processes in large hierarchical organizations with either a single objective [70, 71, 81, 86, 112, 125, 126, 131, 132, 153, 158, 164, 165, 167, 168, 169, 171, 182, 181, 208] or multiple decision criteria [1, 31, 47, 73, 72, 80, 82, 121, 147, 163, 188, 195].

In addition to some of the key methods mentioned in the above references, a seemingly endless number of other variations and extensions can be found in the context of practical applications prominently including problems in resource allocation, regional, and environmental planning [11, 82, 104, 116, 139, 148, 150], production planning and pricing [84, 211], and in connection with MDO for design and systems engineering [29, 183]. Finally, an important although significantly less studied area is the consideration of computational aspects by providing numerical comparisons and benchmarks for existing multilevel optimization techniques [44].

In this section, we present a variety of multilevel formulations and methods that are based on the particular model initially proposed by researchers at the Systems Engineering Department at Case Western University in Cleveland, Ohio (USA) [82, 194], which is chosen because of its general simplicity and continued use by both decision-makers and researchers during the last few years, most recently the Málaga Group of Multicriteria Analysis (MAGMA) at University Málaga (Spain) [21, 76].

$$\text{TOP:} \quad \text{Minimize} \quad \mathbf{F}(\mathbf{f}) = (F_1(\mathbf{f}_1,\ldots,\mathbf{f}_k),\ldots,F_m(\mathbf{f}_1,\ldots,\mathbf{f}_k)) \tag{12.15a}$$

$$\text{MOP}_i\text{:} \quad \underset{\mathbf{x}_i}{\text{Minimize}} \quad \mathbf{f}_i(\mathbf{x}_i,\mathbf{y}_i,\mathbf{z}_i) \tag{12.15b}$$

$$\text{subject to} \quad \mathbf{g}_i(\mathbf{x}_i,\mathbf{y}_i,\mathbf{z}_i) \leq \mathbf{0} \tag{12.15c}$$

$$\text{where} \quad \mathbf{y}_i = \sum_{j=1}^{k} C_{ij}\mathbf{z}_j \text{ and } \mathbf{z}_i = \mathbf{h}_i(\mathbf{x}_i,\mathbf{y}_i) \tag{12.15d}$$

In this formulation, all decision details remain hidden on the top-level problem TOP whose objective vector $\mathbf{F} = (F_1, F_2, \ldots, F_m)$ depends solely on the combined vector of aggregated individual objectives $\mathbf{f} = (\mathbf{f}_1, \mathbf{f}_2, \ldots, \mathbf{f}_k)$ of the multiobjective subproblems MOP_i, $i = 1,\ldots,k$. Independent of any global decision variables, the objectives and constraints of all lower-level problems only include local decision variables $\mathbf{x}_i$ as well as the system inputs and outputs $\mathbf{y}_i$ and $\mathbf{z}_i$ that serve as coupling variables and are related through the coupling relation (12.15d). To reduce notational burden, we denote $\mathbf{x} = (\mathbf{x}_1,\ldots,\mathbf{x}_k)$ and, with the analogous definition of $\mathbf{y}$ and $\mathbf{z}$, let

$$S_i = \{(\mathbf{x}_i,\mathbf{y}_i,\mathbf{z}_i) : \mathbf{g}_i(\mathbf{x}_i,\mathbf{y}_i,\mathbf{z}_i) \leq \mathbf{0},\ \mathbf{h}_i(\mathbf{x}_i,\mathbf{y}_i) - \mathbf{z}_i = \mathbf{0}\}, i = 1,\ldots,k \tag{12.16a}$$

$$S_i(\bar{z}) = \left\{(\mathbf{x}_i,\mathbf{y}_i,\mathbf{z}_i) \in S_i : \mathbf{z}_i = \bar{\mathbf{z}}_i,\ \mathbf{y}_i = \sum_{j=1}^{k} C_{ij}\bar{\mathbf{z}}_j\right\}, i = 1,\ldots,k \tag{12.16b}$$

$$S = \left\{(\mathbf{x},\mathbf{y},\mathbf{z}) : (\mathbf{x}_i,\mathbf{y}_i,\mathbf{z}_i) \in S_i,\ \mathbf{y}_i = \sum_{j=1}^{k} C_{ij}z_j,\ i = 1,\ldots,k\right\} \tag{12.16c}$$

$$S(\bar{\mathbf{z}}) = \{(\mathbf{x},\mathbf{y},\mathbf{z}) : (\mathbf{x}_i,\mathbf{y}_i,\mathbf{z}_i) \in S_i(\bar{\mathbf{z}})\} = \{(\mathbf{x},\mathbf{y},\mathbf{z}) \in S : \mathbf{z} = \bar{\mathbf{z}}\} \tag{12.16d}$$

where the sets S_i and S correspond to the MDO notions of individual and multiple discipline feasible solutions introduced in Section 12.3.1, respectively, while $S_i(\bar{\mathbf{z}})$ and $S(\bar{\mathbf{z}})$ explicitly fix all system outputs $\mathbf{z}$ which implicitly determines the system inputs $\mathbf{y}$ and thus decouples the input-output relation (12.15d) of the individual subproblems.

Based on the actual decision hierarchy and organizational structure of the given problem of interest, we typically distinguish between three decision situations with different numbers of decision makers (DM): (i) one DM for the complete problem with efficient set $E(S,\mathbf{F})$, (ii) k DM for the individual subproblems with efficient set $E(S,\mathbf{f})$, and (iii) $k+1$ DM for the top and the individual subproblems with efficient set $E(S,(\mathbf{F},\mathbf{f}))$. The next Theorem 12.1, however, shows that under a consistency assumption on the overall objectives $\mathbf{F}$ with respect to the individual subproblem objectives $\mathbf{f}$, cases (i) and (iii) can be reduced to (ii) so that we follow most other researchers and in the following restrict our consideration to the second case only.

Theorem 12.1 (Theorem 4.1 in [82], page 81). *Let* $\mathbf{F}$ *be a monotone (increasing) function in* $\mathbf{f}$ *so that* $\mathbf{F}(\mathbf{f}) \leq \mathbf{F}(\mathbf{f}')$ *whenever* $\mathbf{f} \leq \mathbf{f}'$. *Then* $E(S,(\mathbf{F},\mathbf{f})) = E(S,\mathbf{f})$.

Primarily focusing on parts of the research undertaken by the Case Western and MAGMA groups, we now illustrate some of the basic decomposition-coordination schemes, all of which will follow the same general pattern: first, based on an underlying optimality (or coordination) condition for the original overall problem, certain optimization variables are fixed at the upper level and then used as coordination parameters to decouple these conditions and decompose the overall problem into a set of smaller and independent local optimization problems. Second, after these problems have been solved on the lower level either sequentially or in parallel, a coordinating master problem checks the remaining conditions for the top-level problem and accordingly updates the coordination parameters until converging to a final solution at which the coordination conditions are also met for the original problem.

12.3.2.1 Hierarchical Multiobjective Optimization (Case Western Group)

Based on their lasting impact also on many of the more recent decomposition-coordination methods, first we briefly review some of the major schemes developed by researchers at the Systems Engineering Department at Case Western University who seem to be first to propose model (12.15) and supply much of its initial analysis.

Hierarchical Generating Method

This method is developed for the generation of efficient decisions for the overall problem, $E(S,\mathbf{f})$, based on the computation of efficient decisions for the individual subproblems, $E(S_i,\mathbf{f}_i)$, together with the observation that any given set of values for the output variables $\mathbf{z}$ uniquely determines the input variable $\mathbf{y}$ and, therefore, can be used as suitable coordination parameter to decouple the individual subproblems. Hence, after setting all coupling variables to some fixed values $\bar{\mathbf{z}}$, this method solves

$$\text{MOP}_i(\bar{\mathbf{z}}):\ \underset{(\mathbf{x}_i,\mathbf{y}_i,\mathbf{z}_i)\in S_i(\bar{\mathbf{z}})}{\text{Minimize}}\ \mathbf{f}_i(\mathbf{x}_i,\mathbf{y}_i,\mathbf{z}_i) = (\mathbf{f}_{i1}(\mathbf{x}_i,\mathbf{y}_i,\mathbf{z}_i),\dots,\mathbf{f}_{im_i}(\mathbf{x}_i,\mathbf{y}_i,\mathbf{z}_i)) \tag{12.17}$$

and then checks overall efficiency using the coordination condition in Theorem 12.2.

Theorem 12.2 (Theorem 4.4 in [82], page 87). *Let $\mathbf{f}_i$ and $\mathbf{g}_i$ be convex, $\mathbf{h}_i$ be linear, and all functions be continuously differentiable. Under some regularity condition, $(\mathbf{x},\mathbf{y},\mathbf{z}) \in E(S,\mathbf{f})$ if and only if $(\mathbf{x}_i,\mathbf{y}_i,\mathbf{z}_i) \in E(S_i(\mathbf{z}),\mathbf{f}_i)$ and there is $\boldsymbol{\lambda} > 0$ such that*

$$\lambda_i(\boldsymbol{\pi}_i^z)^T + \sum_{j=1}^{k} \lambda_j(\boldsymbol{\pi}_j^y)^T C_{ji} = 0 \text{ for all } i = 1,\dots,k \tag{12.18}$$

where $\boldsymbol{\pi}_i^y$ and $\boldsymbol{\pi}_i^z$ are the Lagrangean multipliers associated with constraints (12.15d).

Hence, if this condition can be verified for a given solution to the subproblems, then it is also an efficient decision for the overall problem and can be presented as candidate solutions to DM. Otherwise, if the condition does not hold, or if the DM is interested in generating additional solutions, then we may interactively vary the output variables $\bar{\mathbf{z}}$ to repeat the process and sequentially influence the continued search for a final preferred decisions for the overall problem. Further technical details with some early applications can be found in the original references [82, 118, 119, 195].

Feasible and Nonfeasible Coordination Methods

While the above generating method gives DM lots of flexibility in choosing appropriate output variables, however, it does not provide any particular information on how to actually modify these coordination parameters as to guarantee that the resulting solution is indeed efficient for the overall problem. In remedy of this drawback, two more sophisticated coordination schemes respectively known as model and goal coordination utilize the interaction (or coupling) variables $\mathbf{z}$ and the Lagrangean multipliers $\boldsymbol{\pi}_i^y$ and $\boldsymbol{\pi}_i^z$ in Theorem 12.2 as coordination parameters. These methods are based on their well-established single-objective analogons also known as model-coordination or feasible (primal, prediction) method [73, 131, 132], and goal-coordination or nonfeasible (interaction balance, price) method [169, 168, 208] which indicates that only the first method guarantees that all intermediate iterates are MDF and satisfy the coupling (12.15d) even before the final coordination is completed.

Feasible Method

Similar to the previous hierarchical generating method, this method also fixes the system outputs $\mathbf{z}$ which then uniquely determine the system inputs $\mathbf{y}$ and thus allows to decouple the individual subproblems as in (12.17). Different from before, however, the subsequent model coordination utilizes an additional trade-off vector $\boldsymbol{\tau}(\mathbf{f}(\mathbf{x},\bar{\mathbf{y}},\bar{\mathbf{z}}))$ that is obtained from a surrogate worth trade-off (SWT) analysis [22, 23, 78, 79, 83] of the partial derivatives of the objective and constraint functions with respect to the coordination parameter $\mathbf{z}_i$ and used to incorporate preference tradeoffs of DM into the following necessary condition for a preferred solution

$$(\boldsymbol{\tau}_i(\mathbf{f}_i))^T \frac{\partial \mathbf{f}_i}{\partial \mathbf{z}_i} + (\boldsymbol{\pi}_i^g)^T \frac{\partial \mathbf{g}_i}{\partial \mathbf{z}_i} - \boldsymbol{\pi}_i^h + \sum_{j=1}^{k} \frac{\tau_{jr}}{\tau_{ir}} (\boldsymbol{\lambda}_j)^T C_{ji} = \mathbf{0},\ i = 1,\ldots,k \tag{12.19}$$

where $\boldsymbol{\pi}^g \geq \mathbf{0}$, $\boldsymbol{\pi}^h$, and $\boldsymbol{\lambda}$ are the Lagrangean multipliers associated with constraints $\mathbf{g}$ and $\mathbf{h}$ in (12.15c) and (12.15d), and the coupling relation in (12.15d) or (12.16a), respectively, and τ_{ir} is the tradeoff with respect to some reference objective $\mathbf{f}_r$. Rather than varying the system outputs $\mathbf{z}$ arbitrarily, a proper set of values can now be found by solving the coordination equation (12.19) and updating its coordination variables using, e.g., a gradient-type method [72]. Moreover, to find multiple solutions, the DM may still interactively change or provide new tradeoffs $\boldsymbol{\tau}$ and thus control or guide the search for a preferred efficient decision for the overall problem until reaching some personal aspirations and goals. Clearly, in the presence of additional subsystem and system constraints, the above condition may become slightly more complicated because total derivatives may not exist and need to be replaced by directional derivatives, discussed in more technical detail in the original references [194, 195].

Nonfeasible Method

Other than in the feasible method, intermediate iterates in the nonfeasible method are typically not feasible for any subproblem and generated based on a Lagrangean relaxation scheme that uses the multipliers $\boldsymbol{\pi}^g$, $\boldsymbol{\pi}^h$, and $\boldsymbol{\lambda}$ as coordination variables, with the corresponding coupling constraints (12.15d) added as relaxation term to the first of each subproblem objective. Like before, this method also uses a given tradeoff vector $\boldsymbol{\tau}(\mathbf{f}(\mathbf{x},\mathbf{y},\mathbf{z}))$ and initial multipliers $\boldsymbol{\lambda}$ to solve the decoupled subproblems

$$\text{MOP}_i(\bar{\boldsymbol{\tau}},\bar{\boldsymbol{\lambda}}): \underset{(\mathbf{x}_i,\mathbf{y}_i,\mathbf{z}_i)\in S_i}{\text{Minimize}} \left(f_{i1} + \sum_{j=1}^{k} \frac{\bar{\tau}_{jr}}{\bar{\tau}_{ir}} (\bar{\boldsymbol{\lambda}}_j)^T C_{ji}\mathbf{z}_i - (\bar{\boldsymbol{\lambda}}_i)^T y_i, f_{i2}, \ldots, f_{im_i} \right) \tag{12.20}$$

and then checks both the coupling constraints (12.15d) and the a priori tradeoffs $\bar{\boldsymbol{\tau}}$ for consistency with the actually obtained tradeoffs $\boldsymbol{\tau}$ at the new optimal objective value

$$\mathbf{y}_i - \sum_{j=1}^{k} C_{ij}\mathbf{z}_j = \mathbf{0} \quad \text{and} \quad \boldsymbol{\tau}_i(\mathbf{f}_i) - \bar{\boldsymbol{\tau}}_i = \mathbf{0}, \quad i = 1,\ldots,k. \tag{12.21}$$

Other Coordination Methods and Applications

In addition to the hierarchical, feasible, and nonfeasible coordination methods, numerous other approaches exist to further extend these basic schemes for decision situations in which tradeoffs also depend on objectives in other subproblems [195] or involve dynamic discrete-time systems with temporal multilevel hierarchies [193, 196]. Prominently including the hierarchical overlapping method, hierarchical holographic modeling, and the envelope approach [117, 119], these and some other methods are described in comprehensive detail in the collective monograph [82] together with several large-scale applications in the context of long-term planning of power systems and energy storage [194], and policy decisions in regional and environmental planning [140]. Finally, for group hierarchies involving multiple decision-makers, for which the above schemes may cause convergence difficulties due to inconsistent specification of tradeoffs, several other extensions also use the formulation of Stackelberg games [177], the inclusion of tradeoff negotiation [83], or the adaption of one of several existing follower-leader schemes [37, 38, 36, 166].

12.3.2.2 Hierarchical Multiobjective Optimization (Málaga Group)

Following some of the methods in Section 12.3.2.1 that are developed by the research group at Case Western, the Málaga Group of Multicriteria Analysis (MAGMA) of the University at Málaga (Spain) has also continued recent work in the area of hierarchical multiobjective optimization leading to several interesting new contributions.

Hierarchical Generating Method

This new generating method is based on the direct observation [21, Theorem 1] that

$$E(S,\mathbf{f}) = E\left(\bigcup\nolimits_{\bar{\mathbf{z}}} E(S(\bar{\mathbf{z}}),\mathbf{f})\right) \tag{12.22}$$

and applies the feasible method from Section 12.3.2.1 to the weighted-sum subproblem

$$V_i(\bar{\mathbf{z}}) = \underset{(\mathbf{x}_i,\mathbf{y}_i,\mathbf{z}_i)\in S_i(\bar{\mathbf{z}})}{\text{Minimize}} \ \mathbf{w}_i^T \mathbf{f}_i(\mathbf{x}_i,\mathbf{y}_i,\mathbf{z}_i). \tag{12.23}$$

Theorem 12.3 (Theorem 2 in [21]). *Let* $\mathbf{f}_i$, $\mathbf{g}_i$, *and* $\mathbf{h}_i$ *be as in Theorem 12.2. Under some regularity condition,* $(\mathbf{x},\mathbf{y},\mathbf{z}) \in E_p(S,\mathbf{f})$ *if and only if* $(\mathbf{x}_i,\mathbf{y}_i,\mathbf{z}_i) \in E_p(S_i(\mathbf{z}),\mathbf{f}_i)$ *and* $\mathbf{z}$ *is a properly efficient solution for the combined weighted-sum coordination problem*

$$\underset{\mathbf{z}}{\text{Minimize}}\, \mathbf{V}(\mathbf{z}) = (V_1(\mathbf{z}), V_2(\mathbf{z}), \ldots, V_k(\mathbf{z})). \tag{12.24}$$

Based on Theorem 12.3 and using the Lagrangean functions for the subproblems $V_i(\mathbf{z})$

$$L_i(\mathbf{x}_i,\mathbf{y}_i,\mathbf{z}_i) = \mathbf{w}_i^T \mathbf{f}_i(\mathbf{x}_i,\mathbf{y}_i,\mathbf{z}_i) + (\boldsymbol{\pi}_i^g)^T \mathbf{g}_i(\mathbf{x}_i,\mathbf{y}_i,\mathbf{z}_i) + (\boldsymbol{\pi}_i^h)^T (\mathbf{h}_i(\mathbf{x}_i,\mathbf{y}_i) - \mathbf{z}_i) \tag{12.25a}$$

$$+ (\boldsymbol{\pi}_i^y)^T (\mathbf{y}_i - \sum_{j=1}^{k} C_{ij}\bar{\mathbf{z}}_j) + (\boldsymbol{\pi}_i^z)^T (\mathbf{z}_i - \bar{\mathbf{z}}_i) \tag{12.25b}$$

the same coordination condition as in Theorem 12.2 (12.18) can be derived for Theorem 12.3.

Lagrangian Duality Approach

While the above method uses the Lagrangean function for a feasible method, this second approach results in a nonfeasible method using the Lagrangean dual problem

$$\phi(\boldsymbol{\lambda}) = \underset{(\mathbf{x}_i,\mathbf{y}_i,\mathbf{z}_i)\in S_i}{\text{Minimize}} \sum_{i=1}^{k} \left(\mathbf{w}_i^T \mathbf{f}_i(\mathbf{x}_i,\mathbf{y}_i,\mathbf{z}_i) + \boldsymbol{\lambda}_i^T (\mathbf{y}_i - \sum_{j=1}^{k} C_{ij}\mathbf{z}_j) \right) \tag{12.26}$$

whose objective function is separable and, hence, decomposable into the k functions

$$\tilde{\mathbf{f}}_i(\boldsymbol{\lambda},\mathbf{x}_i,\mathbf{y}_i,\mathbf{z}_i) = \mathbf{f}_i(\mathbf{x}_i,\mathbf{y}_i,\mathbf{z}_i) + \frac{1}{\mu_i}(\boldsymbol{\lambda}_i^T \mathbf{y}_i - \sum_{j=1}^{k} \boldsymbol{\lambda}_j^T C_{ji}\mathbf{z}_i)\mathbf{e} \tag{12.27}$$

where $\mathbf{e} = \{1,\ldots,1\}$ is the vector of all ones, and $\mu_i = \sum_j w_{ij}$. Different from the nonfeasible method (12.20), however, in this formulation the relaxed coupling constraints do not only contribute to the first objective but are divided among all participating objectives and proportional to the weight μ_i assigned to each resulting subproblem

$$\phi_i(\boldsymbol{\lambda}) = \underset{(\mathbf{x}_i,\mathbf{y}_i,\mathbf{z}_i)\in S_i}{\text{Minimize}} \ \tilde{\mathbf{w}}_i^T \tilde{\mathbf{f}}_i(\boldsymbol{\lambda},\mathbf{x}_i,\mathbf{y}_i,\mathbf{z}_i) \tag{12.28}$$

where the modified weights $\tilde{\mathbf{w}}$ are chosen to satisfy $\mu_i \tilde{w}_i = \mathbf{w}_i$. Like other nonfeasible methods fixing values for $\boldsymbol{\lambda}$ as coordination parameters, the following results hold.

Theorem 12.4 (Theorem 1 in [137]). *Let* $\mathbf{f}_i$*,* $\mathbf{g}_i$*, and* $\mathbf{h}_i$ *be as in Theorem 12.2. Under some regularity assumption, if* $(\mathbf{x},\mathbf{y},\mathbf{z}) \in E_p(S,f)$*, then there exist* $\boldsymbol{\lambda}$ *such that* $(\mathbf{x}_i,\mathbf{y}_i,\mathbf{z}_i) \in E_p(S_i,\tilde{\mathbf{f}}_i(\boldsymbol{\lambda}))$*, where* $\tilde{\mathbf{f}}_i(\boldsymbol{\lambda})$ *denotes the function defined in* (12.27)*, for all* $i = 1,\ldots,k$*.*

The opposite direction of Theorem 12.4, however, is generally not true because a subproblem solution may not satisfy the coupling relationships (12.15d), or ignore the possibility to improve objectives that are dropped in its respective subproblem [60].

Theorem 12.5 (Theorem 2 in [137]). *Let* $\mathbf{f}_i$*,* $\mathbf{g}_i$*, and* $\mathbf{h}_i$ *be as before, and at least one* f_{ij} *be strictly convex. Under some regularity assumption,* $(\mathbf{x},\mathbf{y},\mathbf{z}) \in E_p(S,\mathbf{f})$ *if and only if there exist* $\boldsymbol{\lambda}$ *such that* $(\mathbf{x}_i,\mathbf{y}_i,\mathbf{z}_i) \in E_p(S_i,\tilde{\mathbf{f}}_i(\boldsymbol{\lambda})))$ *for all i, and positive scalars* $\mu_i > 0$ *such that* $\boldsymbol{\lambda}$ *is a solution to the weighted-sum Lagrangean coordination problem*

$$\underset{\boldsymbol{\lambda}}{\text{Maximize}}\, \phi(\lambda) = \sum_{i=1}^{k} \mu_i \phi_i(\boldsymbol{\lambda}). \tag{12.29}$$

Based on Theorem 12.5, the coordination mechanism first assigns the weights $\mu_i > 0$ and fixes the multipliers $\boldsymbol{\lambda}$ associated with the coupling constraints to separate the individual subproblems, which then can be solved independently at the lower level with the weights $\mathbf{w}_i$ for each objective $\mathbf{f}_i$ chosen by the DM. Resulting in an initial set of properly efficient solutions, in an iterative process the upper level continues to adjust $\boldsymbol{\lambda}$ to maximize the coordination function $\phi(\boldsymbol{\lambda})$ before reoptimization on the subproblem level until no further improvement is possible. Taking an active role, the DM may interactively vary the objective weights to generate multiple solutions or to guide the search for a decision that meets given tradeoffs or personal preferences.

Hierarchical Goal Programming Method

Instead of using weights to steer DM preferences, this goal programming method is based on aspiration levels $\boldsymbol{\alpha} = (\alpha_1,\ldots,\alpha_k)$ for the overall objective $\mathbf{F}(\mathbf{f}) \leq \boldsymbol{\alpha}$, and $\boldsymbol{\beta}_i = (\beta_{i1},\ldots,\beta_{im_i})$ for the subproblem objectives $\mathbf{f}_i \leq \boldsymbol{\beta}_i$, resulting in the problem

$$\text{Find} \quad (\mathbf{x},\mathbf{y},\mathbf{z}) \in S \tag{12.30a}$$
$$\text{so that} \quad \mathbf{F}(\mathbf{f}) = \mathbf{F}(\mathbf{f}_1,\mathbf{f}_2,\ldots,\mathbf{f}_k) \leq \boldsymbol{\alpha} \tag{12.30b}$$
$$\mathbf{f}_i = \mathbf{f}_i(\mathbf{x}_i,\mathbf{y}_i,\mathbf{z}_i) \leq \boldsymbol{\beta}_i,\ i = 1,\ldots,k. \tag{12.30c}$$

Denoting the set of feasible solutions to (12.30), (12.30a, 12.30b), and (12.30a, 12.30c) by $S(\boldsymbol{\alpha},\boldsymbol{\beta})$, $S(\boldsymbol{\alpha})$, and $S(\boldsymbol{\beta})$, respectively, we obtain the following result similar to Theorem 12.1.

Theorem 12.6 ([76]). *Let* $\mathbf{F}$ *be monotone (increasing) in* $\mathbf{f}$ *so that* $\mathbf{F}(\mathbf{f}) \leq \mathbf{F}(\mathbf{f}')$ *whenever* $\mathbf{f} \leq \mathbf{f}'$*. If* $\mathbf{F}(\boldsymbol{\beta}) \leq \boldsymbol{\alpha}$*, then* $S(\boldsymbol{\alpha},\boldsymbol{\beta}) = S(\boldsymbol{\beta})$*. Moreover,* $S(\boldsymbol{\alpha}) = \bigcup_{\{\boldsymbol{\beta}:\mathbf{F}(\boldsymbol{\beta})\leq\boldsymbol{\alpha}\}} S(\boldsymbol{\beta})$*.*

The first statement of this theorem shows that a top-level DM is redundant if goals $\boldsymbol{\alpha}$ and $\boldsymbol{\beta}$ are consistent. Moreover, the second statement says that any (exclusive) top-level decision is subsumed in the set of feasible subproblem decisions with consistent goals. Finally, to generate a set of feasible decisions that minimize the deviation from these goals, this method uses a weighted goal-programming (GP) formulation

$$\text{GP}(\boldsymbol{\beta}): \underset{(\mathbf{x},\mathbf{y},\mathbf{z})\in S,\boldsymbol{\varepsilon}\geq \mathbf{0}}{\text{Minimize}} \sum_{i=1}^{k} \mathbf{w}_i^T \boldsymbol{\varepsilon}_i \text{ subject to } \mathbf{f}_i(\mathbf{x}_i,\mathbf{y}_i,\mathbf{z}_i) - \boldsymbol{\varepsilon}_i \leq \boldsymbol{\beta}_i,\ i=1,\ldots,k. \tag{12.31}$$

Essentially a weighted-sum method applied to the deviation variables $\boldsymbol{\varepsilon}$, problem GP($\boldsymbol{\beta}$) can be solved using any suitable optimization technique including the MDO approaches or the single-objective analogons of the feasible or nonfeasible methods.

Feasible coordination

As before, a feasible coordination fixes system outputs $\bar{\mathbf{z}}$ and solves the subproblems

$$V_i(\bar{\mathbf{z}}) = \underset{(\mathbf{x}_i,\mathbf{y}_i,\mathbf{z}_i)\in S_i(\bar{\mathbf{z}}),\boldsymbol{\varepsilon}_i\geq \mathbf{0}}{\text{Minimize}} \quad \mathbf{w}_i^T \boldsymbol{\varepsilon}_i \text{ subject to } \mathbf{f}_i(\mathbf{x}_i,\mathbf{y}_i,\mathbf{z}_i) - \boldsymbol{\varepsilon}_i \leq \boldsymbol{\beta}_i \tag{12.32}$$

analogously to (12.23), then yielding the same result as already stated in Theorem 12.3.

Nonfeasible coordination

For a nonfeasible coordination, similar to problem (12.26) we relax the coupling constraints (12.15d) with associated Lagrangean multipliers $\boldsymbol{\lambda}$ as coordination parameters

$$\phi(\boldsymbol{\lambda}) = \underset{(\mathbf{x}_i,\mathbf{y}_i,\mathbf{z}_i)\in S_i,\ \boldsymbol{\varepsilon}_i\geq 0}{\text{Minimize}} \sum_{i=1}^{k}\left(\mathbf{w}_i^T \boldsymbol{\varepsilon}_i + \boldsymbol{\lambda}_i^T(\mathbf{y}_i - \sum_{j=1}^{k} C_{ij}\mathbf{z}_j)\right) \tag{12.33a}$$

$$\text{subject to} \quad \mathbf{f}_i(\mathbf{x}_i,\mathbf{y}_i,\mathbf{z}_i) - \boldsymbol{\varepsilon}_i \leq \boldsymbol{\beta}_i \tag{12.33b}$$

and then separate the objective function resulting in the k decomposed subproblems

$$\phi_i(\boldsymbol{\lambda}) = \underset{(\mathbf{x}_i,\mathbf{y}_i,\mathbf{z}_i)\in S_i,\ \boldsymbol{\varepsilon}_i\geq 0}{\text{Minimize}} \quad \mathbf{w}_i^T \boldsymbol{\varepsilon}_i + \boldsymbol{\lambda}_i^T \mathbf{y}_i - \sum_{j=1}^{k} \boldsymbol{\lambda}_j^T C_{ji}\mathbf{z}_i \tag{12.34a}$$

$$\text{subject to} \quad \mathbf{f}_i(\mathbf{x}_i,\mathbf{y}_i,\mathbf{z}_i) - \boldsymbol{\varepsilon}_i \leq \boldsymbol{\beta}_i \tag{12.34b}$$

that can be solved independently to update $\boldsymbol{\lambda}$ and maximize its dual function $\phi(\boldsymbol{\lambda})$.

Other Goal Programming Approaches and Applications

A general benefit of goal programming (GP) approaches in comparison to methods that are exclusively based on weighted sums is the DM's additional flexibility to articulate personal preferences in the form of aspiration targets and goals and thus gain additional control on the objective values of the generated candidates and the finally chosen decision [27, 93]. Consequently, although not without critique [149], there also exist several earlier applications of GP to the hierarchical multiobjective analysis of large-scale (engineering) systems [3, 26, 43, 68, 150, 151, 202], as well as to hierarchical management models for the design and development of information systems and decision structures within decentralized organizations [75, 115].

12.4 Coordination of Objective Decompositions

Whereas the major goal of the decision decompositions in Section 12.3 is to efficiently remedy the large numbers of optimization and decision variables to accelerate or otherwise facilitate the computation of efficient decision and thus generate a set of possible candidate solutions, our main focus in this section is the effective handling of large numbers of decision criteria to facilitate the articulation and modeling of preferences, enhance the evaluation of the associated objective tradeoffs, and ultimately support the choice of a final decision that is preferred by the DM. In particular, and among numerous other approaches that are already described in the literature [10, 21, 54, 60, 57, 60, 76, 82, 85, 98, 106, 114, 136, 152, 155, 190, 191, 198, 211], we only focus on some of the most recent contributions that are originally developed in the dissertation by this author [52] and colleagues over the last years.

12.4.1 Tradeoff-Based and 2D Decision-Making

An inherent and arguably the most challenging aspect of any decision-making process is the articulation or specification of preferences between the different decision criteria and their subsequent mathematical representation in the form of consistent value tradeoffs [98]. Intrinsically defined for pairs of objectives only, however, the complexity of this task grows exponentially with an increasing number of criteria, so that our following approach first divides all objectives into smaller groups for which tradeoff decisions are known and can be represented in the form of general preference or domination cones, described in more detail in Section 12.4.1.1. Using a novel coordination mechanism based on the concept of approximate efficiencies, we then explain in Section 12.4.1.2 how these decisions can again be integrated into a consistent overall solution for the combined problem. Finally, in Section 12.4.1.3, we briefly highlight the additional consequence and common benefit of any objective decomposition that due to reduced problem dimensionalities and especially for groups of at most three objectives it is still possible to fully visualize the decomposed Pareto surfaces with significant advantages for overall perception and tradeoff analysis by the DM [4, 5, 48, 124, 123, 128, 133, 184, 185, 186]. Maintaining the same feasible decision set for both original MOP and decomposed subproblems

MOP$_i$, we write

$$\text{MOP: } \underset{\mathbf{x}\in X}{\text{Minimize}}\ \mathbf{f}(\mathbf{x}) = (\mathbf{f}_1(\mathbf{x}), \mathbf{f}_2(\mathbf{x}), \ldots, \mathbf{f}_k(\mathbf{x})) \tag{12.35a}$$

$$\text{MOP}_i\text{: } \underset{\mathbf{x}\in X}{\text{Minimize}}\ \mathbf{f}_i(\mathbf{x}) = (f_{i1}(\mathbf{x}), f_{i2}(\mathbf{x}), \ldots, f_{im_j}(\mathbf{x})) \tag{12.35b}$$

where the subproblem objectives $\mathbf{f}_i$ form subsets of the components of the original objective function **f**, for which it it also possible that the same or multiple objectives participate as a common reference in several or all subproblems. Similar to before, the (weakly) efficient sets of original and subproblems still satisfy the relationships that $E(X,\mathbf{f}) \subseteq E_w(X,\mathbf{f}) \subseteq X$ and $E(X,\mathbf{f}_i) \subseteq E_w(X,\mathbf{f}_i) \subseteq X$, respectively, and furthermore, that $E(X,\mathbf{f}_i) \subseteq E(X,\mathbf{f})$ if $\mathbf{f}_i$ is injective, and $E_w(X,\mathbf{f}_i) \subseteq E_w(X,\mathbf{f})$ generally [60]. The reverse of the last statement, however, is usually not true unless we impose very restrictive technical assumptions on the objective function **f** [51, 144].

12.4.1.1 Tradeoff Coordination Within Subproblems Using Preference Cones

As mentioned earlier, the decomposition (12.35) allows the explicit distinction between two different kinds of tradeoffs, one within and the other among the different decomposed subproblems. The former is particularly useful because although the DM may not be able to fully specify preferences and tradeoffs between each individual pair of objectives, there are many cases in which at least some a priori preferences and tradeoffs are known so that the DM may directly group those objectives together to avoid the costly generation and analysis of other efficient but non-preferred decisions that are already known to be of no true interest. While there are several ways to consider and model such preferences for general MCDM problems [98, 142], the most common approach for MOP is the use of preference or domination cones $D \subseteq \mathbb{R}^n$ based on an a priori tradeoff matrix $T \in \mathbb{R}^{m\times m}$ by DM [54, 89, 206, 213]: if

$$\tau_{ij}(\mathbf{f},\mathbf{x}) = -\frac{\partial f_i(\mathbf{x})}{\partial f_j} \equiv -\frac{d_i}{d_j} \leq T_{ij}, \text{ or equivalently, } d_i + T_{ij}d_j \geq 0 \tag{12.36}$$

then any $d \in D(T) = \{d \in \mathbb{R}^m : d_i + T_{ij}d_j \geq 0\}$ gives a favorable tradeoff vector for the DM resulting in a set of so-called nondominated solutions $\mathbf{f}(\mathbf{x})$, for which there does not exist a favorable tradeoff vector $\mathbf{d} \in D(T) \setminus \{0\}$ and feasible decision $\mathbf{x}' \in X$ such that $\mathbf{f}(\mathbf{x}') = \mathbf{f}(\mathbf{x}) - \mathbf{d}$. The polyhedral cone $D = D(T)$ or its underlying matrix can then be employed during the optimization to restrict the generation of candidates to solutions that remain efficient under this new preference model, or further extended to more general non-polyhedral or variable cones [52, 53, 61]. Clearly, to make optimal use of such approaches, the DM should first identify which tradeoff rates T_{ij} are known a priori, and then decompose the objective function of the original problem accordingly by grouping those objectives together into the same subproblems for which an associated tradeoff matrix can be derived. In particular, because the resulting nondominated set is typically smaller than the full Pareto set so that it is usually sufficient to generate fewer candidate solutions, the overall benefit is then two-fold: first, together with the reduced problem dimensionality we are also able to reduce the computational costs, possibly further enhanced by one of decision decompositions in Section 12.3, and second, the subsequent decision analysis is likely to require fewer resources as well because only relatively few tradeoffs remain to be evaluated to select the preferred solution from among this smaller but more accurate set of candidate solutions, accurate with respect to the true preferences of the DM.

12.4.1.2 Tradeoff Coordination Among Subproblems Using Approx. Efficiencies

Whereas the specification of a priori tradeoffs within subproblems and their coordination using preference cones are a convenient way to facilitate the decision-making process on the subproblem level, all remaining tradeoffs, that occur among the different subproblems, must be accomplished by some other means. In particular, and although it can be shown that all (weakly) efficient decisions for the subproblems are also (weakly) efficient for the overall problem [60], it is well-known that many efficient decisions for the original problem do not belong to the set of (weakly) efficient decision for any subproblem and thus remain unknown to the DM if we only generate decisions that are truly efficient: e.g., in MDO, where different subproblems could correspond to the different disciplines, the best design would typically be the best compromise between all design teams rather than the best design developed with respect to one individual discipline. Consequently, rather than choosing a best decision with respect to an individual subproblem, we typically seek a final decision that can also be slightly suboptimal for each individual subproblem but remains efficient overall with acceptable tradeoffs between its lacks of efficiencies, that can be modeled using the concept of approximate, or $\boldsymbol{\varepsilon}$-efficiency [58, 59, 109, 122, 203]. Precisely, a feasible decision $\mathbf{x} \in X$ is said to be $\boldsymbol{\varepsilon}$-efficient for MOP if there does not exist another feasible decision $\mathbf{x}'$ so that $\mathbf{f}(\mathbf{x}') \leq \mathbf{f}(\mathbf{x}) - \boldsymbol{\varepsilon}$, and the set of $\boldsymbol{\varepsilon}$-efficient decisions is denoted by $E(X, \mathbf{f}, \boldsymbol{\varepsilon})$. With the analogous definition for the individual subproblems MOP$_i$ with objective functions $\mathbf{f}_i$ and efficiency lacks $\boldsymbol{\varepsilon}_i$, it then can be shown [54, 60] that for proper choice of $\boldsymbol{\varepsilon}_i \geq \mathbf{0}$, the sets of $\boldsymbol{\varepsilon}_i$-efficient decisions for the individual subproblems MOP$_i$ contain the full efficient set of the overall problem

$$\bigcup_i E(X, \mathbf{f}_i) \subseteq E_w(X, \mathbf{f}) \subseteq \bigcap_i \bigcup_{\boldsymbol{\varepsilon}_i \geq \mathbf{0}} E(X, \mathbf{f}_i, \boldsymbol{\varepsilon}_i). \tag{12.37}$$

Based on this result, the tradeoff coordination between different subproblems can be accomplished by choosing suitable values for the lacks of efficiency $\boldsymbol{\varepsilon}_i$ for each subproblem MOP$_i$, which then serve as underlying coordination or tolerance parameters that permit to relax efficiency in each individual subproblem to achieve a better compromise and, thus, an overall preferable solution for MOP. An interactive decision-making procedure that supports and guides the DM with this process is explained and illustrated in more detail in the original references [54, 57, 60] and enables the DM to freely navigate between the different subproblems, explore their sets of efficient and $\boldsymbol{\varepsilon}$-efficient decisions, include a priori tradeoffs using the concept of preferences cones, and coordinate any remaining tradeoffs by making educated choices on the coordination parameters based on additional information obtained from a sensitivity analysis. As for the underlying optimization, the original subproblems MOP_i are then modified into a collection of coordination problems (COP)

$$\text{COP}_i(\boldsymbol{\varepsilon},\mathbf{r}):\ \underset{\mathbf{x}\in X}{\text{Minimize}}\ \mathbf{f}_i(\mathbf{x}) = (f_{i1}(\mathbf{x}), f_{i2}(\mathbf{x}), \ldots, f_{im_i}(\mathbf{x})) \tag{12.38a}$$

$$\text{subject to}\ \ \mathbf{f}_j(\mathbf{x}) \leq \mathbf{r}_j + \varepsilon_j,\ j \neq i \tag{12.38b}$$

that depend on an additional reference point $\mathbf{r}_i$ that models the current aspiration levels for each subproblem together with its maximal permissible deviations $\boldsymbol{\varepsilon}_i$ when compromising its associated criteria $\mathbf{f}_i$ with the ones in the respective other subproblems. For the solution set $E(X,\mathbf{f}_i,\boldsymbol{\varepsilon},\mathbf{r})$ of each COP_i, it can be shown [60] that

$$\bigcup_{\boldsymbol{\varepsilon}}\bigcup_{\mathbf{r}} E(X,\mathbf{f}_i,\boldsymbol{\varepsilon},\mathbf{r}) = E_w(X,\mathbf{f}) \tag{12.39}$$

and hence that every efficient decision for any COP_i is weakly efficient for MOP, and moreover, that every efficient decision for MOP can be generated as an efficient decision for any coordination problem COP_i with a suitable choice of reference points $\mathbf{r}$ and coordination parameters $\boldsymbol{\varepsilon}$. In particular, the smaller the values for some coordination parameter $\boldsymbol{\varepsilon}_i$, the smaller the acceptable deviation from its current aspiration level $\mathbf{r}_i$ thus giving the DM additional flexibility and, at the same time, close control on all permissible objective tradeoffs and targets when using the decision-making procedure outlined above. Finally, while the coordination problems (12.38) resemble the epsilon-constraint method (12.3) and the hierarchical goal programming approach (12.31) and thus can be solved using similar techniques, in principle, more specific generating methods for $\boldsymbol{\varepsilon}$-efficient solutions are also presented in [58, 59].

12.4.1.3 2D Decision-Making and Selected Applications

In specialization of the above approach, additional benefits arise if the DM decides to decomposed the original problem into subproblems with at most two or three objectives that enable the immediate visualization of all Pareto sets in the form of a two-dimensional Pareto frontier or three-dimensional Pareto surface for each individual subproblem, together with numerous other information such as reference points, aspiration levels, or decision targets [4, 5, 48, 124, 123, 128, 133, 184, 185, 186]. While this does not only facilitate the interaction between the analyst and DM by making effective use of digital computer graphics, which is particularly important for DM whose training is less analytical and who prefer visual representations over numerical information, especially a decomposition into bicriteria subproblems also makes tradeoff evaluations within each subproblem the simplest possible as there remains only one tradeoff pair that needs to be considered to identify preferred solutions for each individual subproblem. Therefore initially introduced as 2D performance navigation or 2D decision-making [57] and only subsequently extended also for more general objective decompositions [54, 60], several successful applications of this method have already been described for a variety of problems in structural optimization, automotive engineering, vehicle layout configuration, truss topology design, portfolio selection, and financial management [54, 52, 56, 57, 60].

12.4.2 Multiscenario Multiobjective Optimization

This new area in multiobjective optimization has recently been developed in a collaborative effort by engineers and operations researcher in the Departments of Mechanical Engineering and Mathematical Sciences at Clemson University [62, 170, 207]. Motivated by the challenges and increased complexity in today's engineering industry and managerial businesses, it provides an effective new methodology that enables DM to consider multiple decision situations or scenarios simultaneously in a uniform modeling and solution framework. In its most general form, each multiple-scenario multiple-objective problem (MSMOP) can be formulated as

$$\text{MSMOP: Minimize}\,\{\mathbf{F}_s(\mathbf{x}) : s \in S\} \text{ subject to } \mathbf{x} \in X = \bigcup\nolimits_{s \in S} X_s \tag{12.40}$$

where S is a set of scenarios with individual objective functions and decision sets $\mathbf{F}_s$ and X_s, respectively, and a common decision set $X = \bigcup_{s \in S} X_s$. Such problems arise in a host of practical real-life applications and are particularly well-suited for problems involving one or several decision uncertainties, for which individual scenarios can be associated with different data instances and optimal solutions correspond to those commonly feasible decisions $\mathbf{x} \in X$ that are robust with respect to each possible scenario [107]. Other applications include structural optimization and engineering design, where different scenarios are frequently defined as different loading conditions, and multidisciplinary optimization in which different subproblems can be associated with the individual disciplines as discussed in much more detail in the former Section 12.3. Furthermore, MSMOP can be used for product platform design [64, 63] for which scenarios represent different products that are manufactured from a common set of shared components, resulting in modularity and cost benefits but suffering from potential sacrifices in individual product performance, and automotive engineering and vehicle configuration design for which typical scenarios include different steering maneuvers and varying road or driving conditions that require tradeoffs between, e.g., ground clearance, vehicle dynamics, and overall maintainability [56, 62, 74]. To coordinate between the different scenarios, the traditional all-in-one (AiO) approach combines all individual scenario objectives $\mathbf{F}_i$ into one overall objective function and then solves MSMOP as one large-scale MOP

$$\text{AiO: } \underset{\mathbf{x} \in X}{\text{Minimize}}\,(\mathbf{F}_1(\mathbf{x}), \mathbf{F}_2(\mathbf{x}), \ldots, \mathbf{F}_S(\mathbf{x}))\,. \tag{12.41}$$

Clearly, this approach has several major shortcomings including the very large number of objective function components that makes the solution of this problem, the analysis of its efficient decisions, and any assessment and evaluation of tradeoffs and preferences by the DM extremely difficult. Hence, the proposed alternative approach uses the decomposition-coordination method in Section 12.4.1 and formulates the decomposed subproblems MOP_i in (12.35b) according to the specific scenarios s as

$$\text{MOP}_s\text{: } \underset{\mathbf{x}\in X}{\text{Minimize}}\ \mathbf{F}_s. \tag{12.42}$$

The solution and subsequent tradeoff coordination then follows the same decision-making procedure outlined in the previous section, maintaining its benefits also for the facilitated solution of multicriteria decision problems under multiple scenarios.

12.5 Summary

This chapter presents an overview of the most recent advances in decomposition and coordination methods in the context of multi-criteria decision analysis. The overview collects and reports on a subjective selection of some of the most important methods and results that are spread out in the operations research, management sciences, and engineering literature, and provides numerous references to additional journal and conference articles, technical reports, and scholarly theses and dissertations. The reader interested either in theoretical details or more specific applications is referred to those sources for original formulations, complete derivations, proofs, and real-life applications. As a major benefit, however, this overview presents all methods and results in a brief but unified perspective and with common notation to facilitate their further comparison and analysis and to simplify the selection of a particular method of choice if needed. The author thus hopes that this chapter promotes the active use of interactive decomposition-coordination methods for decision making and stimulates further interest and growth in this inspiring area of research.

In spite of the significant amount of new and continuing research contributions, there are mainly two aspects which the author considers most critical to develop and utilize efficient and effective new methods that are capable to support decision-makers in facing today's immense challenges as outlined in the introduction at the beginning of this chapter. First, while the majority of methods uses either decision or objective decompositions to enhance the computation and optimization of efficient decisions, or to facilitate their subsequent analysis and the final decision by dividing the overall decision process into a hierarchical or nonhierarchical sequence of smaller decision problems, it is eventually necessary to develop efficient mechanisms to combine advantages and benefits and, at the same time, remedy individual drawbacks of the different existing approaches. The author hopes that the collection and unified presentation of several existing methods in this chapter enables its reader to undertake future research into this direction. Second, and in addition to new scientific research in theory and methodology, it seems of urgent importance to engage more actively in a dialogue with actual practitioners and potential users of these proposed methods, to educate decision-makers about the potential of operations research and optimization techniques to make better decisions, and to find possible ways that enable the smooth transition from currently established decision-making tools in business, industry, and government to new and hopefully more effective interactive decision-making procedures. If any of the work presented in this chapter raises such interest or motivates such pursuit, then it has met its original objective.

Acknowledgements The material presented in this chapter was collected and developed while the author received support from the Automotive Research Center, a U.S. Army Center of Excellence for Modeling and Simulation of Ground Vehicles led by the University of Michigan in Ann Arbor, USA, and from MITACS, a Center of Excellence for the Mathematical Sciences in Canada. The author would also like to thank his doctoral adviser Margaret M. Wiecek and Georges M. Fadel for their earlier coordination of this work and many fruitful discussions on the topic of decomposition.

References

1. P.L. Abad. Hierarchical optimal control model for co-ordination of functional decisions in a firm. *European Journal of Operational Research*, 32(1):62–75, 1987.
2. M.A. Abo-Sinna. Generating an α-Pareto optimal solution to multiobjective nonlinear programming problems with fuzzy parameters: a decomposition method. *Journal of Fuzzy Mathematics*, 10(2):423–439, 2002.
3. K.B. Afia and W.J. Davis. The hierarchical search model: An experimental tool for programming decision-making hierarchies. *Large Scale Systems - Theory and Applications*, 11(3):243–265, 1986.
4. G. Agrawal, C.L. Bloebaum, and K. Lewis. Intuitive design selection using visualized n-dimensional Pareto frontier. In *Collection of Technical Papers - AIAA/ASME/ASCE/AHS/ASC Structures, Structural Dynamics and Materials Conference*, volume 1, pages 145–158, 2005.
5. G. Agrawal, K. Lewis, K. Chugh, C.-H. Huang, S. Parashar, and C.L. Bloebaum. Intuitive visualization of Pareto frontier for multi-objective optimization in n-dimensional performance space. In *Collection of Technical Papers - 10th AIAA/ISSMO Multidisciplinary Analysis and Optimization Conference*, volume 3, pages 1523–1533, 2004.
6. N.M. Alexandrov and R.M. Lewis. Analytical and computational aspects of collaborative optimization for multidisciplinary design. *AIAA Journal*, 40(2):301–309, 2002.
7. N.M. Alexandrov and R.M. Lewis. Reconfigurability in MDO problem synthesis, part 1. In *10th AIAA/ISSMO Multidisciplinary Analysis and Optimization Conference*, volume 1, pages 73–82, Reston, VA, 2004.
8. N.M. Alexandrov and R.M. Lewis. Reconfigurability in MDO problem synthesis, part 2. In *10th AIAA/ISSMO Multidisciplinary Analysis and Optimization Conference*, volume 1, pages 83–99, Reston, VA, 2004.
9. N.M. Alexandrov and R.M. Lewis. Synthesis of distributed design problems. In *IEEE International Conference on Systems, Man, and Cybernetics*, volume 1, pages 72–77, New York, NY, 2005.
10. S. Azarm and S. Narayanan. Multiobjective interactive sequential hybrid optimization technique for design decision making. *Engineering Optimization*, 32(4):485–500, 2000.
11. N. Baker and J. Freeland. Recent advances in R&D benefit measurement and project selection methods. *Management Science*, 21(10):1164–1175, 1975.
12. J.F. Benders. Partitioning procedures for solving mixed-variables programming problems. *Numerische Mathematik*, 4:238–252, 1962.

13. H.P. Benson and E. Sun. Outcome space partition of the weight set in multiobjective linear programming. *Journal of Optimization Theory and Applications*, 105(1):17–36, 2000.
14. H.P. Benson and E. Sun. A weight set decomposition algorithm for finding all efficient extreme points in the outcome set of a multiple objective linear program. *European Journal of Operational Research*, 139(1):26–41, 2002.
15. C.L. Bloebaum, P. Hajela, and J. Sobieszczanski-Sobieski. Non-hierarchic system decomposition in structural optimization. *Engineering Optimization*, 19(3):171–186, 1992.
16. V.Y. Blouin, J.D. Summers, G.M. Fadel, and J. Gu. Intrinsic analysis of decomposition and coordination strategies for complex design problems. In *Collection of Technical Papers - 10th AIAA/ISSMO Multidisciplinary Analysis and Optimization Conference*, volume 3, pages 1853–1864, 2004.
17. P. Bogetoft and J. Tind. Dual decomposition schemes for multicriteria decision making. *IMA Journal of Mathematics Applied in Business and Industry*, 2(4):269–279, 1989.
18. R.D. Braun. *Collaborative optimization: An architecture for large-scale distributed design*. PhD dissertation, Stanford University, Stanford, CA, 1996.
19. R.D. Braun and I.M. Kroo. Development and application of the collaborative optimization architecture in a multidisciplinary design environment. In N.M. Alexandrov and M.Y. Hussaini, editors, *Multidisciplinary Design Optimization: State of the Art*, pages 98–116. SIAM, Philadelphia, 1997.
20. R.E. Burkard, H.W. Hamacher, and J. Tind. On general decomposition schemes in mathematical programming. *Mathematical Programming Study*, (24):238–252, 1985.
21. R. Caballero, T. Gómez, M. Luque, F. Miguel, and F. Ruiz. Hierarchical generation of Pareto optimal solutions in large-scale multiobjective systems. *Computers & Operations Research*, 29(11):1537–1558, 2002.
22. V. Chankong and Y.Y. Haimes. The interactive surrogate worth trade-off (ISWT) method for multiobjective decision making. In S. Zionts, editor, *Multiple Criteria Problem Solving*, pages 42–67. Springer-Verlag, Berlin, 1978.
23. V. Chankong and Y.Y. Haimes. *Multiobjective Decision Making: Theory and Methodology*. North-Holland, Amsterdam, 1983.
24. V. Chanron and K. Lewis. A study of convergence in decentralized design processes. *Research in Engineering Design*, 16(3):133–145, 2005.
25. V. Chanron, T. Singh, and K. Lewis. Equilibrium stability in decentralized design systems. *International Journal of Systems Science*, 36(10):651–662, 2005.
26. A. Charnes, R.W. Clower, and K.O. Kortanek. Effective control through coherent decentralization with preemptive goals. *Econometrica*, 35(2), 1967.
27. A. Charnes and W. W. Cooper. Goal programming and multiple objective optimizations: Part 1. *European Journal of Operational Research*, 1(1):39–54, 1977.
28. A. Chattopadhyay and N. Pagaldipti. A multidisciplinary optimization using semi-analytical sensitivity analysis procedure and multilevel decomposition. *Computers & Mathematics with Applications*, 29(7):55–66, 1995.
29. J. Chen, Y. Zhong, R. Xiao, and J. Sun. The research of the decomposition-coordination method of multidisciplinary collaboration design optimization. *Engineering Computations*, 22(3):274–285, 2005.
30. G. Coates, R.I. Whitfield, A.H.B. Duffy, and W. Hills. Coordination approaches and systems - part II: An operational perspective. *Research in Engineering Design*, 12(2):73–89, 2000.
31. B. Colson, P. Marcotte, and G. Savard. Bilevel programming: A survey. *4OR*, 3(2):87–107, 2005.
32. A.J. Conejo, E. Castillo, R. Mínguez, and R. García-Bertrand. *Decomposition Techniques in Mathematical Programming*. Springer-Verlag, Berlin, 2006.
33. J. Contreras, A. Losi, M. Russo, and F.F. Wu. Simulation and evaluation of optimization problem solutions in distributed energy management systems. *IEEE Transactions on Power Systems*, 17(1):57–62, 2002.
34. A.B. Cooper, P. Georgiopoulos, H.M. Kim, and P.Y. Papalambros. Analytical target setting: An enterprise context in optimal product design. *Transactions of the ASME*, 128(1):4–13, 2006.

35. E.J. Cramer, J.E. Dennis, Jr., P.D. Frank, R.M. Lewis, and G.R. Shubin. Problem formulation for multidisciplinary optimization. *SIAM Journal on Optimization*, 4(4):754–776, 1994.
36. J.B. Cruz, Jr. Leader-follower strategies for multilevel systems. *IEEE Transactions on Automatic Control*, 23(2):244–255, 1978.
37. J.B. Cruz, Jr. Survey of leader-follower concepts in hierarchical decision-making. In A. Bensoussan and J.L. Lions, editors, *Analysis and Optimization of Systems*, pages 384–396. Springer, Berlin, 1980.
38. J.B. Cruz, Jr., S.V. Drakunov, and M.A. Sikora. Leader-follower strategy via a sliding mode approach. *Journal of Optimization Theory and Applications*, 88(2):267–295, 1996.
39. T. Csergőffy. Benders decomposition for multiobjective programs with cone inclusions. *Central European Journal of Operations Research*, 9(3):267–275, 2001.
40. G.B. Dantzig and M.N. Thapa. *Linear Programming*. Springer-Verlag, New York, 2003.
41. G.B. Dantzig and P. Wolfe. The decomposition algorithm for linear programs. *Econometrica*, 29:767–778, 1961.
42. J.P. Dauer and M.S.A Osman. Decomposition of the parametric space in multiobjective convex programs using the generalized Tchebycheff norm. *Journal of Mathematical Analysis and Applications*, 107(1):156–166, 1985.
43. W.J. Davis and D.T. Whitford. A note on the optimality of the modified generalized goal decomposition model. *Management Science*, 31(5):640–643, 1985.
44. A.J. De Wit and F. Van Keulen. Numerical comparison of multi-level optimization techniques. In *Collection of Technical Papers - AIAA/ASME/ASCE/AHS/ASC Structures, Structural Dynamics and Materials Conference*, volume 3, pages 2339–2369, Reston, VA, 2007.
45. A.V. DeMiguel. *Two decomposition algorithms for nonconvex optimization problems with global variables*. PhD dissertation, Stanford University, Stanford, CA, 2001.
46. A.V. DeMiguel and W. Murray. An analysis of collaborative optimization methods. 8th AIAA/USAF/NASA/ISSMO Symposium on Multidisciplinary Analysis and Optimization, AIAA Paper 004720, 2000.
47. V. DeMiguel and W. Murray. A local convergence analysis of bilevel decomposition algorithms. *Optimization and Engineering*, 7(2):99–133, 2006.
48. J. Eddy and K.E. Lewis. Visualization of multidimensional design and optimization data using cloud visualization. In *Proceedings of the ASME Design Engineering Technical Conference*, volume 2, pages 899–908, 2002.
49. M. Ehrgott. *Multicriteria Optimization*. Springer, Berlin, 2005.
50. M. Ehrgott and X. Gandibleux. *Multiple Criteria Optimization: State of the Art Annotated Bibliographic Surveys*. Kluwer Academic Publishers, Boston, MA, 2002.
51. M. Ehrgott and S. Nickel. On the number of criteria needed to decide Pareto optimality. *Mathematical Methods of Operations Research*, 55(3):329–345, 2002.
52. A. Engau. *Domination and Decomposition in Multiobjective Programming*. PhD dissertation, Clemson University, Clemson, South Carolina, 2007.
53. A. Engau. Variable preference modeling with ideal-symmetric convex cones. *Journal of Global Optimization*, 42(2):295–311, 2008.
54. A. Engau. Tradeoff-based decomposition and decision-making in multiobjective programming. *European Journal of Operations Research*, 199(3):883–891, 2009.
55. A. Engau, M.F. Anjos, and A. Vannelli. On interior-point warmstarts for linear and combinatorial optimization. Technical report, University of Waterloo, 2009.
56. A. Engau, G.M. Fadel, and M.M. Wiecek. A multiobjective decomposition-coordination framework with an application to vehicle layout and configuration design. *Proceedings in Applied Mathematics and Mechanics*, 7(1):2060039–2060040, 2008.
57. A. Engau and M.M. Wiecek. 2D decision-making for multicriteria design optimization. *Structural and Multidisciplinary Optimization*, 34(4):301–315, 2007.
58. A. Engau and M.M. Wiecek. Exact generation of epsilon-efficient solutions in multiple objective programming. *OR Spectrum*, 29(2):335–350, 2007.
59. A. Engau and M.M. Wiecek. Generating epsilon-efficient solutions in multiobjective programming. *European Journal of Operational Research*, 177(3):1566–1579, 2007.

60. A. Engau and M.M. Wiecek. Interactive coordination of objective decompositions in multiobjective programming. *Management Science*, 54(7):1350–1363, 2008.
61. A. Engau and M.M. Wiecek. Introducing nonpolyhedral cones into multiobjective programming. In V. Barichard, M. Ehrgott, X. Gandibleux, and V. T'kindt, editors, *Multiobjective Programming and Goal Programming*, pages 35–45. Springer, Berlin, 2009.
62. G.M. Fadel, I. Haque, V.Y. Blouin, and M.M. Wiecek. Multi-criteria multi-scenario approaches in the design of vehicles. In *Proceeeding of 6th World Congresses of Structural and Multidisciplinary Optimization*, Rio de Janeiro, Brazil, 2005.
63. R. Fellini, M. Kokkolaras, P. Papalambros, and A. Perez-Duarte. Platform selection under performance bounds in optimal design of product families. *Journal of Mechanical Design*, 127(4):524–535, 2005.
64. R. Fellini, M. Kokkolaras, P.Y. Papalambros, and A. Perez-Duarte. Platform selection under performance loss constraints in opgtimal design of product families. In *Proceedings of the 2002 ASME Technical Conference*, volume 2, 2002.
65. E. Fernández and J. Puerto. Multiobjective solution of the uncapacitated plant location problem. *European Journal of Operational Research*, 145(3):509–529, 2003.
66. J. Figueira, S. Greco, and M. Ehrgott. *Multiple Criteria Decision Analysis. State of the Art Surveys*. Springer, New York, 2005.
67. J. Fliege. An efficient interior-point method for convex multicriteria optimization problems. *Mathematics of Operations Research*, 31(4):825–845, 2006.
68. J.R. Freeland and N.R. Baker. Goal partitioning in hierarchical organization. *Omega*, 3(6):673–688, 1975.
69. C. Fulga. Decentralized cooperative optimization for multi-criteria decision making. In *Advances in Cooperative Control and Optimization*, pages 65–80. Springer, Berlin, 2007.
70. A.M. Geoffrion. Elements of large-scale mathematical programming I - Concepts. *Management Science*, 16:652–675, 1970.
71. A.M. Geoffrion. Elements of large scale mathematical programming II - Synthesis of algorithms and bibliography. *Management Science*, 16:676–691, 1970.
72. A.M. Geoffrion, J.S. Dyer, and A. Feinberg. An interactive approach for multicriteria optimization with an application to the operation of an accademic department. *Management Science*, 19:357–369, 1972.
73. A.M. Geoffrion and W.W. Hogan. Coordination of two-level organizations with multiple objectives. In A.V. Balakrishnan, editor, *Techniques of Optimization*, pages 455–466. Academic Press, New York, 1972.
74. M. Gobbi and G. Mastinu. Symbolic multi-objective optimization of the dynamic behaviour of actively suspended road vehicles. *International Journal of Vehicle Design*, 28:189–213, 2002.
75. M.H. Goedhart and J. Spronk. An interactive heuristic for financial planning in decentralized organizations. *European Journal of Operational Research*, 86(1):162– 175, 1995.
76. T. Gómez, M. González, M. Luque, F. Miguel, and F. Ruiz. Multiple objectives decomposition-coordination methods for hierarchical organizations. *European Journal of Operational Research*, 133(2):323–341, 2001.
77. R.T. Haftka and L.T. Watson. Decomposition theory for multidisciplinary design optimization problems with mixed integer quasiseparable subsystems. *Optimization and Engineering*, 7(2):135–149, 2006.
78. Y.Y. Haimes and W.A. Hall. Multiobjectives in water resource systems analysis: The surrogate worth trade off method. *Water Resources Research*, 10(4):615–623, 1974.
79. Y.Y. Haimes and W.A. Hall. Analysis of multiple objectives in water quality. *Journal of the Hydraulics Division*, 101(4):387–400, 1975.
80. Y.Y. Haimes and D. Li. Hierarchical multi-objective analysis for large-scale systems: Review and current status. *Automatica*, 24(1):53–69, 1988.
81. Y.Y. Haimes and D. Macko. Hierarchical structures in water resources systems management. *IEEE Transactions on Systems, Man, and Cybernetics*, 3(4):396–402, 1973.
82. Y.Y. Haimes, K. Tarvainen, T. Shima, and J. Thadathil. *Hierarchical Multiobjective Analysis of Large-scale Systems*. Hemisphere Publishing, New York, 1990.

83. W.A. Hall and Y.Y. Haimes. The surrogate worth trade-off method with multiple decision-makers. In M. Zeleny, editor, *Multiple Criteria Decision Making*, pages 207–233. Springer, Berlin, 1976.
84. G.M. Heal. *The Theory of Economic Planning*. North-Holland, Amsterdam, 1973.
85. M.I. Henig. Interactive dynamic programming: A framework for solving multicriteria decomposable problems. *Control Theory and Advanced Technology*, 5(4):377–389, 1989.
86. D.M. Himmelblau. *Decomposition of Large-Scale System*. North-Holland, 1973.
87. C.-H. Huang and C.L. Bloebaum. Multi-objective Pareto concurrent subspace optimization for multidisciplinary design. *AIAA Paper*, pages 2786–2796, 2004.
88. C.H. Huang, J. Galuski, and C.L. Bloebaum. Multi-objective Pareto concurrent subspace optimization for multidisciplinary design. *AIAA Journal*, 45(8):1894–1906, 2007.
89. B.J. Hunt. *Multiobjective Programming with Convex Cones: Methodology and Applications*. PhD thesis, Clemson University, Clemson, South Carolina, USA, December 2004.
90. M.L. Hussein and M.A. Abo-Sinna. Decomposition of multiobjective programming problems by hybrid fuzzy-dynamic programming. *Fuzzy Sets and Systems*, 60(1):25–32, 1993.
91. M.K. Hyung, N.F. Michelena, P.Y. Papalambros, and Tao J. Target cascading in optimal system design. *Transactions of the ASME*, 125(3):474–480, 2003.
92. M.K. Hyung, D.G. Rideout, P.Y. Papalambros, and J.L. Stein. Analytical target cascading in automotive vehicle design. *Transactions of the ASME*, 125(3):481–489, 2003.
93. J.P. Ignizio. Generalized goal programming: An overview. *Computers & Operations Research*, 10(4):277–289, 1983.
94. J. Jahn. *Vector Optimization: Theory, Applications, and Extensions*. Springer-Verlag, Berlin, 2004.
95. C.D. Jilla. *A multiobjective, multidisciplinary design optimization methodology for the conceptual design of distributed satellite systems*. PhD dissertation, Massachusetts Institute of Technology, Cambridge, MA, 2002.
96. C.D. Jilla and D.W. Miller. Multi-objective, multidisciplinary design optimization methodology for distributed satellite systems. *Journal of Spacecraft and Rockets*, 41(1):39–50, 2004.
97. M.A.E. Kassem. Decomposition of the fuzzy parametric space in multiobjective nonlinear programming problems. *European Journal of Operational Research*, 101(1):204–219, 1997.
98. R.L. Keeney and H. Raiffa. *Decisions with Multiple Objectives: Preferences and Value Tradeoffs*. John Wiley & Sons, New York, 1976.
99. J.E. Kelley, Jr. The cutting-plane method for solving convex programs. *Journal of SIAM*, 8:703–712, 1960.
100. L. Kerbache and J.M. Smith. Multi-objective routing within large scale facilities using open finite queueing networks. *European Journal of Operational Research*, 121(1):105–123, 2000.
101. H.M. Kim. *Target Cascading in Optimal System Design*. PhD dissertation, University of Michigan, Ann Arbor, 2001.
102. Y. Kim, Y.-H. Jeon, and D.-H. Lee. Multi-objective and multidisciplinary design optimization of supersonic fighter wing. *Journal of Aircraft*, 43(3):817–824, 2006.
103. S. Kodiyalam and J. Sobieszczanski-Sobieski. Multidisciplinary design optimization - some formal methods, framework requirements, and application to vehicle design. *International Journal of Vehicle Design*, 25(1-2):3–22, 2001.
104. N.I. Komkov and A.S. Pleshchinskiy. The decomposition method of allocating resources in a multiobjective program. *Izvestiya Akademii Nauk SSSR, Tekhnicheskaya Kibernetika*, (6):26–31, 1987.
105. M.B. Korchemkin. Decomposition and aggregation for solving transportation and transshipment problems. *Modelling, Simulation & Control C: Environmental, Biomedical, Human & Social Systems*, 6(2):25–35, 1986.
106. P. Korhonen. Interactive methods. In J. Figueira, S. Greco, and M. Ehrgott, editors, *Multiple Criteria Decision Analysis: State of the Art Surveys*, pages 641–666. Springer, New York, 2005.
107. P. Kouvelis and G. Yu. *Robust Discrete Optimization and Its Applications*. Kluwer Academic Publishers, Dordrecht, 1997.

108. I. Kroo. Decomposition and collaborative optimization for large scale aerospace design. In N. Alexandrov and M.Y. Hussaini, editors, *Multidisciplinary Design Optimization: State of the Art*. SIAM, 1996.
109. S.S. Kutateladze. Convex ε-programming. *Soviet Mathematics*, 20(2):391–393, 1979.
110. K.F. Lam and E.U. Choo. Goal programming in preference decomposition. *Journal of the Operations Research Society*, 46(2):205–213, 1995.
111. A.H. Land and A.G. Doig. An automatic method of solving discrete programming problems. *Econometrica*, 28:497–520, 1960.
112. L.S. Lasdon. *Optimization Theory for Large Systems*. Dover, Mineola, NY, 2002. Reprint of the 1970 original.
113. J.B. Lassiter, M.M. Wiecek, and K.R. Andrighetti. Lagrangian coordination and analytical target cascading: Solving ATC-decomposed problems with Lagrangian duality. *Optimization and Engineering*, 6(3):361–381, 2005.
114. R. Lazimy. Solving multiple criteria problems by interactive decompositions. *Mathematical Programming*, 35(3):334–361, 1986.
115. S.M. Lee, M. Gen, and B.H. Rho. A revised iterative algorithm for decomposition goal programming. *International Journal of Systems Science*, 14(12):1383–1393, 1983.
116. S.M. Lee and B.H. Rho. Multicriteria decomposition model for two-level, decentralized organizations. *International Journal on Policy and Information*, 9(1):119–133, 1985.
117. D. Li and Y.Y. Haimes. The envelope approach for multiobjective optimization problems. *IEEE Transactions on Systems, Man, and Cybernetics*, 17(6):1026–1038, 1987.
118. D. Li and Y.Y. Haimes. Hierarchical generating method for large-scale multiobjective systems. *Journal of Optimization Theory and Applications*, 54(2):303–333, 1987.
119. D. Li and Y.Y. Haimes. Correction to: "The envelope approach for multiobjective optimization problems". *IEEE Transactions on Systems, Man, and Cybernetics*, 18(2):332, 1988.
120. D. Li, J.B. Yang, and M.P. Biswal. Quantitative parametric connections between methods for generating noninferior solutions in multiobjective optimization. *European Journal of Operational Research*, 117(1):84–99, 1999.
121. E.R. Lieberman. Hierarchical multiobjective programming: An overview. In A. Goicoechea, L. Duckstein, and S. Zionts, editors, *Multiple Criteria Decision Making: Theory and Applications in Business, Industry and Government*, pages 211–225. Springer, Berlin, 1992.
122. P. Loridan. Necessary conditions for ε-optimality. *Mathematical Programming Study*, (19):140–152, 1982.
123. A.V. Lotov, L.V. Bourmistrova, R.V. Efremov, V.A. Bushenkov, A.L. Buber, and N.A. Brainin. Experience of model integration and Pareto frontier visualization in the search for preferable water quality strategies. *Environmental Modelling and Software*, 20(2):243–260, 2005.
124. A.V. Lotov, V.A. Bushenkov, and G.K. Kamenev. *Interactive Decision Maps*. Kluwer Academic Publishers, Boston, MA, 2004.
125. D. Macko and Y.Y. Haimes. Overlapping coordination of hierarchical structures. *IEEE Transactions on Systems, Man, and Cybernetics*, 8(10):745–751, 1978.
126. M.S. Mahmoud. Multilevel systems control and applications: A survey. *IEEE Transactions on Systems, Man, and Cybernetics*, 7(3):125–143, 1977.
127. E. Matthias and M.M. Wiecek. Multiobjective programming. In J. Figueira, S. Greco, and M. Ehrgott, editors, *Multiple Criteria Decision Analysis: State of the Art Surveys*, pages 667–722. Springer, New York, 2005.
128. C.A. Mattson and A. Messac. Pareto frontier based concept selection under uncertainty, with visualization. *Optimization and Engineering*, 6(1):85–115, 2005.
129. C.D. McAllister, T.W. Simpson, K. Hacker, K. Lewis, and A. Messac. Integrating linear physical programming within collaborative optimization for multiobjective multidisciplinary design optimization. *Structural and Multidisciplinary Optimization*, 29(3):178–189, 2005.
130. A.F. Mehr and I.Y. Tumer. A multidisciplinary and multiobjective system analysis and optimization methodology for embedding Integrated Systems Health Management (ISHM) into NASA's complex systems. In *Proceedings of IDETC2006: ASME 2006 Design Engineering Technical Conferences*, 2006.

131. M.D. Mesarović, D. Macko, and Y. Takahara. *Theory of Hierarchical, Multilevel, Systems*. Academic Press, New York, 1970.
132. M.D. Mesarović, D. Macko, and Y. Takahara. Two coordination principles and their application in large scale systems control. *Automatica*, 6:261–270, 1970.
133. A. Messac and X. Chen. Visualizing the optimization process in real-time using physical programming. *Engineering Optimization*, 32(6):721–747, 2000.
134. N.F. Michelena and P.Y. Papalambros. Hypergraph framework for optimal model-based decomposition of design problems. *Computational Optimization and Applications*, 8(2):173–196, 1997.
135. K. Miettinen. *Nonlinear Multiobjective Optimization*. Kluwer Academic Publishers, Dordrecht, 1999.
136. K. Miettinen. Interactive nonlinear multiobjective procedures. In M. Ehrgott and X. Gandibleux, editors, *Multiple Criteria Optimization: State of the Art Annotated Bibliographic Surveys*, pages 227–276. Kluwer, Boston, MA, 2002.
137. F. Miguel, T. Gomez, M. Luque, F. Ruiz, and R. Caballero. A decomposition-coordination method for complex multi-objective systems. *Asia-Pacific Journal of Operational Research*. Accepted, forthcoming.
138. J. Montusiewicz and A. Osyczka. A decomposition strategy for multicriteria optimization with application to machine tool design. *Engineering Costs and Production Economics*, 20(2):191–202, 1990.
139. P. Nijkamp and P. Rietveld. Multi-objective multi-level policy models: An application to regional and environmental planning. *European Economic Review*, 15(1):63–89, 1981.
140. P. Nijkamp and P. Rietveld. Multiple objective decision analysis in regional economics. In P. Nijkamp, editor, *Handbook of Regional and Urban Economics*, volume 1, pages 493–541. North-Holland, Amsterdam, 1986.
141. Alexandrov N.M. and Lewis R.M. Algorithmic perspectives on problem formulations in MDO. Technical report, 2000.
142. M. Öztürk, A. Tsoukiàs, and Ph. Vincke. Preference modelling. In J. Figueira, S. Greco, and M. Ehrgott, editors, *Multiple Criteria Decision Analysis: State of the Art Surveys*, pages 29–71. Springer, Berlin, 2005.
143. H. Park, N. Michelena, D. Kulkarni, and P. Papalambros. Convergence criteria for hierarchical overlapping coordination of linearly constrained convex design problems. *Computational Optimization and Applications*, 18(3):273–293, 2001.
144. N. Popovici. Pareto reducible multicriteria optimization problems. *Optimization*, 54(3):253–263, 2005.
145. M.E. Posner. The decomposition of nonlinear problems. *Mathematical Programming*, 15(1):360–362, 1978.
146. S. Rabeau, Ph. Depince, and F. Bennis. COSMOS: Collaborative optimization strategy for multi-objective systems. In *Proceedings of TMCE 2006*, pages 487–500, Ljubljana, Slovenia, 2006.
147. U. Reimers. Zur koordination von entscheidungen in hierarchischen organisationen bei mehrfacher zielsetzung. In *Operations Research Proceedings 1981 (Göttingen)*, pages 408–413. Springer, Berlin, 1982.
148. P. Rietveld. *Multiple Objective Decision Methods and Regional Planning*. North-Holland, Amsterdam, 1980.
149. R.E. Rosenthal. Goal programming - A critique. *New Zealand Operational Research*, 11(1):1–7, 1983.
150. T.W. Ruefli. A generalized goal decomposition model. *Management Science*, 17:505–518, 1971.
151. T.W. Ruefli. Linked multi-criteria decision models. In J.L. Cochrane and M. Zeleny, editors, *Multiple Criteria Decision Making*, pages 406–415. University of South Carolina Press, Columbia, SC, 1973.
152. T.L. Saaty. *The Analytic Hierarchy Process*. McGraw-Hill, New York, 1980.
153. A.P. Sage. *Methodology for Large-scale Systems*. McGraw-Hill, New York, 1977.

154. M. Sakawa. Fuzzy multiobjective and multilevel optimization. In M. Ehrgott and X. Gandibleux, editors, *Multiple Criteria Optimization: State of the Art Annotated Bibliographic Surveys*, pages 171–226. Kluwer, Boston, MA, 2002.
155. M. Sakawa and F. Seo. Interactive multiobjective decisionmaking for large-scale systems and its application to environmental systems. *IEEE Transactions on Systems, Man, and Cybernetics*, 10(12):796–806, 1980.
156. M. Sakawa and H. Yano. Trade-off rates in the hyperplane method for multiobjective optimization problems. *European Journal of Operational Research*, 44(1):105–118, 1990.
157. M. Sakawa and H. Yano. An interactive fuzzy decomposition method for large-scale multiobjective nonlinear programming problems. In G. Fandel and T. Gal, editors, *Multiple Criteria Decision Making*, pages 393–402. Springer, Berlin, 1997.
158. N.R. Sandell, Jr., P. Varaiya, M. Athans, and M.G. Safonov. Survey of decentralized control methods for large scale systems. *IEEE Transactions on Automatic Control*, 23(2):108–128, 1978.
159. J.L. Sanders. A nonlinear decomposition principle. *Operations Research*, 13(2):266–271, 1965.
160. G.N. Saridis and H.E. Stephanou. A hierarchical approach to the control of a prosthetic arm. *IEEE Transactions on Systems, Man, and Cybernetics*, 7(6):407–420, 1977.
161. S. Sayin. An algorithm based on facial decomposition for finding the efficient set in multiple objective linear programming. *Operations Research Letters*, 19(2):87–94, 1996.
162. S. Sayin. A procedure to find discrete representations of the efficient set with specified coverage errors. *Operations Research*, 51(3):427–436, 2003.
163. J. Shankar, C.J. Ribbens, R.T. Haftka, and L.T. Watson. Computational study of a nonhierarchical decomposition algorithm. *Computational Optimization and Applications*, 2(3):273–293, 1993.
164. T. Shima. New overlapping coordination methodologies for large-scale optimization problems. *Automatica*, 27(6):1011–1024, 1991.
165. T. Shima and Y.Y. Haimes. The convergence properties of hierarchical overlapping coordination. *IEEE Transaction on Systems, Man, and Cybernetics*, 14(1):74–87, 1984.
166. M.A. Sikora, S.V. Drakunov, and Jr. J.B. Cruz. A sliding mode based leader-follower strategy for multi-level hierarchies. In R.P. Agarwal, editor, *Recent Trends in Optimization Theory and Applications*, pages 393–408. World Scientific, River Edge, NJ, 1995.
167. M.G. Singh. *Dynamical Hierarchical Control*. North-Holland, Amsterdam, 1977.
168. M.G. Singh and A. Titli, editors. *Handbook of Large Scale Systems Engineering Applications*. North-Holland, Amsterdam, 1979.
169. M.G. Singh and A. Titli. *Systems: Decomposition, Optimization and Control*. Pergamon Press, Oxford, 1984.
170. V. Singh. Multi-scenario multi-criteria optimization in engineering design. Master's thesis, Clemson University, Clemson, South Carolina, USA, 2001.
171. N.J. Smith and A.P. Sage. An introduction to hierarchical systems theory. *Computers and Electrical Engineering*, 1(1):55–71, 1973.
172. J. Sobieszczanski-Sobieski. Sensitivity of complex, internally coupled systems. *AIAA Journal*, 28(1):153–160, 1990.
173. J. Sobieszczanski-Sobieski and R.T. Haftka. Multidisciplinary aerospace design optimization: Survey of recent developments. *Structural Optimization*, 14(1):1–23, 1997.
174. J. Sobieszczanski-Sobieski, B.B. James, and A.R. Dovi. Structural optimization by multilevel decomposition. *AIAA Journal*, 23(11):1775–1782, 1985.
175. J. Sobieszczanski-Sobieski and S. Kodiyalam. Bliss/s: A new method for two-level structural optimization. *Structural and Multidisciplinary Optimization*, 21(1):1–13, 2001.
176. R.S. Solanki and J.L. Cohon. Approximating the noninferior set in linear biobjective programs using multiparametric decomposition. *European Journal of Operational Research*, 41(3):355–366, 1989.
177. von H. Stackelberg, editor. *The Theory of the Market Economy*. Oxford University, Oxford, 1952.

178. W. Stadler, editor. *Multicriteria Optimization in Engineering and in the Sciences.* Plenum Press, New York, 1988.
179. W. Stadler. Caveats and boons of multicriteria optimization. *Microcomputers in Civil Engineering*, 10(4):291–299, 1995.
180. W. Stadler and J. Dauer. Multicriteria optimization in engineering: A tutorial and survey. In M.P. Karmat, editor, *Structural Optimization: Status and Future*, pages 209–249. American Institute of Aeronautics and Astronautics, Washington, DC, 1992.
181. T. Stoilov and K. Stoilova. *Noniterative Coordination in Multilevel Systems.* Kluwer Academic Publishers, Dordrecht, 1999.
182. T. Stoilov and K. Stoilova. Goal and predictive coordination in two level hierarchical systems. *International Journal of General Systems*, 37(2):181–213, 2008.
183. C. Stummer and R. Vetschera. Decentralized planning for multiobjective resource allocation and project selection. *Central European Journal of Operations Research*, 11(3):253–279, 2003.
184. G.M. Stump, T.W. Simpson, M. Yukish, and E.N. Harris. Design space visualization and its application to a design by shopping paradigm. In *Proceedings of the ASME Design Engineering Technical Conference*, volume 2B, pages 795–804, 2003.
185. G.M. Stump, T.W. Simpson, M.A. Yukish, and L. Bennet. Multidimensional visualization and its application to a design by shopping paradigm. In *9th AIAA/ISSMO Symposium on Multidisciplinary Analysis and Optimization Conference*, 2002. AIAA 2002-5622, Atlanta, GA.
186. G.M. Stump, M.A. Yukish, J.D. Martin, and T.W. Simpson. The ARL trade space visualizer: An engineering decision-making tool. *Collection of Technical Papers - 10th AIAA/ISSMO Multidisciplinary Analysis and Optimization Conference*, 5:2976–2986, 2004.
187. H. Tadghighi. Aero/acoustic optimization model - A multiobjective, multilevel decomposition based optimization technique. *Annual Forum Proceedings - American Helicopter Society*, 1:197–215, 1998.
188. K. Tammer. The application of parametric optimization and imbedding to the foundation and realization of a generalized primal decomposition approach. In J. Guddat, H.Th. Jongen, B. Kummer, and F. Nožička, editors, *Parametric Optimization and Related Topics*, pages 376–386. Akademie-Verlag, Berlin, 1987.
189. R.V. Tappeta and J.E. Renaud. Multiobjective collaborative optimization. *Journal of Mechanical Design*, 119(3):403–411, 1997.
190. R.V. Tappeta and J.E. Renaud. Interactive multiobjective optimization procedure. *AIAA Journal*, 37(7):881–889, 1999.
191. R.V. Tappeta, J.E. Renaud, A. Messac, and G.J. Sundararaj. Interactive physical programming: Tradeoff analysis and decision making in multicriteria optimization. *AIAA Journal*, 38(5):917–926, 2000.
192. R.V. Tappeta, J.E. Renaud, and J.F. Rodriguez. An interactive multiobjective optimization design strategy for decision based multidisciplinary design. *Engineering Optimization*, 34(5):523–544, 2002.
193. K. Tarvainen. Nonfeasible hierarchical multicriteria methods. In A. Straszak, editor, *Large Scale Systems: Theory and Applications.*
194. K. Tarvainen. *Hierarchical multiobjective optimization.* PhD dissertation, Case Western Reserve University, Cleveland, OH, 1981.
195. K. Tarvainen and Y.Y. Haimes. Coordination of hierarchical multiobjective systems: Theory and methodology. *IEEE Transactions on Systems, Man, and Cybernetics*, 12(6):751–764, 1982.
196. K. Tarvainen, Y.Y. Haimes, and I. Lefkowitz. Decomposition methods in multiobjective discrete-time dynamic problems. *Automatica*, 19(1):15–28, 1983.
197. C. Tribes, J-F. Dube, and J-Y. Trepanier. Decomposition of multidisciplinary optimization problems: Formulations and application to a simplified wing design. *Engineering Optimization*, 37(8):775–796, 2005.
198. M. Verma, G. Rizzoni, D.A. Guenther, and L. James. Modeling, simulation and design space exploration of a MTV 5.0 ton cargo truck in MSC-ADAMS. SAE2005-01-0938, 2005.

199. T.L. Vincent. Renewable resource management. In W. Stadler, editor, *Multicriteria Optimization in Engineering and in the Sciences*, pages 161–186. Plenum, New York, 1988.
200. T.C. Wagner and P.Y. Papalambros. Decomposition analysis and optimization of an automotive powertrain design model. *Engineering Optimization*, 31(1-3):273–299, 1999.
201. J. Wallenius, J.S. Dyer, P.C. Fishburn, R.E. Steuer, S. Zionts, and K. Deb. Multiple criteria decision making, multiattribute utility theory: Recent accomplishments and what lies ahead. *Management Science*, 54(7):1336–1349, 2008.
202. M. Weitzmann. Iterative multilevel planning with iterative targets. *Econometrica*, 38(1), 1970.
203. D.J. White. Epsilon efficiency. *Journal of Optimization Theory and Applications*, 49(2):319–337, 1986.
204. Coates G. Whitfield, R.I., A.H.B. Duffy, and W. Hills. Coordination approaches and systems - part I: a strategic perspective. *Research in Engineering Design*, 12(1):48–60, 2000.
205. R.I. Whitfield, A.H.B. Duffy, G. Coates, and W. Hills. Distributed design coordination. *Research in Engineering Design*, 13(3):243–252, 2002.
206. M.M. Wiecek. Advances in cone-based preference modeling for decision making with multiple criteria. *Decision Making in Manufacturing and Services*, 1(1-2):153–173, 2007.
207. M.M. Wiecek, V.Y. Blouin, G.M. Fadel, A. Engau, B.J. Hunt, and V. Singh. Multi-scenario multi-objective optimization with applications in engineering design. In V. Barichard, M. Ehrgott, X. Gandibleux, and V. T'kindt, editors, *Multiobjective Programming and Goal Programming*, pages 283–298. Springer, Berlin, 2009.
208. D.A. Wismer. *Optimization Methods for Large-Scale Systems ... with Applications*. McGraw-Hill, New York, 1971.
209. L.A. Wolsey. *Integer Programming*. John Wiley & Sons, New York, 1998.
210. G.A. Wrenn and A.R. Dovi. Multilevel decomposition approach to the preliminary sizing of a transport aircraft wing. *Journal of Aircraft*, 25(7):632–638, 1988.
211. J.B. Yang, C. Chen, and Z.J. Zhang. The interactive decomposition method for multiobjective linear programming and its applications. *Information and Decision Technologies*, 14(4):275–288, 1988.
212. M. Yoshimura, K. Izui, and Y. Fujimi. Optimizing the decision-making process for large-scale design problems according to criteria interrelationships. *International Journal of Production Research*, 41(9):1987–2002, 2003.
213. P.L. Yu. Cone convexity, cone extreme points, and nondominated solutions in decision problems with multiobjectives. *Journal of Optimization Theory and Applications*, 14:319–377, 1974.
214. P.L. Yu and M. Zeleny. Linear multiparametric programming by multicriteria simplex method. *Management Science*, 23(2):159–170, 1976.
215. M. Zeleny. *Linear Multiobjective Programming*. Springer-Verlag, Berlin, 1974.

Part IV
Applications

Chapter 13
Applying the EPISSURE Approach for the Evaluation of Business Sponsorship Performance

Stéphane André and Bernard Roy

Abstract This paper presents the application of an approach designed to evaluate non-financial performance in companies. Within a defined perimeter, the approach called EPISSURE produces an "evaluation of non-financial performance with a hierarchical set of synthesis indicators co-constructed during a process of framed dialogue." The paper discusses how the EPISSURE approach was tested and set up within several companies for the purpose of evaluating sponsorship projects and deciding on their follow-up. Test results seem to indicate that the EPISSURE approach is decidedly appropriate for evaluating non-financial performance.

13.1 Introduction

This paper examines the application of an approach designed to evaluate non-financial performance in companies. The application is applied to sponsorship projects. The approach was designed within a much broader vision [4]. The purpose was to meet the ever-growing need of business to take account of non-financial performance [7, 8, 14, 31, 40, 18], from the perspective of decision aiding. Synthesising non-financial performance with an operational perspective involved designing an approach that could:

- Adjust to special cultural, sociopolitical and environmental traits that, within the given company, delineate the perimeter of the relevant non-financial performance components

Stéphane André
LAMSADE, Université Paris-Dauphine, 120/122 rue Réaumur - 75002 Paris, France e-mail: stephane.andre@dauphine.fr

Bernard Roy
LAMSADE, Université Paris-Dauphine, Place du Maréchal De Lattre de Tassigny, F-75 775 Paris Cedex 16, France e-mail: roy@lamsade.dauphine.fr

C. Zopounidis and P.M. Pardalos (eds.), *Handbook of Multicriteria Analysis*, Applied Optimization 103, DOI 10.1007/978-3-540-92828-7_13, © Springer-Verlag Berlin Heidelberg 2010

- Take account of the multiple factors involving heterogeneous aspects that can often only be grasped in qualitative terms, entailing some share of ill determination that should not be overlooked
- Produce a few synthesis evaluations that the various stakeholders will consider legitimate and appropriate to the set goal.

The approach (presented in Section 13.2) is designed to produce an "**evaluation of non-financial performance with a hierarchical set of synthesis indicators co-constructed within a process of framed dialogue**", within a defined perimeter. The approach is called EPISSURE (splice), which is a nautical term meaning a joint made by splicing, i.e., to join or unite (ropes or rope ends) by weaving together the end strands (adapted from Webster's New World Dictionary, 1976). The term reflects one of the concerns that helped guide the design of the approach, i.e., to interweave the separate components of performance. Section 13.3 examines how the EPISSURE approach was tested and set up in several companies for the evaluation sponsorship projects (with a view to deciding on their follow-up). Section 13.4 describes the main results at the end of the tests. Finally, the conclusion addresses the main hurdles that have to be overcome for the application of the EPISSURE approach.

13.2 The EPISSURE Approach

After outlining the approach, we describe the tools that were used to define the synthesis indicators, the dialogue process that is an integral part of the approach, and the set-up of the EPISSURE approach. This procedure clears the way for the users' proper appropriation of the tools and legitimises the resulting evaluations for the main stakeholders.

13.2.1 The Outline of the Approach

Two normative principles to ground the approach were laid down *ex-ante*.

- Principle no. 1: The suggested evaluation is hierarchical, i.e., classified into successive synthesis levels. These levels match a hierarchy of responsibilities into the organization. The importance of the hierarchical agglomerative clustering principle is advocated in numerous research papers on performance evaluation [1, 2, 9, 22, 27]. About evaluation in a context of decision aiding, authors as Keeney [23] and Saaty [39] highlighted the interest of hierarchical levels. Nevertheless they did not establish a link with the hierarchy of responsibilities into an organization.
- Principle no. 2: At each hierarchical level, the evaluations (except perhaps for some at the lowest levels) rely on purely ordinal verbal scales. The number of

degrees on the scales must be adjusted to its matching level; also, the number of degrees must be high enough to mirror evolutions and to be understandable by the stakeholders operating at the said level.

Four levels were selected. The purpose of the first set-up stage of EPISSURE was to detail how the levels were to be understood and tailored within the company where the approach was applied. The four levels are called, and characterised as follows:

- *Level 1 or elementary indicator level:* It involves the elementary indicators that are considered relevant for evaluating the performance of actions [33, 34](chapter 8).
- *Level 2 or criteria level:* It concerns the major points of view for defining the groups of elementary indicators: groups wherein each indicator is related to the same significance axis, which matches a criterion [36] (page 46).
- *Level 3 or level of the actions to evaluate:* We use this term to refer to the typical entity whose performance has to be evaluated. In each concrete case of application, the term must be replaced with the one that designates what - viz., a project, an investment, a procedure, an operational mode, etc. - must be evaluated in the company.
- *Level 4 or level of organisation:* It involves the structure in charge of the different actions.

The elementary indicators of level 1 are what may be considered as the primary data providing the basis for working out all the other evaluations. The elaboration involves different syntheses, each one taking into account the evaluations of the level immediately below. Three syntheses are required (see Figure 13.1):

- The **Operational** synthesis that clears the way for going from the evaluations on the elementary indicators to the evaluations on the criteria.
- The **Intermediate** synthesis that makes it possible to go from the evaluations on the criteria to the evaluations of the actions.
- The **Managerial** synthesis that involves going from the evaluations of the actions to the evaluation of the organisation.

Each synthesis uses tools that are described in the next subsection. Compliant with the second principle (see beginning of this subsection), the tools solely involve purely ordinal scales (save certain exceptions). For the following reasons, these types of scales [35] are deemed the most appropriate for non-financial performance evaluations:

1. The significance of each degree, defined by a verbal description, is easily understandable by the different stakeholders, corporate in-house and external players alike.
2. Any errors of interpretation due to a definition of degrees solely based on numeric data are avoided. These errors stem from individuals' tendency to attribute - automatically and without good reason - significance to the arithmetic relations

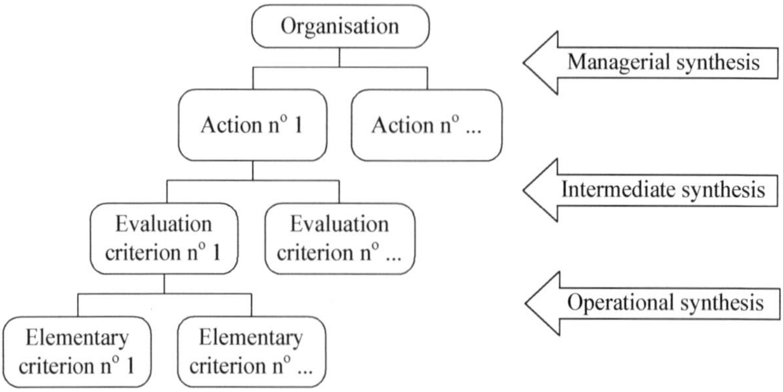

Fig. 13.1 Hierarchical Structure and Corresponding Syntheses

existing between numbers, i.e., a ratio equal to two translates into twice as much satisfaction or intensity of preference; equal differences reflect equal satisfaction or preference variations. This is only true if the numbers express a quantity or a variation of quantity; said quantity being understood as based on a well-defined unit. Non-financial performance cannot usually e grasped in these terms.

3. Stating that one degree rather than the next closest must be used to evaluate the relevant object (whatever the level) may seem arguable in many instances. This ill determination may be properly taken into account by indifference or preference thresholds linked to the scale.

13.2.2 The Tools for the Approach

They tools are dedicated to the given synthesis, and so are described below in succession for each synthesis. The grounds for the choices we have made can be found in André [4].

However, the three types of syntheses do share a common trait. Save certain exceptions, the evaluations they aggregate refer to very heterogeneous, not easily commensurable data. Accordingly, resorting to aggregation procedures (save for the operational synthesis in some cases) that strictly limit the possibility of systematically compensating for poor evaluations with good evaluations seemed appropriate. For non-financial performance, such compensations do not seem warranted especially since, with purely ordinal scales, grounding the equivalences implied by the said compensations is hard to do based on clearly understandable considerations. Actually, this is only feasible at the cost of recoding the scales that - more often than not - entails the loss of significance of the degrees.

13.2.2.1 The Operational Synthesis

At the lowest level (level 1), EPISSURE involves indicators, called elementary indicators whose purpose for each of the actions (sponsorship projects in this case) is to describe an aspect, characteristic, factor, expected consequence, etc. that ought to be taken into consideration in the evaluation process. The basic data, and often the indicators per se, exist within the organisation (the company or association in charge of the sponsorship project). In the operational synthesis, the point is to group them per point of view so as to define a coherent family of criteria [33]. In the case of sponsorship (see 13.3.2), each criterion c was defined as the synthesis of elementary indicators with a common scale (not necessarily the same from one criterion to the next). The evaluation of a project did not have any quantitative significance for any of the elementary indicators. When the evaluation was not naturally verbal and was in other words numeric, the numbers only served to reflect an order of preference. In these conditions, defining what we call the **partial evaluation of a project on one criterion**consisted in aggregating elementary verbal evaluations that were all located on the same purely ordinal scale (see 13.3.2).

We were able to use an unique tool in each business that is dealt with in this paper. This tool is the **rule-based weighted median** which we are going to describe below. This tool was especially well tailored to the aggregation of purely ordinal data with a view to restricting compensation possibilities (a goal that proved appropriate to the studied situations).

Let us denote $g_c(a)$ the partial evaluation of project a on any one of the criteria c that the operational synthesis has to build. Let $h_1, ..., h_k, ..., h_{H_c}$ be the set of elementary indicators that the synthesis must take into account. Evaluations $h_k(a)$ are on the same purely ordinal scale defined by the ordinate series $E_1, ..., E_j, ..., E_m$ of the set of degrees it comprises with E_1 designating the best and E_m the worst. For the application presented in this paper, the selected scale did not depend on the criteria c, it was defined (see 13.3.2.2) by 16 degrees reflecting the verbal description of a position according to a set goal. To define the median of the evaluations of a on the relevant H_c elementary indicators, the evaluations $h_k(a)$ had to be arranged in an order starting from the best to the worst. The degree located in the middle of the resulting ordinate series $s(a)$ is, by definition, the median of the evaluations.

If one wants to differentiate the role that the elementary indicators should play in calculating the median, a set of weights characterising the relative importance to attribute to each indicator may be taken into account in the manner explained below. Let P_k be the weight attributed to indicator h_k Here, the weight must be defined by an integer (a high number if indicator h_k is to play a more preponderant role in the synthesis). Let $S(a)$ be the series of degrees deducted from $s(a)$ by having degree $h_k(a)$ come into play not once but P_k times. The weighted median is defined by the degree that is located in the middle of the ordinate series $S(a)$.

The partial evaluation of the project a on criterion c may be defined by median $M(a)$ of the ordinate series $S(a)$. As will be seen in 13.3.4.2, the evaluation of action a according to the point of view that criterion c should reflect may show some "defects", in some cases. The defects may be remedied by introducing rules that take account of all the evaluations $h_k(a)$ on the elementary indicators and thus replace $M(a)$ with a somewhat different degree $\hat{M}(a)$. The rule base that should be introduced obviously depends on the defect to correct. The type of observed defects was as follows: $M(a)$ could lead to a partial evaluation on criterion c that was judged as too favourable considering that there were poor evaluations in $S(a)$. The said rules involve boundaries B (or milestones) defined as follows:

- If E_m is in $S(a)$, then $g_c(a)$ must not exceed B_m
- if E_{m-1} is in $S(a)$, then $g_c(a)$ must not exceed B_{m-1}
- and so forth
- if E_t is in $S(a)$, then $g_c(a)$ must not exceed B_t

For obvious reasons of coherence, the boundaries must be defined so that $B_m \leq B_{m-1} \leq ... \leq B_t$ The rule-based weighted median is written as follows:

$$g_c(a) = \min\{M(a), B_m(a), B_{m-1}(a), ..., B_t(a)\}$$

where

$$B_j(a) = \begin{cases} B_j \text{ if } E_j \text{ is in } S(a) \\ E_1 \text{ if } E_j \text{ is not in } S(a) \end{cases}$$

In these conditions, the partial evaluation $g_c(a)$ of project a on criterion c is necessarily one of the degrees of the ordinate series $E_1, ..., E_m$ defining the common scale for all the indicators h_k $(k = 1, ..., h_{H_c})$ that have be taken into consideration to build criterion c (see Figure 13.2 for example). As the scale is not necessarily the same from one criterion to the next, strictly speaking the degrees should be written $E_1^c, ..., E_m^c$. Up to now, exponent c was omitted for simpler notations. From now on, we will write E^c for the scale.

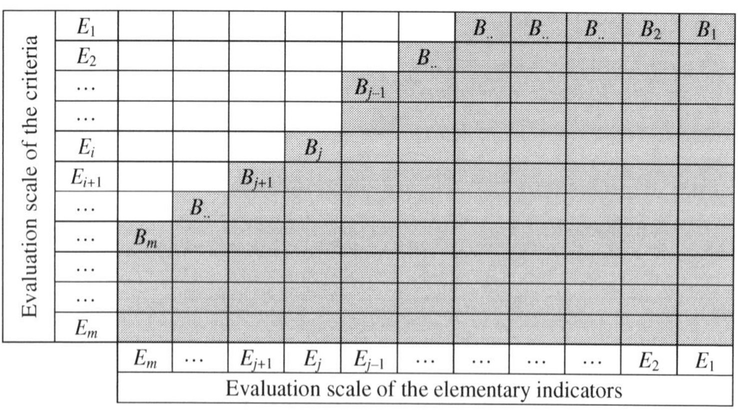

Fig. 13.2 Example of the Rule Base

13.2.2.2 The Tools for the Intermediate Synthesis

The purpose of this synthesis is to provide the means of appreciating the performance of each project a by associating what we call a **comprehensive evaluation** $g(a)$ with each project. Compliant with the second principle (see 13.2.1), $g(a)$ must be a verbally described degree on a necessarily purely ordinal scale $\boldsymbol{E}$. The comprehensive evaluation $g(a)$ must be regarded as a synthesis of partial evaluations $g_c(a), \forall c \in F$. To build the synthesis, EPISSURE implements a multi-criteria aggregation procedure. We felt that is was requisite for the aggregation procedure to take into account the following points.

i) The purely ordinal character of the scales E^c on which the partial evaluations $g_c(a)$ are located.
ii) The fact that two partial evaluations on different criteria may, even if their verbal description is the same, concern totally heterogeneous aspects of the actual project (in other words, non-commensurable aspects).
iii) The virtual impossibility of grounding rules that systematically compensate for poor partial evaluations with good partial evaluations, on clearly understandable bases.
iv) The fact that two projects with very similar partial evaluations on criterion c may not signify a preference in favour of the project with the best evaluation of the two.
v) The need to differentiate the importance of the role played by the different criteria in determining comprehensive project evaluation.

The ELECTRE methods [15] (chapter 5) and (see also chapter 3 of this book) take these requirements into account very well. As shown below, the ELECTRE TRI method is highly suitable (although it was not designed with this object in mind) for defining the comprehensive aggregation $g(a)$ based on partial aggregations, by taking into account the five above-mentioned requirements. That is why, we choose ELECTRE TRI method for the intermediate synthesis.

ELECTRE TRI [15] (chapter 3) and [36] was designed to assign actions (in this paper, we will still refer to them as projects) to pre-defined ordinate categories. In ELECTRE TRI, each project a is characterised by partial evaluations $g_c(a)$ on a family F of criteria, where the evaluations are on purely ordinal scales E^c. To each of these scales, indifference and preference thresholds could be associated. The purpose of these thresholds is to take into account the relative meaningfulness of the preference gap separating two different evaluations (see iv above). The criteria with their thresholds are usually called **pseudo-criteria**. For a precise definition of the thresholds and of the pseudo-criterion tool, see Roy and Bouyssou [36] and Roy and Vincke [37]. The importance of the role played by the criteria (see v above) may be differentiated to define the category to which a project must be assigned. To do so, a weight and a possible veto power is linked to each criterion. The greater the weight of a criterion, the greater the role that the criterion plays in defining the category of assignment. The veto makes it possible to block the assignment to certain categories due to the partial evaluation of the project on the relevant criterion.

Applying the ELECTRE TRI method, EPISSURE used the same scale **E** to assign a category to the relevant projects. Scale $\boldsymbol{E}$ (selected for the application in this paper) was defined (see 13.3.2.2) by the following degrees:

$\boldsymbol{E}_1$: projects largely exceeding set goals
$\boldsymbol{E}_2$: projects reaching or slightly exceeding set goals
$\boldsymbol{E}_3$: projects falling short of goals
$\boldsymbol{E}_4$: projects falling far short of goals

In ELECTRE TRI, categories are defined based on what is called their limit profiles. By definition, the limit profiles are projects that border on two consecutive categories. Such a project is defined by all the partial evaluations that are justifiably assigned to the project so that it effectively characterises the border. The project defined by the boundary between two categories is not only the lower limit profile of the best of these two categories but also the upper limit profile of the worse of these two categories (see Figure 13.3).

To assign a project to a category, ELECTRE TRI compares its partial evaluations with limit profile evaluations criterion by criterion. Based on the result of the comparisons, a project is assigned to one category rather than to another, i.e., this project is evaluated by one degree rather than another. We would like to point out that ELECTRE TRI has two possible assignment procedures. For EPISSURE, we have used the so-called pseudo conjunctive procedure. With the procedure, a project a can only be assigned to a category $\boldsymbol{E}_h$ if a sufficient majority of criteria (coalition of weights is high enough) evaluate project a at least as well as the lower limit profile of $\boldsymbol{E}_h$ and providing that none of the criteria (which do not evaluate the project as well) veto the assignment to the category (for a complete description and justification of this assignment procedure, see above-mentioned references).

13.2.2.3 Managerial Synthesis Tools

The purpose of the synthesis is to provide an overview of the performance of all the projects, to the general management of the organisation (level 4 of Figure 13.1). The main aim of the overview [13] (p. 44) and [1] (p. 314) is to enable the managers to:

- Identify the widest gaps compared to the set goals and to find the origin of the gaps.
- Communicate on the performance of the various projects and highlight its main traits, within the organisation.
- Identify the projects whose follow-up may pose a problem and help select new projects, if appropriate.

EPISSURE does not suggest any mathematical tool for the overview. The only suggested tools are graphics. The type of graphic components that could be selected may draw on tools such as Dashboards [1] (p. 314) or Balanced Scorecards [21, 22]. Obviously, the components depend on the context, and specifically on the reporting

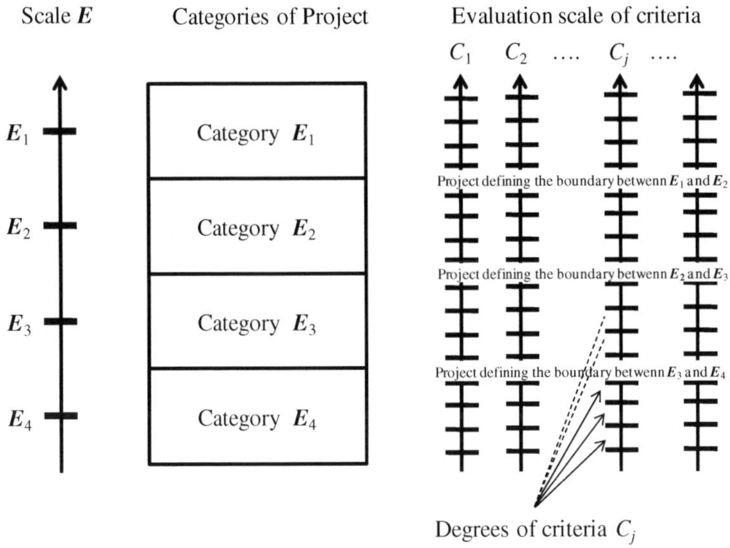

Fig. 13.3 Links between scale E, categories of projects and degrees of criteria

practices in the relevant organisation. Thanks to the hierarchical structure of the approach (principle no. 1), highlighting the characteristics and factors with the greatest impact on overall project performance is easily done. With a computer tool, it is easy to click on the graphics and scroll down from comprehensive performance to partial performance on any given criterion, and from the criterion to elementary evaluations on the various indicators taken into account.

13.2.3 The Framed Dialogue[1] Process

The reasons that prompted us to introduce a dialogue process as an integral part of the EPISSURE approach are not discussed in this paper; please refer to Andre [4] and Andre and Roy [38]. As any other dialogue approach, the objective is that the different stakeholders involved in the evaluation reach a common vision. We deliberately chose framed dialogue because we considered that the framework of the shared vision had to be defined by the company. Actually for the business, performance evaluation is not a goal per se; it must serve the aim of creating value for all the stakeholders (shareholders, employees, environment and civil society). Consequently, performance evaluation must lead to an operational evaluation system for performance management. Framed dialogue is defined as a structured dialogue pro-

[1] The word dialogue is used as a translation of the French word "Concertation" for which there is no equivalent in English.

cess which take place within a delimited space called framework. This framework is built on:

- The two normative design principles (see 13.2.1).
- The synthesis tools selected for each of the three levels (see 13.2.2).
- The identified stakeholders involved in the evaluation (see 13.2.4.1).

In these conditions, dialogue must address the components that still need to be defined so that the synthesis tools become operational within this framework. The components involve what is called Characteristic to be Identified by Dialogue (CID). It should be pointed out that a group of experts could have identified all the components (without any dialogue). To provide readers immediately with a quick overview of what CID comprise, we have broken them down per major type; we list what the characteristics pertain to in each type (content or the value to attribute to certain parameters), for each of the relevant synthesis:

- The CID pertaining to performance definition:
  - Operational synthesis: The elementary indicators linked to the criteria.
  - Intermediate synthesis: The criteria.
  - Managerial synthesis: The graphic representation of the managerial synthesis.
- The CID contextualise the performance components and are instrumental in drawing up the different synthesis:
  - Operational synthesis: Relative importance of the elementary indicators and the rules involved in the aggregation.
  - Intermediate Synthesis: Relative importance of the criteria, information pertaining to the modelling of poor knowledge and veto thresholds.
  - Managerial synthesis: No generalisation is possible, but characteristics have to be determined according to the selected mode of graphic display.
- The CID determining performance goals.

We would like to point out that dialogue and framing may be perceived as conflicting. The designed framework was defined accurately enough so that the company could pursue its goals of value creation while leaving enough leeway for real discussions between the stakeholders. Consequently, the set-up of framed dialogue requires several precautions that we detail during the different set-up stages of the EPISSURE approach.

13.2.4 The Set-up Stages of the EPISSURE Approach

The set-up of the EPISSURE approach must be tailored to each evaluation context to become effectively operational. There are three set-up stages in the following order:

- Stage no. 1: Defining the components of the evaluation framework and tailoring the synthesis tools.
- Stage no. 2: Dialogue phase per se for identifying CID content or the value of their parameters, depending on the case.
- Stage no. 3: Dialogue phase per se for identifying CID content or the value of their parameters, depending on the case.

To ensure the coherence of the approach, it is advisable that the same individual, called a facilitator (consult Maystre and Bollinger [28] for additional information), conduct all the stages. His/her function is to implement the synthesis tools properly as well as facilitate dialogue:

- Present the approach so that it is understood by all.
- Facilitate discussions between the stakeholders while checking that each stakeholder takes the floor to express his/her point of view and goals, and that he/she is heard [36].

The neutrality of the facilitator is important to forestall any stakeholder from challenging the process [10, 24].

The detailed content of the stages depends on the evaluation context. This is discussed in Section 13.3. Nevertheless, the main components of stages 1 and 2 can now be detailed.

13.2.4.1 Stage no. 1

During the first step (stage no. 1.1), the purpose is to defined the framework for dialogue. Company appointed stakeholders (for instance, the organisational managers or performance control managers) with support from the facilitator construct the framework. These stakeholders form the steering group. After a phase where the approach is presented and clarified (specifically what the CID encompass and their *raison d'etre* (see 13.2.3), the group details the following items:

- Evaluation goal and the decisions the evaluation will aid.
- The four hierarchical levels of the evaluation.
- The ordinal verbal scales of the evaluations matching each evaluation level.
- The stakeholders that should be included in the dialogue; the latter are organised into what is called the working group (WG).
- The components that still need to be identified so that the synthesis tools are operational, i.e., the components related to the Characteristics to be Identified by Dialogue (CID).

During the second step (stage no. 1.2), the facilitator tailors the synthesis tools to the framework that was designed earlier (see 13.3.3 - stage no. 1.2). This includes:

- Selecting the synthesis tool that is the most appropriate for the operational synthesis.
- Tailoring ELECTRE TRI to the corporate evaluation context.
- Suggesting a graphics display for the managerial synthesis.

13.2.4.2 Stage no. 2

This is the dialogue phase per se. It should be organised within a working group whose content is decided during stage no. 1. The purpose of this stage is: one, to set-up the framework and two, foster a constructive discussion within the working group. This stage is incremental and iterative for each CID:

- Incremental: the target of each working group meeting (2 to 3 hours) is to reach an agreement on the content of an CID or on the value of its parameters. The definition of criteria c of F (see 13.2.2.1) is usually the target of the first meeting.
- Iterative: each meeting begins with the review and possible amendment of an CID that was on the agenda of the previous meeting. EPISSURE provides for the possibility of going over the work of the previous meeting. The iterative method may seem risky as it might question the progress made at the earlier WG meeting. However, we will see that this option fosters the construction of a common vision where each stakeholder feels that he/she is part of the process (see 13.4). This point corroborates what Joerin and Rondier [20] (p. 19) established about cognitive learning processes during which it is common to review an earlier stage to modify, polish, or complete results.

In conclusion to this Section, we would like to draw the reader's attention to two special features of EPISSURE:

- Dialogue only pertains to concrete proposals and not to general evaluation principles.
- The point is not to incorporate alternatives but to build a common vision for all the actors involved in performance evaluation. For instance, the working group members have to agree on one set of values to assign to the weights (see 13.2.2.1 and 13.2.2.2) and not on several sets according to the different value systems of the members [6]. We will see that this deliberate choice did not cause any particular problems.

Here we conclude the general presentation of the EPISSURE approach that proposes an innovative approach in an attempt to provide answers to the issue of broadening the value concept. In the next Section, we describe a series of operational implementations with a view to evaluating the sponsorship project.

13.3 Testing and Implementing EPISSURE in Companies, for Sponsorship Projects

Thanks to *IMS- Entreprendre pour la Cité* (an institute for sponsorship and solidarity)[2], we were able to contact about ten companies likely to be interested in

[2] Created in 1986, *IMS-Entreprendre pour la Cité* is the backbone for a network of 200 firms. Its purpose is to help the businesses integrate innovative societal commitment approaches into their

testing and possibly implementing EPISSURE to evaluate sponsorship project performance. After a first series of contacts, four companies were selected. The four companies suggested testing seven different sponsorship projects. In this Section, we first present the companies. Then, we describe the set-up stages of the EPISSURE approach.

13.3.1 Presentation of the Selected Companies

The four selected corporate foundations were:

- The Petzl Foundation[3] (Grenoble) with a project for a mountain school in Nepal
- The Kronenbourg Foundation[4] (Strasbourg) with two projects:
  - A project for a restaurant promoting social integration.
  - A project with the association *Jardin de la montagne verte* (green mountain garden), where fruit and vegetable crops are grown by adults with disabilities.
- The Olympique Lyonnais / Cegid Foundation[5] (Lyons) with two projects:
  - The Immersion project whose goal is to enable young people with disabilities to discover the corporate world.
  - The Doctor Clown project providing entertainment for sick children in hospital.
- The RATP Foundation[6] (Paris) with two projects:
  - The *T'as trouvé un job* (you've found a job) project for the professional integration of young people.
  - The *Rencontre sur Tatamis* (the tatami mat encounters) project teaching civic education through sports to young people.

An association (which was different for each project) ran each of the seven projects and the companies were project partners. Accordingly, the companies were seeking an evaluation approach tailored to their evaluation contexts:

- The evaluation required taking into account very heterogeneous data (see 13.2.2.2 ii).
- More often than not, the basic data was solely qualitative. Although the data underwent a quantitative evaluation, a great deal of the data was only meaningful for ordinal comparisons (see 13.2.2.2 i).

---

social responsibility policy, approaches that meet their development challenges as well as society's expectations.

[3] http://fr.petzl.com/petzl/frontoffice/static/fondation/fondation-petzl.htm

[4] http://www.brasseries-kronenbourg.com/_corporate/entreprise_engagee

[5] http://www.olweb.fr/index.php?lng=fr&pid=910101

[6] http://www.fondation-ratp.fr

- The data often also contained a share of arbitrariness, thus small gaps were devoid of meaning (see 13.2.2.2 iv).
- The favourable data for a good evaluation did not have to compensate systematically for the data that was unfavourable to a good evaluation (see 13.2.2.2 iii).

Building synthesis indicators based on dialogue was a first for the different stakeholders and fuelled huge expectations.

We would like to remark that the goals of the different evaluated projects covered very different fields, i.e., national vs. international, young people vs. adults, business vs. civil society. The companies sponsoring the projects were also of different sizes (ranging from small enterprises to major groups) and worked in different businesses. Last, it should be underscored that some associations had poor knowledge of evaluation practices and were worried that the topic might jeopardise their autonomy unlike others associations that already had a solid experience of evaluation practices.

13.3.2 Framing the Dialogue Approach (Stage no. 1.1)

The stage was conducted by a group of stakeholders called the "steering group" (see 13.2.4.1). In the cases studied in this paper, the groups included the head of the foundation, an IMS representative and a facilitator (see 13.2.4). They had to define (see 13.2.4.1):

- The evaluation goals and the decisions they would aid.
- The four hierarchical levels of the evaluation.
- The ordinal verbal scales of evaluations matching each evaluation level.
- The stakeholders that should be included in the dialogue, i.e., the composition of the working group (WG).
- The components pertaining to Characteristics to be Identified by Dialogue (CID).

The framing stage was special to each company. However, the features common to each company can be presented.

13.3.2.1 Evaluation Goal

The evaluation had to enlighten the foundation's decision (usually by the Board on a yearly basis) on project follow-up: continue the investment in a project conditionally or unconditionally, or stop the investment. The evaluation did not only consist in judging whether a project was good or bad but in evaluating whether the project had reached the set goals. In the different cases, the companies had invested in the projects with a view to results: general interest for society and the company's own interests ("sponsorship is no longer an act of pure philanthropy, but a form of investment that business legitimately expects to produce a positive return"). Consequently, the set goals were impact goals and not resource-based goals. Although this may seem a standard discourse in the corporate world, it seemed quite new within the context of the cases we studied. Actually, although the idea of a goal was in the air, the terms of the goals were not always clearly identified within the companies and associations that we met.

13.3.2.2 The Hierarchical Levels of the Evaluation

The following levels were used for all the corporate cases presented in this paper (see Figure 13.4):

- Level 4 corresponded to the foundation. At this level, an overview of the different projects funded by the company had to be feasible.
- Level 3 was selected as the level for each project. At this level, the approach had to situate a project vis-à-vis its goals.
- Level 2 was selected as the intermediate evaluation level of a project, i.e., what we called the success criteria. The criteria were defined as the ex-post requisites according to which the stakeholders involved in the project would consider it successful. The methodology for identifying the success criteria was detailed during stage no. 2 (see 13.3.4).
- Last, Level 1 was defined as the elementary indicator level.

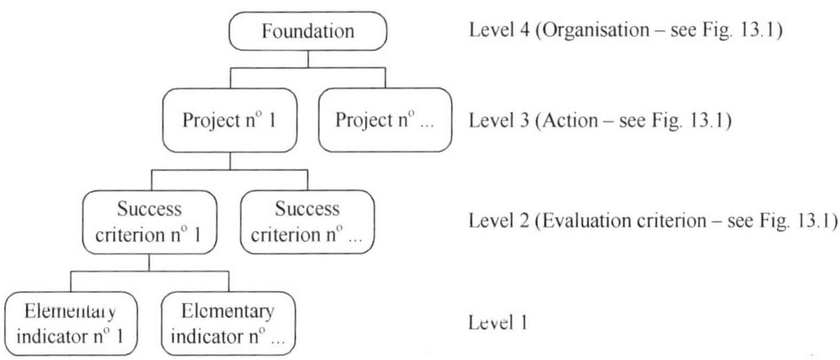

Fig. 13.4 The 4 hierarchical levels of evaluation

13.3.2.3 Ordinal Verbal Scales

The same scales were selected for all the corporate cases. For each scale, the principle consisted in taking a position compared to the goals (the above-mentioned evaluation goal).

For Level 1, a scale, which was the same for all the elementary indicators, was a 16 degree-scale arranged into 8 main degrees where each degree was broken down into two ratings "high" and "low" (see Figure 13.5). This scale is also to be found automatically in level 2, because of the way the operational synthesis was defined (rule base weighted median). Consequently, level 2 is not mentioned below.

E_1	Goal very largely exceeded	High
E_2		Low
E_3	Goal largely exceeded	High
E_4		Low
E_5	Goal exceeded	High
E_6		Low
E_7	Goal reached	High
E_8		Low
E_9	Goal partially reached	High
E_{10}		Low
E_{11}	Goal very partially reached	High
E_{12}		Low
E_{13}	Goal not reached	High
E_{14}		Low
E_{15}	Goal that the project did not attempt to reach	High
E_{16}		Low

Fig. 13.5 Scale linked to levels 1 and 2

On the basis of the initial evaluations on elementary indicators, three reference situations will have to be defined:

- Satisfactory (level E_8) i.e., the value indicating that the stakeholders estimated that the project had reached its goals based on this one indicator (identifying goals was part of the dialogue phase, see 13.3.4).
- Ideal (level E_1) i.e., the value indicating that the stakeholders estimated that the project could not go any further, based on this one indicator.
- Floor value (level E_{16}) i.e., the value indicating that the stakeholders estimated that the project had not even attempted to reach its goal, based on this one indicator.

The detail of this correspondence can be found in 13.3.4.3.

Based on the same principle again, a levelled scale for level 3 was considered fine enough to evaluate the projects. The levels were defined as indicated in Figure 13.6.

At level 4, the selected graphic representation (see 13.3.3.3) took into account two ordinal scales jointly: the level 3 scale and another scale reflecting the size of the company's investment in the project. Investment size is not only financial as it

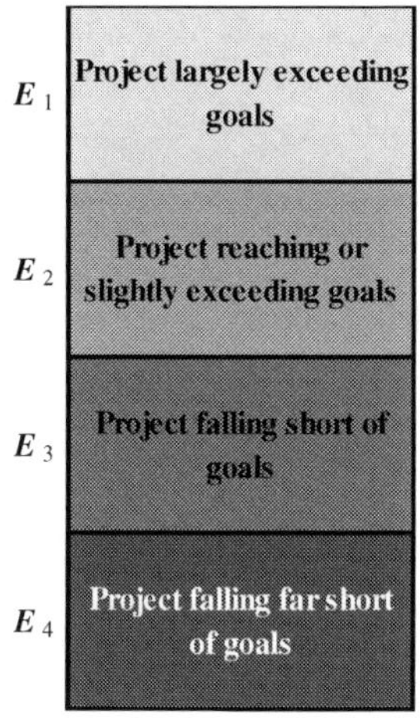

Fig. 13.6 Ordinal scale of evaluation for sponsorship projects

comprises other types of investment (for instance, number of employees involved in the project). The scale had two levels:

- Low Investment.
- High Investment.

13.3.2.4 Composition of the Working Group

Each working group had to count members which are legitimate and representative of the different goals of sponsorship projects. Therefore, the number of members of working group depends on each evaluation context and each of these members had to have leeway over their organization. The composition of the working groups was different in each company, ranging from 4 people (at Petzl) to 12 (at Kronenbourg). However, aside from the facilitator, each group included stakeholders outside the company (mainly from the association in charge of the project) and in-house players.

13.3.2.5 Elements Linked to the Characteristics to be Identified by Dialogue (CID)

So as to incorporate the different components of the evaluation context of a sponsorship action (emerging, not well structured issue, association and corporate stakeholders are not well acquainted, and so on, see 13.3.1), we suggested a mainly ascending process broken down into three sub-stages (see , and):

- Stage no. 2.1 - Identification, formalisation, and ranking of the success criteria (Level 2) corresponding to the identification of the parameters required for the Intermediate Synthesis. The CID at this stage were:
  - The success criteria.
  - The weights of the criteria.
  - The thresholds of discrimination (indifference, preference) and the veto possibilities.
- Stage no. 2.2 - Identification of the elementary indicators (Level 3) linked to the success criteria selected during stage no. 2.1. The CID at this stage were:
  - The elementary indicators.
  - The weights of the elementary indicators.
  - The rule base.
- Stage no. 2.3 - Identification of the goals linked to the elementary indicators. The CID at this stage were performance goals, related to the definition of the reference situations (see 13.3.2.3).
  - Satisfactory.
  - Ideal.
  - Floor value.

For each stage, the dialogue was focused on the CID (Characteristics to be Identified by Dialogue). For the presentation of the EPISSURE process, we would like to underscore the incremental and iterative approach (see 13.2.4.2). The incremental approach was exemplified by the fact that the CID were discussed synthesis operation per synthesis operation. The CID at stage no. 2.1 corresponded to the Intermediate Synthesis (IS) while the CID at stages no. 2.2 and no. 2.3 corresponded to the Operational Synthesis (OS). The iterative approach was exemplified by the fact that each working group meeting was an opportunity to rework points that had been examined earlier (see 13.3.4 for the conditions). Thus defined, the dialogue process required about 4 meetings with the working group. Figure 13.7 below shows how the dialogue principle unfolded.

Choosing the iterative approach was done deliberately to leave room for discussion, enable the issues to mature, and allow the stakeholders to appropriate the approach (evaluation was a new topic for numerous stakeholders - see 13.3.1). We describe the process in detail further down (see 13.3.4).

13.3.3 Tailoring the Synthesis Tools (Stage no. 1.2)

The purpose of this stage is to allow the facilitator to tailor the synthesis tool to each business.

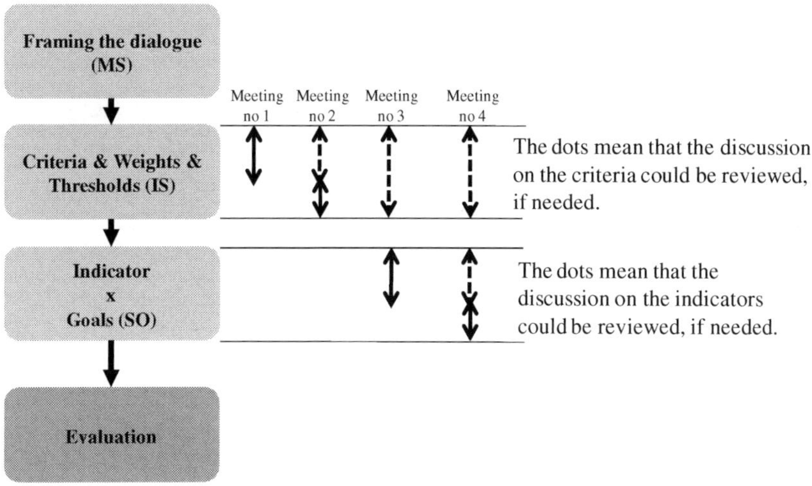

Fig. 13.7 Principle of the dialogue approach

13.3.3.1 Operational Synthesis

The necessary adjustments concern both the way the ordinal 16 levelled scale was used and the rule base adjustment.

a) Positioning Principle for the Evaluation of the Elementary Indicators on the Ordinal Scale

We have to keep in mind that the elementary indicators already existed. Adaptation consisted to link the way the project was previously evaluated to its position on the 16 degrees verbal scale. For the elementary numeric indicators, each degree was demarcated by a high limit and a low limit. For an indicator of growing satisfaction, the value of the low limit corresponded to the value that had to be determined during the dialogue process (see 13.3.4.3; stage no. 2.3). By definition, the high limit corresponded to the low limit of the degree above. In conclusion, project *a* was positioned on a degree if its evaluation ranged within the two limits demarcating the said degree. For elementary indicators on a verbal scale, positioning was done by hand during the evaluation phase.

b) Adjustment of the Rule Base

Based on the principles defined for the rule base (see 13.2.2), the adjustment provided a default rule base (see Figure 13.8), which comprised two rules that fully identify the base:

- If, on all the elementary indicators linked to the criterion, the weakest evaluation of project a ranged between E_1 and E_4, then the value of project a evaluation on the success criterion was not marked with a boundary.
- In the other cases, if the weakest evaluation was equal to degree E, on all elementary indicators linked to the success criterion, then the value of the criterion evaluation was at best equal to degree E_{i-3}.

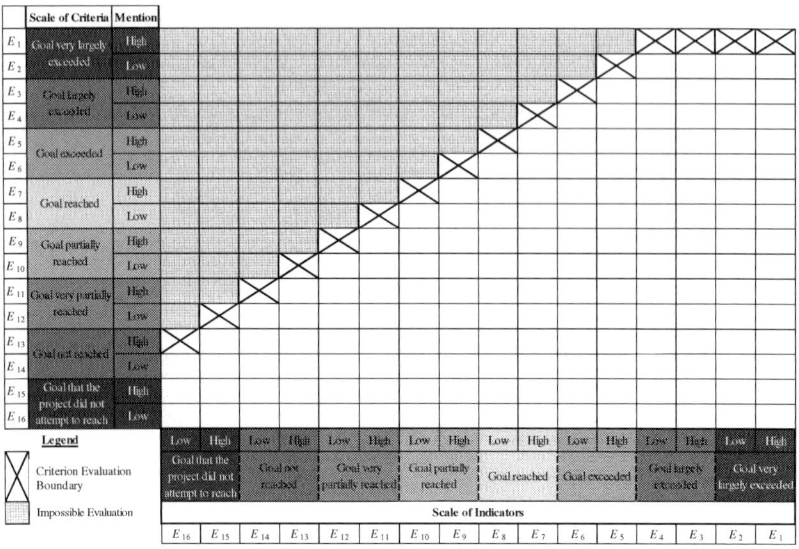

Fig. 13.8 Default Rule Base

This leads to adopt $B_j = E_{j-3}$ for $j > 4$ (see 13.2.2.1).

13.3.3.2 Intermediate Synthesis: ELECTRE TRI Implementation

In this paragraph, we present the outline of how ELECTRE TRI was adjusted. The data taken into account by the synthesis tool were the evaluation degrees linked to each success criterion during a given period (written j; usually the year) for the project (a_j, *which will be simplify as* a). Based on this data, the tool was used to identify the most appropriate degree to which each project was properly assigned.

Four limit profiles were defined for the synthesis of the success criteria. The profiles set the boundary limits for each category:

- "Project largely exceeding its goals" ($\boldsymbol{E}_1$)
- "Project reaching or slightly exceeding goals" ($\boldsymbol{E}_2$)
- "Project falling short of goals" ($\boldsymbol{E}_3$)
- "Project falling far short of goals" ($\boldsymbol{E}_4$)

The identification of the profiles was based on a simple logic. Each category included 4 degrees on the scale used for the success criteria (see Figure 13.9).

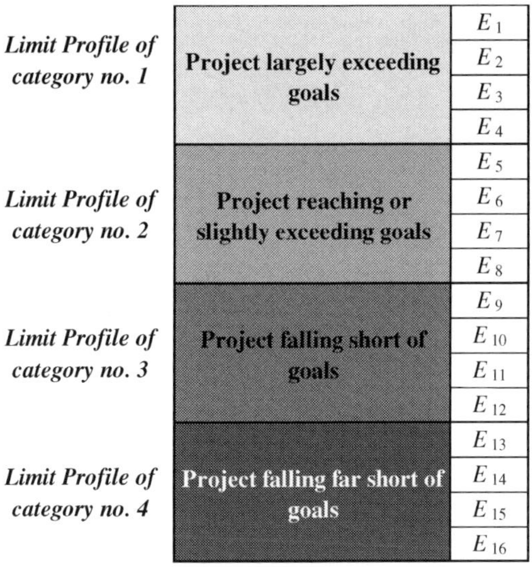

Fig. 13.9 Profile Limits

The different discrimination thresholds were set as follows:

- The value of the indifference threshold set at one degree
- The value of the weak preference threshold set at three degrees

In specific cases and for some criteria, these two thresholds could be choosen as equal to zero. The veto power (when chosen for a criteria) was set to prevent the evaluation on the degrees:

- "Project largely exceeding goals" ($\boldsymbol{E}_1$), if one criterion was evaluated between degrees E_{13} and E_{16}
- "Project reaching or slightly exceeding goals" ($\boldsymbol{E}_2$), if one criterion was evaluated at degrees E_{15} or E_{16}
- No veto threshold prevented the evaluation on the last two degrees: "Project falling short of goals" ($\boldsymbol{E}_3$) and "Project falling far short of goals" ($\boldsymbol{E}_4$).

13.3.3.3 Managerial Synthesis

The adjustment consisted in designing a graphic display providing an overview of the performance of the projects funded by a company (or corporate foundation). For the display, particular care had to be paid to the graphic components of the synthesis overview: shape, type, size, and colour of the graphics. The graphic choices depended on context, and specifically on the reporting practices of the relevant organisations. The care paid to the graphics was very similar to the one involved in Dashboard designs [1] (p. 314).

The figure consisted in crossing the result of the Intermediate Synthesis,- with the size of the investment in each project (see 13.3.2.2), thus making it possible to identify the number of projects crossing both axes. The company assessed the evaluation of the size of the investment. Figure 13.10 below provides an example of the display.

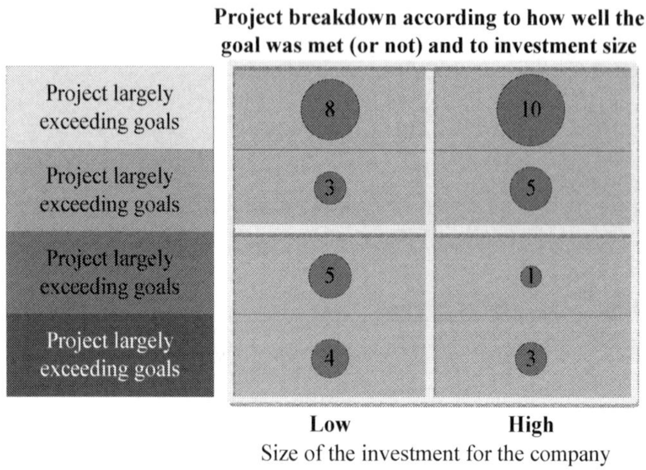

Fig. 13.10 Example of a display of managerial synthesis results

The purpose of the display was to enlighten corporate or foundation authorities' decision by allowing them to identify the projects that needed reviewing (for instance, projects for which the company had made substantial investments and that clearly did not reach their goals). Once the projects were identified, the criteria basing the evaluation and then the related elementary indicators could be detailed thanks to hierarchical interleaving. Consequently, the stakeholders had all the facts for managing the performance of the sponsorship projects funded by the company.

13.3.4 Dialogue Sequence (Stage no. 2)

13.3.4.1 Identification, Formalisation and Ranking of the Success Criteria (Stage no. 2.1)

The CID (Characteristic to be Identified by Dialogue) at this stage cleared the way for implementing the Intermediate Synthesis. The CID corresponded to the parameters required for ELECTRE TRI implementation (see 13.2.3 and 13.3.2.3):

- IS (Intermediate Synthesis) - CID no. 1: the success criteria
- IS - CID no. 2: the weights of the criteria
- IS - CID no. 3: the veto powers

The discussions were usually held during the first two meetings of the working group (WG).

a) Success criteria (IS - CID no.1)

The identification of the success criteria (Level 2) was on the agenda at the first WG meeting. After a general presentation of the EPISSURE approach, the facilitator organised the dialogue. He/she first asked all the members of the working group the following question, "From your standpoint, and only from your standpoint, what are your success criteria for the project?" In other words, "From your standpoint, what would prompt you to say, at project completion, that it was a success?" The purpose of taking a position solely from the standpoint of one's role was to enable each stakeholder to express his/her point of view as well as understand the other stakeholders' points of view.

Each participant was given time to think about the question. During this time, the participants had to jot down their ideas on post-its (a legibly written idea on a post-it), which were in different colours according to origin (foundation, company, association in charge of the project, and others).

Once they had thought it over, each participant presented his/her success criteria one by one. The facilitator grouped the criteria into broad categories. The only questions that were allowed during the presentation phase were questions to clarify understanding.

Once every idea was expressed, they were regrouped or even reformulated. At this stage of the discussion, each participant could express a broader point of view than his/her own position. The recommended reformulation was to use an action verb to express each success criterion. For instance, the project was a success if it managed to "strengthen young people's ties with the corporate world", or "improve the behaviours of young beneficiaries." This type of formulation might seem somewhat restrictive but it allowed each stakeholder to take a position within an impact evaluation logic and not within a resource evaluation logic, which is linked to the evaluation goal (see 13.3.2.1).

During this phase, participants could express their opinions freely. The facilitator made sure that everyone spoke. At the end of the phase, each working group in the different corporate cases managed to set down the terms of the first version of the success criteria. The version changed in the course of the discussions.

b) Weights of the Criteria (IS - CID no. 2)

Before the discussion got started, the facilitator reminded the participants of the significance of the weights in terms of veto power. The purpose of the first question was to find out if the stakeholders conferred the same importance to all the success criteria.

If the answer was no, a discussion was held to identify the weights, using J. Simos's revised card method [16]. To get the discussion going, the facilitator provided one white card per criterion. The facilitator then asked the stakeholders to rank the criteria cards by order of importance, explaining that criteria could be tied, and that one or several white cards could be inserted in the ranking order. He/she went on to explain that the purpose of the white cards was to reflect the size of the gap separating the relative importance assigned to two consecutive criteria in the ranking order.

By default, all the success criteria were put on the same level. Once an agreement had been reached about ranking, the facilitator asked the different stakeholders the following question so that they would identify a range to indicate the ratio between the most important and the least important criterion, "Indicate an interval for the ratio between the weight of criterion (name of the most important criterion) and the weight of criterion (name of the least important criterion)."

During this phase, the only condition laid down by the facilitator was to reach an agreement on the sets of weights. Actually keeping several sets of weights was not envisaged. This choice can be explained by the evaluation goal (see 13.3.2.1) that is to enlighten the company's decision on project follow-up. The set of weights was defined using the SRF software [16].

c) Discrimination threshold and veto power linked to Each Criterion (IS - CID no. 3)

In the first two cases, the facilitator suggested that the working group identify: one, the discrimination thresholds (so as to take into account poor knowledge) and two, the veto thresholds (so as to prevent a good evaluation of the project if one of the criteria received a bad evaluation). Both discussions (on discrimination and veto thresholds) were quite long, because the stakeholders had trouble understanding the different underlying these two notions. Grouping in a same CID of the discrimination threshold and veto power (which do not have the same function) may have facilitated comprehension. Once this was understood, the actual discussion dealt with giving a veto power to different success criteria, or not. In the third case, default values were set for the discrimination thresholds (see 13.3.3.2) and the discussion only dealt with whether or not to give a veto to the different criteria.

During the discussions, the stakeholders often questioned about the criteria of success (definition and/or formulation) and of the weights.

13.3.4.2 Link between Elementary Indicators and Success Criteria (Stage no. 2.2)

The CID at this stage enabled the implementation of the operational synthesis via the rule-based weighted median. The Characteristics to be Identified by Dialogue corresponded to the choice of the elementary indicators that had to be considered for each criterion, for project evaluation. They also corresponded to the values to attribute to the parameters required for method implementation:

- OS (Operational synthesis) - CID no. 1: the elementary indicators
- OS - CID no. 2: the weights of the elementary indicators
- OS - CID no. 3: the rule base

These discussions were often held during the third WG meeting.

a) Elementary Indicators (OS - CID no. 1)

In the course of the first discussions, the associations shared their fear of increased reporting assignments. Although they had understood the advantage of the suggested evaluation approach, they did not want to spend too much time on reporting. From the second corporate case, the facilitator noticed that, although the discussions on the elementary indicators would begin based solely on what existed, needs (save the odd exception) were covered. The approach also provided several advantages. First, it cleared the way for addressing the evaluation faster, because defining and setting up new indicators was not required. Two, it provided a base for the discussion on the goals (see 13.3.4.3 - stage no. 2.3).

The facilitator organised the dialogue in the following manner, "On the criterion (name of the criterion), what extant indicators could be used to evaluate whether goals were reached on this success criterion?" The discussion on the elementary indicators often prompted new questions and discussions on the success criteria. The discussions lead to reformulation or often to the merge of two criteria that were felt to be quite similar after analysing the related elementary indicators.

b) Weights of the Elementary Indicators (OS - CID no. 2)

The purpose of the first question was to find out whether all the stakeholders conferred the same importance to all the indicators. In the different cases and unlike the weights of the success criteria (see 13.3.4.1), the stakeholders agreed on equitable weighting. If this hadn't been the case, the facilitator would have used the same method (J. Simos's cards) and the same dialogue process as for the weights of the success criteria (see 13.3.4.1).

If the stakeholders preferred equal weighting, it was to simplify things. They found it complicated to have different types of weights, something which puzzled most of the stakeholders.

c) Rule Base (OS - CID no. 3)

Before starting the discussion, the facilitator used a numeric example to present the principle of the rule-based median. He/she specifically explained that the rules served to limit the evaluation on a criterion in the case where an indicator had a very bad evaluation whereas all the other evaluations were good. At this stage of the discussion, the facilitator asked the WG members to agree on the need to set up the rule base. In every corporate case the answer was yes. The facilitator suggested a default rule base (see 13.3.3.1). In every corporate case, the default base was accepted.

13.3.4.3 Identification of the Goals Linked to the Elementary Indicators (Stage no. 2.3)

The CID at this stage cleared the way for contractualising the performance goals on each elementary indicator. For each elementary indicator, the CID concerned the matching between the way they were first used to evaluate a project and the scale E_1 to E_16 on which they were eventually located. This matching (see 13.3.2.1) was linked to three reference situations (Satisfactory, Ideal and Floor) that set the meaning attributed to degrees E_8, E_1 and E_{16}, linked to the goals.

These discussions were often held during the fourth and last WG meeting. In one corporate case, a fifth meeting was needed. The dialogue process depended on the type of elementary indicator (described further down). The discussions on identifying the goals often prompted new questions and discussions on the elementary indicators or even on the associated criteria. The discussions led to modifying elementary indicators (one indicator added, merge of two indicators, removal of an indicator) or even changing the associated success criterion (reformulation of the name, merge of two criteria that were, in the end, judged to be quite similar).

a) Dialogue Process for an Elementary Indicator (in numeric form, at first)

The facilitator asked the WG members for at least three reference numeric values, by asking the three following questions in the order below:

- According to you, what is the target value to reach within a year for you to consider that, from the sole standpoint of this indicator, the project could be considered a success? (identification of the value setting the meaning of degree E_8).
- According to you, which value would you consider indicates that the project, from the sole standpoint of this indicator, is a success beyond all expectation? (identification of the value setting the meaning of degree E_1).
- According to you, below which value would you consider that the project, from the sole standpoint of this indicator, is a total failure? (identification of the value setting the meaning of degree E_{16}.

For each of these questions, the stakeholders who originally suggested the elementary indicator answered first. Then, an open discussion began until an agreement was reached. On all the elementary indicators in the seven corporate cases, a consensus on these performance goals was rapidly reached. According to the stakeholders, the main reason for this was that the discussion on the performance goals took place at the end of the process. The stakeholders knew each other well and could have constructive discussions. For instance, Figure 13.11 below illustrates the result for a indicator monitoring the number of young people to train.

			No. of graduates benefiting from arrangement	
			Goal	**Eval.**
E_1	Goal very largely exceeded	High	**1300**	
E_2		Low		
E_3	Goal largely exceeded	High		
E_4		Low		
E_5	Goal exceeded	High		
E_6		Low		
E_7	Goal reached	High		
E_8		Low	**800**	
E_9	Goal partially reached	High		
E_{10}		Low		
E_{11}	Goal very partially reached	High		
E_{12}		Low		
E_{13}	Goal not reached	High		
E_{14}		Low		
E_{15}	Goal that the project did not attempt to reach	High		
E_{16}		Low	**600**	

Fig. 13.11 Identification of the performance goals for the elementary indicators

Once an agreement was reached on the three reference values, linear interpolation was used to calculate the values corresponding to the other degrees. The facilitator presented the result and asked the WG members if they agreed with the values calculated for degrees $(E_2,...,E_7,E_9,E_{10},...E_{15})$. Going back to the same example, Figure 13.12 illustrates what the facilitator presented.

In cases where the proposal was not validated by the working group, the stakeholders could choose to set other reference values for other degrees (supplemental to degrees E_1, E_8 and E_{16}). Another linear interpolation calculation was then carried out to set the value of the degrees between the reference values. Based on the same example as before, Figure 13.13 illustrates the case where the value matching degree E_4 was selected as the reference.

			No. of graduates benefiting from arrangement	
			Goal	**Eval.**
E_1	Goal very largely exceeded	High	**1300**	
E_2		Low	1229	
E_3	Goal largely exceeded	High	1157	
E_4		Low	1086	
E_5	Goal exceeded	High	1014	
E_6		Low	943	
E_7	Goal reached	High	871	
E_8		Low	**800**	
E_9	Goal partially reached	High	775	
E_{10}		Low	750	
E_{11}	Goal very partially reached	High	725	
E_{12}		Low	700	
E_{13}	Goal not reached	High	675	
E_{14}		Low	650	
E_{15}	Goal that the project did not attempt to reach	High	625	
E_{16}		Low	**600**	

Fig. 13.12 Calculation of the value of the performance goals based on the three values identified by the stakeholders of the evaluation process

b) Dialogue Process for an Ordinal Verbal Elementary Indicator

The same process was applied except that the different degrees of the scale were described with precise definitions. First, the facilitator asked the WG members to describe precisely the degree matching the relevant reference situations (satisfactory, ideal and floor). Then, the facilitator asked for the description of at least two additional degrees (see illustrations in Figures 13.13 and 13.14).

For instance, figure 13.14 below illustrates an indicator monitoring the set-up of a local organisation.

For each situation, the stakeholders that came up with the elementary verbal indicator were the first to submit a formulation. An open discussion then began until an agreement was reached. For every of the seven corporate cases, a consensus was rapidly reached on the verbal description of the reference situations which contractualised the performance goals.

			No. of graduates benefiting from arrangement	
			Goal	**Eval.**
E_1	Goal very largely exceeded	High	**1300**	
E_2		Low	1233	
E_3	Goal largely exceeded	High	1167	
E_4		Low	1100	
E_5	Goal exceeded	High	1014	
E_6		Low	943	
E_7	Goal reached	High	871	
E_8		Low	**800**	
E_9	Goal partially reached	High	775	
E_{10}		Low	750	
E_{11}	Goal very partially reached	High	725	
E_{12}		Low	700	
E_{13}	Goal not reached	High	675	
E_{14}		Low	650	
E_{15}	Goal that the project did not attempt to reach	High	625	
E_{16}		Low	**600**	

Fig. 13.13 Calculation of the value of the performance goals based on the four values identified by stakeholders of the evaluation process

13.3.5 Validation and Implementation of the EPISSURE approach (Stage no. 3)

The evaluation approach that was built then had to be formally validated by the companies. The foundation authority usually validated the approach. Notably, once EPISSURE was set up, it could be used in a contract situation between the different stakeholders. Although dialogue was critical during the set-up phase, it then had to make way for a conformation tool whose goal was to prompt the stakeholders to adopt the desired type of behaviour [29].

13.4 Observed Results

Out of seven corporate cases EPISSURE approach was stopped for three of them (two Kronenbourg projects ant Petzl project) mainly for lack of follow-up. However, EPISSURE approach is still in place in four of them (RATP and Olympique Lyonnais projects). Several people were trained to act as facilitators within the IMS association and several corporate foundations. Accordingly, implementation has now

			Set-up of a durable local organisation	
			Goal	**Eval.**
E_1	Goal very largely exceeded	High	Financial autonomy of the system	
E_2		Low		
E_3	Goal largely exceeded	High		
E_4		Low		
E_5	Goal exceeded	High	The entire team is in place and operates autonomously	
E_6		Low		
E_7	Goal reached	High		
E_8		Low	The entire team is in place but is not autonomous vis-à-vis the foundation	
E_9	Goal partially reached	High		
E_{10}		Low		
E_{11}	Goal very partially reached	High	A mixes team Népal / Europe is in place	
E_{12}		Low		
E_{13}	Goal not reached	High	A technical manager is appointed	
E_{14}		Low		
E_{15}	Goal that the project did not attempt to reach	High		
E_{16}		Low	No contact, or at the study stage	

Fig. 13.14 Illustration of the evaluation degrees for a verbal indicator, at first

been broadened to include the evaluation of other sponsorship projects, specifically within the RATP foundation. The facts lead us to believe that the advantage of the EPISSURE approach has been validated for the evaluation of sponsorship action performance. Below, we discuss a series of results that were established within the different corporate cases.

a) Contribution of the Approach to Sponsorship Action Evaluation

In sponsorship action evaluation, there is a very strong demand for tailored, relevant tools enabling effective evaluation and providing an answer to companies' standard question, "were we right to invest in this project?" Although it is not expressed as such, the greatest expectation is to have synthesis providing a better understanding of the available information. Actually, the task is tricky not because of any lack of information but because of an overabundance of data and of the multi-criterion aspect of the data (see 13.3.1). The strong expectation is reflected by the fact that the WG members were readily available. They all wanted to reach concrete satisfactory results. We organised several dozen working groups and met diverse stakeholders (a total of fifty different stakeholders; companies, foundations, and associations). In every case, attendance was close to 100% at every single meeting.

Although the working groups were held independently, they led to the identification of success criteria (Level 2) that belong to three families of the same type:

- Social criteria evaluating the social impact (or general interest) of the project (for instance, introducing the corporate world to young people with disabilities).
- Association related criteria, specifically the association's ability to become self-reliant (for instance, reach financial autonomy).
- Corporate related criteria (for instance, strengthen employee motivation, clear the way for recruitment, provide ideas for new products).

The name of the criteria and the related elementary indicators were obviously particular to each company. Interestingly, compared to the literature on sponsorship action evaluation [12, 30, 32, 41], association related criteria are innovations. This undoubtedly bears witness to the fact that the construction of sponsorship action evaluation is still underway.

b) Ability of the Framed Dialogue Process to Foster Genuine Dialogue

The actual problem consisted in finding a process that was serious and reliable enough to reassure evaluation "regulars", but not too "sophisticated" to scare the "novices" (see 13.3.1). The process also had to prompt each stakeholder to become a responsive listener. As a result, virtually all the stakeholders involved in the dialogue process expressed their satisfaction. The only actual problem was addressing the case of an association representative who was unable to attend the first two WG meetings. He was uncooperative when work began on the elementary indicators (see 13.3.4.2 - stage no. 2.2). He was afraid that the company would use the data to lower financial aid. The positive contribution of the framed dialogue process is clear, provided that the process is experienced from beginning to end.

The seven corporate cases highlight that the process fostered individual and collective learning mechanisms contributing to the design of shared solutions. Consequently, we can talk about a cognitive side of the EPISSURE approach. Actually, the agreements on the different CID (Characteristic to be Identified by Dialogue) were not reached based on pre-existing components, but on components that the WG members constructed together. An instance of the cognitive aspect is the list of success criteria (IS - CID no. 1 - see 13.3.4.1). The first meeting was an opportunity for each participant to list his/her success criteria (solely from his/her point of view; see 13.3.4.1). The list of criteria was then reworked during the entire process (iterative aspect of the approach; see 13.3.4.2 and 13.3.4.3). At the end of the process, in each corporate case, the working groups were able to draw up substantially revised lists of success criteria. The lists were shorter, better defined, and above all were common to all, unlike the lists that had been put together at the start of the first meeting. Several factors facilitated the collective learning process. First, there was the dialogue framework, which was never experienced as being overly restrictive. On the contrary, it fostered discussions within a common space. Because the framework was established during the framing phase of the dialogue process (meaning it was tailored to the special features of the context; see 13.3.2.3), it was well accepted. The iterative and incremental aspects (see 13.3.1) of the EPISSURE approach also fostered ties and trust within the working group. The WG members actually had time to get to know one another. They felt much freer to discuss as the process allowed them to review the result of their discussions at each working group meeting. Last, there was the facilitator; one of his/her assignments was to stimulate a common thought process.

c) Ability of the Synthesis Tools to Take into Account the Special Features of Sponsorship Action Performance

Identifying synthesis indicators was not a foregone undertaking, see Alfsen and Moe [3] (p. 10), Henderson [19], Anielski [5], and Faucheux & Nicolaï [14] to find numerous reserves about synthesis indicators. Nevertheless, the synthesis indicators, which were designed thanks to the selected synthesis tools (see 13.2.2) were very productive for evaluating non-financial performance. These good observed results were possible here thanks to the deliberate choices of the EPISSURE approach (see 13.2.1):

- The hierarchical structure of the synthesis indicators enabled stakeholders to have a good grasp on the components making up the synthesis.
- The use of ordinal scales meant that only the ordinal significance of the evaluation data was taken into account.
- The use of synthesis tools limited compensation and incorporated poor knowledge.

The positive feedback was also due to the fact that the approach was easily tailored to the different situations thanks to:

- The existence of different tools for the operational and managerial synthesis, the choice was quite free, meaning that a tool close to the field and to the standard practices of the relevant stakeholders could be selected.

- The choice of evaluation compared to goals - this fairly standard idea made it possible to set goals tailored to each situation. As a result, the synthesis addressed the gaps with the goals listed on an ordinal scale, which had been built based on simple ideas, viz., goal has not been reached, has been reached, has been exceeded. This choice meant that indicators whose evaluations were on a quantitative or qualitative basis could be incorporated.

d) Advantage of the EPISSURE Approach for Decision-Aid

The EPISSURE approach seems to have a real advantage within the context of the decision-aid process by promoting collective action twofold. One, it contributes to improving stakeholders' representation of the relevant issue (cognitive aspect of the approach mainly experienced during stage no. 2). Two, it prompts stakeholders to take action to reach the goals they are assigned (compliance aspect of the approach mainly experienced during use -see stage no. 3). Here we find the two sides of management tools [29] (p. 42.): the cognitive side and the conformation side. During stage no. 2, the cognitive aspect prevails. We have highlighted this through the observed learning processes (see 13.4b). The cognitive components provide an answer to the complexity of non-financial performance evaluation. During this phase, it should be pointed out that the conformation aspect is not missing. The fact that stakeholders agree to conduct dialogue within a given framework is a point in case. The stakeholders' representation of the context can evolve thanks to the alliance of both aspects. These findings are similar to the findings of the research conducted by Lebreton et al. [26]. Actually, in the course of participating and designing a system of indicators, a stakeholder becomes aware of what he/she perceives of the workings of the actual systems and learns from the other stakeholders' different points of view. The evolution involves a sizeable cognitive load, relying essentially on the acquisition of structured information [20]. Faced with new information, stakeholders find themselves in one of two possible situations [17]. Either the information is "coherent" with their representation of the system and enriches them, or the information creates "incoherence" and they have to deconstruct/reconstruct their representation of the system. The dialogue process then enables stakeholders to learn about the sit uation they are facing, better understand it, and gradually feel they are in a position to act. Here, we clearly find the foundation of our framed incremental and iterative dialogue process (see 13.2.4). Once this phase is over, the conformation side prevails to the extent where the stakeholders are "urged" to act to reach the goals they have been assigned. As the goals regularly come up for discussion (usually every year), conformation will again make way for dialogue. Here too, we find results that have been highlighted at the Canada Research Chair on territorial decision aid. Actually, once the indicator system is implemented, it provides a formalised representation of the dynamics of the evaluated actions, thus enabling the different stakeholders to communicate and exchange on the complexity of the evaluation.

The last point is the fact that the EPISSURE approach seems to help legitimise decisions. Actually, managing increasingly means legitimising, i.e., producing arguments likely to make a corporate decision acceptable by its stakeholders [25] (p. 81). Yet everything happens as if the said legitimacy is now the result of management systems able to build both objects and collective agreements [11] (p. 4). Thanks to the framed dialogue process, the EPISSURE approach makes it possible to build the objects and collective agreements via the identification of the CID (Characteristic to be Identified by Dialog), among others. The evaluation stemming from the EPISSURE approach is the end-result of collaborative work and does not seem to be some sort of hat trick. From this standpoint, the EPISSURE approach does not legitimise the decision per se but the components underpinning the decision.

13.5 Conclusion

For the successful set-up of EPISSURE, for sponsorship or, more broadly, for non-financial performance evaluation, some obstacles must be overcome.

The first obstacle is to find a facilitator who is sufficiently knowledgeable about the different multi-criteria synthesis tools to select the most appropriate (operational and managerial synthesis tools) and to tailor the use of ELECTRE TRI to the evaluation context.

The second obstacle involves the framed dialogue process. The EPISSURE approach requires constructive behaviour during the work sessions. This supposes that one, the stakeholders regularly attend meetings to share their opinions and two, they agree to let their guard down enough to foster a discussion on the "real" issues. For the purpose of facilitating the stakeholders' constructive behaviour, the role of the facilitator is again decisive. To do so, he/she must have teaching skills to explain the approach, be a proficient group organiser, and have a knack for mediation to foster consensus on tricky points.

The third obstacle is undoubtedly the most complex. It concerns incorporating the tools within the organisation. Actually, the EPISSURE approach can only enable the management of non-financial performance if it is implemented over time. The first stage is the successful set-up of the EPISSURE approach. This supposes one, receiving strong support from management and two, tailoring the approach to the customs and practices of the organisation (for instance, adjusting the technical vocabulary of the synthesis tools to the corporate language). The next step is to ensure the durability of the approach within the organisation. This assignment is trickier and, for the time being, we do not have enough perspective to observe durability in the different cases. However, we can assume that durability supposes that the EPISSURE approach must:

- Enrich extant practices without adding overly strict constraints, the goal is to prompt organisation stakeholders to appropriate the approach gradually.
- Remain coherent with the corporate strategy; any policy moving toward a solely financial strategy would void the EPISSURE approach.
- Be incorporated into the reporting processes that already exist in the company.

Although we believe we have validated the advantage of the approach, the different companies still have to be monitored to analyse the durable integration of EPISSURE within an organisation. The analysis of coherence compared to other management tools and the analysis of the capacity to change behaviours are the main points requiring special attention.

References

1. F. Adam and J.-C. Pomerol. Critical factors in the development of executive systems - leveraging the dashboard approach. In M. Mora, G. Forgionne, and J. Gupta, editors, *Decision Making Support Systems: Achievements and Challenges for the New Decade*, pages 305–330. IGI Publishing, Hershey, PA, 2002.
2. F. Adam and J.-C. Pomerol. Developing practical support tools using dashboards of information. In F. Burstein and C.W. Holsapple, editors, *Handbook on Decision Support Systems 2*, pages 151–173. Springer-Verlag, 2008.
3. K. Alfsen and T. Moe. An international framework for constructing national indicators for policies to enhance sustainable development. In *Expert Group Meeting on Indicators of Sustainable Development*, New York, 13-15 December 2005. UN Department of Economic and Social Affairs, Division for Sustainable Development.
4. S. André. Evaluation de la performance non financière des entreprises: Apport des méthodes multicritère d'aide à la décision. Thèse d'Informatique / Aide à la Décision, Université Paris Dauphine, Paris, France, 2005.
5. M. Anielski. Alberta sustainability trends 2000: The genuine progress indicators report 1961-1999. Pembina Institute, Drayton Valley, 2001.
6. V. Belton and J. Pictet. A framework for group decision using a MCDA model: Sharing, aggregating or comparing individual information. *Revue des Systèmes de Décision*, 6(3):283–303, 1997.
7. CEE. Direction générale de l'emploi et des affaires sociales: Promouvoir un cadre européen pour la responsabilité sociale des entreprises. Commission Européenne, 2001.
8. CERES. Dix principes de protection de l'environnement et de transparence. Coalition for Environmentally Responsible Economies, 1989.
9. R. Chenhall and K. Langfield. Multiple perspective of performance measures. *European Management Journal*, 25(4):266–282, 2007.
10. S. Damart and B. Roy. The uses of cost-benefit analysis in public transportation decision-making in France. *Transport Policy*, 16(4):200–212, 2009.
11. A. David, A. Hatchuel, and R. Laufer. Sciences de gestion et sicences sociales: Un déficité d'identité. In A. David, A. Hatchuel, and R. Laufer, editors, *Les Nouvelles Fondations des Sciences de Gestion*, pages 1–12. Vuibert, Paris, 2001.
12. F. Debiesse. *Le Mécénat*. Presses Universitaires de France, Paris, 2007.
13. C. Dover. How dashboards can change your culture. *Strategic Finance*, 86(4):43–48, 2004.
14. S. Faucheux and I. Nicolaï. Quels indicateurs pour la responsabilité sociales des entreprises? Une application à l'industrie de l'aluminium. *Revue Gérer et Comprendre*, 76:42–54, 2004.
15. J.R. Figueira, B. Roy, and V. Mousseau. ELECTRE methods. In J.R. Figueira, S. Greco, and M. Ehrgott, editors, *Multiple Criteria Decision Analysis: State of the Art Surveys*, pages 133–162. Springer, New York, 2005.
16. J. Figuiera and B. Roy. Determining the weights of criteria in the ELECTRE type methods with a revised Simos' procedure. *European Journal of Operational Research*, 139:317–326, 2002.
17. F. Flourentzos. Constructivisme piagétien dans l'aide à la décision - Contribution au développement durable en architecture. Thèse no 2418, Ecole Polytechnique Fédérale de Lausanne, 2001.

18. World Bank Group. Pollution prevention and abatement handbook, toward cleaner production. The World Bank Group in collaboration with the United Nations Environment Programme and the United Nations Industrial Development Organization, Washington, D.C, 1999.
19. H. Henderson. What's next in the great debate about measuring wealth and progress? *Challenge*, 39:50–56, 1996.
20. F. Joerin and P. Rondier. Indicateurs et décision territoriale: Pourquoi? quand? comment? In G. Sénécal, editor, *Les Indicateurs Socio-territoriaux et les Métropoles*, pages 9–36. Presses del'Université Laval, Québec, 2007.
21. R. Kaplan and D. Norton. The balanced scorecard: Measures that drive performance. *Harvard Business Review*, 70(1):71–79, 1992.
22. R. Kaplan and D. Norton. Using the balanced scorecard as a strategic management system. *Harvard Business Review*, 74(1):75–85, 1996.
23. R. Keeney. *Value-Focused Thinking: A Path to Creative Decisionmaking*. Harvard University Press, Cambridge, MA, 1996.
24. S. Labbouz, B. Roy, Y. Diab, and M. Christen. Implementing a public transport line: Multi-criteria decision-making methods that facilitate concertation. *Operational Research: An International Journal*, 8(1):5–31, 2008.
25. R. Laufer. Les institutions du management: Légitimité, organisation et nouvelle rétorique. In A. David, A. Hatchuel, and R. Laufer, editors, *Les Nouvelles Fondations des Sciences de Gestion*, pages 45–81. Vuibert, Paris, 2001.
26. M. Lebreton, G. Desthieux, and P. Rondier. Chaire de recherche du Canada en aide à la décision territoriale. Université de Laval, available at http://www.adt.chaire.ulaval.ca/2_recherche/programme_recherche.php#systemes_indicateurs, 2009.
27. H. L¨ning, V. Malleret, J. Meric, Y. Pesqueux, E. Chiapello, D. Michel, and A. Solé. *Le Contrôle de Gestion: Organisation et Mise en Oeuvre*. Edition Dunod, Paris, 3rd edition, 2008.
28. L. Maystre and D. Bollinger. *Aide à la Négociation Multicritère: Pratique et Conseils - Collection Gérer l'Environnement*. Presses Polytechniques et Eniversitaires Romandes, Lausanne, 1999.
29. J.C. Moisdon. *Du Mode d'Existence des Outils de Gestion*. Edition Seli Arslam, Paris, 1997.
30. K. Nielsen. *Le Mécénat Mode d'Emploi*. Economica, Paris, 2007.
31. ONU. Action 21: Document publié par les Nations Unies à la suite du sommet de Rio. Available at www.un.org/french/ga/special/sids/agenda21/index.html, 1992.
32. M.E. Porter and M.R. Kramer. The competitive advantage of corporate philantropy. *Harvard Business Review*, 80(12):56–67, 2002.
33. B. Roy. *Méthodologie Multicritère d'Aide à la Décision*. Economica, Paris, 1985.
34. B. Roy. *Multicriteria Methodology for Decision Aiding*. Kluwer Academic Publishers, Dordrecht, 1996.
35. B. Roy. L'aide à la decision ajourd'hui: Que devrait-on en attendre? In A. David, A. Hatchuel, and R. Laufer, editors, *Les Nouvelles Fondations des Sciences de Gestion*. Vuibert, Paris, 2000.
36. B. Roy and D. Bouyssou. *Aide Multicritère à la Décision: Méthodes et Cas*. Economica, Paris, 1993.
37. B. Roy and P. Vincke. Pseudo-orders: Definition, properties and numerical representation. *Mathematical Social Sciences*, 14:263–274, 1987.
38. André S. and Roy B. Conception et mise en place d'un outil d'évaluation de la performance environnementale - Le cas des raffineries de la société Total. *Journal of Decision Systems*, 16(3):335–367, 2007.
39. T.L. Saaty. *Decision Making for Leaders: The Analytical Hierarchy Process for Decisions in a Complex World*. Wadsworth, Belmont, MA, 1982.
40. S. Schmidheiny. *Changing Course: A Global Business Perspective on Development and the Environment*. MIT Press, Cambridge, MA, 1992.
41. V. Seghers. *Ce qui Motive les Enterprises Mécènes*. Autrement, Paris, 2007.

Chapter 14
Optimal Capital Structure
Reflections on Economic and Other Values

Marc B.J. Schauten and Jaap Spronk

Abstract Despite a vast literature on the capital structure of the firm there still is a big gap between theory and practice. Starting with the seminal work by Modigliani and Miller, much attention has been paid to the optimality of capital structure from the shareholders' point of view. Over the last few decades studies have been produced on the effect of other stakeholders' interests on capital structure. Another area that has received considerable attention is the relation between managerial incentives and capital structure. Furthermore, the issue of corporate control and, related, the issue of corporate governance, receive a lion's part of the more recent academic attention for capital structure decisions. From all these studies, one thing is clear: The capital structure decision (or rather, the management of the capital structure over time) has to deal with more issues than the maximization of the firm's market value alone. In this paper, we give an overview of the different objectives and considerations that have been proposed in the literature. We show that capital structure decisions can be framed as multiple criteria decision problems which can then benefit from multiple criteria decision support tools that are widely available.

14.1 Introduction

Despite a vast literature on the capital structure of the firm (see [10, 22], for overviews) there still is a big gap between theory and practice (see e.g. [6, 18]). Starting with the seminal work by Modigliani and Miller [35, 36], much attention

Marc B.J. Schauten
Erasmus University, P.O. Box 1738, 3000DR Rotterdam, The Netherlands
e-mail: schauten@ese.eur.nl

Jaap Spronk
Erasmus University, P.O. Box 1738, 3000DR Rotterdam, The Netherlands
e-mail: jspronk@rsm.nl

C. Zopounidis and P.M. Pardalos (eds.), *Handbook of Multicriteria Analysis*, Applied Optimization 103, DOI 10.1007/978-3-540-92828-7_14, © Springer-Verlag Berlin Heidelberg 2010

has been paid to the optimality of capital structure from the shareholders' point of view.

Over the last few decades studies have been produced on the effect of other stakeholders' interests on capital structure. Well-known examples are the interests of customers who receive product or service guarantees from the company (see e.g. [19]). Another area that has received considerable attention is the relation between managerial incentives and capital structure (Ibid.). Furthermore, the issue of corporate control[1] (see [27]) and, related, the issue of corporate governance[2] (see [50]), receive a lion's part of the more recent academic attention for capital structure decisions.

From all these studies, one thing is clear: The capital structure decision (or rather, the management of the capital structure over time) involves more issues than the maximization of the firm's market value alone. In this paper, we give an overview of the different objectives and considerations that have been proposed in the literature. We make a distinction between two broadly defined situations. The first is the traditional case of the firm that strives for the maximization of the value of the shares for the current shareholders. Whenever other considerations than value maximization enter capital structure decisions, these considerations have to be instrumental to the goal of value maximization. The second case concerns the firm that explicitly chooses for more objectives than value maximization alone. This may be because the shareholders adopt a multiple stakeholders approach or because of a different ownership structure than the usual corporate structure dominating finance literature. An example of the latter is the co-operation, a legal entity which can be found, in among others, many European countries. For a discussion on why firms are facing multiple goals, we refer to Hallerbach and Spronk [20, 21].

In Section 14.2 we will describe objectives and considerations that, directly or indirectly, clearly help to create and maintain a capital structure which is "optimal" for the value maximizing firm. In Section 14.3 we describe other objectives and considerations. Some of these may have a clear negative effect on economic value, others may be neutral and in some cases the effect on economic value is not always completely clear. Section 14.4 shows how, for both cases, capital structure decisions can be framed as multiple criteria decision problems which can then benefit from multiple criteria decision support tools. Section 14.5 gives a brief summary.

[1] Corporate Control is defined by Jensen and Ruback [27] as the rights to determine the management of corporate resources - that is, the rights to hire, fire and set the compensation of top-level managers.

[2] According to Shleifer and Vishney [50] corporate governance deals with the ways in which suppliers of finance to corporations assure themselves of getting a return on their investment. A broader definition is given by the OECD: "Corporate governance is the system by which business corporations are directed and controlled. The corporate governance structure specifies the distribution of rights and responsibilities among different participants in the corporation, such as, the board, managers, shareholders and other stakeholders, and spells out the rules and procedures for making decisions on corporate affairs. By doing this, it also provides the structure through which the company objectives are set, and the means of attaining those objectives and monitoring performance".

14.2 Maximizing Shareholder Value

According to the neoclassical view on the role of the firm, the firm has one single objective: maximization of shareholder value. Shareholders possess the property rights of the firm and are thus entitled to decide what the firm should aim for. Since shareholders only have one objective in mind - wealth maximization - the goal of the firm is maximization of the firm's contribution to the financial wealth of its shareholders. The firm can accomplish this by investing in projects with a positive net present value.[3] Part of shareholder value is determined by the corporate financing decision.[4] Two theories about the capital structure of the firm - the trade-off theory and the pecking order theory - assume shareholder wealth maximization as the one and only corporate objective. We will discuss both theories including several market value related extensions. Based on this discussion we formulate a list of criteria that is relevant for the corporate financing decision in this essentially neoclassical view.

The original proposition I of Modigliani and Miller [35] states that in a perfect capital market the equilibrium market value of a firm is independent of its capital structure, i.e. the debt-equity ratio.[5] If proposition I does not hold then arbitrage will take place. Investors will buy shares of the undervalued firm and sell shares of the overvalued firm in such a way that identical income streams are obtained. As investors exploit these arbitrage opportunities, the price of the overvalued shares will fall and that of the undervalued shares will rise, until both prices are equal.

When corporate taxes are introduced, proposition I changes dramatically. Modigliani and Miller [35, 36] show that in a world with corporate tax the value of firms is among others a function of leverage. When interest payments become tax deductible and payments to shareholders are not, the capital structure that maximizes firm value involves a hundred percent debt financing. By increasing leverage, the payments to the government are reduced with a higher cash flow for the providers of capital as a result. The difference between the present value of the taxes paid by an unlevered firm (G_u) and an identical levered firm (G_l) is the present value of tax shields ($PVTS$). Figure 14.1 depicts the total value of an unlevered and a levered firm. The higher leverage, the lower G_l, the higher $G_u - G_l (= PVTS)$.[6]

---

3 This view is seen as an ideal by many; see for example [23].

4 Financial decisions that influence the value of the firm are the capital budgeting decision and the corporate financing decision. In this paper we focus on the corporate financing decision made by the firm.

5 As Miller and Modigliani [35] formulate their proposition I in a perfect capital market: "The market value of any firm is independent of its capital structure and is given by capitalizing its expected return (i.e. cash flows) at the ρ_k (i.e. capitalization rate) appropriate to its class." With as a result of the former "That is, the average cost of capital to any firm is completely independent of its capital structure and is equal to the capitalization rate of a pure equity stream of its class." ([35], p. 268–269).

6 See Schauten and Tans [49] for a derivation of the cost of tax for the government.

Balance sheet of the unlevered firm

Pre-tax firm value	PV government's claim (G_u)
	PV residual claim equityholders (E_u)
Total value (TV)	Total value (TV)

Balance sheet of the levered firm

Pre-tax firm value	PV government's claim (G_l)
	PV residual claim equityholders (E_l)
	Debt (D)
Total value (TV)	Total value (TV)

Fig. 14.1 Pre-tax value of the firm
This figure presents the expanded balance sheet of the unlevered and the levered firm with on the left hand side the pre-tax value of the firm and on the right hand side the present value of the tax payments to the government by the unlevered firm (G_u) and the levered firm (G_l), the market value of equity of the unlevered firm (E_u) and the levered firm (E_l) and the market value of debt of the levered firm (D).

In the traditional trade-off models of optimal capital structure it is assumed that firms balance the marginal present value of interest tax shields [7] against the marginal direct costs of financial distress or direct bankruptcy costs. [8] Additional factors can be included in this trade-off framework. Other costs than direct costs of financial distress are agency costs of debt [26]. Often cited examples of agency costs of debt are the underinvestment problem [37], [9] the asset substitution problem ([16, 26]), the "play for time" game by managers, the "unexpected increase of leverage (combined with an equivalent pay out to stockholders to make to increase the impact)," the "refusal to contribute equity capital" and the "cash in and run" game [7]. These problems are caused by the difference of interest between equity and debt holders and could be seen as part of the indirect costs of financial distress. Another benefit of debt - besides the PVTS - is the reduction of agency costs between managers and external holders of equity ([23, 24, 26]). Jensen and Meckling [26] argue that debt, by allowing larger managerial residual claims because the need for external equity is reduced by the use of debt, increases managerial effort to work. In addition, Jensen [23] argues that high leverage reduces free cash (flow) with less resources to waste

[7] Miller [33] argued that under certain conditions, the corporate tax advantage of debt may be offset by tax disadvantages at the personal level, making leverage from a tax shield perspective irrelevant.

[8] Direct bankruptcy costs are the costs of the use of the legal mechanism allowing creditors to take over a firm when it defaults [7]. Direct bankruptcy costs consist of administrative costs and legal fees. Robichek and Myers [46] and Baxter [3] suggest that the cost associated with bankruptcy might represent the missing element in the theory of Miller and Modigliani. However, Modigliani and Miller [35] already remark that reorganization involves costs and might have unfavorable effects on earnings prospects, with a discount on the value of heavily indebted companies as a result, see Ibid. footnote 18.

[9] The underinvestment problem is sometimes referred to as the debt overhang problem [19] (p.563).

on unprofitable investments as a result.[10] The agency costs between management and external equity are often left out the trade-off theory since it assumes managers not acting on behalf of the shareholders (only) which is an assumption of the traditional trade-off theory.

In Myers' [38] and Myers and Majluf's [42] pecking order model there is no optimal capital structure.[11] Instead, because of asymmetric information and signaling problems associated with external financing, firm's financing policies follow a hierarchy, with a preference for internal over external finance, and for debt over equity.[12] A strict interpretation of this model suggests that firms do not aim at a target debt ratio. Instead, the debt ratio is just the cumulative result of hierarchical financing over time. (See [51]). Original examples of signaling models are the models of Ross [47] and Leland and Pyle [30]. Ross [47] suggests that higher financial leverage can be used by managers to signal an optimistic future for the firm and that these signals cannot be mimicked by unsuccessful firms.[13] Leland and Pyle [30] focus on owners instead of managers. They assume that entrepreneurs have better information on the expected cash flows than outsiders have. The inside information held by an entrepreneur can be transferred to suppliers of capital because it is in the owner's interest to invest a greater fraction of his wealth in successful projects. Thus the owner's willingness to invest in his own projects can serve as a signal of project quality. The value of the firm increases with the percentage of equity held by the entrepreneur relative to the percentage he would have held in case of a lower quality project. (See [10]).

The stakeholder theory formulated by Grinblatt and Titman [19] suggests that the way in which a firm and its *non-financial* stakeholders interact is an important determinant of the firm's optimal capital structure. Non-financial stakeholders are those parties other than the debt and equity holders.[14] Non-financial stakeholders include

---

[10] Jensen predicts a positive relation between leverage and profitability if the market for corporate control is effective and forces firms to commit to paying out cash by levering up. However, if this market is ineffective, i.e. managers prefer to avoid the disciplining role of debt, a negative relation between profitability and leverage could be expected [45]. The free cash flow theory of Jensen could then be presented as separate theory that assists the trade-off theory in explaining why managers do not fully exploit the tax advantages of borrowing (as suggested by Myers [41], p.99).

[11] In 1984, the pecking order story was not new. Donaldson [12, 13] for example observed pecking order behavior in case studies. However, the pecking order until then was viewed as managerial behavior - possibly to avoid the discipline of capital markets.

[12] The pecking order theory assumes that managers know more about their companies' prospects, risks and values than do outside investors.

[13] Such unsuccessful firms do not have sufficient cash flow. This concept is easily applied to dividend policy as well. A firm that increases dividend payout is signalling that it has expected future cash flows that are sufficiently large to meet debt payments and dividend payments without increasing the probability of bankruptcy. (See [10]). Miller and Rock [34] develop a financial signalling model founded on the concept of "net dividends." An unexpected increase in dividends will increase shareholders' wealth and an unexpected issue of new equity or debt will be indebted as bad news about the future prospects of the firm.

[14] The stakeholder theory is probably inspired by, among others, Baxter [3] and Kim [29] who discuss indirect costs of financial distress.

firm's customers, employees, suppliers and the overall community in which the firm operates. These stakeholders can be hurt by a firm's financial difficulties. For example customers may receive inferior products that are difficult to service, suppliers may lose business, employees may lose jobs and the economy can be disrupted. Because of the costs they potentially bear in the event of a firm's financial distress, non-financial stakeholders will be less interested ceteris paribus in doing business with a firm having a high(er) potential for financial difficulties. This understandable reluctance to do business with a distressed firm creates a cost that can deter a firm from undertaking excessive debt financing even when lenders are willing to provide it on favorable terms (Ibid., p.598). These considerations by non-financial stakeholders are the cause of their importance as determinant for the capital structure. This stakeholder theory could be seen as part of the trade-off theory (see [7], p.481, although the term "stakeholder theory" is not mentioned) since these stakeholders influence the indirect costs of financial distress.[15]

As the trade-off theory (excluding agency costs between managers and shareholders) and the pecking order theory, the stakeholder theory of Grinblatt and Titman [19] assumes shareholder wealth maximization as the single corporate objective.[16]

Based on these theories, a huge number of empirical studies have been produced. See e.g. Harris and Raviv [22] for a systematic overview of this literature.[17] More recent studies are e.g. Shyum-Sunder and Myers [51], testing the trade-off theory against the pecking order theory, Kemsley and Nissim [28] estimating the present value of tax shields, Andrade and Kaplan [1] estimating the costs of financial distress and Rajan and Zingales [45] investigating the determinants of capital structure in the G-7 countries. Rajan and Zingales [45][18] explain differences in leverage of individual firms with firm characteristics. In their study leverage is a function of tangibility of assets, market-to-book ratio, firm size and profitability.[19] Barclay and Smith [2] provide an empirical examination of the determinants of corporate debt maturity. Graham and Harvey [18] survey 392 CFOs about among others capital structure. We come back to this Graham and Harvey study in Section 14.3.[20]

---

[15] The stakeholder theory could also explain observed pecking order behaviour in the market. See [19], p.613.

[16] In the Modigliani and Miller world, where agency problems are absent, maximizing the value of the firm is identical to maximizing shareholder's wealth. When agency problems exist there are ways to increase shareholder wealth at the expense of other stakeholders. (See e.g., [9], p.261).

[17] Harris and Raviv divide the evidence into four categories: i) evidence of general capital structure trends; ii) event studies that measure the impact on share prices of an announcement of a capital structure change, iii) studies that relate firm/industry characteristics to capital structure, iv) studies that measure the relationship between capital structure and factors associated with corporate control.

[18] Examples of other cross sectional studies before 1991 are: Bradley et al. [5], Long and Malitz [32], and Titman and Wessels [54].

[19] See Lemmon et al. [31] for empirical evidence against the explanatory power of determinants of capital structure such as size, market-to-book, profitability, and industry.

[20] For European firms Brounen et al. [8] did a similar survey as Graham and Harvey did for U.S. firms.

Cross sectional studies as by Titman and Wessels [54], Rajan and Zingales [45], Barclay and Smith [2] and Wald [55] model capital structure mainly in terms of leverage and then leverage as a function of different firm (and market) characteristics as suggested by capital structure theory.[21] We do the opposite. We do not analyse the effect of several firm characteristics on capital structure (c.q. leverage), but we analyse the effect of capital structure on variables that co-determine shareholder value. In several decisions, including capital structure decisions, these variables may get the role of decision criteria. Criteria which are related to the trade-off and pecking order theory are listed in Table 14.1. We will discuss these criteria using a simplified example in Section 14.4. Figure 14.2 illustrates the basic idea of our approach.

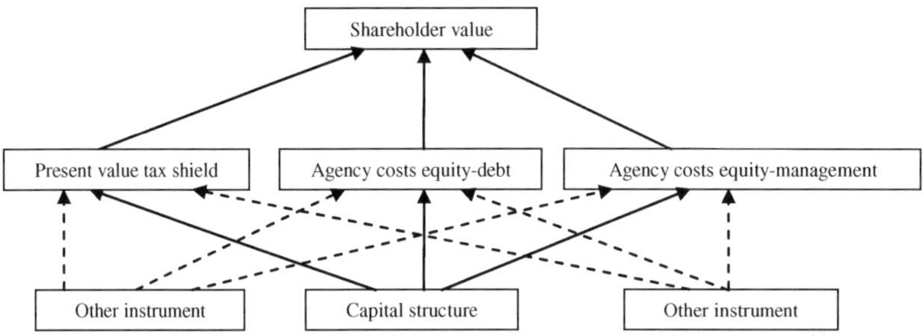

Fig. 14.2 Example of the basic idea of assumed relations within the neoclassical view

Figure 14.2 shows that shareholder value is related to the present value of tax shields and agency costs (both listed in Table 14.1 as determinants of shareholder value). The financing decision or "capital structure choice" now is an instrument that influences the value of these determinants. For example, the higher the leverage, the higher the present value of the tax shield. However, besides the financing decision, "other instruments" could have an influence (reflected with dotted arrows in Figure 14.2) on the value of these determinants as well. For example the decision to acquire assets that could be written of fast, influences the tax benefits of the interest deductibility. Of course, the financing decision influences the agency costs as well. For example, it could be argued that the agency costs between equity and debt increase with leverage. However, the tangibility of assets influences these agency costs as well. If a firm decides to invest in tangible assets this could have a negative impact on the magnitude of these agency costs. Put differently, agency costs are not minimized using one instrument only. Instead, a multiplicity of instruments is involved.

[21] In cross-sectional research, capital structure theories are tested by analyzing the relation between leverage (as endogenous variable) and some firm (and or country/institutional) characteristics (as exogenous variables). For example the static trade-off theory predicts that firms with a high profitability have higher leverage. A positive cross-sectional relation between the determinant profitability and leverage will be analysed. Proxies are used to measure leverage on the one hand and profitability on the other. If proxies are perfect indicators for the determinants then econometric tests reveal whether a relation between the variables exists. See e.g. [9].

The financing problem - even in a neoclassical context - is complex, because i) relevant "value determinants" are not influenced by capital structure only and ii) most if not all of these determinants cannot be translated into clearly quantifiable costs or benefits, even if we neglect the possible effect of other instruments on the selected determinants.

Table 14.1 Multiple criteria or determinants of capital structure

Category	#	Multiple criteria	References
Economic values	1	*Tax shield*	
		- corporate level	[35, 36]
		- personal level	[33]
	2a	*Direct costs of financial distress*	[35]
	2b	*Agency costs equity-debt*	
		- underinvestment	[37]
		- asset substitution (risk shifting)	[7, 16, 26]
		- refusing to contribute equity capital	Ibid.
		- cash in and run	Ibid.
		- playing for time	Ibid.
		- bait and switch	Ibid.
	2c	*Non-financial stakeholders*	
		- customers	[19]
		- employees	Ibid.
		- suppliers	Ibid.
		- community	Ibid.
	3	*Agency costs equity-management*	
		- residual claim	[26]
		- reduction free cash flow (overinvestment);	[23]
		corporate control shareholders,	[27]
		corporate governance	[50]
	4	*Following hierarchy and flexibility* (real options)	[38, 42]
	5	*Signaling*	[30, 47]
	6	*Subsidy*	[15]

14.3 Other Objectives and Considerations

A lot of evidence suggests that managers act not only in the interest of the shareholders (see [41]). Neither the static trade-off theory nor the pecking order theory can fully explain differences in capital structure. Myers [41] (p.82) states that "Yet even 40 years after the Modigliani and Miller research, our understanding of these firms financing choices is limited."[22] Results of several surveys (see [8, 9, 18]) reveal that CFOs do not pay a lot of attention to variables relevant in these shareholder wealth maximizing theories. Given the results of empirical research, this does not come as a surprise.

The survey by Graham and Harvey finds only moderate evidence for the trade-off theory. Around 70% have a flexible target or a somewhat tight target or range. Only 10% have a strict target ratio. Around 20% of the firms declare not to have an optimal or target debt-equity ratio at all.

In general, the corporate tax advantage seems only moderately important in capital structure decisions. The tax advantage of debt is most important for large regulated and dividend paying firms. Further, favorable foreign tax treatment relative to the U.S. is fairly important in issuing foreign debt decisions.[23] Little evidence is found that personal taxes influence the capital structure.[24] In general potential costs of financial distress seem not very important although credit ratings are. According to Graham and Harvey this last finding could be viewed as an (indirect) indication of concern with distress. Earnings volatility also seems to be a determinant of leverage, which is consistent with the prediction that firms reduce leverage when the probability of bankruptcy is high. Firms do not declare directly that (the present value of the expected) costs of financial distress are an important determinant of capital structure, although indirect evidence seems to exist. Graham and Harvey find little evidence that firms discipline managers by increasing leverage. Graham and Harvey [18] (p.227) explicitly note that "1) managers might be unwilling to admit to using debt in this manner, or 2) perhaps a low rating on this question reflects an unwillingness of firms to adopt Jensen's solution more than a weakness in Jensen's argument."

The most important issue affecting corporate debt decisions is management's desire for financial flexibility (excess cash or preservation of debt capacity). Furthermore, managers are reluctant to issue common stock when they perceive the market is undervalued (most CFOs think their shares are undervalued). Because asymmetric information variables have no power to predict the issue of new debt or equity, Harvey and Graham conclude that the pecking order model is not the true model of the security choice.[25]

The fact that neoclassical models do not (fully) explain financial behavior could be explained in several ways. First, it could be that managers do strive for creating shareholder value but at the same time also pay attention to variables other than the variables listed in Table 14.1. Variables of which managers think, that they are

[22] These firms are public, non-financial corporations with access to U.S. or international capital markets.

[23] According to Graham and Harvey the most popular reason to issue foreign debt is that it provides a natural hedge against foreign currency devaluation.

[24] Graham [17] argues that companies do not make full use of interest rate tax shields.

[25] For European firms [8], p.99) find moderate support for the static trade-off theory. The results of the pecking order theory, the desire for financial flexibility and pecking order behavior are important considerations but as Graham and Harvey [18] conclude, asymmetric information is not the driving force behind this behavior.

(justifiably or not) relevant for creating shareholder value. Second, it could be that managers do not (only) serve the interest of the shareholders but of other stakeholders as well.[26] As a result, managers integrate variables that are relevant for them and or other stakeholders in the process of managing the firm's capital structure. The impact of these variables on the financing decision is not per definition negative for shareholder value. For example if "value of financial rewards for managers" is one of the goals that is maximized by managers - which may not be excluded - and if the rewards of managers consists of a large fraction of call options, managers could decide to increase leverage to lever the volatility of the shares with an increase in the value of the options as a result. The increase of leverage could have a positive effect on shareholder wealth (e.g. the agency costs between equity and management could be lower) but the criterion "value of financial rewards" could (but does not have to) be leading. Third, shareholders themselves do possibly have other goals than shareholder wealth creation alone. Fourth, managers rely on certain (different) rules of thumb or heuristics that do not harm shareholder value but can not be explained by neoclassical models either.[27] Fifth, the neoclassical models are not complete or not tested correctly (see e.g. [31, 51]).

Either way, we do expect that variables other than those founded in the neoclassical property rights view are or should be included explicitly in the financing decision framework. To determine which variables *should* be included we probably need other views or theories of the firm than the neoclassical alone. Zingales [56] argues that "... *corporate finance theory, empirical research, practical implications, and policy recommendations are deeply rooted in an underlying theory of the firm*" (Ibid., p.1623.). Examples of attempts of new theories are "the stakeholder theory of the firm" (see e.g. [14]), "the enlightened stakeholder theory" as a response (see [25]), "the organizational theory" (see [39, 40, 41]) and "the stakeholder equity model" (see [52]).

We introduce an organizational balance sheet which is based on the organizational theory of Myers [39]. The intention is to offer a framework to enhance a discussion about criteria that could be relevant for the different stakeholders of the firm. In Myers' organizational theory employees (including managers) are included as stakeholders; we integrate other stakeholders as suppliers, customers and the community as well. Figure 14.3 presents the adjusted organizational balance sheet.

Note that pre-tax value of the existing assets and the growth opportunities is the value of the firm including the present value of all stakeholders' surplus. The present value of the stakeholders' surplus (ES plus OTS) is the present value of future costs of perks, overstaffing, above market prices for inputs (including above

[26] Block [4] finds that on average 56% of his surveyed CFOs of Fortune 1,000 companies has stockholder wealth maximization as predominant goal. This percentage is much lower than 100% but higher than the results of Petty et al. [44] and Stanley and Block [53] where this percentage was only 11% (of their sample of Fortune 500 Companies) and 21% (of their sample of Fortune 1,000 companies) respectively.

[27] Miller [33] (p.272) states that 'harmful heuristics, like harmful mutations in nature die out. Neutral mutations that serve no function, but do no harm, can persist indefinitely.' Miller [33] (p.273) further argues that a pool of neutral mutations could be of value when the environment changes.

Balance sheet of the levered firm

Pre-tax value existing assets	(PTA)	PV residual claims equityholders	(E)
Pre-tax value growth opportunities	(PVGO)	Debt	(D)
		Employees' Surplus	(ES)
		Other stakeholders' Surplus	(OTS)
		PV government claims	(GI)
Pre-tax value	(PTV)	Pre-tax value	(PTV)

Fig. 14.3 Adjusted organizational balance sheet in market values

market wages), above market services provided to customers and the community etc.[28] Depending on the theory of the firm, the pre-tax value can be distributed among the different stakeholders following certain "rules." Note that what we call "surplus" in this framework is still based on the "property rights" principle of the firm. Second, only distributions in market values are reflected in this balance sheet. Neutral mutations are not.[29]

Based on the results of Graham and Harvey [18] and common sense we formulate a list of criteria or heuristics that could be integrated into the financing decision framework. Some criteria lead to neutral mutations others do not. We call these criteria "quasi non-economic criteria"; non-economic because the criteria are not based on the neoclassical view. Quasi, because the relations with economic value are not always clear cut. We include criteria that lead to neutral mutations as well, because managers might have good reasons that we overlook or are relevant for other reasons than financial wealth.

The broadest decision framework we propose in this paper is the one that includes both the economic and quasi non-economic variables. Figure 14.4 illustrates the idea. The additional quasi non-economic variables are listed in Table 14.2. This list is far from complete. Relevant variables to be included depend on i) the theory of the firm, ii) characteristics of the particular firm/industry/country and iii) judgment and the preferences of the manager(s).

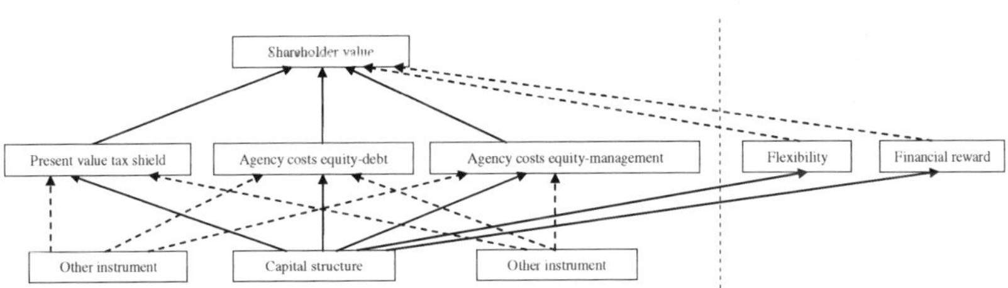

Fig. 14.4 Example of basic idea of possible relations

[28] To a certain extent - as long as debt is not risk free - the firm can expropriate wealth from the debt holders which would result in a broader definition.

[29] Myers [38] defines - after Miller [33] - neutral mutations as financing patterns or habits which have no material effect on firm value and makes managers feel better.

Financial flexibility (excess cash), the first variable in Table 14.2 is valued by managers because it increases their independence from the capital market. Managers may invest more often in projects that do not create shareholder value when they have excess cash or unused debt capacity. For this reason financial flexibility could be relevant for at least employees and the suppliers of resources needed for these projects. As long as managers only would invest in zero net present value projects this variable would have no value effect in the organizational balance sheet. But if it influences the value of the sum of the projects undertaken this will be reflected in this balance sheet. Of course, financial flexibility is also valued for economic reasons, see Section 14.2 and 14.4.

The probability of bankruptcy influences job security for employees and the duration of a "profitable" relationship with the firm for suppliers, customers and possibly the community. For managers (and other stakeholders without diversified portfolios) the probability of default could be important. The cost of bankruptcy is for them possibly much higher than for shareholders with diversified portfolios. As with financial flexibility, the probability of default influences shareholder value as well. In Section 14.2 and 14.4 we discuss this variable in relation to shareholder value. Here the variable is relevant, because it has an effect on the wealth or other "valued" variables of stakeholders other than the equity (and debt) holders.

We assume owner-managers dislike sharing control of their firms with others. For that reason, debt financing could possibly have non-economic advantages for these managers. After all, common stock carries voting rights while debt does not. Owner-managers might prefer debt over new equity to keep control over the firm. Control is relevant in the economic framework as well, see Section 14.2 and 14.4.

In practice, earnings dilution is an important variable effecting the financing decision.[30] Whether it is a neutral mutations variable or not, the effect of the financing decision on the earnings per share is often of some importance.[31] If a reduction in the earnings per share (EPS) is considered to be a bad signal, managers try to prevent such a reduction. Thus the effect on EPS becomes an economic variable. As long as it is a neutral mutation variable, or if it is relevant for other reasons, we treat EPS as a quasi non-economic variable.

The reward package could be relevant for employees. If the financing decision influences the value of this package this variable will be one of the relevant criteria for the manager. If it is possible to increase the value of this package, the influence on shareholder value is ceteris paribus negative. If the reward package motivates the manager to create extra shareholder value compared with the situation without the package, this would possibly more than offset this negative financing effect.

Other criteria that might be relevant are maturity matching, because of liquidity reasons, and the influence of the capital structure on the credit rating of debt.

---

[30] E.g., Block [4] finds that on average 28% of his surveyed CFOs of Fortune 1,000 companies have 'growth in earnings per share' defined as predominant goal of the firm, and Brav et al. [6] state that three-fourth of their survey respondents (financial executives) indicate that increasing earnings per share is an important factor affecting their share repurchase decision.

[31] In perfect capital markets earnings dilution does not influence the value of equity. This is often misunderstood, see [7], Chapter 32.

Table 14.2 Multiple criteria of capital structure

Category	#	Multiple Criteria
Quasi non-economic values	1	Flexibility (self sufficiency)
	2	Job security
	3	Control
	4	Earnings dilution
	5	Financial reward
	6	Maturity matching, Credit rating etc.

14.4 Capital Structure as Multiple Criteria Decision Problem

Ideally, capital structure decisions are embedded in a capital structure management process, with 1) periodic planned evaluations (e.g. around reporting dates and connected with dividend decisions), 2) events or anticipated events concerning the assets of the company (large investments, mergers and acquisitions, unexpected results) or 3) concerning the liabilities side (changing financial market circumstances, new products offered by the financial industry, refinancing loans etc.). Given the multiplicity of considerations, the large variety of choices and the presence of many contingent claims, both real and financial, make many capital structure decisions unfit for being framed as an optimization problem. In such cases, it does make sense to solicit a variety of solutions by advisors, banks and other providers of capital, which can then be compared in terms of their impact on the criteria considered to be important for the firm concerned.

The factors considered to be important are determined by firm, industry, environmental, country or institutional characteristics. For example, profitability, risk, tangibility of assets, size, growth opportunities of the firm, the competition within and concentration of the industry, the legal system and corporate governance regulations are all more or less important in the selection and weighting of the appropriate criteria.

As an example of capital structure as a multiple criteria decision problem, consider the 100% equity financed firm "OCS." In the coming year OCS has to make an investment and financing decision.[32] Let:

x = new investment in millions of euros;
y = new issue of debt, in millions of euros;
z = new issue of equity, in millions of euros.

Assume the investment generates a perpetual free cash flow of 1 million. Assume for simplicity there are only two financing solutions: 100% debt (plan 1) and 100% equity financing (plan 2). OCS is a listed firm. Managers own 10% of total equity. Assume the unlevered cost of capital is 10% and x is €10 million. The corporate tax rate is 30%. Taxes on a personal level are 0%. OCS has to decide whether she goes

[32] This example is based on Myers and Pogue [43]. For another example, where different financing proposals for an M&A financing problem are compared, see Schauten and Spronk [48].

ahead with the project and if so, whether $y = €10$ million or $z = €10$ million. To support the financing decision OCS evaluates both financing solutions on the criteria listed in Table 14.1 and Table 14.2. If possible, the influence of the financing plans on the criteria is measured in euros. If this is not possible, we only make a qualitative statement. The scores on the economic and quasi non-economic criteria are given in Table 14.3. In this example we choose to score the quasi non-economic variables from the perspective of the manager.[33] The economic variables are scored from the perspective of the shareholders.

Tax shield. The main advantage of debt financing is the reduction of the present value of the government's claim. In general, the higher the proportion interest bearing debt, the higher the PVTS. However, the level of non-debt tax shields [11] and, among others a low level and high variability of earnings could have a negative impact on the PVTS of additional debt. If we assume the profits are high enough to realize the tax shields then the tax shield score on the corporate level of plan 1 is corporate tax rate times the amount of debt, i.e. $0.3 \times €10$ million $= €3$ million.[34] If on the personal level income tax for received interests is higher than for equity income, the advantage on the corporate level could be offset by the disadvantage on the personal level. For now, we assume there are no personal taxes. This implies there is no difference on the criterion "Tax shield on a personal level."

Direct costs of financial distress or the direct bankruptcy costs are the costs of the legal mechanism that allows creditors taking over the assets of a firm when a firm defaults (see [7]). If a firm increases leverage, it increases the probability of default and the present value of the direct costs of bankruptcy. Lenders foresee these costs and foresee that they will pay them if default occurs. Therefore lenders will charge a higher interest rate which reduces both equity cash flows and equity value as a result. If we assume that the risk of the assets in place of OCS is low, and the size of the investment is small relative to the expected free cash flow, the expected probability of default is low. The impact of plan 1 on the direct costs of financial distress then is limited. Of course, plan 2 scores better on this criterion than plan 1.

Agency costs equity-debt. If OCS is not in financial distress, the probability that OCS will play games with the debt holders is small. But if the FCFs are unexpectedly low, it could be that managers on behalf of the existing shareholders try to expropriate wealth from the debt holders. Therefore the agency costs equity-debt are low but positive. Of course the agency costs equity-debt are zero if the investment is financed with an issue of shares.

Non-financial stakeholders. If stakeholders foresee that - because of a higher leverage - the probability of default exceeds acceptable levels, stakeholders could e.g. charge higher prices or buy less products. If the products need a lot of services the value of the assets in place and the value of the new project could be reduced by using an excessive amount of debt. If OCS chooses plan 1 we assume customers will buy less products and employees will charge higher wages. We assume that neither suppliers nor the community is impacted by the financing decision.

---

[33] It is possible to score the criteria from the perspective of other stakeholders as well.

[34] We assume that the additional amount of debt is fixed and the assets of the project serve as collateral.

Table 14.3 Example scores simplified example "OCS"

Category	#	Multiple Criteria	Scores plan 1 and 2		
			Plan 1	Plan 2	Preference for plan
Panel A: Economic values	1	*Tax shield*			
		- corporate level	€3 million	€0	1
		- personal level	€0	€0	–
	2a	*Direct costs financial distress*	👎	👍	2
	2b	*Agency costs equity-debt*	👎	👍	2
	2c	*Non-financial stakeholders*			
		- customers	👎	👍	2
		- employees	👎	👍	2
		- suppliers	–	–	–
		- community	–	–	–
	3	*Agency costs equity-management*			
		- residual claim	👍	👎	1
		- free cash flow	👍	👎	1
		- control	👎	👍	2
	4	*Following hierarchy*	👍	👎	1
		Flexibility	👎	👍	2
	5	*Signaling*	👍	👎	1
	6	*Subsidy*	–	–	–
Panel B: Quasi non-economic values	1	*Flexibility*	👎	👍	2
	2	*Job security*	👎	👍	2
	3	*Control*	👍	👎	1
	4	*Earnings dilution*	👍	👎	1
	5	*Financial reward*	👍	👎	1

Agency costs equity-management. Under plan 1 the residual claim managers hold remains the same. That means that the price of shirking for the managers remains the same as well. Under plan 2 this price decreases, which means the agency costs caused by a reduction in the residual claim for the managers increases. Under plan 1 free cash flows (FCFs) are reduced because of the promised interest payments. Under plan 2 these FCFs are not reduced. This means that plan 1 scores better on both criteria; residual claim and free cash flow. Given the stake managers have, under plan 1 they could prevent harder possible bidders to take-over the firm. If plan 2 is chosen the stake of the managers dilutes and - we assume - the power of the market for corporate control increases. Plan 2 scores better than plan 1 on the criterion control. The governance structure of the firm, e.g. the way the firm rewards their managers influences the importance of the FCF problem.

Following hierarchy / flexibility. If debt is issued instead of equity the negative impact of mispricing caused by information asymmetry is reduced. However, plan 1 also has a possible negative effect: plan 1 reduces the FCFs, which may negatively influence the future flexibility of the firm. Financial flexibility (excess cash or the preservation of debt capacity) is valued positively because it prevents firms from not investing in positive net present value projects. For example if the net present value of a new project is 1.5 million and the firm has - because of a lack of excess cash, i.e. a lack of financial flexibility - to issue shares to collect 10 million but are really worth 12 million, the firm will not pursue. It only goes ahead if the net present value of the project is at least 2 million. (See [38], p.584.) The score for plan 1 is relatively good for the aspect hierarchy and bad for expected flexibility.

Signaling. Given information asymmetry it could be argued that if managers have the incentive to always issue the correct signal (that is to tell the truth) an issue of debt could be interpreted as a positive signal about future cash flows [47]. The score for plan 1 then is better than the score for plan 2.

Subsidy. There is no subsidy.

The first quasi non-economic variable flexibility is reduced if managers select plan 1. As under panel A FCF is reduced if debt is issued. If the new project generates positive FCFs then expected flexibility will increase due to an accumulation of free cash.

Job security increases inversely with the probability of default. If the new project contributes to stability of the firm's cash flows the new project could increase job security.

We assume that the managers do not like their stake to dilute. Managers prefer plan 1. This is also in accordance with the control score in Panel A of Table 14.3 where we assume that external shareholders prefer plan 2.[35]

Earnings dilution is higher if new shares are issued. If managers prefer higher earnings per share, plan 1 is favored by managers. Expected earnings increase due to the profitability of the new project, while the number of shares remains the same.

If the financial reward exists - besides the equity stake - of call options, plan 1 again is best. It increases the volatility of equity with a relative positive effect on call options as a result. If plan 2 is implemented the volatility remains the same.[36]

The next step is that the manager evaluates the relative scores on all the criteria and gives his/her own weighting factors to the relevant criteria and then decides which plan is optimal.[37] If the perceived value of all the side effects under the favored plan is positive the manager will go ahead with this project.[38] This simplified "numerical" example shows how complex capital structure problems can be. Even, if we only take the economic criteria into account.

14.5 Summary

The capital structure decision (or rather, the management of the capital structure over time) is never a goal on its own, but should be instrumental to the goal of the

[35] Management could prefer Plan 2 if for instance the power of certain active monitoring shareholders is reduced by a placement of new shares to minority shareholders.

[36] We assume the volatility of the assets remains the same.

[37] MCDA methods that allow the incorporation of quantitative and qualitative criteria could support this decision problem. See Zopounidis [57] for arguments that could justify the use of MCDA methods in investment decisions and portfolio management decisions.

[38] We assume the present value without side effects equals €1 million/0.1=€10 million.

firm. In the traditional case of the firm that strives for the maximization of the value of the shares for the current shareholders, all choices concerning the capital structure should be evaluated in terms of their effect on the firm's market value. No wonder that so much research effort is devoted to the value effects of capital structure decisions. The capital structure decision is often pictured as an optimization problem in which a value function including all costs and benefits is to be maximized, possibly subject to some hard constraints.

We have shown that the management of the firm's capital structure is not that easy at all. The reason is that a number of considerations that enter the capital structure decision and have value implications, cannot be translated into clearly quantifiable costs or benefits that can be entered into the value function or be transformed into hard constraints. Examples discussed include agency costs between equity holders and management (including corporate control and corporate governance), costs of financial distress, benefits and costs for other financial stakeholders, flexibility and even the tax shield. Still these considerations cannot be ignored in the capital structure decision and its economic value implications. Therefore, we propose to translate some of these considerations as separate criteria, which can be traded off against the hard and quantifiable criterion of market value.

Many firms exist that explicitly choose for more objectives than value maximization alone. This may be because the shareholders adopt a multiple stakeholders approach or because of a different ownership structure than the usual corporate structure dominating finance literature. An example of the latter is the co-operation, a legal entity which can be found in, among others, many European countries. So in addition to the criteria that capture the value implications of capital structure decisions, this kind of firms may have other criteria as well. An example is bankruptcy risk and its implications for various stakeholders.

Ideally, capital structure decisions are embedded in a capital structure management process, with 1) periodic planned evaluations (e.g. around reporting dates and connected with dividend decisions), 2) events or anticipated events concerning the assets of the company (large investments, mergers and acquisitions, unexpected results) or 3) concerning the liabilities side (changing financial market circumstances, new products offered by the financial industry, refinancing loans). Given the multiplicity of considerations, the large variety of choices (e.g. all the specifications that can be connected with a loan or with a leasing contract) and the presence of many contingent claims, both real and financial, makes many capital structure decisions unfit for being framed as an optimization problem. In such cases, it does make sense to solicit a variety of solutions by banks and advisors, which can then be compared in terms of their impact on the criteria considered to be important for the firm concerned. The definition of the criteria and the study of the impact of the decision alternatives on these criteria is thus a sine qua non for financial structure decisions.

Acknowledgements We thank the participants of various meetings of the EURO Working Group on Financial Modelling, for their constructive remarks. In addition, we thank J.O. Soares of the Technical University in Lisbon for his very useful suggestions.

References

1. G. Andrade and S.N. Kaplan. How costly is financial (not economic) distress? Evidence from highly leveraged transactions that became distressed. *Journal of Finance*, 43:1443–1493, 1998.
2. M.J. Barclay and C.W. Smith. The maturity structure of corporate debt. *Journal of Finance*, 50:609–631, 1995.
3. N.D. Baxter. Leverage, risk of ruin and the cost of capital. *Journal of Finance*, 22(3):395–403, 1967.
4. S. Block. Are there differences in capital budgeting procedures between industries? An empirical study. *The Engineering Economist*, 50:55–67, 2005.
5. Jarrell G. Bradley, M. and E.H. Kim. On the existence of an optimal capital structure: Theory and evidence. *Journal of Finance*, 42:1245–1260, 1984.
6. A. Brav, J.R. Graham, C.R. Harvey, and R. Michaely. Payout policy in the 21st century. *Journal of Financial Economics*, 77:483–527, 2005.
7. R.A. Brealey, S.C. Myers, and F. Allen. *Principles of Corporate Finance*. McGraw-Hill, New York, 2006.
8. D. Brounen, A. De Jong, and K. Koedijk. Corporate finance in Europe: Confronting theory with practice. *Financial Management*, 33(4):71–101, 2004.
9. K. Cools. *Capital structure choice*. PhD thesis, Katholieke Universiteit Brabant, 1993.
10. T.E. Copeland, J.F. Weston, and K. Shastri. *Financial Theory and Corporate Policy*. Pearson Addison Wesley, New York, 4th edition, 2005.
11. H. DeAngelo and R. Masulis. Optimal capital structure under corporate and personal taxation. *Journal of Financial Economics*, 8(1):3–29, 1980.
12. G. Donaldson. *Strategy for Financial Mobility*. Irwin, Homewood, IL., 1971.
13. G. Donaldson. *Managing Corporate Wealth*. Preager Publ., New York, 1984.
14. T. Donaldson and L.E. Preston. The stakeholder theory of the corporation: Concpets, evidence, and implications. *The Academy of Management Review*, 20(1):65–91, 1995.
15. D. Galai. Taxes, M-M propositions and government's implicit cost of capital in investment projects in the private sector. *European Financial Management*, 4:143–157, 1998.
16. D. Galai and R. Masulis. The option pricing model and the risk factor of stock. *Journal of Financial Economics*, 3(1-2):53–81, 1976.
17. J.R. Graham. How big are tax benefits of debt? *Journal of Financial Economics*, 55(5):1901–1941, 2000.
18. J.R. Graham and C.R. Harvey. The theory and practice of corporate finance: Evidence from the field. *Journal of Financial Economics*, 60:187–243, 2001.
19. M. Grinblatt and S. Titman. *Financial markets and corporate strategy*. Irwin/McGraw Hill, Boston, 2002.
20. W.G. Hallerbach and J. Spronk. A multidimensional framework for financial-economic decisions. *Journal of Multi-Criteria Decision Analysis*, 11:111–124, 2002.
21. W.G. Hallerbach and J. Spronk. The relevance of MCDM for financial decisions. *Journal of Multi-Criteria Decision Analysis*, 11:187–195, 2002.
22. M. Harris and A. Raviv. The theory of capital structure. *Journal of Finance*, 46:297–356, 1991.
23. M.C. Jensen. Agency costs of free cash flow, corporate finance and takeovers. *American Economic Review*, 76:323–329, 1986.
24. M.C. Jensen. The eclipse of the public corporation. *Harvard Business Review*, 67:61–74, 1989.
25. M.C. Jensen. Value maximization, stakeholder theory, and the corporate objective function. *Journal of Applied Corporate Finance*, 14(3):8–21, 2001.
26. M.C. Jensen and W. Meckling. The theory of the firm: Managerial behavior, agency costs and the ownership structure. *Journal of Financial Economics*, 3:305–360, 1976.
27. M.C. Jensen and R.S. Ruback. The market for corporate control. *Journal of Financial Economics*, 11:5–50, 1983.

28. D. Kemsley and D. Nissim. Valuation of the debt tax shield. *Journal of Finance*, 57:2045–2073, 2002.
29. E.H. Kim. A mean variance theory of optimal capital structure and corporate debt capacity. *Journal of Finance*, 33:45–63, 1978.
30. H. Leland and D. Pyle. Informational asymmetries, financial structure, and financial intermediation. *Journal of Finance*, 32:371–387, 1977.
31. M.L. Lemmon, M.R. Roberts, and J.F. Zender. Back to the beginning: Persistence and the cross-section of corporate capital structure. *Journal of Finance*, 63:1575–1608, 2008.
32. M. Long and I. Malitz. The investment financing nexus: Some empirical evidence. *Midland Corporate Finance Journal*, 3:53–59, 1985.
33. M.H. Miller. Debt and taxes. *Journal of Finance*, 32:261–275, 1977.
34. M.H. Miller and K. Rock. Dividend policy under asymmetric information. *Journal of Finance*, 40(4):1031–1051, 1985.
35. F. Modigliani and M.H. Miller. The cost of capital, corporation finance and the theory of investment. *American Economic Review*, 48:261–297, 1958.
36. F. Modigliani and M.H. Miller. Corporate income taxes and the cost of capital: A correction. *American Economic Review*, 53:433–443, 1963.
37. S.C. Myers. Determinants of corporate borrowing. *Journal of Financial Economics*, 19:147–176, 1977.
38. S.C. Myers. The capital structure puzzle. *Journal of Finance*, 39:575–592, 1984.
39. S.C. Myers. Still searching for optimal capital structure. *Journal of Applied Corporate Finance*, 1(6):4–14, 1993.
40. S.C. Myers. Outside equity. *Journal of Finance*, 55:1005–1037, 2000.
41. S.C. Myers. Capital structure. *Journal of Economic Perspectives*, 15(2):81–102, 2001.
42. S.C. Myers and N. Majluf. Corporate financing and investment decisions when firms have information that investors do not have. *Journal of Financial Economics*, 13(2):187–221, 1984.
43. S.C. Myers and G.A. Pogue. A programming approach to corporate financial management. *Journal of Finance*, 29:579–599, 1974.
44. J.W. Petty, Jr. Scott, D.F., and M.M. Bird. The capital budgeting decision process of large corporations. *The Engineering Economist*, 20(3):159–172, 1975.
45. R.G. Rajan and L. Zingales. What do we know about capital structure? Some evidence from international data. *Journal of Finance*, 50:1421–1460, 1995.
46. A. Robichek and S. Myers. Problems in the theory of optimal capital structure. *Journal of Financial and Quantitative Analysis*, 1:1–35, 1966.
47. S.A. Ross. The determination of financial structure: The incentive signalling approach. *Bell Journal of Economics*, 8(1):23–40, 1977.
48. M.B.J. Schauten and J. Spronk. Optimal capital structure decision in a multi-criteria framework: Solutions for an m&a case as proposed by practicing financial experts. *The Journal of Financial Decision Making*, 5(1):69–98, 2009.
49. M.B.J. Schauten and B. Tans. Cost of capital of government's claims and the present value of tax shields. available at SSRN: http://ssrn.com/abstract=627862, 2006.
50. A. Shleifer and R.W. Vishny. A survey of corporate governance. *Journal of Finance*, 52(2):737–783, 1997.
51. L. Shyam-Sunder and S.C. Myers. Testing static tradeoff against pecking order models of capital structure. *Journal of Financial Economics*, 51:219–244, 1999.
52. A. Soppe. *Finance as an instrument to a sustainable company*. PhD thesis, Erasmus University Rotterdam, 2006.
53. M.T. Stanley and S.B. Block. A survey of multinational capital budgeting. *Financial Review*, 19:36–51, 1984.
54. S. Titman and R. Wessels. The determinants of capital structure choice. *Journal of Finance*, 43:1–20, 1988.
55. J.K. Wald. How firm characteristics affect capital structure: An international comparison. *Journal of Financial Research*, 22:161–187, 1999.
56. L. Zingales. In search of new foundations. *Journal of Finance*, 55(4):1623–1653, 2000.
57. C. Zopounidis. Multicriteria decision aid in financial management. *European Journal of Operational Research*, 119:404–415, 1999.

Chapter 15
Applications of MCDA in Marketing and e-Commerce

Stelios Tsafarakis, Kleanthi Lakiotaki, and Nikolaos Matsatsinis

Abstract This chapter emphasizes on the major components under which MCDA applications in marketing and e-commerce have been developed and describes characteristic examples of research works that apply MCDA methodologies in marketing and e-commerce. The chapter is divided into two main sections separating the MCDA applications in the marketing discipline from those that appear in the e-commerce field. In each section fundamental notions of marketing and e-commerce are discussed accordingly and some characteristic examples of research works are analytically mentioned. The aim of this work is to endow candidate researchers that are interested in applying MCDA methodologies in marketing and e-commerce with adequate background information to further develop their scopes and ideas.

15.1 Introduction

The application of multi-criteria methods in marketing is mainly focused on two wide areas of research: the measurement, analysis and description of customer preferences regarding products and services, as well as the modeling and prediction of consumer purchasing behavior. The interest for these areas was raised by the establishment of consumer research as a distinct field of marketing science in the late '50. Managers discovered that individuals react differently to the same marketing campaigns, offers, and new product introductions. This was the beginning of the rejection of the theory of sales (first develop the product and then locate the potential customers). Nowadays, satisfying individual consumer needs has received much more attention in marketing science. Companies try to identify customer preferences by studying the way a consumer evaluates, chooses and uses a particular product. Multi-Attribute Utility Theory constitutes the long-established approach for mod-

Technical University of Crete, Department of Production and Management Engineering, Decision Support Systems Laboratory, Chania, Greece e-mail: tsafarakis@isc.tuc.gr, {klio, nikos}@ergasya.tuc.gr

C. Zopounidis and P.M. Pardalos (eds.), *Handbook of Multicriteria Analysis*, Applied Optimization 103, DOI 10.1007/978-3-540-92828-7_15, © Springer-Verlag Berlin Heidelberg 2010

eling consumer behavior, along with advanced heuristic algorithms that simulate individual purchasing process, while disaggregation methods are used for revealing customer preferences.

Since the 1990s, the socio-economic context has generally been referred to as the information and knowledge age or "the dot com/Internet revolution" and the attractiveness of using websites as an integrated marketing media has increased as more users have access to the Internet. Thus, traditional marketing applications adopted the "e-" trend and new application areas of marketing research revealed. In the late 90's when online stores emerged and e-commerce became a research topic on its own the foundations of decision theories such as these found in "Theory of Games and Economic Behavior," by von Neumann and Morgenstern [62] and later in "Decisions with Multiple Objectives: Preferences and Value Trade-Offs," by Keeney and Raiffa [29] inspired computer scientists to apply them in various e-commerce problems.

Following the explosion of the internet, Multiple Criteria Decision Analysis (MCDA), a well established field of Decision Science has penetrated into new application areas, such as electronic commerce. Nevertheless, even though significant work has been undertaken towards this application field and many MCDA tools have been integrated into web applications, often in an ad hoc manner and with varying degrees of success, there exist still much ongoing work as well as new trends and unexplored aspects of MCDA applications in e-commerce.

The purpose of this chapter is to survey the main applications that have been hitherto referred in the literature and to identify the principal directions under which these applications have been developed. To achieve that, this chapter is divided into two main sections, the first devoted to the applications of MCDA in marketing and the second to the applications of MCDA in e-commerce. Although e-commerce employs principles and aspects from marketing science and research and e-marketing is considered a fundamental element for a successful e-commerce site, marketing is regarded in this chapter as a more traditional application area than e-commerce, which, despite its exponential development, it is still considered in its infancy.

Following the introduction (section 15.1) which aims to clarify the theme and the scopes of this chapter, section 15.2.1 is devoted to a fundamental aspect of both marketing and e-commerce applications, the measurement of customer preferences. Basic notions like utility function are briefly discussed and examples of MCDA that measure customer preferences are analyzed. Subsequently (section 15.2.2) fundamental concepts of strategic marketing like segmentation, targeting and positioning are presented and explained by means of Multi-criteria preference disaggregation methods applied to marketing problems. Section 15.2 of the chapter concludes by describing how MCDA methodologies have been applied to choice models.

Applications of MCDA in e-commerce are described in the second part of this chapter (section 15.3) after a brief mention in the basic definitions and concepts of e-commerce (section 15.3.1). Focus is given to applications of MCDA in the B2C e-commerce (section 15.3.2) and specifically to recommender systems. The necessity and importance of applying MCDA methodologies in these systems is discussed and some works of initial attempts in this direction are described. Furthermore, in section 15.3.3, characteristic research works are referred that demonstrate the potentiality of applying MCDA methodologies in e-commerce.

In the last part (section 15.4) a summary of the basic conclusions that arise from this chapter is provided that identifies the major points of this work. The keystones under which MCDA methodologies have been applied in marketing and e-commerce are mentioned to provide the reader with a state of the art and possible extensions in this research direction.

15.2 Marketing Applications

Measuring consumer preferences concerning product criteria constitutes an important success factor in new product design and development. When a consumer reaches the stage of the purchasing decision, he usually faces a number of competing products. This initial set of products, which is called the *awareness* set, is then reduced to a smaller one, the *consideration* set, with the use of some basic criteria (usually the products that do not have a certain feature are excluded from the consideration set). Next, the customer collects some extra information, in order to further reduce the number of the alternative products to the *choice* set, from which he will finally choose the one to purchase (here the products whose features are within a specific range of values are included in the choice set). This final stage is the most difficult for the customer to arrive to a decision, as he has to process all the available information in order to select the product that he will finally purchase. At this point, the consumer is assumed to evaluate each alternative according to set of criteria, which are determined by his rationality and knowledge. The customer decomposes every product into a number of criteria, and implicitly assigns different importance to each, according to his preferences and needs. Lancaster [34] and Fishbein [20], motivated by the fields of psychology and economy respectively, were the first to introduce the notion that customers evaluate a product according to a set of criteria. Each product category is related to different criteria and every individual assigns a different level of importance to each. In marketing research it is assumed that all consumers evaluate a specific product category using a common set of criteria.

Hauser and Urban [27] showed that the utility theory of von Neumann and Morgenstern [62] is consistent with the descriptive models used to represent consumer behavior. Such models relate the customer preferences with his final selection, using a utility function for each criterion. Through this function the individual expresses the level of satisfaction he expects to obtain from each criterion's value. The combination of the marginal utilities for each product's criterion results in a total utility value for the product. In such a way we can quantify the level of satisfaction a customer expects to obtain from each product, as a function of the values of the criteria that form the product. The higher the utility value a consumer assigns to a product, the higher the probability to purchase it.

15.2.1 Consumer Preference Measurement

The first step in a consumer preference measurement procedure is the selection of the appropriate set of criteria that customers use in order to evaluate a product belonging to a certain category. The criteria can be either qualitative, for which a discrete number of feasible levels is determined, or quantitative, for which a specific range of possible values is determined. Next, customer evaluations with regard to the specified criteria for a number of alternatives must be elicited. This is usually implemented with the use of questionnaires which are collected during the conduction of a market survey. The respondents evaluate each of the alternatives according to the set of criteria. In some cases, photos of or even the alternatives themselves may be present at the time of the interview. The interviewer can make a trial use, in case for instance of a cell phone or a computer, or even taste the alternatives, in case of food products. Depending on the method that will be used for the estimation of the criteria utility functions, the respondent will be further asked to make some kind of comparison between the alternatives. If for example UTASTAR is going to be employed, the interviewer will be asked to rank the alternatives from the most to the least preferred. An example of a multi-criteria table that each customer that participates in such a survey fills in is shown in Table 15.1.

Table 15.1 A customer's multi-criteria matrix

	Criterion 1	Criterion 2	Criterion 3	*Ranking*
Product 1	3	1	2	2
Product 2	2	2	1	5
Product 3	1	2	3	3
Product 4	2	3	2	4
Product 5	3	2	1	6
Product 6	4	2	3	1

The table's rows contain the alternatives included in the study, and the columns contain the criteria used for the alternatives assessment. Each cell of the table contains the evaluation of a certain alternative in a certain criterion. The last column contains the rank order of the alternatives, where the alternative that the respondent considers as best is ranked first. The data contained in a respondent's multi-criteria table is entered into a preference disaggregation method like UTASTAR, which estimates an additive utility function of the form:

$$u(\mathbf{g}) = w_1 u_1(g_1) + w_2 u_2(g_2) + \ldots + w_n u_n(g_n)$$

where $u_i(g_i)$ is the marginal utility, and w_i the importance weight of criterion g_i.

Baourakis et al. [6] employed UTASTAR to measure the preference of two hundred consumers regarding Cretan wine brands. Each respondent evaluated four wines according to eight criteria, and ranked the wines from the most to the least preferred. The importance weights that resulted from the application of UTASTAR were used to estimate the percentage of consumers that consider each criterion as important, as illustrated in Table 15.2.

Table 15.2 Criteria that customers consider important in Cretan wine brands

Criteria	Consumers (%)
Odor	72
Quality	48
Image	29
Package	28
Authenticity	10
Environmental influence	8
Price	7
Advertisement	6

15.2.2 Modeling Consumer Behavior

Modeling consumer behavior constitutes a basic process in strategic marketing. According to Kotler and Keller [31], the heart of modern strategic marketing can be described as STP marketing, where companies are moving away from *mass marketing* and apply methods of identifying market segments (segmentation), selecting one or more of them (targeting), and developing products and marketing offerings (positioning) tailored to each.

15.2.2.1 Market Segmentation

Kotler and Armstrong [30] define segmentation as the division of market into distinct groups of buyers with different needs, characteristics, or behavior who might require separate products or marketing mixes. One of the basic elements of market segmentation is the *segmentation base*, the set of variables or characteristics according to which consumers are assigned to different segments. Frank et al. [21] classify segmentation bases into *general*, which are independent of products, services or circumstances, and *product-specific*, which are related to both the consumer and the product. Wedel and Kamakura [64], further classify the bases into *observable*, which are measured directly, and *unobservable*, which are inferred, proposing the classification scheme shown in Figure 15.1.

Unobservable product-specific bases have been used for over forty years, since Yankelovich [66] employed buyers' perceptions of product criteria as segmentation variables, based on the assumption that a customer's behavior is formed by his individual perceptions about the product features, instead of the real features themselves. Hence, each individual forms a different attitude towards the product fea-

	General	Product-specific
Observable	Cultural, geographic, demographic and socioeconomic variables	User status, usage frequency, store loyalty and patronage, situations
Unobservable	Psychographics, values, personality and life-style	Benefits, perceptions, elasticities, attributes, preferences, intention

Fig. 15.1 Classification of segmentation bases

tures and follows his personal point of view, while the features remain actually the same. In case some customers share similar perceptions, we can form groups of customers (market segments), where similar behavior is observed. People that belong to the same segment can be addressed with targeted marketing offers and appropriate products. In 1968, Haley introduced *benefit segmentation*, where the market is divided into groups of customers that seek similar benefits from the product [25]. In a research conducted in the toothpaste market, he identified four benefit segments: *economy* (people who seek for a low price), *medicinal* (people who seek for decay prevention), *cosmetic* (people who seek for bright teeth), and *taste* (people who seek for good tasting). Criteria importance weights are the variables used in most benefit segmentation studies, as the segments revealed are easy to be described, and small in number, in contrast to criteria marginal utilities, where the large number of variables may produce too many segments to be effectively managed.

Another basic element of market segmentation is the method employed. Wind [65] classifies segmentation techniques into *a priori*, where the type and number of segments are determined in advance by the researcher, and *post hoc*, where the results of the data analysis determine the type and number of segments. Segmentation methods are also classified into *descriptive*, which analyze the associations across a single set of segmentation bases with no distinction between dependent or independent variables, and *predictive*, which analyze the association between two sets of variables, where one set consists of dependent variables to be explained/predicted by the set of independent variables (Wedel and Kamakura, 2000).

Multi-criteria preference disaggregation methods have been used to provide the basis for product-specific descriptive segmentation applications, both in a priori and post hoc way. In an a priori approach the manager may, for instance, have decided in some way that he wants to produce three segments for a wine market; one that prefers cheap wines, one that emphasizes the taste and odor of the wine, and one that focuses on the design of the wine's bottle. In this case after applying the preference disaggregation method to each customer, the researcher will produce one group that assigns the highest importance weight to the criterion price, one that assigns higher importance to the criteria taste and odor, and one that assigns the highest weight to criterion package. If the same sample is to be segmented post hoc, the attribute importance weights derived from UTASTAR will be used as variables in a clustering algorithm. The segments' number and type (described by highest weighted criteria within segment) will be determined by the clustering algorithm.

Tsafarakis et al. [60] applied a post hoc descriptive segmentation using the criteria importance weights derived from the application of UTASTAR to the results of a survey conducted in Paris, for the development of a new Cretan olive oil product, specifically designed for the French market requirements [56]. Two hundred and four olive oil consumers were interviewed providing personal information (demographic characteristics) and information concerning their oil usage patterns (type, frequency etc). They also tasted 6 different olive oils, evaluated them in 6 criteria and rank them according to their purchase probability. UTASTAR was applied to each customer separately using as inputs the evaluations he provided for each criterion as well as the final ranking of the products. Next, cluster-based benefit segmentation was conducted using the criteria importance weights. First, hierarchical clustering was conducted in order for the number of segments as well as an initial estimation of the cluster centroids to be determined. The results constituted the input to a k-means algorithm, which optimized the clustering results and provided the final segment centroids, as well as the customer membership. The Euclidean distance was used as a function of measuring the similarity between every pair of objects (5-dimensional vectors of criteria weights) in the data set (customers), and the Ward algorithm (inner squared distance) for grouping the objects into clusters. The application of hierarchical clustering resulted in an initial solution of 3 different segments. The final solution provided by k-means is shown on Table 15.3.

Table 15.3 Results of the segmentation using importance weights

Segment			Criteria				
No	Name		Image	Color	Odor	Taste	Package
1	Gourmet	27	0.154	0.099	0.211	0.423	0.113
2	Balanced	99	0.256	0.213	0.198	0.207	0.126
3	Aesthetic	78	0.234	0.188	0.087	0.172	0.319

As it emerges from Table 15.3, the customers belonging to first segment give very high importance to the Taste of the oil, and high importance to its Odor, while they are not very interested in the rest criteria. We can define them as the "Gourmet" segment, mainly concerned about the quality of the product. The analysis of the second segment reveals that none criterion dominates the others. This group of customers is the most "Balanced" of the three giving almost equal importance to all criteria. Customers who belong to the last segment give priority to the Package, Image and Color of the oil, while not considering its Odor and Taste as significant criteria. This group of customers is defined as the "Aesthetic" segment, since they favor mostly the product's appearance. The authors suggest that the firm conducting the survey should not consider the Gourmet segment, since it constitutes the smallest one. The second segment is probably the best choice, since it is the largest one. The targeting of this segment requires the development of a product that competes well to all criteria. If the firm decides to adopt a more concentrated marketing strategy, then it should target the third segment, giving emphasis to the packaging and the image of the oil.

Matsatsinis and Siskos [43] applied to the same data set another descriptive post hoc segmentation, where the basis was the criteria that consumers considered as significant. A criterion is considered as a significant one when it overcomes a significance threshold. As significance threshold they consider the mean value of the criterion's utilities across the whole customer sample. The authors used an overlapping clustering scheme where a customer may belong to multiple segments. Four segments were identified as shown in Table 15.4, where the sign "+" indicates the significance of the criterion.

Table 15.4 Results of the segmentation using criteria significance

		Criteria					
		Color	Image	Price	Package	Taste	Odor
1	34%	+				+	+
2	29%	+		+			
3	46%			+	+		
4	37%		+	+	+		

15.2.2.2 Market Targeting

Market targeting, the step that follows market segmentation, is the procedure of assessing the attractiveness of each market segment and choosing one or more of them to enter. Evaluation of target markets relies on criteria such as segment's size, profitability and growth rates, among others.

Tsafarakis et al. [60] proposed a methodology for target market selection, which assists firms in identifying those customers that are more critical to switch among brands, the *undecided* customers. By applying this method, a company has high probabilities of winning the largest possible number of new customers, avoiding a typical targeting approach that may look effective in theory, but result in a failure in practice. The model's main advantage is that it does not rely solely on consumers' elasticities of the product's price, but adopts a more qualitative approach, taking into account the consumers' complete preference structure, The authors identify the undecided customers through the study of the conjunction of preferences they show about the whole set of products, as described next.

It is widely accepted, that the decision maker focuses on the absolute difference in attractiveness among the alternatives provided before making a choice [17]. The range of the utilities that the customer assigns to a set of products ($r = U_{\max} - U_{\min}$), is a measure of his ability to define a well-established order of preferences [42]. A customer who assigns similar utilities to the whole set of products (low values of r) expects to gain the same amount of benefits regardless of his final decision.

He feels unable to discriminate or incapable of weighting the relevant differences among the available products, and will probably choose randomly the one that he will finally purchase (Figure 15.2, Case 1). Medium values of r (Figure 15.2, Case 2) indicate a customer who has established a more clear order of preferences, feeling that some products are better than others. Yet, the relatively small differences in the utility values are potentially reversible, and the customer expresses his difficulty to select from the set of products. Large difference in attractiveness among alternatives (Figure 15.2, Case 3) makes it easier for the customer to arrive at a decisive choice. He has established a strong order of preferences, by separating the products into different classes of importance. This is a clear situation, indicating a determined decision maker.

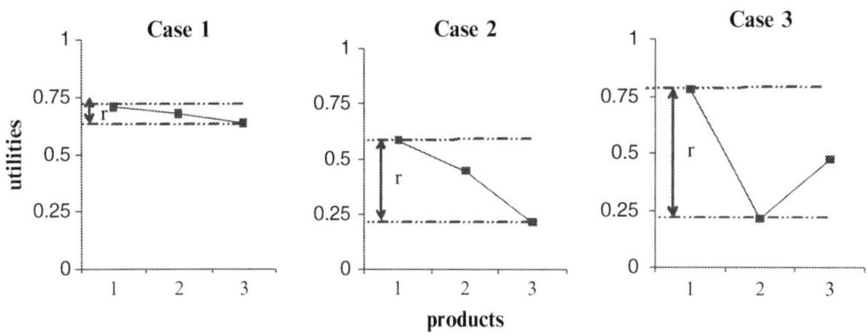

Fig. 15.2 Utilities distributions for different values of the r coefficient

Although the range of the utilities distribution is a strong indicator of the choice pattern a customer follows (random or not), it is not a sufficient measure of his confidence. Cases 4 and 5 (Figure 15.3) both show distributions with a wide range of utilities, representing customers who are able to segregate the acceptable products from the unacceptable ones.

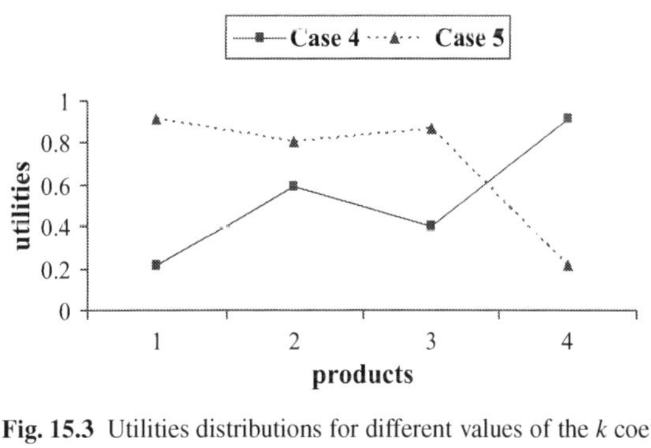

Fig. 15.3 Utilities distributions for different values of the k coefficient

However in Case 4 the customer is capable of separating the whole set, while in Case 5 he cannot distinguish the products with assigned utilities in the range between 0.8 and 0.9. While the first customer expresses a clear preference structure, we cannot assume the same for the second one. Tversky and Shafir [61], argue that the availability of competing alternatives of comparable attractiveness, creates a conflict to the decision maker, since he has not an immediate reason to choose either alternative over the others. The kurtosis coefficient (k) of the utilities distribution is another measure of the difficulty a customer expresses to select among a set of products. A flat utilities distribution (low kurtosis) represents a determined customer who feels certain about his preferences, while a peaked one (high kurtosis) describes a confused one who seems more elastic in making his final decision. Now we consider Cases 6 and 7 (Figure 15.4).

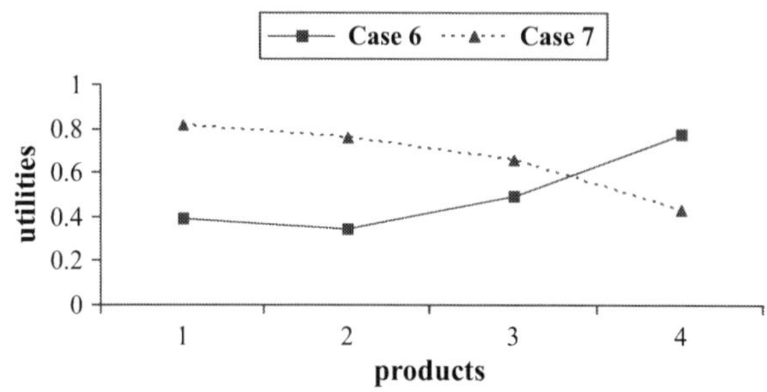

Fig. 15.4 Utilities distributions for different values of the s coefficient

Both situations have approximately the same value for the range of utilities and the kurtosis coefficient, but differ considerably in the skewness (s) of the distribution. The two customers express a difficulty in making a choice among products, since r has a medium value and the distribution is peaked. According to Hsee and Leclerc [28], when a number of attractive products are presented jointly to a consumer, they will look less attractive and will be less likely to be chosen, than when presented separately. The opposite rule applies for non attractive alternatives. While in Case 7 the majority of the products constitute good choices (positive skewness), their comparison confuses the customer and makes his selection process more difficult. On the contrary, the consumer who evaluates most of the products as poor options (negative skewness) faces an easier choice situation (Case 6), since he is able to distinguish the 4th alternative as an acceptable one.

Summarizing the preceding analysis, a customer is defined as *undecided*, if the distribution of the utilities he assigns to the set of products under consideration is narrow, peaked and positively asymmetric. Specifically, the condition holds for the following values of the coefficients: $r < 0.6$, $k > 0.5$, and $s < 0.25$. These values came up from extensive simulation tests conducted by Matsatsinis and Samaras [41], who studied different types of consumer buying behavior for certain utilities distributions. Two approaches for targeting the undecided customers are proposed, both of which constitute a two-staged hybrid descriptive segmentation procedure using two different bases. In the first approach, an analysis of the shape of the assigned product utilities distribution takes place for each customer separately. The range of the distribution is calculated, which constitutes the dominant factor in defining his decision making pattern. In addition, the kurtosis and skewness coefficients are estimated, which describe different preference structures within the same range of utilities. Then, an a-priori segmentation is conducted, where the sample is partitioned on the base of the values of r, k and s. The segment that arises for $r < 0.6$, $k > 0.5$, and $s < 0.25$, is identified as the undecided part of the customers base. The second stage of the model consists of a post-hoc clustering-based approach using an unobservable, product-specific base. Benefit segmentation is conducted within the undecided set of customers, using their criteria importance weights. The study of the emerging segments, which constitute the final results of the model, reveals valuable information regarding the preferences of the undecided customers. The second approach comprises the same steps in the opposite order. Initially, benefit segmentation takes place with the use of the criteria importance weights. Among the segments that arise the firm may select the one (or more) that contains the most undecided customers. Depending on the scope of the application (product design, positioning etc.), a firm can utilize the importance consumers give to each attribute of the product, to develop an effective marketing strategy.

The authors applied the first approach to the olive oil data set. Initially the utilities that consumers expect to gain from each product were estimated, with the use of UTASTAR, as well as the importance weights they assign to each attribute. Next, the undecided customers were identified, who constitute the target sample. The number of consumers whose utilities distribution exhibits the appropriate conditions was 68. Finally, k-means clustering was applied in order for the undecided customers' preferences to be revealed (Table 15.5).

Table 15.5 Characteristics of the resulting segments

Segment			Criteria				
No	Name		Image	Color	Odor	Taste	Package
1	Gourmet	9	0.154	0.099	0.211	0.423	0.113
2	Balanced	33	0.256	0.213	0.198	0.207	0.126
3	Aesthetic	26	0.234	0.188	0.087	0.172	0.319

The firm should consider targeting segments 2 or 3, which contain the largest number of undecided customers.

15.2.2.3 Market Simulation

Nowadays the economic environment where firms operate has become more competitive than ever. The globalization of the markets and the shorter product life cycles, due to the rapid technology development, put high pressure on companies' profitability. In order to survive under such circumstances a firm must continuously introduce new products or redesign its existing ones. However, such procedures entail risk, due to the high cost of the development of a new product, and the serious problems a new product's failure may cause to the company. In order to minimize uncertainty in a new product launch, managers try to assess its market penetration before it enters the production stage. This is accomplished through simulating the market with the use of choice models, which take as input the utilities a customer assigns to products and convert them to choice probabilities. Popular choice models usually applied in market simulations are the first choice rule, where the customer is assumed to deterministically select the product with the highest utility, the Bradley-Terry-Luce [10, 37], a constant utility model which uses a probabilistic choice rule, $P_{ij} = U_{ij} / \sum_{j=1}^{n} U_{ij}$, and the MultiNomial Logit [44], a random utility model with a deterministic choice rule, $P_{ij} = e^{U_{ij}} / \sum_{j=1}^{n} e^{U_{ij}}$, where P_{ij} the probability that customer i chooses product j, U_{ij} the utility that customer i assigns to product j, and n the number of competing products.

When simulating a market, the manager formulates a scenario where x customers have to select one product among m alternatives. For each customer i a vector of product choice probabilities $[P_{i1}, P_{i2}, ..., P_{im}]$ is calculated with the use of a choice model, and the total choice likelihood for a product j results from the integration of its choice probabilities across the whole customer base: $CP_j = \sum_{i=1}^{x} P_{ij}$. Finally, the simulated market share for each product j is estimated: $MS_j = 100 \times CP_j / \sum_{k=1}^{m} CP_k \%$.

Numerous simulations can be conducted, by altering the configuration of one or more products and observe the relative change in market shares. In this way, the outcome of different strategic moves in a market with competition can be anticipated. Market simulations can capture cannibalism or cross-elasticity effects between different brands or attributes, answer what-if questions about new product introductions, product modifications, or product line extensions given a current competitive environment, reveal price/sales elasticities and guide pricing strategy [48]. Individual brand switching behavior can be revealed, enabling a company to design new products that take share mostly from its competitors without cannibalizing its existing product line. As Baier and Gaul [5] state, the determination of an adequate choice model is the most cumbersome task in market simulation situations. Allenby et al. [4] stress the complexity of human behavior, and propose the adjustment of choice models to better predict it. An appropriate and theoretically justified method for tuning simulated results to more closely fit actual market shares is the adjustment of the exponent used in choice models [49].

Matsatsinis and Samaras [41] indicate that a very difficult problem in marketing decision making situations is the selection of the most appropriate choice model for analyzing a customer's buying behavior, and propose the selection of a different choice model for each consumer, through the study of the distribution of the total utilities he assigns to the set of products. Particularly, they consider the distribution's range, kurtosis and skewness, for selecting the choice model that better describes each consumer's purchasing pattern. The values of the three coefficients constitute the input that trigger a total of 27 if-then rules, which comprise a knowledge base containing 8 different brand choice models. This knowledge base is one of the three subsystems used in the MARKEX (MARKet Expert) system [42], which integrates decision support methodologies and intelligent systems technologies for conducting market simulations. Two of the models used arise from the calibration at the individual level of the Pessemier and the MNL model, with the use of r as the exponent (they call them Width of Utilities-1and Width of Utilities-2 respectively).

Tsafarakis et al. [59] propose a method for maximizing a choice model's predicting validity regarding customer purchasing behavior, through calibrating the choice model's exponent. They extend the approach of Matsatsinis and Samaras [41] by using an exponent a on the Pessemier model [51] that depends not only to the range of the product utilities distribution but also to its kurtosis and skewness. A linear relationship is proposed between the exponent a_i that models customer i and the range (r), the kurtosis (k), and the skewness (s) of his total utilities distribution:

$$a_i = R \times r_i + K \times k_i + S \times s_i$$

where the three parameters R, K, and S are not individually based, but have differential effects depending on each customer's r, k, and s coefficients. The values of the R, K, and S are optimized using an optimization algorithm, so that the model gives the closest overall fit to actual market shares. A Monte Carlo simulation was conducted, in order for the proposed approach to be compared with the state of the art model and three classic approaches. The proposed model gave the best results in respect to market shares predictive validity in all data sets.

15.3 e-Commerce Applications

15.3.1 Some Interesting Remarks on e-Commerce

The evolution of e-commerce can be attributed to a combination of society reorganization and technological innovation. Through Internet which played a major role in this evolution and appeared in the late 1960s, e-commerce took off with the arrival of the World Wide Web and browsers in the 1990s.

The advancement of the information technology (IT) has enabled a variety of products and services to be displayed without consideration of the physical aspects of space and time. As a result, the "e-market", a virtual place for online transactions, has become a center of attention and great efforts have been put into identifying the major success factors in e-commerce, the electronic commerce of business in the e-market [58]. The amount of trade conducted electronically has grown extraordinarily with widespread Internet usage and two of the most representative e-commerce sites launched in 1995 Amazon.com and eBay.com prove the aforementioned statement.

In order to understand electronic commerce it is important to identify the different terms that are used. Based on the parties involved in the business transaction, e-commerce can be mainly classified as Business-to-Customer (B2C) which covers customer interactions with an e-business site where users may collect information, order products, and use support services through the Web site and Business-to-Business (B2B) which is conducted through industry-sponsored marketplaces and through private exchanges set up by large companies for their suppliers and customers. Common examples of popular B2C e-commerce include Amazon.com or BestBuy.com, while most representative B2B cases include Tradekey.com and Alibaba.com [9].

E-commerce was first developed in the early 1970s with innovations like Electronic Funds Transfer (EFT) where funds can be routed electronically from one organization to another, Electronic Data Interchange (EDI) which was used to electronically transfer routine documents that expanded electronic transfers from financial transactions to other types of transaction processing, or Interorganizational System (IOS), a system, which allows the flow of information to be automated between organizations in order to reach a desired supply chain management system, which enables the development of competitive organizations [36].

The number of online e-tailers (electronic retailers) continues to grow each and every year, and new marketing strategies are constantly being developed by businesses to ensure that their products are ultimately being found online and purchased by the end-user.

Chu et al. [15] provide an evolutionary perspective of e-commerce Web sites by identifying four different eras, the Pre-Web, the Reactive Web, the Interactive Web and the Integrative Web (as shown in Figure 15.5), where not only traditional e-commerce activities can take place but also e-business processes, such as e-supply chain management, e-collaboration, e-reengineering, and e-procurement.

Through time, various definitions have been attributed to the term e-commerce; from the simple "doing business over the internet" or "doing business online" to more sophisticated like the one mentioned in [22] "... a modern business methodology that addresses the needs of organizations, merchants, and consumers to cut costs while improving the quality of hoods and services and increasing the speed of service delivery, by using the internet". The only fundamental difference of electronic commerce from the traditional commerce is that business in the first case is done electronically without the need of physical existence. Nevertheless, elementary aspects of commerce like the craving desire of consumers to buy with satisfaction and the obsession of companies to increase their gains remain indelible over time and technology changes.

As the amount of data that needs to be processed through a web interface increases, companies rely on more automated decision aids. One of the first researchers that proposed the use of intelligent software agents to reduce information overload was Maes [38] while an early example of a "man-machine interactive"

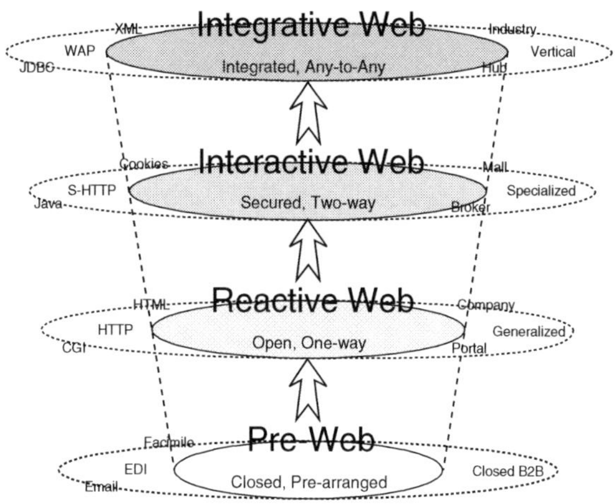

Fig. 15.5 The evolution of e-commerce web sites

decision support tool for selecting among alternatives based on trade-offs among multiple attributes is provided by [18]. Within this new environment, there is an obvious motivation for marketers to enhance the shopping experience for consumers in order to increase sales and profitability. One way of automating the decision making task is to add "intelligence" to the decision support tools used for analyzing the information generated in B2C and B2B electronic commerce applications.

15.3.2 MCDA in B2C e-Commerce

The main objective of some of business-to-consumer (B2C) tools is to collect valuable consumer data from real decision makers. Apart from demographic or browsing behavioral data, preference data is an important aspect of customer modeling. There are many different approaches to modeling preferences [13, 50], and several different strategies have been incorporated into e-commerce sites.

Wallenius et al. [63] in their work "Multiple Criteria Decision Making, Multiattribute Utility Theory: Recent Accomplishments and What Lies Ahead" identified two major application areas of MCDA in e-commerce. These are multi-attribute online auctions and comparison shopping agents both lying in the B2C area of e-commerce. In this chapter we extend the second category into a broader approach to include not only comparison shopping agents, agents that identify the "best deal" for each customer, but also any agent that helps customer to identify the "best buy" meaning the closest to her preferences product. This means that price constitutes just one of the attributes that outlines customer preference schema. These agents are commonly referred as recommendation agents, a new, but enormously evolved research field in e-marketing.

In order to construct and utilize a recommendation agent featuring with MCDA approaches, the user must be able to express preferences for different levels of performance on criteria. Individuals with low levels of knowledge regarding a product class may attain a difficulty in making the necessary trade-offs in terms of product attributes, and therefore feel uncomfortable with both the process and the results of the ranking that is offered.

15.3.2.1 MCDA in Recommender Systems

Recommender systems (or recommendation agents) are software applications that attempt to reduce information overload, a natural consequence of information technology development due to its ability to produce more information more quickly and to disseminate this information to a wider audience than ever before. The goal of a Recommender System is to recommend items of interest to end users based on their preferences. A recommender system (RS) will potentially suggest to the end user to watch or not a movie, to buy or not an item, to listen or not to a song and so forth. In this sense, an accurate Recommender System will ideally be able to act on behalf of the user [54]. To achieve its goal, it must gain knowledge of the user's value system and decision policy.

Suppose u be a utility function that measures usefulness of item s to user c, i.e., $u : C \times S \rightarrow R$, where R is a totally ordered set. Then, for each user, $c \in C$ we want to choose such item s that maximizes the user's utility. More formally:

$$\forall c \in C, s'_c = \arg\max u(c,s), s \in S$$

Most existing recommender systems use the so called collaborative filtering approach, some others are based on the content-based approach and many attempts combine these two methods into hybrid frameworks. Several reviews exist on Recommender Systems [46, 54], one of the most representative being that of Adomavicius and Tuzhilin [2].

A major concept in RS is the notion and degree of personalization. "Personalization is the ability to provide content and services that are tailored to individuals based on knowledge about their preferences and behavior" is just one of the various definitions attributed to the concept of personalization [2]. Personalization and customization are considered increasingly important elements of marketing applications. These terms usually refer to exploiting information about a user (that may be a customer, an individual, or a group) to better design products and services targeted to that user.

The majority of existing Recommender Systems obtains an overall numerical rating r_{ij}, as input information for the recommendation algorithm. This overall rating depends only on one single criterion that usually represents the overall preference of user i on item j. However, the key to more effective personalization services is the ability to develop a system able to understand not only what people like, but why they like it. In other words, an accurate modeling of a users' value system and thus an effective preference representation schema, will potentially lead to the design of a recommendation algorithm with increased performance. A system can understand how users think about items, by considering the knowledge about the underlying attributes that attract users to choose this particular item and hence recognize preferences, not just patterns, ensuring a more sophisticated understanding of a user.

Recently, some works have started employing multi-criteria ratings directly in their recommendation algorithms [40], or indirectly in related to RS's tasks such as user profiling [32]. Also, in the commercial sector, Yahoo Movies has launched a recommendation service that employs user-specific multi-criteria ratings for each movie. However, to incorporate multi-criteria rating in an existing recommendation process or to design new recommendation techniques, careful consideration is necessary, to achieve maximum accuracy [1]. Some approaches of user preference modeling have already been applied to recommender systems and mainly adopt techniques and methodologies from the greater field of artificial intelligence, knowledge engineering or data mining, like ontological user profiling [45] or ubiquitous user modeling [8]. Nevertheless, although multiple criteria analysis methodologies have been extensively studied in operations research community, not much has been done yet though, in the field of recommender systems [2].

Manouselis and Costopoulou [39] identified a set of dimensions that distinguish, describe and categorize multi-criteria recommender systems, based on existing taxonomies and categorizations and integrated these dimensions into an overall framework that was used for the analysis and classification of 37 different multi-criteria recommender systems.

15.3.3 Examples of MCDA Applications in e-Commerce

Lakiotaki et al. [33] studied the application of Multiple Criteria Decision Analysis methodologies in Recommender Systems and presented UTARec, a Recommender System that incorporates MCDA techniques. UTARec system was based on a four step methodological framework: 1) Data acquisition and representation, 2) Data set configuration 3) Modeling user preferences and 4) Recommendation process. UTARec's performance was demonstrated through a movie recommendation application. The corresponding movie data set included ratings for 25 different movies. These ratings provided information on four criteria in addition to an overall rating. The four criteria, upon which each user was asked to evaluate a movie, were: a) story (C1), b) acting (C2), c) direction (C3) and d) visuals (C4). A characteristic multi-criteria matrix had the form of Table 15.6.

To model user preferences the authors employed the UTASTAR algorithm [57], an improved version of the original UTA (UTilités Additives) method [55]. This algorithm, adopts the preference disaggregation principle, the philosophy of which, is to assess/ infer preference models from given preferential structures and is shown in Figure 15.6.

Table 15.6 A sample of multi-criteria data input matrix

User ID	Ranking order	C1	C2	C3	C4	Movie ID
	1	13	12	13	11	1
	2	10	13	9	13	4
	3	9	11	9	13	25
	4	10	10	9	9	23
	5	6	9	7	13	9
...	...	...	...	...	...	
	1	13	11	11	13	9
	2	10	9	9	9	18
	2	11	11	13	9	2
	...	...	...	...	...	...

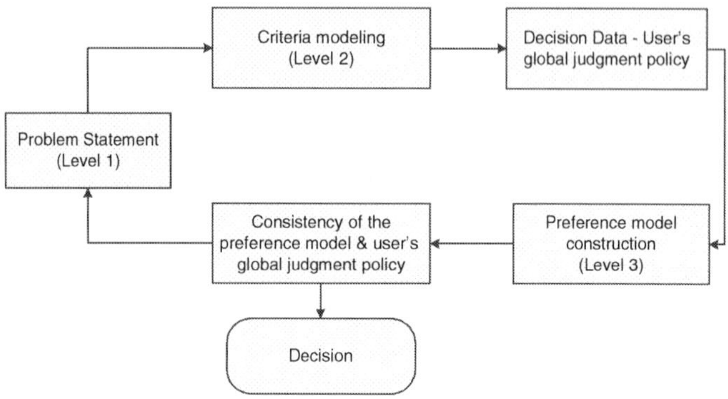

Fig. 15.6 The disaggregation - aggregation approach

By applying a simple additive model on the performances of the movies upon all four criteria, the authors calculated an overall utility score for every movie. They evaluated UTARec's accuracy by performing validity analysis and calculated Kendall's tau, a measure of correlation between two ordinal-level variables. Their model's prediction accuracy was additionally evaluated in terms of classification performance. To do so, they calculated True positive rate, the actual number of positives correctly classified by their model over the total positives and False Positive Rate, the result of the number of negatives incorrectly classified, divided by the total negatives. They plotted an ROC graph which depicts relative trade-offs between true positives and false positives and found a result of 0.81 for the Area Under Curve.

Pu and Faltings [53] introduced and developed a travel planning system, called SmartClient, where new techniques for better adapting interaction of users with an electronic catalog system to actual buying behavior where explained. Their model was based on a conversation that supported the buyer in formulating his or her needs and in deciding which criteria to apply in selecting a product to buy. Later, further work in the technology of example critiquing agents confirmed that example critiquing significantly reduces users' task time and error rate while increasing decision accuracy [52]. Li Chen in her PhD thesis [12] describes two primary technologies: one is called example critiquing agents aimed to stimulate users to conduct trade off navigation and freely specify feedback criteria to example products; another termed as preference-based organization interfaces designed to take two roles: explaining to users why and how the recommendations are computed and displayed, and suggesting critique suggestions to guide users to under-stand existing trade off potentials and to make concrete decision navigations from the top candidate for better choices.

Bartelt and Lamersdorf [7] exploit multi-criteria decision analysis through a different approach, as a tool to present a multi-criteria taxonomy of e-business models. They build up a taxonomy using multiple criteria to present individual business models. Their taxonomy can be used to analyze and enhance existing systems and business models as well as to develop new internet strategies for companies.

Al-Aomar and Dweiri [3] describe a web-based Decision support System (DSS) that provides a straightforward method for making product selection decisions in online environment. By segmenting prospective online customers into three levels (High: H, Medium: M, and Low: L) and directing them to use three different decision modules Analytical Hierarchy Process (AHP), Simple Multi-Attribute Rating Technique (SMART), or Direct Scaling Method (DSM) according their level of product knowledge, they manage to enhance the front-end of company's business portal.

Denguir-Rekik et al. [16] proposed a multi-criteria decision making support system, called a "Feedback Based Diagnosis System" (FBDS), to aid the marketing team of an e-commerce (EC) organisation in its activities. The FBDS database was composed of customers' satisfaction measures. These measures were related to the different services an EC offers to its customers. Thus, they constitute a multi-criteria (MC) evaluation of EC performances. In the general framework of recommender systems, these available MC evaluations are considered as useful information for other customers to help them to objectively, rationally and exhaustively assess and compare the numerous ECs among the ones likely to meet their needs. FBDS is not concerned with improving or automating a recommendation process for customers but it is merely EC management team oriented.

Grigoroudis et al. [23] presented a framework for analyzing changes of e - customer preferences. Their presented results mainly focused on demonstrating how several tools, like perceptual maps, may be used in order to analyze changes of customer preferences. For their study they conducted and used two independent customer satisfaction surveys in different time periods on behalf of one of the major Internet Service Provider (ISP) in Greece. Their analyses were based on non-parametric statistical techniques, as well as on the multi-criteria satisfaction analysis (MUSA) method, which is actually formulated according to the multi-criteria disaggregation approach [24].

Chiu et al. [14], provide a theoretically and empirically based model for evaluating and selecting an e-commerce strategy. They adopted a fuzzy multiple criteria decision-making method in their research and applied their model by demonstrating an empirical study of the Taiwan ISP industry.

Hämäläinen [26] presents Decisionarium (www.decisionarium.hut.fi), the first public site for interactive multi-criteria decision support with tools for individual decision-making as well as for group collaboration and negotiation. Decisionarium includes a family of software like:

1. Web-HIPRE (www.hipre.hut.fi) which supports value tree and AHP analysis including group models.
2. The RICH methodology (www.rich.hut.fi) which allows the decision maker to provide incomplete ordinal preference statements when considering the relative importance of attributes in a value tree.
3. Opinions-Online (www.opinion.hut.fi) which is a platform for surveys voting and group collaboration. There are different ways for voting, multi-attribute scoring, surveys as well as interactive viewing of the results.
4. Joint Gains (www.jointgains.hut.fi) which applies the method of improving directions to support multiparty negotiations in a multi-criteria setting.
5. Smart Swaps which offers an implementation of the even swaps procedure (www.smart-swaps.hut.fi).
6. Windows software WINPRE and PRIME-Decisions for value tree analysis under incomplete information.

All of the tools above are web-based, so global interaction is natural and links can be utilized for multimedia information support. Decisionarium also offers access to complete e-learning modules based on the use of the software.

Butler et al. [11] describe potential applications of multi-attribute preference models (MAPM) in e-commerce and offer some guidelines for their implementation. MAPM are methodologies for modeling complex preferences that depend on more than one attribute or criterion, and include multi-attribute utility theory, conjoint analysis, and the Analytic Hierarchy Process. They focused on applications of MAPM models in B2C and B2B websites, where preferences of consumers are assessed for the purpose of identifying products or services that closely match their needs. In their paper they focused on the approaches that feature additive, compensatory value functions but many of the ideas can be also applied to outranking approaches like ELECTRE [19].

Lenz and Ablovatski [35] apply some main MCDA techniques (SAW, TOPSIS, AHP, PROMETHEE, DEA) to a decision problem using a software developed for this purpose. They review the aforementioned techniques and they propose a hybrid technique called GiUnTa that reconciles the different rankings obtained from each MCDA technique. They demonstrate their idea by applying and comparing the various MCDA methods to a hypothetical company named "DotNet" which plans to launch a new product in a foreign market or country. The decision problem is to identify the optimal country of interest from an alternative set of 20 countries. These alternatives are evaluated by considering 5 partially conflicting criteria.

With the widespread use of mobile devices and the pervasiveness of wireless networks, a new type of commerce, called mobile commerce, emerged. First attempts to apply MCDA techniques in m-commerce can be found in the work of Ondrus and Pigneur [47] who propose two disruption analyses to draw the disruptiveness profile of mobile payment solutions compared to other payment instruments. They tried to discover what factors have hindered the technical and commercial development by using a DSS based on the ELECTRE I multi-criteria decision making method.

15.4 Conclusions

MCDA methods have played an important role in marketing research, and are widely used in commercial applications. The aggregation-disaggregation approach and the Multi-Attribute Utility Theory constitute fundamental tools for marketing applications like market segmentation, product design and positioning etc. However, the other two research approaches of MCDA (outranking methods, goal programming) have not yet been applied to marketing problems. This may constitute an area of future research.

In the e-commerce sector, albeit Multi-Criteria Decision Analysis tools have been integrated into various web applications, there is still need to develop an overall framework for the researchers to follow. So far, various attempts of enabling MCDA methodologies in e-commerce web sites have approach the issue in a diverse, often purpose dependent manner. For example, MCDA methodologies have been applied in Recommender Systems, in User Modeling or to the design of Intelligent Interfaces. On the whole, MCDA methodologies have been mainly exploited in the broader field of consumer behavior modeling and scattered approaches are encountered in other related to e-commerce objectives. For example, MCDA tools have been applied to classify e-business models, to aid the marketing team of an e-commerce organization, or to evaluate and select an e-commerce strategy. Nevertheless, the necessity of establishing a coherent methodology to apply MCDA techniques in e-commerce becomes imperative as the rate of traditional companies that incorporate the suffix "e-" in their business philosophy is increasing over time.

References

1. G. Adomavicius and Y. Kwon. New recommendation techniques for multicriteria rating systems. *IEEE Intelligent Systems*, 22(3):48–55, 2007.
2. G. Adomavicius and A. Tuzhilin. Towards the next generation of recommender systems: A survey of the state-of-the-art and possible extensions. *IEEE Transactions on Knowledge and Data Engineering*, 17(6):734–749, 2005.
3. R. Al-Aomar and F. Dweiri. A customer-oriented decision agent for product selection in web-based services. *International Journal of Information Technology and Decision Making*, 7(1):35–52, 2008.
4. G. Allenby, G. Fennell, J. Huber, T. Eagle, T. Gilbride, D. Horsky, J. Kim, P. Lenk, R. Johnson, E. Ofek, B. Orme, T. Otter, and J. Walker. Adjusting choice models to better predict market behavior. *Marketing Letters*, 16(3-4):197–208, 2005.
5. D. Baier and W. Gaul. Market simulation using a probabilistic ideal vector model for conjoint data. In A. Gustafsson and A. Hermann, editors, *Conjoint Measurement, Methods and Applications*, pages 97–120. Springer, Berlin, 2nd edition, 2001.

6. G. Baourakis, N.F. Matsatsinis, and Y. Siskos. Agricultural product development using multi-dimensional scaling and multicriteria analyses: The case of wine. *European Journal of Operational Research*, 94(2):321–334, 1996.
7. A. Bartelt and W. Lamersdorf. A multi-criteria taxonomy of business models in electronic commerce. In *International Conference on Distributed Systems Platforms in Electronic Commerce*, 2001.
8. S. Berkovsky. Ubiquitous user modeling in recommender systems. *Lecture Notes in Computer Science*, 3538:496–498, 2005.
9. J. Botha, C.H. Bothma, and P. Geldenhuys. *e-Commerce*. Juta and Company Limited, 2004.
10. R.A. Bradley and M.E. Terry. Rank analysis of incomplete block designs I: The method of paired comparisons. *Biometrika*, 39:324–345, 1952.
11. J.C. Butler, J.S. Dyer, J. Jia, and K. Tomak. Enabling e-transactions with multi-attribute preference models. *European Journal of Operational Research*, 186:748–765, 2008.
12. L. Chen. *User decision improvement and trust building in product recommender systems*. PhD thesis, EPFL, Lausanne, 2008.
13. L. Chen and P. Pu. Survey of preference elicitation methods. Technical report, EPFL, 2004.
14. Y.C. Chiu, J.Z. Shyu, and G.H. Tzeng. Fuzzy MCDM for evaluating the e-commerce strategy. *International Journal of Computer Applications in Technology*, 19(1):12–22, 2004.
15. S.C. Chu, L.C. Leung, Y.V. Hui, and W. Cheung. Evolution of e-commerce web sites: A conceptual framework and a longitudinal study. *Information & Management*, 44:154–164, 2007.
16. A. Denguir-Rekik, J. Montmain, and G. Mauris. A possibilistic-valued multi-criteria decision-making support for marketing activities in e-commerce: Feedback based diagnosis system. *European Journal of Operational Research*, 195(3):876–888, 2009.
17. R. Dhar. Consumer preference for a no choice option. *Journal of Consumer Research*, 24(2):215–231, 1997.
18. J.S. Dyer. A time-sharing computer program for the solution of the multiple criteria problem. *Management Science*, 19(12):1379–1383, 1973.
19. J. Figueira, V. Mousseau, and B. Roy. ELECTRE methods. In J. Figueira, S. Greco, and M. Ehrgott, editors, *Multiple Criteria Decision Analysis: State of the Art Surveys*, pages 133–153. Springer, Boston, 2005.
20. M. Fishbein. *Readings in Attitude Theory and Measurement*. John Wiley, New York, 1967.
21. R.E. Frank, W.F. Massy, and Y. Wind. *Market Segmentation*. Prentice Hall, Englewood Cliffs, New Jersey, 1972.
22. R. Goel. *e-Commerce*. New Age International, 2007.
23. E. Grigoroudis, P. Kyriazopoulos, Y. Siskos, A. Spyridakos, and D. Yannacopoulos. Tracking changes of e-customer preferences using multicriteria analysis. *Managing Service Quality*, 17(5):538–562, 2007.
24. E. Grigoroudis and Y. Siskos. Preference disaggregation for measuring and analysing customer satisfaction: The MUSA method. *European Journal of Operational Research*, 143(1):148–170, 2002.
25. R.I. Haley. Benefit segmentation: A decision-oriented research tool. *Journal of Marketing*, 32:30–35, 1968.
26. R.P. Hämäläinen. Decisionarium: Aiding decisions, negotiating and collecting opinions on the web. *Journal of Multi-Criteria Decision Analysis*, 12(2-3):101–110, 2003.
27. J.R. Hauser and G.L. Urban. Consumer preference axioms: Behavioural postulates for describing and predicting stochastic choice. *Management Science*, 24(13):1331–1341, 1978.
28. C.K. Hsee and F. Leclerc. Will products look more attractive when presented separately or together? *Journal of Consumer Research*, 25(2):175–186, 1998.
29. R. Keeney and H. Raiffa. *Decision with Multiple Objectives: Preference and Value Tradeoffs*. Cambridge University Press, New York, 1993.
30. P. Kotler and G. Armstrong. *Principles of Marketing*. Prentice Hall, New Jersey, 12th edition, 2007.
31. P. Kotler and K.L. Keller. *Marketing Management*. Prentice Hall, Pearson, NJ, 13th edition, 2009.

32. K. Lakiotaki, P. Delias, V. Sakkalis, and N.F. Matsatsinis. User profiling for recommender systems: The role of utility functions. *Operational Research: An International Journal*, 9(1):3–16, 2009.
33. K. Lakiotaki, S. Tsafarakis, and N.F. Matsatsinis. UTA-Rec: A recommender system based on multiple criteria analysis. In *ACM Conference on Recommender Systems*, 2008.
34. K. Lancaster. A new approach to consumer theory. *Journal of Political Economy*, 74(2):132–157, 1966.
35. H.J. Lenz and A. Ablovatski. MCDA: Multi-criteria decision making in e-commerce. In G.D. Riccia, D. Dubois, R. Kruse, and H.-J. Lenz, editors, *Decision Theory and Multi-Agent Planning*, pages 31–48. Springer, New York, 2006.
36. J. Loke. The history & evolution of e-commerce. http://ecommercesite.wordpress.com/2008/06/11/the-history-evolution-of-e-commerce (accessed 25 May 2009), 2008.
37. R.D. Luce. *Individual Choice Behavior: A Theoretical Analysis*. John Wiley, New York, 1959.
38. P. Maes. Agents that reduce work and information overload. *Communications of the ACM*, 37(7):30–40, 1994.
39. N. Manouselis and C. Costopoulou. Analysis and classification of multi-criteria recommender systems. *World Wide Web: Internet and Web Information Systems*, 10(4):415–441, 2007.
40. N.F. Matsatsinis, K. Lakiotaki, and P. Delias. A system based on multiple criteria analysis for scientific paper recommendation. In *11th Panhellenic Conference on Informatics*, 2007.
41. N.F. Matsatsinis and A.P. Samaras. Brand choice model selection based on consumers' multicriteria preferences and experts' knowledge. *Computers and Operations Research*, 27(7-8):689–707, 2000.
42. N.F. Matsatsinis and Y. Siskos. MARKEX: An intelligent decision support system for product development decisions. *European Journal of Operational Research*, 113(2):336–354, 1999.
43. N.F. Matsatsinis and Y. Siskos. *Intelligent Support Systems for Marketing Decisions*. Kluwer, Heidelberg, 2003.
44. D. McFadden. Conditional logit analysis of qualitative choice behaviour. In P. Zaremka, editor, *Frontiers in Econometrics*, page 105142. Academic Press, New York, 1974.
45. S.E. Middleton, N.R. Shadbolt, and D.C.D. Roure. Ontological user profiling in recommender systems. *ACM Transactions on Information Systems*, 22(1):54–88, 2004.
46. M. Montaner, B. Lopez, and J.L. Rosa. A taxonomy of recommender agents on the internet. *Artificial Intelligence Review*, 19:285–330, 2003.
47. J. Ondrus and Y. Pigneur. Towards a holistic analysis of mobile payments: A multiple perspectives approach. *Electronic Commerce Research and Applications*, 5:246–257, 2006.
48. B. Orme. *Getting Started with Conjoint Analysis: Strategies for Product Design and Pricing*. Research Publishers, Madison, WI, 2006.
49. B. Orme and R. Johnson. External effects adjustments in conjoint analysis. Research Paper Series, Sawtooth Software, Sequim, WA, 2006.
50. M. Öztürk, A. Tsoukiàs, and P. Vincke. Preference modelling. In *Dagstuhl Seminar Proceedings*, 2003.
51. E.A. Pessemier, P. Burger, and R. Teach. Using laboratory brand preference scales to predict consumer brand purchases. *Management Science*, 17:371–385, 1971.
52. P. Pu, L. Chen, and P. Kumar. Evaluating product search and recommender systems for e-commerce environments. *Electronic Commerce Research*, 8(1-2):1–27, 2008.
53. P. Pu and B. Faltings. Enriching buyers' experiences: The smartclient approach. In *Proceedings of the SIGCHI Conference on Human Factors in Computing Systems*, New York, NY, USA. ACM.
54. P. Resnick and H.R. Varian. Recommender systems. *Communications of the ACM*, 40(3):56–58, 1997.
55. Y. Siskos, E. Grigoroudis, and N.F. Matsatsinis. UTA methods. In J. Figueira, S. Greco, and M. Ehrgott, editors, *Multiple Criteria Decision Analysis: State of the Art Surveys*, pages 297–344. Springer, Boston, 2005.
56. Y. Siskos, N.F. Matsatsinis, and G. Baourakis. Multicriteria analysis in agricultural marketing: The case of french olive oil market. *European Journal of Operational Research*, 130(2):315–331, 2001.

57. Y. Siskos and D. Yannacopoulos. UTASTAR: An ordinal regression method for building additive value functions. *Investigação Operacional*, 5(1):39–53, 1985.
58. C. Standing, P. Love, R. Stockdale, and D. Gengatharen. Examining the relationship between electronic marketplace strategy and structure. *IEEE Transactions on Engineering Management*, 53(2):297–311, 2006.
59. S. Tsafarakis, A. Doulamis, and N.F. Matsatsinis. Individual choice model calibration. In *Proceedings of the 38th EMAC Conference*, 2009.
60. S. Tsafarakis, E. Grigoroudis, and N.F. Matsatsinis. Targeting the undecided customer. In *Proceedings of the 37th EMAC Conference*, 2008.
61. A. Tversky and E. Shafir. Choice under conflict: The dynamics of deferred decision. *Psychological Science*, 3:358–361, 1992.
62. J. von Neumann and O. Morgenstern. *Theory of Games and Economic Behaviour*. Princeton University Press, New Jersey, 1947.
63. J. Wallenius, J.S. Dyer, P.C. Fishburn, R.E. Steuer, S. Zionts, and K. Deb. Multiple criteria decision making, multiattribute utility theory: Recent accomplishments and what lies ahead. *Management Science*, 54(7):1336–1349, 2008.
64. M. Wedel and W.A. Kamakura. *Market Segmentation: Conceptual and Methodological Foundations*. Kluwer, Boston, 2nd edition, 2000.
65. Y. Wind. Issues and advances in segmentation research. *Journal of Marketing Research*, 15:317–337, 1978.
66. D. Yankelovich. New criteria for market segmentation. *Harvard Business Review*, 42:83–90, 1964.

Index

Made in the USA
Monee, IL
21 January 2021

8438c05b-d45a-400c-a9d9-84d196af6c5aR01